# Lecture Notes in Electrical Engineering

## Volume 1004

The book series *Lecture Notes in Electrical Engineering* (LNEE) publishes the latest developments in Electrical Engineering—quickly, informally and in high quality. While original research reported in proceedings and monographs has traditionally formed the core of LNEE, we also encourage authors to submit books devoted to supporting student education and professional training in the various fields and applications areas of electrical engineering. The series cover classical and emerging topics concerning:

- Communication Engineering, Information Theory and Networks
- Electronics Engineering and Microelectronics
- Signal, Image and Speech Processing
- Wireless and Mobile Communication
- Circuits and Systems
- Energy Systems, Power Electronics and Electrical Machines
- Electro-optical Engineering
- Instrumentation Engineering
- Avionics Engineering
- Control Systems
- Internet-of-Things and Cybersecurity
- Biomedical Devices, MEMS and NEMS

For general information about this book series, comments or suggestions, please contact leontina.dicecco@springer.com.

To submit a proposal or request further information, please contact the Publishing Editor in your country:

**China**

Jasmine Dou, Editor (jasmine.dou@springer.com)

**India, Japan, Rest of Asia**

Swati Meherishi, Editorial Director (Swati.Meherishi@springer.com)

**Southeast Asia, Australia, New Zealand**

Ramesh Nath Premnath, Editor (ramesh.premnath@springernature.com)

**USA, Canada**

Michael Luby, Senior Editor (michael.luby@springer.com)

**All other Countries**

Leontina Di Cecco, Senior Editor (leontina.dicecco@springer.com)

**** This series is indexed by EI Compendex and Scopus databases. ****

Chandan Giri · Takahiro Iizuka · Hafizur Rahaman ·
Bhargab B. Bhattacharya
Editors

# Emerging Electronic Devices, Circuits and Systems

## Select Proceedings of EEDCS Workshop Held in Conjunction with ISDCS 2022

Springer

*Editors*
Chandan Giri
Department of Information Technology
Indian Institute of Engineering Science
and Technology, Shibpur
Howrah, West Bengal, India

Hafizur Rahaman
Department of Information Technology
Indian Institute of Engineering Science
and Technology, Shibpur
Howrah, West Bengal, India

Takahiro Iizuka
HiSIM Research Center
Hiroshima University
Higashihiroshima, Japan

Bhargab B. Bhattacharya
Indian Institute of Technology (IIT)
Kharagpur, West Bengal, India

ISSN 1876-1100 ISSN 1876-1119 (electronic)
Lecture Notes in Electrical Engineering
ISBN 978-981-99-0054-1 ISBN 978-981-99-0055-8 (eBook)
https://doi.org/10.1007/978-981-99-0055-8



This Springer imprint is published by the registered company Springer Nature Singapore Pte Ltd.
The registered company address is: 152 Beach Road, #21-01/04 Gateway East, Singapore 189721, Singapore

# Contents

**Hardware-Efficient *Q*-Learning Accelerator for Robot Path Planning** . . . . . . . . 1
Harsh Advani, Jimmy Patel, and Tapas Kumar Maiti

**Performance Analysis of Temperature on Wireless Performance for Vertically Stacked Junctionless Nanosheet Field Effect Transistor** . . . 11
Sresta Valasa, Shubham Tayal, and Laxman Raju Thoutam

**Analysis of the $NH_3$ Adsorption on Boron-Arsenic Co-doped Monolayer Graphene: A First Principle Study** . . . . . . . . 19
Aditya Tiwari, Naresh Bahadursah, Sandip Bhattacharya, and Sayan Kanungo

**Quantum Fault-Tolerant Implementation of a Majority-Based 4-Bit BCD Adder** . . . . . . . . 33
Laxmidhar Biswal, Niladri Pratap Maity, and Hafizur Rahaman

**CNTFET-Based Universal Filter Using DO-CCII** . . . . . . . . 47
Mohd Yasir and Naushad Alam

**Designing of Energy-Efficient XOR Gate Implementing DWM Spintronics** . . . . . . . . 61
Afreen Khursheed and Kavita Khare

**Short-Channel Effects in Independently Controlled MG-MOSFET** . . . . . 75
Soumajit Ghosh, Mitiko Miura-Mattausch, Hafizur Rahaman, Takahiro Iizuka, and H. J. Mattausch

**Reduction of Interconnect Delay and Resistance While Minimizing Grid Area in GNR-Based VLSI Routing Problem** . . . . . . . . 85
Subrata Das, Debesh Kumar Das, and Soumya Pandit

**Modeling of Pristine and Intercalation Doped Multilayer Graphene Nanoribbon Conductors with Energy-per-Layer Screening** ... 99
Santasri Giri Tunga, Sandip Bhattacharya, Subhajit Das, and Hafizur Rahaman

**Enhancing Lifetime of Non-volatile Memory Caches by Write-Aware Techniques** ... 109
S. Sivakumar, Mani Mannampalli, and John Jose

**Microfluidic Dilution by Recycling Arbitrary Stock Solutions Using Various Mixing Models** ... 125
Abhishek Ghosh, Debraj Kundu, Sudip Poddar, Shigeru Yamashita, Robert Wille, and Sudip Roy

**Design and Analysis of Posit Processing Engine with Embedded Activation Functions for Neural Network Applications** ... 139
Pranose J. Edavoor, Aneesh Raveendran, Vivian Desalphine, and David Selvakumar

**Multinet Global Routing Algorithm for On-Chip Optical Interconnects to Minimize Optical Signal Loss** ... 155
Anik Saha, Subhajit Chatterjee, Supriyo Srimani, Tuhina Samanta, and Hafizur Rahaman

**Performance Enhancement of Dielectric Engineered Doping Less InGaN Tunnel FET for Low Power Analog/Radio Frequency Applications** ... 171
Arnab Som and Sanjay Kumar Jana

**Voltammetric Detection and Controlled Inhibition of Decarboxylation of Gallic Acid (GA) in Green Tea Using Eugenol** ... 181
Aindrila Roy, Debopam Bhattacharya, Chirantan Das, Basudev Nag Chowdhury, Anupam Karmakar, and Sanatan Chattopadhyay

**Impact of a Tubular Dielectric Medium on Peak Noise and Crosstalk Delay in a Coaxial TSV** ... 191
Maya Chandrakar and Manoj Kumar Majumder

**Novel Approach for the Reduction of Critical Paths in Static Timing Analysis Without Degradation in QOR** ... 209
K. Ranjit Kannan and G. Lakshminarayanan

**Coupling Transition Reduction on On-Chip Buses Using Adaptive Bus Encoding (ABE)** ... 217
Sumantra Sarkar, Ayan Biswas, Anindya Sundar Dhar, and Rahul M. Rao

**Modelling and Analysis of Confluence Attack by Hardware Trojan in NoC** ........ 231
Sachin Bagga, Ruchika Gupta, and John Jose

**Investigating the Impact of Ge-Quantum Well Width in Si/$SiO_2$/Ge/$SiO_2$/Pt Resonant Tunneling Device with NEGF Formalism** ........ 247
Nilayan Paul, Basudev Nag Chowdhury, and Sanatan Chattopadhyay

**Comparative Analysis of Normal and Anemic RBC by Employing Impedimetric and Voltammetric Studies** ........ 259
Debopam Bhattacharya, Aindrila Roy, Chirantan Das, Basudev Nag Chowdhury, Anupam Karmakar, and Sanatan Chattopadhyay

**Differential Fault Analysis of Trivium Using Artificial Neural Network on SoC Platform** ........ 269
Arijit Tewary, Swagata Mandal, Amlan Chakrabarti, Debasri Saha, and Avishek Adhikari

**Investigation of Adders for Retinal Neuromorphic Circuits** ........ 281
Payal Shah and Surendra Singh Rathod

**Sputtered $HfO_2$/$ZrO_2$ Induced Interfacial Ferroelectric HZO Layer for Negative Capacitance Applications** ........ 297
Ankita Sengupta, Basudev Nag Chowdhury, Bodhishatwa Roy, Subhrajit Sikdar, and Sanatan Chattopadhyay

**Gain Flattening of Erbium-Doped Fiber Amplifier Using an In-Line M-S-M Fiber Structure** ........ 307
Protik Roy and Partha Roy Chaudhuri

**Experimental Demonstration of Electric Field Sensing Using Sagnac Loop Based Fiber Cantilever Configuration** ........ 313
Isha Sharma and Partha Roy Chaudhuri

**An SMT-Based Reverse Engineering of Register Allocation in High-Level Synthesis** ........ 319
Mohammed Abderehman and Chandan Karfa

**Hardware Primitives-Based Accelerator Architecture for NTRU-HRSS Scheme** ........ 331
J. Mervin, Shabbir Darbar, and David Selvakumar

**Distributed Agent-Based Voltage Control Approach for Active Distribution Systems** ........ 349
Ahmed Bedawy, Naoto Yorino, Yutaka Sasaki, Yoshifumi Zoka, Kihembo Samuel Mumbere, and Ryuta Kubo

**Application Mapping of Fully Connected 3D NoC Using Latency Prediction Model** ........ 363
Ramesh Sambangi, B. Hari Krishnan, Kanchan Manna, Santanu Chattopadhyay, and Sudipta Mahapatra

**Smart Device and Mobile Application for Remote Health Monitoring and Alarming** ........ 375
Mrittika Ghosh, Ankan Ghosh, Aditi Bhattacharya, Abir Kapat, Soumyadeep Mandal, Sambit Prasad, and Ananya Banerjee

**Designing a Silicon-on-Insulator (SOI) Waveguide with an Aim of Studying Nonlinear Pulse Reshaping** ........ 385
Hemant, Somen Adhikary, and Mousumi Basu

**FaceDig: A Deep Neural Network-Based Fake Image Detection Scheme** ........ 395
Simantini Ghosh, Suman Kayal, Manab Malakar, Anirbit Sengupta, Supriyo Srimani, and Abhijit Das

**Self-heating Effects on Power Loss of SiC-Based General-Purpose Inverter-Stack Circuit** ........ 405
Subhajit Das, Takahiro Iizuka, and Mitiko Miura-Mattausch

**A Novel Approach to Model and Analyze Wafer–Wafer Hybrid Bonding** ........ 419
Debika Chaudhuri, Hafizur Rahaman, and Tamal Ghosh

**Successive Approximation Register Analog-to-Digital Converter—A Tutorial** ........ 427
Shruti Konwar, Utkarsh Jaiswal, and Bibhu Datta Sahoo

**Origin of Hump in $I_{ds}$ for Body-Tied SOI-MOSFET and Its Influence on Circuit Performance** ........ 445
T. Iizuka, M. Miura-Mattausch, H. Kikuchihara, and S. Ghosh

**IR-LED Using Electroluminescence in PbS Quantum Dot** ........ 455
Abhigyan Ganguly, Siddhartha S. Nath, and Viranjay M. Srivastava

**100X Increase in Industrial and Personal Productivity Augmenting the State-of-the-Art Technologies AI/ML, Edge Computing, and 5G Network** ........ 463
Biswajit Patra

# About the Editors

**Chandan Giri** received B.Tech. degree in Computer Science & Engineering from Calcutta University, Kolkata, India, in 2000 and subsequently Master of Engineering (M.E) in Computer Science & Engineering from Jadavpur University, Kolkata, India, in 2002 and the Ph.D. degree from the Department of Electronics & Electrical Communication Engineering, Indian Institute of Technology Kharagpur, in 2008. He is currently Associate Professor at the Department of Information Technology, Indian Institute of Engineering Science and Technology, Shibpur, India. His research interests are wireless sensor networks, testing and design-for-testability of integrated circuits (especially 3D and multicore chips), microfluidic biochip design, and testing.

**Takahiro Iizuka** received Ph.D. degree from Hiroshima University, Higashi-Hiroshima, Japan, in 2013. From 1986 to 2012, he was with NEC Corporation, NEC Electronics Corporation, and Renesas Electronics Corporation, where he was involved in carrier transport modeling for TCAD and led a SPICE modeling team during 2003–2012. During these years, he was also with the Semiconductor Technology Academic Research Center, Yokohama, Japan, where he was involved with the development of the HiSIM family models interacting with Hiroshima University as Visiting Researcher. Since 2012, he has been with the HiSIM Research Center at Hiroshima University, where he is involved in compact modeling, including maintaining, improving, and developing the HiSIM family models. Dr. Iizuka is Member of IEEE, The Institute of Electronics, Information and Communication Engineers, Japan (IEICE), and The Japan Society of Applied Physics (JSAP), respectively.

**Hafizur Rahaman** is Full Professor at the Indian Institute of Engineering Science and Technology (IIEST), Shibpur, India. He was Royal Society Postdoctoral Research Fellow (2006–2009) at the University of Bristol, UK. During 2013–2015, he was Visiting Professor at the University of Bremen, Germany, under the DST-DAAD Research Scheme. His research interests include the design and testing of integrated circuits, nano-biochips, and emerging nanotechnologies including quantum computing. He has published over 350 research articles in journals and refereed conference proceedings. To his credit, he has supervised 17 doctoral and 46 master theses besides guiding several projects at the undergraduate level. Eight more doctoral theses are now in progress. He leads the VLSI design and test group at IIEST, Shibpur, India.

**Bhargab B. Bhattacharya** is currently Distinguished Visiting Professor of Computer Science and Engineering at the Indian Institute of Technology (IIT) Kharagpur. Before that, he had been on the faculty of the Indian Statistical Institute, Kolkata, for over 35 years. He received his B.Sc. degree in physics from the Presidency College, Calcutta, and B.Tech., M.Tech., and Ph.D. degrees from the University of Calcutta, India. Dr. Bhattacharya served as Visiting Professor at the University of Nebraska-Lincoln and Duke University, USA; University of Potsdam, Germany; Kyushu Inst. of Tech., Japan; Tsinghua University, Beijing, China; National Tsing Hua University, Taiwan; and IIT Guwahati, India. He is Fellow of the Indian National Academy of Engineering (INAE), Fellow of the National Academy of Sciences (India), and Fellow of the IEEE.

# Hardware-Efficient *Q*-Learning Accelerator for Robot Path Planning

**Harsh Advani, Jimmy Patel, and Tapas Kumar Maiti**

**Abstract** This work aims to reduce the actual processing time, i.e., latency of a mobile robot, required to perform any action in any particular state. Hence, we proposed an efficient hardware architecture to implement *Q*-learning testing algorithm which is suitable for real-time robotics applications. The developed architecture improves the performance and accuracy, which in our case is robot path planning. Major focus lies on the two parameters, i.e., learning and recognition speeds.

**Keywords** *Q*-learning · Mobile robot · Reinforcement learning · Agent · State · Action · *Q*-value

## 1 Introduction

Machine learning (*ML*) is a subset of artificial intelligence (AI) that provides capability to a computer to learn without being explicitly programmed by users [1]. From the name, we realized that it makes the computer more similar to humans by the ability to learn. Basic *ML* classification is depicted in Fig. 1. *ML* is based on the idea that the machine can learn from data, identify patterns and then make decisions accordingly. Reinforcement learning (*RL*) is a method in machine learning that makes a sequence of decisions in particular environment [2, 3]. It is an algorithm which learns based on the observations of environment. The advantage of using *RL* is that, it provides rewards that lead to success even when the environment is too large and complex. *RL* allows an agent to learn by exploring the environment and observing the results and rewards. The goal is to find the suitable action that would result in maximum reward. In this technique, there is no requirement to provide training examples, which is actually a large dataset, and this is the main advantage of this technique. Real-time learning is possible in *RL* which implies that it can provide outcomes while improving at the same time. Figure 2 illustrates the action-reward feedback loop of a general *RL* model. *RL* allows an agent to learn by exploring the

H. Advani · J. Patel · T. K. Maiti (✉)
Dhirubhai Ambani Institute of ICT, Gandhinagar, Gujarat 382007, India
e-mail: tkm.iitkgp@gmail.com

 
C. Giri et al. (eds.), *Emerging Electronic Devices, Circuits and Systems*,
Lecture Notes in Electrical Engineering 1004,
https://doi.org/10.1007/978-981-99-0055-8_1

environment and observing the results and rewards. The goal is to find the suitable action that would result in maximum reward. In this technique, there is no requirement to provide training examples, which is actually a large dataset and this is the main advantage of this technique. Real-time learning is possible in *RL* which implies that it can provide outcomes while improving at the same time. Figure 2 illustrates the action-reward feedback loop of a general *RL* model.

*Q*-learning is a *RL* which uses *Q*-values, also known as action values, to improve the actions of the learning agent, iteratively. *Q* in *Q*-learning stands for quality which represents how for a given action is useful in acquiring future rewards [4]. *Q*-learning is an off-policy reinforcement learning algorithm used to find the optimal action to be taken in the given current state. Off-policy means that the *Q*-learning function learns from random actions and hence the policy is not required. The algorithm works in the following fashion: (i) the agent starts in a state, takes action and receives reward; (ii) for the next action, the agent will have two choices, i.e., referring *Q*-table and select the action that is having the highest value or to take a random action [5]; (3) after above steps, the agent updates the *Q*-values, i.e., *Q*[State, Action]. The *Q*-learning equation, which is also known as Bellman's equation, is used to train the agent in particular environment, as shown below [4],

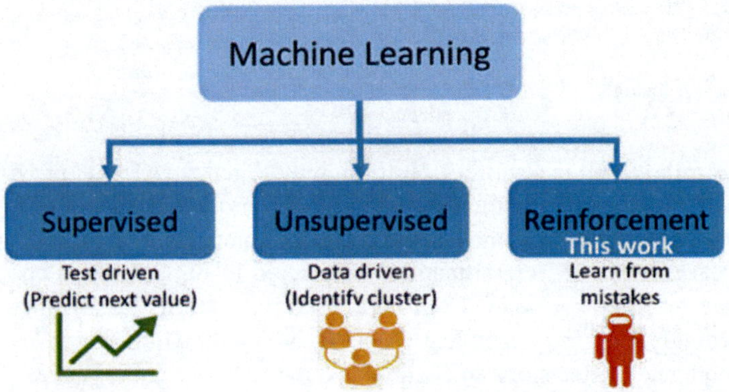


**Fig. 1** Classification of machine learning. This work focused on *Q*-learning which is a subset of reinforcement learning (*RL*)

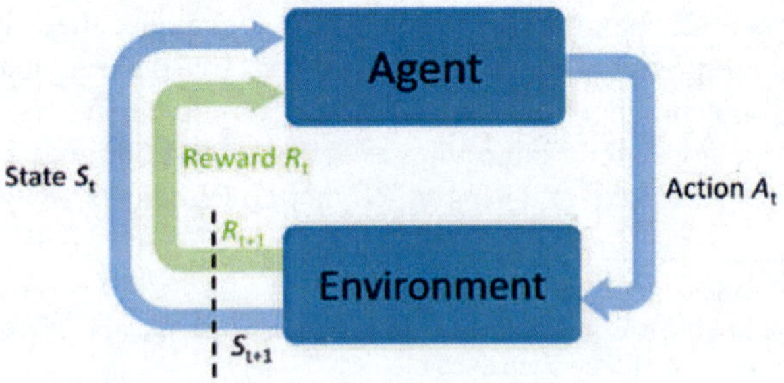


**Fig. 2** Action-reward feedback loop [3]. Here, $A_t$, $R_t$ and $S_t$ denote the action, reward and sate, respectively

$$\left\{Q(S_t, A_t) \leftarrow (1-\alpha) * Q(S_t, A_t) + \alpha * \left(R_t + \gamma * \max_A(S_{t+1}, A_t)\right)\right\} \quad (1)$$

Here, $\alpha$ is learning rate at which the model will be trained, $R_t$ is the reward for each action taken, $\gamma$ is the discount factor; $A_t$, $R_t$ and $S_t$ denotes the action, reward and sate, respectively. Currently, research efforts have been focused on improving the *RL* algorithms. This in turn will provide a smooth operation of *RL* applications such as controlling robotics arm, autonomous vehicle and humanoid robots. We have a different perspective in terms of research.

Currently, Python/MATLAB is used to implement these types of algorithms and perform their training and testing. Here, testing refers to the real-time use of the algorithm incorporated in applications such as humanoid robot and autonomous vehicle [6]. This has a processing time in milliseconds. Hence, we aim to reduce the processing time, i.e., latency of the robot by implementing the *Q*-learning testing algorithm in Verilog-HDL, and thereby providing an efficient hardware architecture that will control the robot [7–9]. This method will reduce the processing time from milliseconds to microseconds. Thus, we are focusing on implementing 2D environment in Verilog-HDL and perform testing based on that.

## 2 Deep Reinforcement Learning for Walking Robot

*RL* a type of machine learning is all about how the virtual agent thinks to take actions in an environment in order to maximize the total reward. Figure 3 shows the *RL* model used in this project. The environment is the world where the actions need to be taken in order to reach the target. Here, the environment is the lower body of the robot, i.e., two limbs. Here, the actor-critic agent model is chosen, DDPG algorithm. There are two neural networks in this project one called actor, which has state observations as input and provides actions as output. The second is the critic, which estimates the value of the state of the environment based on the actions taken by the actor. This helps to update the weight values of both the neural networks. This project aims to provide the lower body of robot the ability to balance and walk. The architecture used for this work is shown in Fig. 4.

The major aspect to discuss here is the extremely high computational cost of training the agent using deep reinforcement learning, which can even take few days to complete. Figure 5a shows that the training was carried out for around 2000 episodes. The results obtained after training the model are shown in Fig. 5b–d. Figure 6 shows the simulation of the project, i.e., robot walking along the path toward the final destination. In this work, the most important thing is the interaction of the robot with the floor. The main aim of this project was to achieve a movement which is similar to human walking, which here is forward motion along with balancing and improving the movement. As the robot movement is a very complicated task and requires a lot of resources, we divided in two parts, one of which is training and second is testing.

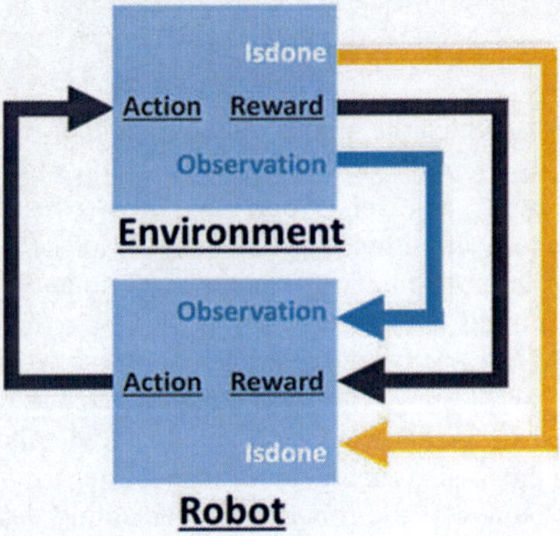


**Fig. 3** Reinforcement learning (*RL*) model implemented for humanoid robot simulation

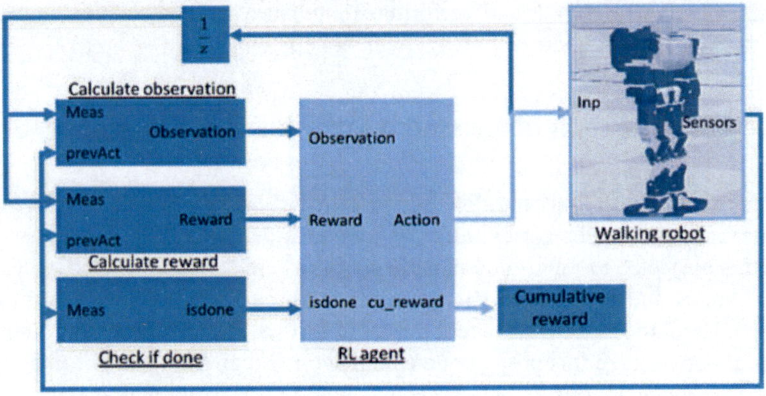


**Fig. 4** Block diagram architecture of reinforcement learning model for robot path planning

The major aspect to discuss here is the extremely high computational cost of training the agent using deep reinforcement learning, which can even take few days to complete. Figure 5a shows that the training was carried out for around 2000 episodes. The results obtained after training the model are shown in Fig. 5b–d. Figure 6 shows the simulation of the project, i.e., robot walking along the path toward the final destination. In this work, the most important thing is the interaction of the robot with the floor. The main aim of this project was to achieve a movement which is similar to human walking, which here is forward motion along with balancing and improving the movement. As the robot movement is a very complicated task and requires a lot of resources, we divided in two parts, one of which is training and second is testing.

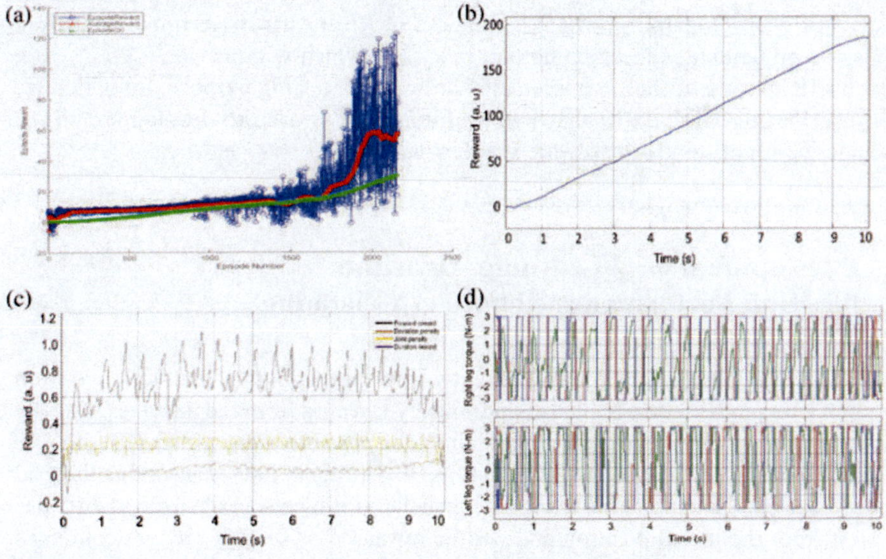


**Fig. 5** **a** Training information, episode reward for walking robot with *RL*, **b** cumulative reward, **c** Individual rewards for different aspects, and **d** left and right legs movements actions (torque in N-m)

**Fig. 6** Simulation of the lower body in the environment after training the model. The training was performed for left and right legs movement of a humanoid robot

Further, more focus is given on testing. We implemented an efficient hardware for it, which in turn will provide a faster execution compared to its current implementation. The area in which we are focusing has a very negligible or have no research papers that aim to implement a hardware for this purpose. Some research work focused on improving the *RL* algorithms that would make smooth simulations of the application [7–9]. We started to explore from an initial level to reduce the complexity and to provide better results, we development hardware implementation of *Q*-learning instead of deep *Q*-learning, which is the base of *RL* for hardware implementation. Thus, we implemented the training part in Python which will provide training data, which will be used further to implement hardware for testing purpose, for which we will use Verilog-HDL. Note that Python framework is used to develop simplified *Q*-lingering algorithm for hardware implementation.

## 3 Development of *Q*-Learning Algorithm for High-Performance Robot Path Planning

In Python, we considered 2D environment where we taught the agent to move from one block to another and also it will optimize by learning from the mistakes. The 2D environment is a $4 \times 4$ grid which contains four possible areas—start (S), tile (T), obstacle (O) and end (E), as illustrated in Fig. 7. The agent moves around in the grid until it reaches the end or meets with an obstacle. If it meets an obstacle, then it has to start from the initial position and will be rewarded as 0. This process continues till it learns from all the mistakes and reaches the end point. Here is visual look of 2D environment.

The agent will have four possible actions to perform—left (0), right (1), forward (2) and backward (3). We implemented *Q*-learning which is the basic form of RL. When the agent moves in a wrong direction, i.e., the direction in which the target position cannot be reached, it will take a random action. Initially, the agent starts to learn we need to allow random movements, and then, gradually its probability should be reduced. It minimizes the error by minimizing the loss. There are 16 possible locations where the agent can be at any given time. In the present state, the agent will

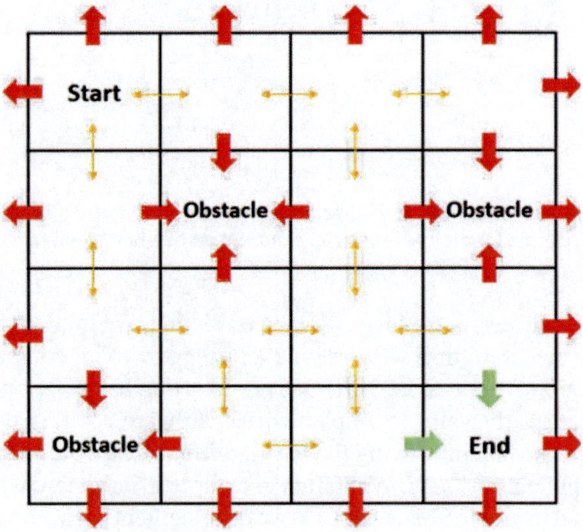


**Fig. 7** Illustrates the 2D environment which is considered to verify *Q*-learning algorithm

have four possible actions that needs to be decided for next state. Hence, 16 possible locations, having 4 possible actions which generates a $Q$-table of $16 \times 4$ matrix form. We trained the agent to perform a number of episodes, which updates $Q$-table with reward values. Figure 8a shows average reward values per thousand episodes, and Fig. 8b shows $Q$-table with updated reward values after training. Using these $Q$-table values, the agent was tested in the 2D environment for five episodes, the output of which is shown in Fig. 9.

We observed that the run time of the testing part in 6.08 ms. As mentioned earlier, we focused on the reduction of processing time, required by the agent to take the

(a)

```
*****Average rewards per thousand episodes*****

1000 : 0.05400000000000004

2000 : 0.19800000000000015

3000 : 0.40900000000000003

4000 : 0.55300000000000004

5000 : 0.62000000000000004

6000 : 0.66000000000000005

7000 : 0.67200000000000005

8000 : 0.67600000000000005

9000 : 0.69100000000000005

10000 : 0.66400000000000005


Score over time: 0.5197
```

(b)

```
*****Q-table*****

[[0.54280691 0.49507268 0.49046761 0.4879708 ]
 [0.34076411 0.38259701 0.34038782 0.50364463]
 [0.41623005 0.4275343  0.39877404 0.47476726]
 [0.33092483 0.27176289 0.30464656 0.44931632]
 [0.55549163 0.19025993 0.23049143 0.41713372]
 [0.         0.         0.         0.        ]
 [0.25712276 0.12218955 0.15739433 0.10350397]
 [0.         0.         0.         0.        ]
 [0.36276422 0.36806197 0.46927636 0.59200631]
 [0.41750877 0.61508794 0.46192973 0.33323423]
 [0.54586577 0.36901888 0.33970338 0.23047793]
 [0.         0.         0.         0.        ]
 [0.         0.         0.         0.        ]
 [0.54680973 0.56743614 0.74351255 0.55522483]
 [0.70158606 0.86730397 0.73039726 0.72871119]
 [0.         0.         0.         0.        ]]
```

**Fig. 8** **a** Average rewards per thousand episodes obtained using developed $Q$-learning algorithm, **b** $Q$-table obtained using developed algorithm which is used for RTL-level implementation in FPGA hardware

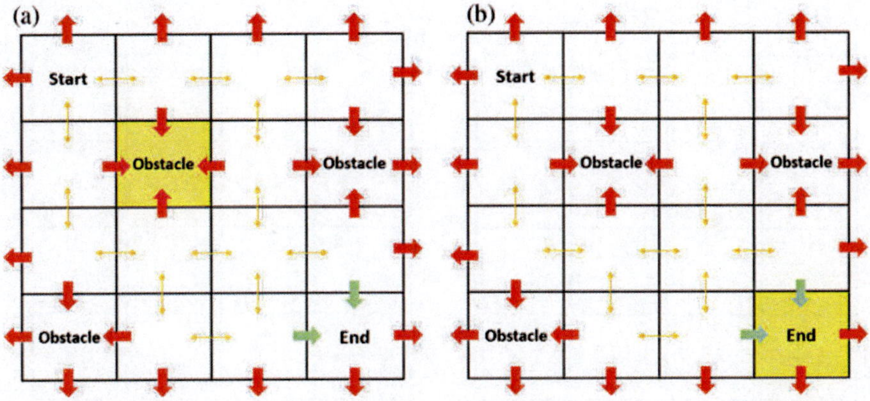


**Fig. 9** Output of testing based on Q-values obtained after training, **a** output window for reward = 0 and **b** output window for reward = 1

**Table 1** Simulation results are obtained using Python

| Episode no. | Last tile | No. of steps | Reward |
|---|---|---|---|
| 0 | Obstacle (O) | 19 | 0 |
| 1 | Obstacle (O) | 28 | 0 |
| 2 | End (E) | 16 | 1 |
| 3 | End (E) | 25 | 1 |
| 4 | End (E) | 32 | 1 |

decision during the testing part. Hence, we trained our agent in Python and will use the $Q$-values obtained from it in Verilog-HDL for the testing part (Table 1).

## 4 Efficient Hardware Architecture in Verilog-HDL

The $Q$-table which is generated after completing the training of agent in Python is saved as a memory file. This memory file will be accessed during the testing part which is implemented in Verilog-HDL. The environment is a $4 \times 4$ grid which contains four possible areas—start (00), tile (01), obstacle (10) and end (11). The environment is similar to that of Python, except 00, 01, 10, 11 (see Fig. 11) in binary are used as states instead of S, T, O, E alphabets. Rest of the specifications of the environment remains same as that were in Python. We simulated the $Q$-learning model for five episodes, the output of which is shown in Fig. 10b.

The run time of the code is observed in the figure above, i.e., 30 ns, which is a huge reduction in comparison with that of Python which was 6.08 ms. Thus, we partially succeeded to achieve our goal, as the main aim of this work is to design a chip which will perform this functionality. Also, in the simulation in Verilog, the agent follows

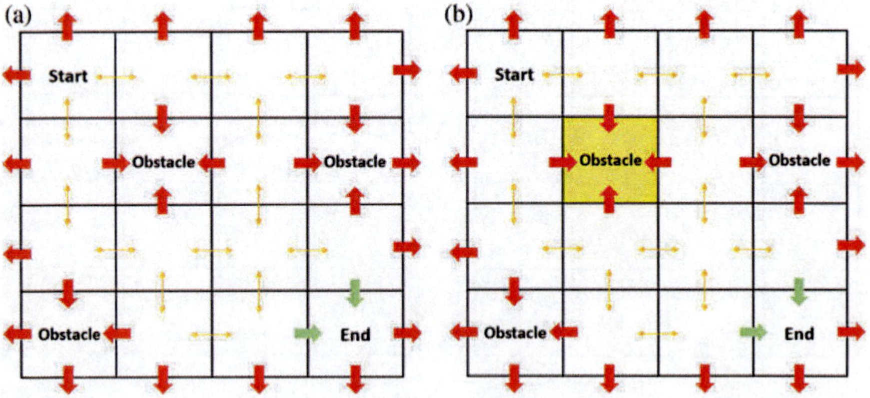


**Fig. 10** **a** Environment in Verilog-HDL similar to Python (refer to Fig. 7), **b** simulated output obtained using Verilog-HDL

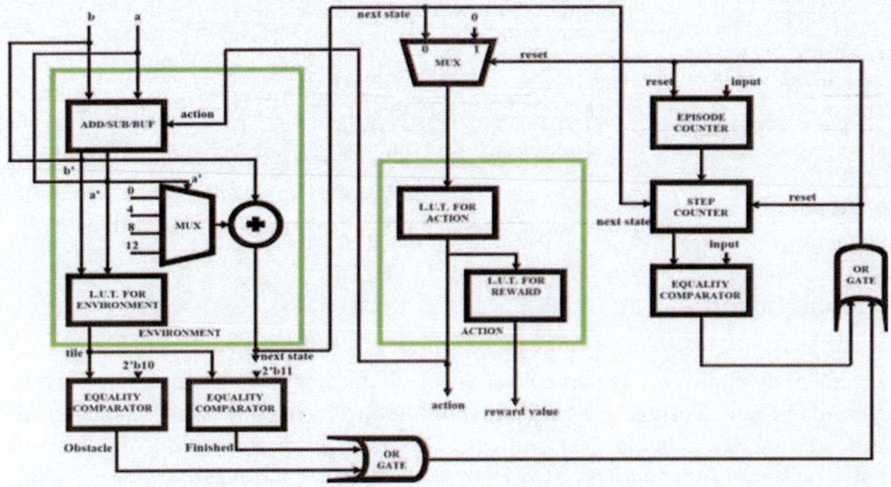


**Fig. 11** Developed RTL-level architecture for *Q*-learning testing based on proposed algorithm

the same step in each episode, which is not the case in Python during training; this is due to the dynamic coding style in Python. As discussed earlier, we have started from scratch, so initially we have followed static coding style, as the implementation of the testing part is itself a complex task. We will try to make it dynamic in future. The code is written in synthesizable way so that we can generate the Register Transfer Level (RTL schematic for our application. We used Quartus Prime Lite to generate the RTL schematic. The proposed efficient hardware architecture for *Q*-learning testing algorithm is shown in Fig. 11. The internal architecture of ADD/SUB/BUF block in our proposed architecture is shown in Fig. 12, corresponding Verilog-HDL based simulation results are depicted in Table 2.

**Fig. 12** ADD/SUB/BUF blocks internal architecture of *Q*-learning circuit

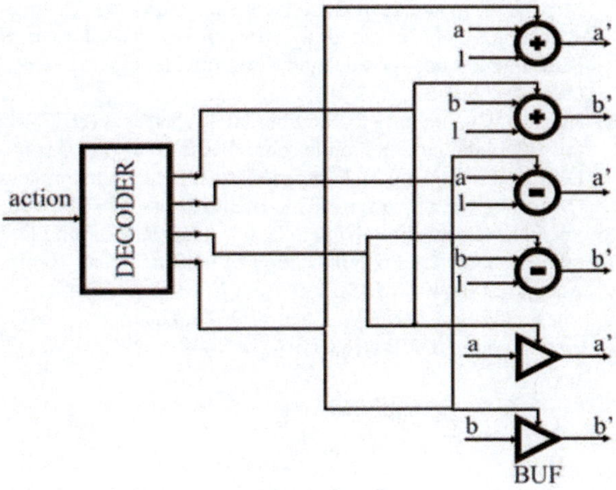

**Table 2** Simulation results are obtained using Verilog-HDL

| Episode no | Last tile | No. of steps | Reward |
|---|---|---|---|
| 0 | Obstacle (O) | 5 | 0 |
| 1 | Obstacle (O) | 5 | 0 |
| 2 | Obstacle (O) | 5 | 0 |
| 3 | Obstacle (O) | 5 | 0 |
| 4 | Obstacle (O) | 5 | 0 |

## 5 Conclusion

Implement dynamic coding part so that the agent does not take same action in every repeated episode. Designing a chip based on the RTL schematic which is generated from Verilog-HDL coding. The run time of the developed algorithm is 30 ns at hardware level in comparison with Python software which was 6.08 ms. Adding more functionality to the design, the robot performed more actions in addition to the existing ones.

## References


1. Wang C, Gong L, Yu Q, Li X, Xie Y, Zhou X (2017) DLAU: a scalable deep learning accelerator unit on FPGA. IEEE Trans Comput Aided Des Integr Circuits Syst 36(3):513–517
2. Watanabe H, Tsukada M, Matsutani H (2021) An FPGA-based on-device reinforcement learning approach using online sequential learning. arXiv:2005.04646v3
3. Gu S, Holly E, Lillicrap T, Levine S (2020) Deep reinforcement learning for robotic manipulation with asynchronous off-policy updates. In: IEEE international conference on robotics and automation (ICRA), pp 1–6, Singapore
4. Jang B, Kim M, Harerimana G, Kim J (2019) Q-learning algorithm: a comprehensive classification and applications. IEEE Access 7:133653–133667
5. Meng Y, Kuppannagari S, Rajat R, Srivastava A, Kannnan R, Prasanna V (2020) QTAccel: a generic FPGA based design for Q-table based reinforcement learning accelerators. In: IEEE international parallel and distributed processing symposium workshops (IPDPSW), New Orleans, LA, USA, pp 1–4
6. Konar A, Chakraborty G, Singh S, Jain L, Nagar A (2013) A deterministic improved Q-learning for path planning of a mobile robot. IEEE Trans Syst Man Cybern Syst 43(5):1141–1153
7. Da Silva L, Torquato M, Fernandes M (2018) Parallel implementation of reinforcement learning Q-learning technique for FPGA. IEEE Access 7:2782–2798
8. Spano S, Cardarilli G, Nunzio L, Fazzolari R, Giardino D, Matta M, Nannarelli A (2019) An efficient hardware implementation of reinforcement learning: the Q-learning algorithm. IEEE Access 7:186340–186351
9. Lin J, Hwang K, Jiang W, Chen Y (2016) Gait balance and acceleration of a biped robot based on Q-learning. IEEE Access 4:2439–2449

# Performance Analysis of Temperature on Wireless Performance for Vertically Stacked Junctionless Nanosheet Field Effect Transistor

**Sresta Valasa, Shubham Tayal, and Laxman Raju Thoutam**

**Abstract** This paper investigates the effect of temperature on the wireless performance characteristics, i.e., linearity and harmonic distortion for the vertically stacked junctionless field effect transistor (JL-NSFET) at gate length ($l_g$) = 16 nm. The transfer characteristics curve ($I_d$-$V_g$), transconductance ($g_m$), and its second- and third-order derivatives, i.e., $g_{m2}$ and $g_{m3}$ performances are explored. The detailed analysis reveals that the temperature shows a profound influence on VIP3 and IIP3 giving the best linearity for the temperature range 77–200 °K with better gate drive. However, the second- and third-order derivatives of distortion, i.e., HD2 and HD3 are not much affected by temperature giving least distortions with negligible change. Therefore, the simulated JL-NSFET optimizes the device linearity and harmonic distortion performances with best suitable for RF and wireless applications.

**Keywords** JL-NSFET · Harmonic distortion · Linearity · Transconductance ($g_m$) · Wireless application · Temperature

## 1 Introduction

The exceptional upsurge in today's communication networks has accelerated the demand for devices featuring exquisite RF properties [1, 2]. While most of RF research findings gave a strong interpretation on various analog parameters involving transconductance ($g_m$), cut-off frequency ($f_T$), maximum oscillation frequency ($f_{MAX}$), yet definite figure of merits reinforcing linearity and harmonic distortion are often overlooked. Linearity has emerged as one of the most critical concerns in RF circuit design. Devices employing wireless communication networks should operate linearly regardless receiving a poor signal in the midst of a strong interfering signal. Else the strong interferer signal may inundate the targeted weak signal or introduce cross-modulation [4–7]. The linearity of the device substantially characterizes and governs the performance of various communication networks. A device's

S. Valasa (✉) · S. Tayal · L. R. Thoutam
Department of Electronics and Communication Engineering, SR University, Warangal, India
e-mail: shreshtavalasa@gmail.com

 
C. Giri et al. (eds.), *Emerging Electronic Devices, Circuits and Systems*,
Lecture Notes in Electrical Engineering 1004,
https://doi.org/10.1007/978-981-99-0055-8_2

excellent linearity results in the reduction of intermodulation distortion and higher order harmonics which can be minimized at both the system and transistor levels [8–10]. High linearity is extremely desirable for designing of low power amplifiers, even in the existence of severe interferences and poor signals.

CMOS devices are widely employed in wireless, and satellite communications, medical, and military equipment, automobiles, and nuclear power plants. It has also been revealed that while operating at lower temperatures, the performance of MOS transistors greatly increases in terms of sub-threshold swing, noise performance, on-currents, short-channel effects, cut-off frequency, carrier mobilities, and gain [11, 12]. Also, there has been an immense requirement for portable and huge battery backup smart devices. In order to cope up with the growing demand, a large increase in transistor density in the IC should be made. With increasing transistor density, the heat dissipation in the IC increases substantially ensuing a rise in operating temperature [13, 14]. These higher temperatures show a crucial impact on the device operation as the device dimensions shrink below 20 nm or perhaps even end up damaging the device [15, 16]. A device's high-temperature reliability assures its accelerated endurance and stability [17]. For enhanced reliability, it is therefore pivotal to investigate temperature effect variation on the wireless performance.

Multi-gate FET devices, including FinFET, nanowire FET, and nanosheet (NS) FET have been investigated by researchers in the recent years to minimize the problem of SCEs [18–21]. Because of the excellent gate control over charge carriers, NSFET is the considered as the most phenomenal contender for upcoming ULSI technology nodes among the multi-gate FETs. This paper explores the wireless performance of junctionless (JL) NSFET for different operating temperatures utilizing linearity and harmonic distortion parameters such as VIP3, IIP3, and higher order coefficients of transconductance ($g_{m2}$ and $g_{m3}$).

## 2 Device Portraiture and Simulation Setup

The simulated three-dimensional geometrical view of the junctionless nanosheet FET is portrayed in Fig. 1. The desired simulations have been carried out using one of the most eminent industrial tools named mixed-mode sentaurus TCAD simulator. The device specifications used for the designing of JL-NSFET are organized in Table 1.

The gate length has been set at 16 nm to design the device. The width, thickness, and effective oxide thickness (EOT) used is 18 nm, 5 nm, and 0.7 nm. A metal gate work function which is equal to 4.8 eV is utilized to develop a junctionless device to alleviate the poly depletion discrepancies. The conventional gate dielectric constant of 3.9 ($SiO_2$) is deployed. A heavy doping level of $1 \times 10^{19}$ cm$^{-3}$ (n-type for arsenic) is used to deter the formation of junction. Two nanosheets are vertically stacked to further analyze the performance of the device. The spacing between these nanosheets is set at 6 nm. A temperature range of 77–400 °K is engaged to compare the device performance. The various simulation models availed for the investigation of device performance is discussed below.

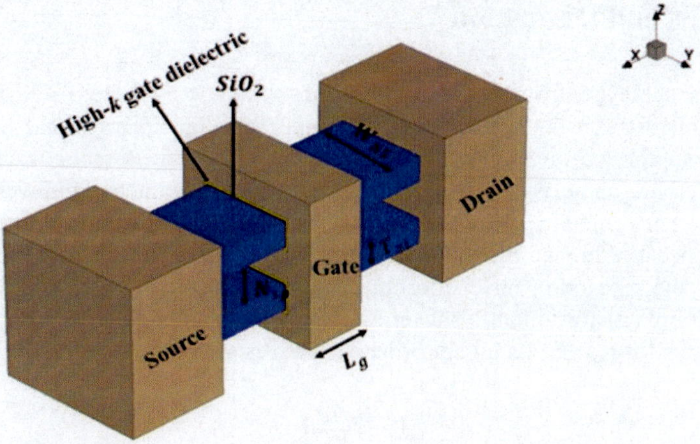


**Fig. 1** 3D geometrical portraiture of JL-NSFET

**Table 1** Device specifications of JL-NSFET

| Device specifications | Values |
|---|---|
| Temperature (T) | 77–400 °K |
| Gate length ($L_g$) | 16 nm |
| Width of the nanosheet ($W_{ns}$) | 18 nm |
| Thickness of silicon ($T_{si}$) | 5 nm |
| Gate dielectric constant (*k*) | 3.9 ($SiO_2$) |
| Work function ($\phi_{ms}$) | 4.8 eV |
| Doping concentration | $1 \times 10^{19}$ cm$^{-3}$ |
| Effective oxide thickness (EOT) | 0.7 nm |
| No. of nanosheets | 2 |
| Spacing between nanosheets ($N_{sp}$) | 6 nm |

To analyze the device performance of strongly doped JL-NSFET, Fermi Dirac statistics simulation model is utilized. To characterize the radiative generation and recombination processes, the SRH recombination/generation along with the band-to-band auger recombination models is utilized. The Lombardi mobility simulation model is also incorporated for a wide range of scattering phenomena, including acoustic phonons and surface roughness. The bandgap narrowing concept is also applied in JL-NSFETs because of their higher doping profile. Further, the quantum density gradient model is used for the quantum confinement phenomenon.

## 3 Results and Discussion

The results achieved from the modeling and simulation of the JL-NSFET are discussed in this section. The impact of temperature on linearity and harmonic distortion is explored here. The transconductance ($g_m$) and its second- and third-order derivatives, namely $g_{m1}$ and $g_{m2}$ are the paramount parameters in investigating the wireless performances such as linearity and harmonic distortion of several FET devices. Since the higher order (second order and third order) transconductance determines the restriction on distortion, they ($g_{m2}$ and $g_{m3}$) must be very less to achieve minimal distortion and is a serious reliability challenge at the device level. The equations for $g_m$ and its higher order derivatives are given by,

$$g_m = \frac{I_d}{V_{gs}} \tag{1}$$

$$g_{m2} = \frac{1}{2}\frac{\partial g_m}{\partial V_{gs}} \tag{2}$$

$$g_{m3} = \frac{1}{3}\frac{\partial g_{m2}}{\partial V_{gs}} \tag{3}$$

The transfer characteristics ($I_d$-$V_g$), $g_m$, $g_{m2}$, and $g_{m3}$ for various temperatures are depicted in Fig. 2.

The transfer characteristics curve for $I_d$-$V_g$ in linear scale at different temperatures are plotted in plotted in Fig. 2a. It is visualized that the increase in gate voltage ($V_g$) increases the drain current at various temperatures. It is also found that the transconductance ($g_m$) rises with an increase in gate voltage ($V_g$) for different temperatures as portrayed in Fig. 2b. Figure 2c, d visualizes that as the temperature of the JL-NSFET increases, the peak amplitude of $g_{m2}$ and $g_{m3}$ increases for temperature ranging between 77 and 200 °K. Later, they start reducing when the operating temperature is raised above 300 °K indicating the peak approaching nearer to the lower gate voltages. This indicates that a lower gate drive is necessary in order to ensure good linearity. It is well known that the linearity of the device is excellent at lower amplitudes of $g_{m2}$ and $g_{m3}$ [6]. However, it is noticed that there is only a slight change in $g_{m2}$ and $g_{m3}$ which is negligible when the temperature is raised from 77 to 400 °K. Thus, it is assured that the linearity of the designed device is good as the operating temperature increases.

VIP3 and IIP3 denote the ingested input voltage (power), where the third harmonic amplitude and power equal the first harmonic amplitude (power) which are mathematically presented as:

$$VIP3 = \sqrt[2]{\left|24 \times \frac{g_m}{g_{m3}}\right|} \tag{4}$$

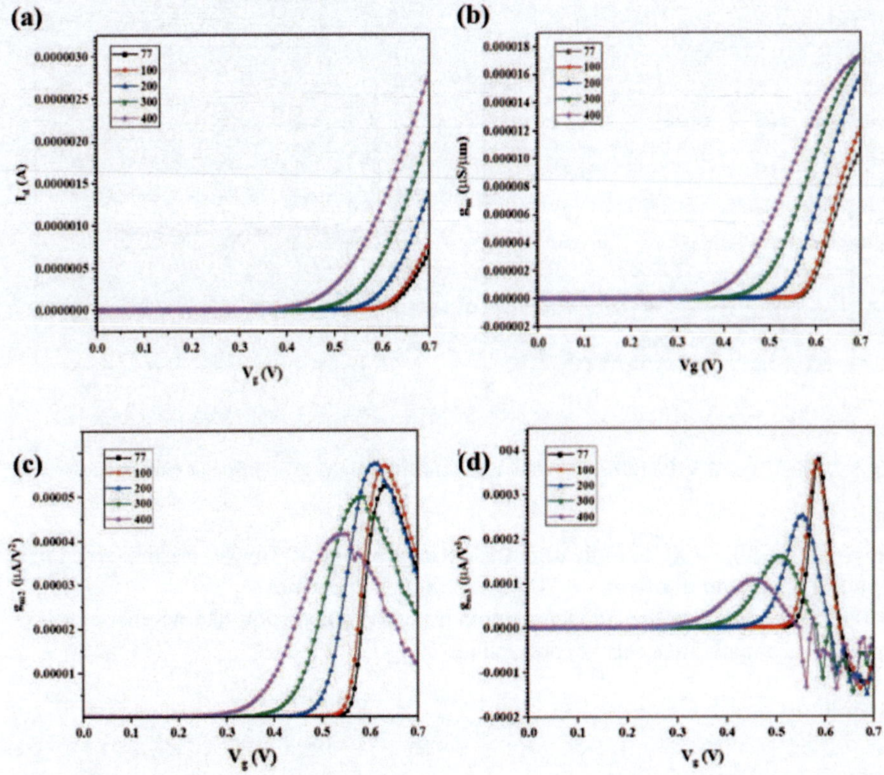


**Fig. 2** Variation of **a** $I_d$-$V_g$ transfer characteristics **b** transconductance ($g_m$) **c** second-order transconductance ($g_{m2}$) **d** third-order transconducatnce ($g_{m3}$) for various temperatures

$$IIP3 = \frac{2}{3} \times \frac{g_m}{g_{m3} \times R_s} \tag{5}$$

where the Rs = 50Ω for most of the RF-based applications.

For analog/RF and microwave functional applications where excellent linearity and minimal distortion values are of sheer importance, VIP3 and IIP3 must be maintained high. From Fig. 3a, it is observed that as the temperature increases from 77 to 200 ˚K, the peak of VIP3 occurs at lower gate voltages. However, when the temperature is increased above the room temperature from 300 ˚K, the peak of VIP3 gets shifted toward higher gate voltages. This might be attributed to the relative change in $g_{m3}$. Thus, we can say that the JL-NSFET offers great linearity within the temperature range 77–200 ˚K offering a good gate control. Furthermore, the same trend is observed for IIP3 too as picturized in Fig. 3b. Peak values of IIP3 are noticed between the temperature range 77–200 ˚K at lower gate voltages indicating excellent gate control, whereas the peak values occur at higher gate voltages for the

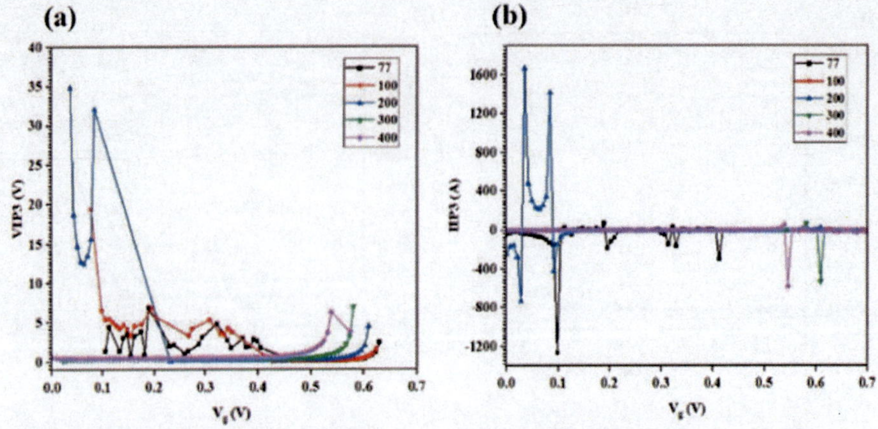


**Fig. 3** Variation of **a** VIP3 **b** IIP3 in relation to gate voltage ($V_g$) for different temperatures

temperatures 300–400 °K indicating the poor gate control. Hence, it is observed that JL-NSFET is good to use at 77–200 °K for better linearities.

Further, to analyze the impact of temperature on distortion characteristics, HD2 and HD3 are mathematically expressed as:

$$\text{HD2} = 0.5 \times V_a \times \frac{g_{m2}}{2! \times g_m} \tag{6}$$

$$\text{HD3} = 0.25 \times V_a^2 \times \frac{g_{m3}}{3! \times g_m} \tag{7}$$

Here, $V_a$ represents the AC amplitude and is chosen to be 50 mV for each and every instance [3]. HD2 and HD3 must be maintained very less to attain minimal distortion. Figure 4a, b picturizes that the second-order harmonic distortion (HD2) and third-order harmonic distortion (HD3) increase between the temperature range 77–200 °K and gradually starts decreasing for 300–400 °K assuring least harmonic distortion. However, deeply analyzing the results, the change is negligible indicating that the JL-NSFET device is the suitable device to obtain less distortions at the specified temperature range.

## 4 Conclusion

The effect of temperature on linearity and harmonic distortion parameters for RFIC and wireless applications has been studied in this paper. The result analysis reveals that the transconductance ($g_m$) increases with an increase in temperature with respect to gate voltage. It is also revealed that the second- and third-order derivatives of transconductance show a reduced peak above the nominal room temperature range

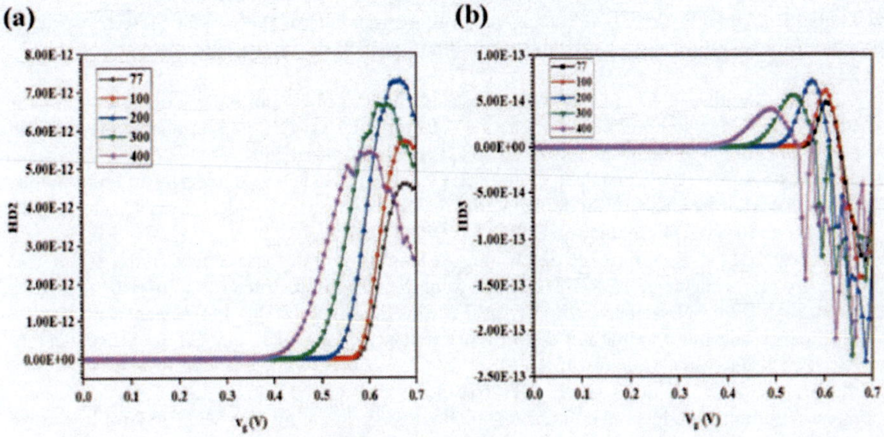


**Fig. 4** Variation of **a** HD2 **b** HD3 in relation to gate voltage ($V_g$) for different temperatures

indicating a good gate control. However, the change is slightly negligible compared with the temperature range 77–200 ˚K. Further, it is observed that the JL-NSFET offers great linearity within the temperature range 77–200 ˚K offering excellent gate control. Besides, it is also noticed that the second- and third-order harmonic distortions (HD2 and HD3) are less for all temperatures ensuring that the JL-NSFET is the most promising candidature for future wireless applications.

## References


1. Emona D, Avik C, Abhijit M (2020) Relative study of analog performance, linearity, and harmonic distortion between junctionless and conventional SOI FinFETs at elevated temperatures. J Electron Mater 49(5):3309–3316
2. Barman KR, Baishya S (2020) Study of temperature effect on analog/RF and linearity performance of dual material gate (DMG) vertical super-thin body (VSTB) FET. Silicon, pp 1–10
3. Tayal S, Bhattacharya S, Jena B, Ajayan J, Muchahary D, Singla P (2021) Linearity performance and harmonic distortion analysis of IGE junctionless silicon nanotube-FET for wireless applications. Silicon, pp 1–6
4. Tayal S, Mittal V, Jadav S, Gupta S, Nandi A, Krishan B (2020) Temperature sensitivity analysis of inner-gate engineered JL-SiNT-FET: an Analog/RF prospective. Cryogenics 108:103087
5. Awadhiya B, Pandey S, Nigam K, Kondekar PN (2017) Effect of ITC's on linearity and distortion performance of junctionless tunnel field effect transistor. Superlattices Microstruct 111:293–301
6. Gupta N, Chaujar R (2016) Investigation of temperature variations on analog/RF and linearity performance of stacked gate GEWE-SiNW MOSFET for improved device reliability. Microelectron Reliab 64:235–241
7. Valasa S, Tayal S, Thoutam LR (2022) Optimization of design space for vertically stacked junctionless nanosheet FET for analog/RF applications. Silicon, pp 1–10

8. Madan J, Chaujar R (2016) Interfacial charge analysis of heterogeneous gate dielectric-gate all around-tunnel FET for improved device reliability. IEEE Trans Device Mater Reliab 16(2):227–234
9. Valasa S, Shinde JR, Ramji DR, Avunoori S (2021) A power and delay efficient circuit for CMOS phase detector and phase frequency detector. In: 2021 6th international conference on communication and electronics systems (ICCES), pp 77–82. IEEE
10. Kumar SP, Agrawal A, Chaujar R, Gupta RS, Gupta M (2011) Device linearity and intermodulation distortion comparison of dual material gate and conventional AlGaN/GaN high electron mobility transistor. Microelectron Reliab 51(3):587–596
11. Kumar A (2017) Effect of trench depth and gate length shrinking assessment on the analog and linearity performance of TGRC-MOSFET. Superlattices Microstruct 109:626–640
12. Kumar A, Gupta N, Tripathi SK, Tripathi MM, Chaujar R (2020) Performance evaluation of linearity and intermodulation distortion of nanoscale GaN-SOI FinFET for RFIC design. AEU-Int J Electron Commun 115:153052
13. Rewari S, Goel A, Verma S, Gupta RS (2019) Linearity and intermodulation distortion assessment of underlap engineered cylindrical junctionless surrounding gate MOSFET for low noise CMOS RFIC design. In: 2019 IEEE 16th India council international conference (INDICON), pp 1–4. IEEE
14. Tayal S, Nandi A (2018) Study of temperature effect on junctionless Si nanotube FET concerning analog/RF performance. Cryogenics 92:71–75
15. Saha R, Bhowmick B, Baishya S (2018) Temperature effect on RF/analog and linearity parameters in DMG FinFET. Appl Phys A 124(9):1–10
16. Lee SY, Lee YS, Jeong YH (2006) A novel phase measurement technique for IM3 components in RF power amplifiers. IEEE Trans Microw Theory Tech 54(1):451–457
17. Yu C, Yuan JS, Yang H (2004) MOSFET linearity performance degradation subject to drain and gate voltage stress. IEEE Trans Device Mater Reliab 4(4):681–689
18. Yum TY, Chiu L, Chan CH, Xue Q (2006) High-efficiency linear RF Amplifier-a unified circuit approach to achieving compactness and low distortion. IEEE Trans Microw Theory Tech 54(8):3255–3266
19. Kumar B, Chaujar R (2021) TCAD temperature analysis of gate stack gate all around (GS-GAA) FinFET for improved RF and wireless performance. Silicon, pp 1–13
20. Ajayan J, Nirmal D, Tayal S, Bhattacharya S, Arivazhagan L, Fletcher AA, Murugapandiyan P, Ajitha D (2021) Nanosheet field effect transistors-A next generation device to keep Moore's law alive: an intensive study. Microelectron J 105141
21. Ajayan J, Nirmal D (2015) A review of InP/InAlAs/InGaAs based transistors for high frequency applications. Superlattices Microstruct 86:1–19

# Analysis of the $NH_3$ Adsorption on Boron-Arsenic Co-doped Monolayer Graphene: A First Principle Study

**Aditya Tiwari, Naresh Bahadursah, Sandip Bhattacharya, and Sayan Kanungo**

**Abstract** Two-dimensional (2D) graphene has drawn significant attention for its potential application in the detection of inorganic gas molecules when doped with appropriate dopants. As of yet, these effects of non-metallic co-doping at the different sub-lattice sites are yet to be observed systematically from a theoretical perspective for gas-molecule detection on graphene. The study investigates molecular adsorption of ammonia ($NH_3$) on boron/arsenic (B/As) monolayer graphene using density functional theory (DFT). In this paper, we evaluate the influence of arsenic impurity on the molecular adsorption of boron-doped graphene in the same and different sub-lattice sites. In the present context, three doping configurations are identified that possess distinct electronic properties and respond characteristically to individual gas molecules. Due to orbital overlaps from the adsorbed gas molecules, molecular adsorption has a considerable impact on the spatial distribution of electronic states along the band edges, resulting in large modulations in the energy bandgaps and effective masses of the co-doped lattices. Co-doping techniques appear to be particularly suitable for electrochemical gas sensing because they result in a significant semiconducting bandgap opening while preserving the inherent nature of graphene. For co-doped lattices, the subsequent molecular adsorptions result in substantial charge transfer between the gas molecules and the host lattice, as well as a significant increase in the density of electronic states around the Fermi level.

**Keywords** First principal calculation · DFT · Graphene · Doping · Co-doping · Molecular adsorption · Gas sensing

---

A. Tiwari · N. Bahadursah · S. Kanungo (✉)
Electrical and Electronics Engineering Department, Birla Institute of Technology and Science-Pilani, Hyderabad Campus, Hyderabad 500078, India
e-mail: sayan.kanungo@hyderabad.bits-pilani.ac.in

S. Bhattacharya
Electronics and Communication Engineering Department, SR University, Aanthsagar, Hsanparthy, Warangal, Telangana 506371, India

S. Kanungo
Materials Center for Sustainable Energy and Environment, Birla Institute of Technology and Science-Pilani, Hyderabad Campus, Hyderabad 500078, India

 
C. Giri et al. (eds.), *Emerging Electronic Devices, Circuits and Systems*,
Lecture Notes in Electrical Engineering 1004,
https://doi.org/10.1007/978-981-99-0055-8_3

# 1 Introduction

Environmental monitoring, pollution control, food safety and industrial automation applications are all benefiting from the development of miniaturized, extremely sensitive and selective gas sensors with quick reaction times [1]. Due to their exceptionally high surface-to-volume ratio and a large number of reactive surface sites, crystalline two-dimensional (2D) materials have shown tremendous potential in this setting, where a small number of gas-molecule adsorption can result in a large change in electronic conductivity [1–3]. As a result, numerous 2D materials have been investigated for gas sensor design, both experimentally and theoretically [4, 5]. Graphene is the earliest and one of the most widely investigated 2D materials. Because of its single atomic layer thickness, strong electrical conductivity and chemical stability, graphene is also a promising material for gas detection [6–8]. However, owing to the unavailability of a natural semiconducting energy bandgap and a lower sensitivity to practically every ambient gas, the performance of pristine graphene-based gas sensors is extremely limited [9, 10].

In this respect, it has already been demonstrated that adding proper substitutional dopants to the graphene lattice improves the interaction between the gas molecule and the host lattice as compared to pristine graphene [8, 11–15]. It should be highlighted that both metallic and non-metallic impurities can be doped into graphene. The introduction of metallic dopants into a semi-metal such as graphene, on the other hand, drastically changes the electrical characteristics of the lattice and tends to induce metallic nature [16]. This has prompted the researcher to investigate several non-metallic dopants in graphene to specifically boost gas-molecule adsorption on the material while maintaining an appropriate energy bandgap. Boron (B), nitrogen (N), phosphorus (P) and arsenic (As) are determined to be the most effective non-metallic dopants for establishing a sufficient semiconducting bandgap while assuring superior gas-molecule adsorption on the dopant site of the lattice [7, 11, 16, 17].

However, for boron (trivalent) and arsenic (pentavalent) doping, respectively, degraded p- and n-type doping must be suffered to open an appreciable bandgap in graphene [7, 16]. The Fermi level jumps into the conduction band or valence band before gas-molecule adsorption in such degenerated doped graphene [7, 16], which is counterproductive for electrochemical gas sensing applications. Moreover, no attempts have been made to establish a semiconducting bandgap in graphene utilizing chemical co-doping while preserving the inherent nature of such material systems for gas-molecule detection too far. However, no attempts have been made to establish a semiconducting bandgap in graphene utilizing chemical co-doping while preserving the inherent nature of such material systems for gas-molecule detection so far. Also, only a few theoretical studies of the effects of non-metallic co-doping on graphene for gas sensing applications have been published [18–20]. These works suggest the catalytic reduction of nitric oxides in the presence of both boron and nitrogen co-doping on the graphene surface, but there are no works currently performed which suggest the use of boron and arsenic co-doping for ammonia gas detection.

## 2 Computational Methodology

In pure graphene, each carbon atom creates three $sp^2$ bonds with its nearest neighbors, which forms a hexagonal honeycomb lattice structure. The $p_z$-orbitals that are out of plane overlap and create π-bonds. The electrical characteristics of graphene's conduction band and valence band edges are determined by delocalized electrons in π-bands and π*-bands, respectively. Furthermore, the $sp^2$ hybridization and consequent production of π-bonds provide graphene with a high degree of in-plane stability as well as a flawlessly flat atomic layer.

The two carbon (C) atoms in the graphene unit cell represent two different sublattices, as shown in Fig. 1a. The graphene is considered in a 4 × 4 × 1 supercell configuration for studying gas-molecule adsorption by repeating the unit cell in the in-plane direction, as shown in Fig. 1b. Two and three C atoms are substituted in the supercell to create 3 separate co-doped systems which account for 6.25 and 9.375% of doping concentrations, respectively.

In the case of single doping configurations, only one C atom of the supercell is replaced with the B and As atom in the graphene lattice. To avoid artificial interaction from its periodic images in an out-of-plane direction, each supercell is given a vacuum of 40 Å. Moreover, single gas-molecule adsorption on the dopant atom site is studied throughout this work, and only the geometrical configuration associated with the lowest binding energy for a certain gas-molecule and lattice interaction is considered. It should also be emphasized that the associated conductivity modulation for any gas-molecule detection is mostly attributed to molecular doping due to charge transfer and a change in the electronic band structure. The Atomistix Tool Kit (ATK) and Virtual Nano Lab (VNL) simulation software packages from Synopsys Quantum-Wise are used in this study to perform density functional theory (DFT)-based first principle calculations.

The Brillouin zone is sampled using a linear combination of atomic orbital (LCAO) double-zeta polarized basis set with a density mesh cut-off energy of 125 Hartree and a Monkhorst–Pack grid of 3 × 3 × 1. Firstly, the geometries are

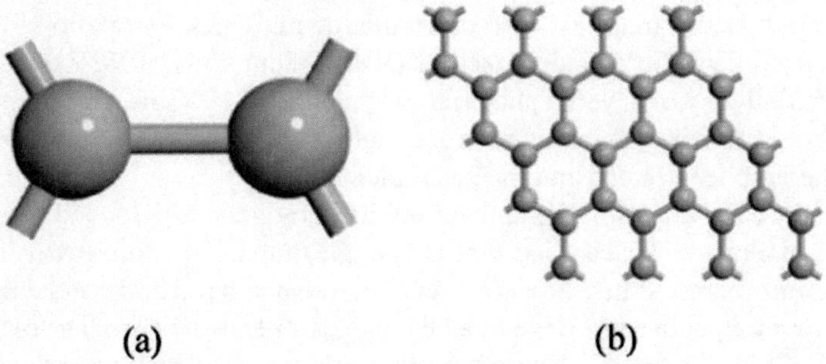


**Fig. 1** Schematic representation of **a** unit cell (1 × 1 × 1) and **b** 4 × 4 × 1 supercell of graphene layer (Gr) from top view

relaxed using the Limited memory Broyden Fletcher Goldfarb Shanno (LBFGS) algorithm, which has pressure and forces tolerances of 0.0001 eV/Å3 and 0.01 eV/Å, respectively [21].

The monolayer of pure graphene has a lattice vector of 2.456 Å and a C-C bond length of 1.42 Å. The monolayer of pure graphene has a lattice vector of 2.456 Å and a C-C bond length of 1.42 Å after optimization. The computed values are quite close to the experimentally determined lattice vector of 2.46 and C-C bond length of 1.42 [22, 23]. When two subsystems such as the host lattice and the gas molecule are involved, the inherent incompleteness of the Linear Combination of Atomic Orbitals (LCAO) basis causes Basis Set Superposition Error (BSSE) [24]. As a result of the artificial interaction between the subsystems, the energy of the lattice/gas-molecule system is overestimated [24]. To overcome this BSSE constraint, the counterpoise (CP) adjustment is applied during the binding energy calculation [21].

The binding energy of individual gas-molecules adsorption on the co-doped graphene is calculated as follows:

$$E_{\text{Binding}} = E_{\text{Gas\_and\_ Host}} - (E_{\text{Host}} + E_{\text{Gas}}) \tag{1}$$

where $E_{\text{Gas\_ and\_ Hosts}}$, $E_{\text{Gas}}$, and $E_{\text{Host}}$ are the ground-state energies of gas-molecule/co-doped graphene binding system after including CP correction, isolated gas molecule and co-doped graphene, respectively. For any stable gas-molecule adsorption, $E_{\text{Binding}}$. After this, the charge transfers between the gas-molecule and doped graphene lattice are given by

$$Q_{\text{Transfer}} = Q_{\text{gas\_after\_adsoption}} - Q_{\text{gas\_bofore\_adsoption}} \tag{2}$$

where $Q_{\text{gas\_after\_adsoption}}$ and $Q_{\text{gas\_bofore\_adsoption}}$ are the Mulliken charges in gas molecule after and before adsorption in co-doped graphene surface. If $Q_{\text{Transfer}} < 0$, the electron is transferred from gas molecule to lattice, which suggests the gas molecule acts as donor type of impurity, whereas $Q_{\text{Transfer}} > 0$ suggests the gas molecule acts as an acceptor type of impurity.

After the geometrical optimization of the system was done using LDA, the various properties calculations were performed using GGA correlation interaction and Revised Purdew Burke and Ernzerhof (RPBE) functional. RPBE facilitates the energy calculations for layered materials [19, 25]. At last, the Grimme DFT-D3 empirical correction is also included to account for the Van der Waals (VdW) forces between the graphene lattice and the gas molecule.

The energy band diagram of graphene and its homogeneously doped counterparts are shown in Fig. 2. We can see that the single doping of boron (trivalent) and nitrogen (pentavalent) shifts the Fermi level in valance band and conduction band, respectively, along with some degenerated bandgap opening in the range of 0.105 eV to 0.199 eV [7, 16, 17, 26, 27]. Therefore, the work progresses with other pentavalent non-metallic dopants like arsenic in order to modulate the electrical and structural properties of pure graphene lattice.

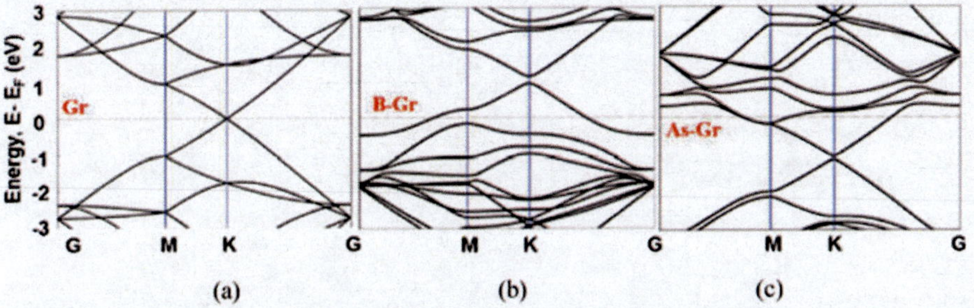


**Fig. 2** Energy band (E-K) structure of **a** pristine graphene, **b** as-doped graphene, **c** B-doped graphene

# 3 Results and Discussion

## *3.1 Effects of B/A Co-doping on Structural and Electronic Properties*

Two distinct dopant species are used as substitutional impurities in the supercell of the pristine graphene lattice in this study. As a result, three different configurations for boron/arsenic (B/As) co-doped lattice are identified and depicted in Fig. 3 based on the locations and number of the two dopant atoms in a hexagonal graphene ring. The numbers '1' and '2' represent the amount of boron (B) and arsenic (As) in the doped lattice, respectively. The structural stability of any 2D material system is best preserved by the size compatibility of the dopant atoms for substitutional doping. The typical atomic radius of C, B and As are 0.70 Å, 0.85 Å and 1.85 Å, respectively. Consequently, after B/As co-doping, the different interatomic bond lengths are observed to be in the range of 1.40–1.73 Å, compared to the C-C bond length of 1.42 Å in pristine graphene. Furthermore, the unit cell lattice vector increases from 2.46 Å to 2.56 Å in one of the B/As co-doped systems.

These indicate a slight change in structural stability in the B/As system compared to the un-doped system. These changes in structural parameters are also evident with the energy band structure modification. The boron has one less electron in the outer shell whereas nitrogen possesses one more than carbon atom which is the reason behind the Fermi shift when these atoms are substituted into the same hexagonal ring of graphene ring. It is interesting to note that when both boron and arsenic are doped together into the graphene lattice the nature of bandgap opening and the position of the Fermi level becomes intrinsic which is a huge departure from the Fermi level position in the case of single individual doping's of B and As atoms which showed strong degeneracies (see Fig. 2). The atomic orbital projection indicates that the p-orbitals of the C atom primarily populated the CBM and VBM in co-doped graphene. For the B/As co-doped system, small contributions from p-orbitals of B and As are observed near the valence band and conduction band edges, respectively. The band structures of the host lattice along with the orbital projections are shown in Fig. 4.

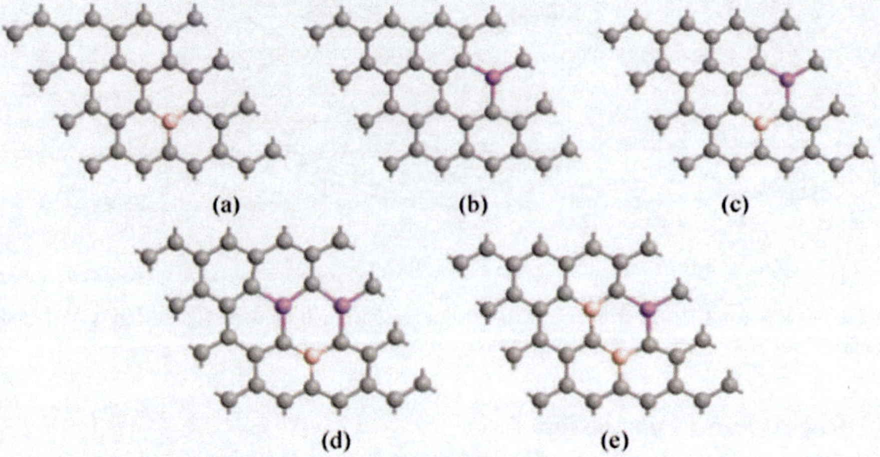


**Fig. 3** Schematic representation of 4 × 4 × 1 super cell of different host configurations from top view, **a** 1B, **b** 1As, **c** 1B1As, **d** 1B2As, **e** 2B1As

## *3.2 Gas-Molecule Adsorption on Homogenously Doped and Co-doped Graphene Systems*

This research looks at the most energetically stable configuration of gas-molecule adsorption on a given single-doped or co-doped graphene lattice. Each gas molecule is first examined in various positions/orientations on the lattice, with the highest binding energy configurations eventually being discovered. The schematic representation of 4 × 4 × 1 super cell of most stable $NH_3$ gas-molecule adsorption configuration of single B-doped, As-doped and B/As co-doped systems is schematically depicted in Fig. 5.

The preferential site for $NH_3$ gas-molecule adsorption is the doped boron atom due to it being electronegative which facilitates the proper transfer of electrons from the nitrogen of ammonia to the boron of the doped lattice. This is the reason why the boron showed the most energetically stable site for both mono-doped and co-doped graphene lattice. The effect of gas-molecule adsorption on the electronic and structural integrity of doped graphene lattice can also be solidified by the visible planar distortion of lattice structures in various systems once the gas molecule interacts with the host layer.

The interaction of gas molecules with doped/co-doped lattice systems is then quantified in terms of binding energies, charge transfer and equilibrium gas-molecule/lattice distance, as shown in Table 1. From Table 1, it should be noted that the binding energies greater than −1.0 eV usually denotes chemosorption, and a negative sign in charge transfer signifies donor-type nature of $NH_3$ gas molecule [27]. Also, we can see that the boron-doped lattice bonds strongly with the $NH_3$ gas molecule which is not the case with pentavalent impurity like arsenic, the boron doping is thus more effective for donor-type gas molecules [7].

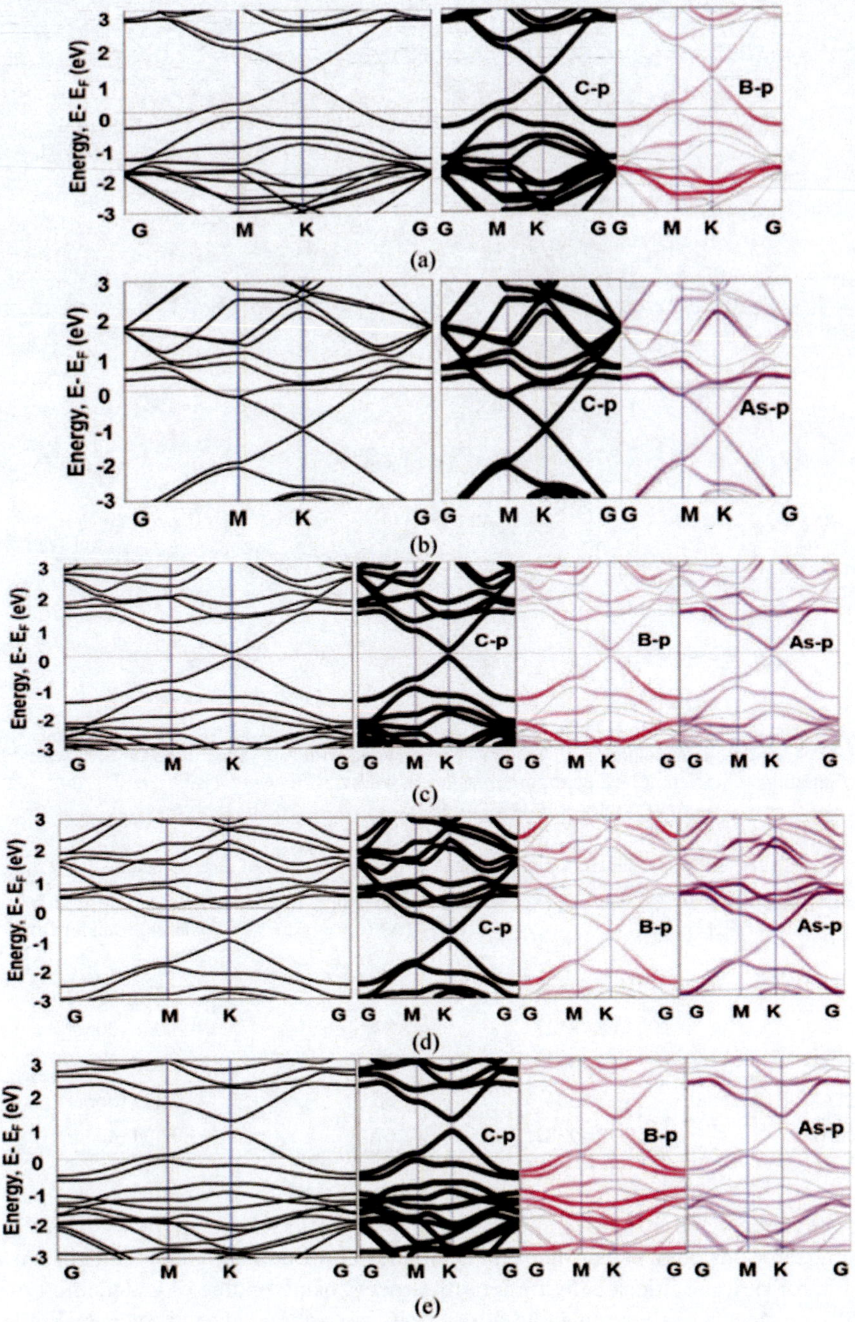


**Fig. 4** Energy band structure along with its orbital projections of different host configurations, **a** 1B, **b** 1As, **c** 1B1As, **d** 1B2As, **e** 2B1As

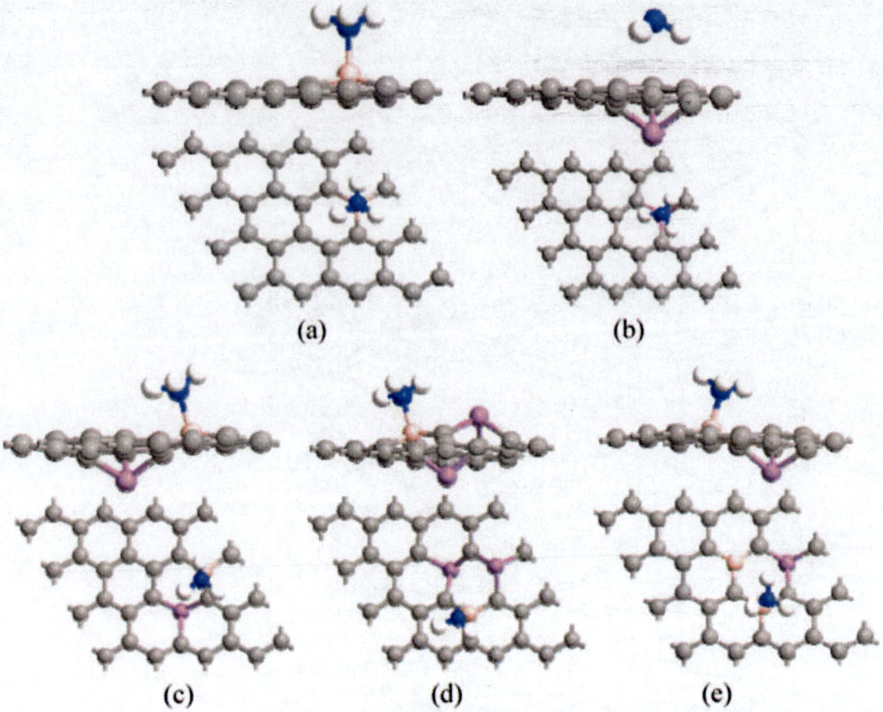


**Fig. 5** Schematic representation of 4 × 4 × 1 host lattice with $NH_3$ adsorption in its most stable configurations of different host configurations, **a** 1B, **b** 1As, **c** 1B1As, **d** 1B2As, **e** 2B1As

**Table 1** Structural properties of doped graphene lattice in presence of gas molecule

| Co-doped lattice in presence of $NH_3$ gas | Binding energy (eV) | Gas/lattice distance (Å) | Electron (Mulliken charge) transfer from gas |
|---|---|---|---|
| B | – 0.24355 | 1.68 | – 0.323 (donor) |
| As | – 2.78127 | 2.82 | – 0.068 (donor) |
| 1B1As | – 1.45649 | 1.68 | – 0.336 (donor) |
| 2B1As | – 1.4264 | 1.66 | – 0.337 (donor) |
| 1B2As | – 3.49382 | 1.67 | – 0.304 (donor) |

The energy band (E-k) structure is used to examine the relative influence of molecular adsorptions on the electronic characteristics of different co-doped lattices. As a result, for various lattice configurations, the energy band topologies and their atomic orbital projections after molecule adsorptions are addressed, as shown in Fig. 6. Also, the total density of states and the respective bandgap change in the host-gas-molecule systems are also a good indications of the effect of gas-molecule adsorptions in different host lattice (see Table 2).

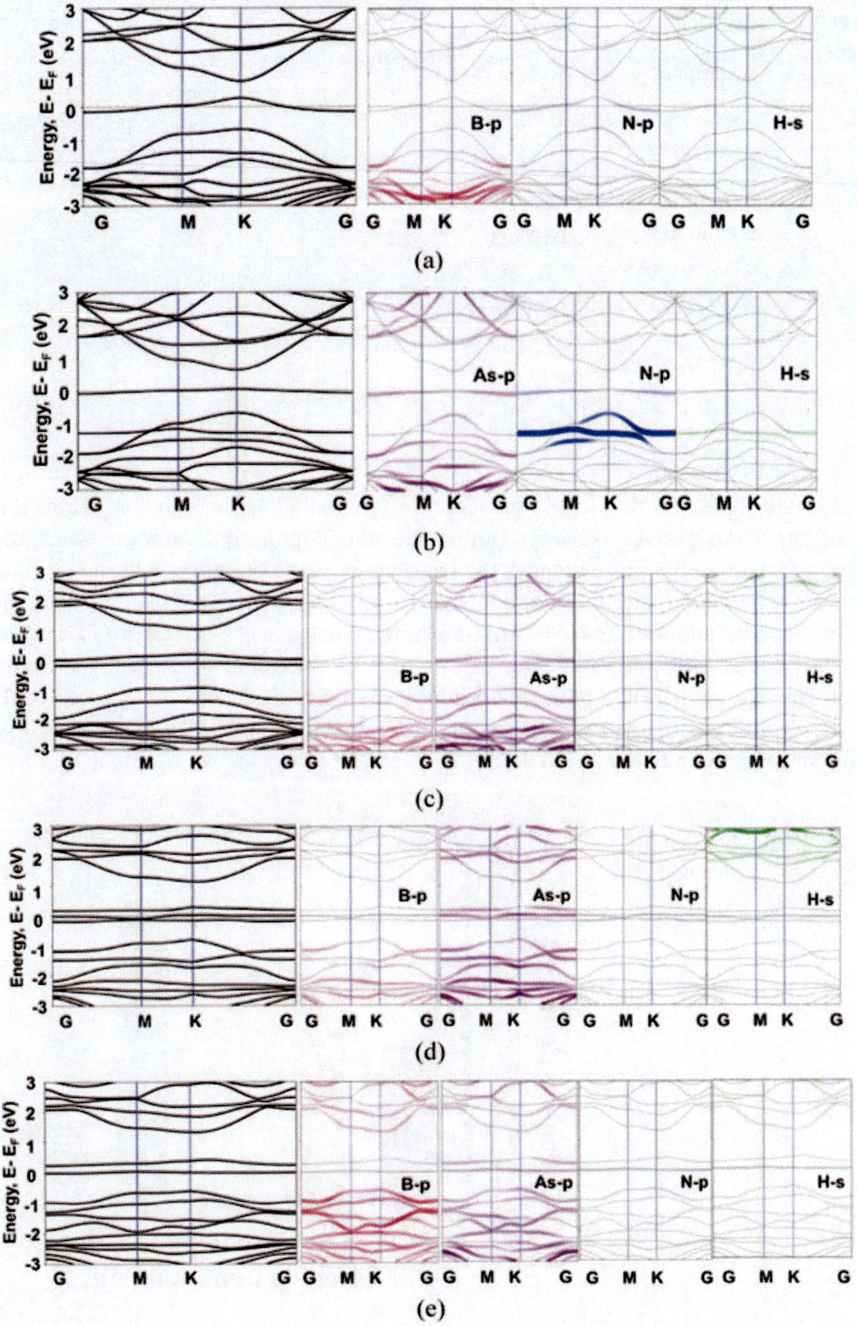


**Fig. 6** Energy band structure along with its orbital projections of different host configurations after $NH_3$ adsorption for **a** 1B, **b** 1As, **c** 1B1As, **d** 1B2As, **e** 2B1As configurations

**Table 2** Bandgap of doped graphene for $NH_3$ gas-molecule adsorption

| Gas molecule and doping configurations | Bandgap (eV) |
|---|---|
| B-Gr | 0.29 |
| $NH_3$-B-Gr | 0.51 (+ 0.32) |
| As-Gr | 0.09 |
| NH3-As-Gr | 0.58 (+ 0.49) |
| 1B1As-Gr | 0.21 |
| $NH_3$-1B1As-Gr | 0.13 (−0.08) |
| 1B2As | 0.22 |
| $NH_3$-1B2As | 0.91 (+ 0.69) |
| 2B1As | 0.36 |
| $NH_3$-2B1As | 0.25 (+ 0.11) |

It should be noted that in the presence of ammonia gas the bandgap of the co-doped lattice (Gr-1B1As) is reduced unlike the other doping configurations where the presence of ammonia gas enhances the bandgap of the host lattice material system. Figure 7 shows a magnitudes of bandgap change in eV for different material systems in presence of ammonia gas. Furthermore, the total density of states (TDOS) of material system is a good indicator of presence of electronic states near the Fermi level. The change in TDOS of the material system indicates the effect of gas molecule on the overall electronic property of material system. The TDOS of various host configurations with and without ammonia gas molecule is shown in Fig. 8.

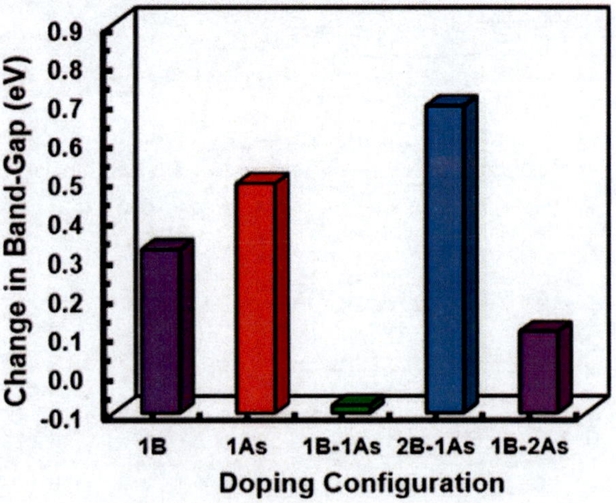


**Fig. 7** Change in bandgap of host systems in presence of $NH_3$ gas in eV

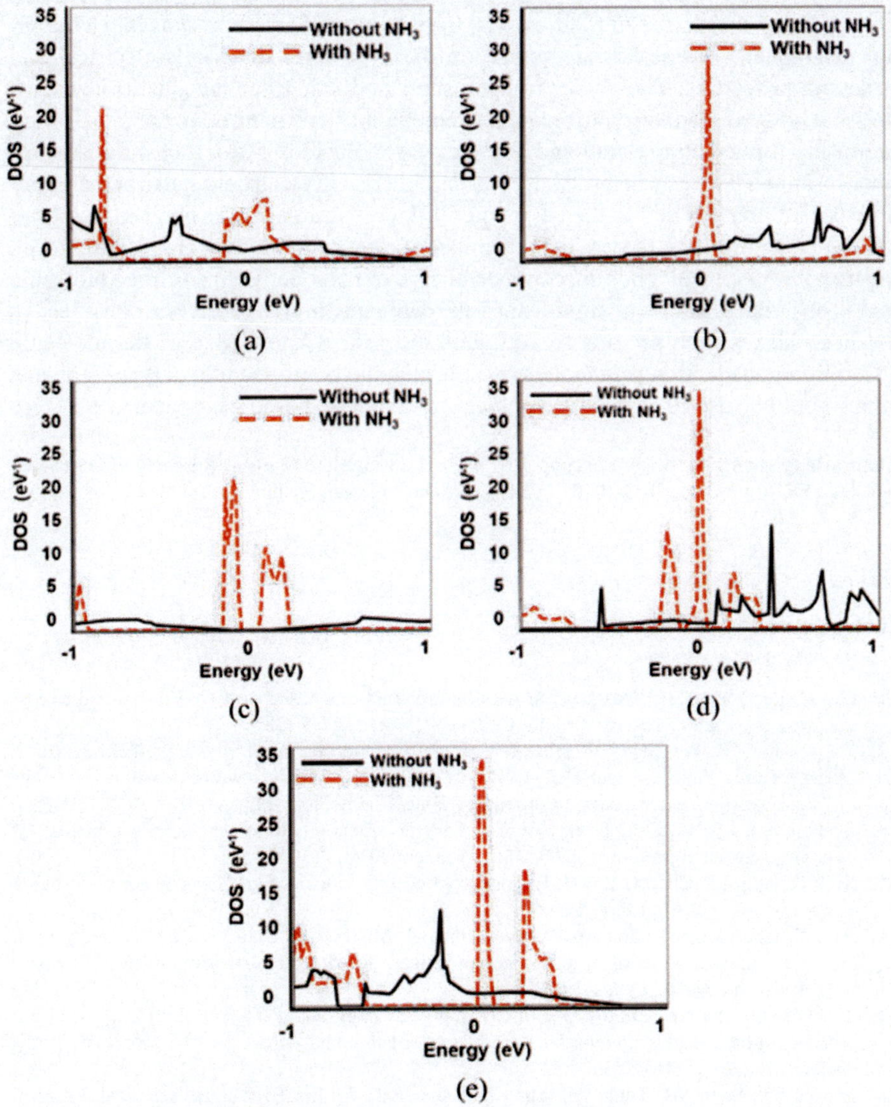


**Fig. 8** Total density of state profiles before and after $NH_3$ adsorption for **a** 1B, **b** 1As, **c** B1As, **d** 1B2As and **e** 2B1As configurations

# 4 Conclusion

The work offers a thorough analysis of B/As co-doping effects in graphene for $NH_3$ gas-molecule detection. The electrical and sensing properties of graphene are highly influenced by the atomic specifications, as well as the relative locations of the dopant species in the sub-lattice. The introduction of arsenic atoms into a boron-doped graphene lattice has an impact on the host layer material's electrical and structural properties. The ability to open semiconducting bandgaps before adsorption in co-doped graphene lattices while keeping intrinsic natures appears to be very promising for detecting donor and acceptor gas molecules. After molecular adsorption, co-doped lattices often show a considerable rise in electronic states at the Fermi level, which is attributed to the augmented E-k curvature near the band edges. After molecular adsorptions, the rise in the number of electronic states near the Fermi level, together with molecular doping owing to charge transfer between adsorbed molecule and host lattices, shows a significant improvement in charge carrier densities. In essence, the research provides a complete theoretical knowledge of the molecule adsorption transduction mechanism on non-metallic co-doped monolayer graphene, which could be useful in the development of graphene-based gas sensors.

**Acknowledgements** The work is supported by part through the Start-up Research Grant (SRG) by DST-SERB (grant no. SRG/2020/000547) awarded to Sayan Kanungo.

## References

1. Bag A, Lee NE (2019) Gas sensing with heterostructures based on two-dimensional nanostructured materials: a review. J Mater Chem C 7(43):13367–13383
2. Yang S, Jiang C, Wei SH (2017) Gas sensing in 2D materials. Appl Phys Rev 4(2):021304
3. Zhang J, Liu L, Yang Y, Green B, Li D, Zeng D (2021) A review on two-dimensional materials for chemiresistive-and FET-type gas sensors. Phys Chem Chem Phys
4. Zuckerkandl E, Pauling L (1965) Evolutionary divergence and convergence in proteins. In: Evolving genes and proteins (pp 97–166). Academic Press (1965)
5. Xu Z, Zheng QS, Chen G (2007) Elementary building blocks of graphene-nanoribbon-based electronic devices. Appl Phys Lett 90(22):223115
6. Basu S, Chatterjee S, Saha M, Bandyopadhay S, Mistry KK, Sengupta K (2001) Study of electrical characteristics of porous alumina sensors for detection of low moisture in gases. Sens Actuators B Chem 79(2–3):182–186
7. Zhang YH, Chen YB, Zhou KG, Liu CH, Zeng J, Zhang HL, Peng Y (2009) Improving gas sensing properties of graphene by introducing dopants and defects: a first-principles study. Nanotechnology 20(18):185504
8. Fowler JD, Allen MJ, Tung VC, Yang Y, Kaner RB, Weiller BH (2009) Practical chemical sensors from chemically derived graphene. ACS Nano 3(2):301–306
9. Gao H, Liu Z (2017) DFT study of NO adsorption on pristine graphene. RSC Adv 7(22):13082–13091
10. Leenerarts O, Partoens B (2008) Adsorpton of H2O, NH3, CO, NO2 and NO2 on graphene: a first principle study. Phys Rev B 77(12):1–6
11. Dai J, Yuan J, Giannozzi P (2009) Gas adsorption on graphene doped with B, N, Al, and S: a theoretical study. Appl Phys Lett 95(23):232105
12. Rad AS (2015) First principles study of Al-doped graphene as nanostructure adsorbent for NO2 and N2O: DFT calculations. Appl Surf Sci 357:1217–1224
13. Ali SMU, Nur O, Willander M, Danielsson B (2010) A fast and sensitive potentiometric glucose microsensor based on glucose oxidase coated ZnO nanowires grown on a thin silver wire. Sens Actuators B Chem 145(2):869–874

14. Liu W, Liu Y, Wang R, Hao L, Song D, Li Z (2014) DFT study of hydrogen adsorption on Eu-decorated single-and double-sided graphene. Phys Status Solidi (b) 251(1):229–234
15. Li Y, Chopra N (2015) Graphene encapsulated gold nanoparticle-quantum dot heterostructures and their electrochemical characterization. Appl Surf Sci 344:27–32
16. Sahithi A, Sumithra K (2020) Adsorption and sensing of CO and NH 3 on chemically modified graphene surfaces. RSC Adv 10(69):42318–42326
17. Yutomo EB, Noor FA, Winata T (2021) Effect of the number of nitrogen dopants on the electronic and magnetic properties of graphitic and pyridinic N-doped graphene–a density-functional study. RSC Adv 11(30):18371–18380
18. Esrafili MD (2019) Boron and nitrogen co-doped graphene nanosheets for NO and NO2 gas sensing. Phys Lett A 383(14):1607–1614
19. Mandado M, Blockhuys F, Van Alsenoy C (2006) On the applicability of QTAIM, Hirshfeld and Mulliken delocalisation indices as a measure of proton spin–spin coupling in aromatic compounds. Chem Phys Lett 430(4–6):454–458
20. Sahithi A, Sumithra K (2021) New insights in the electronic structure of doped graphene on adsorption with oxides of nitrogen. Mater Today Commun 27:102417
21. QuantumATK version Q-2019.12, Synopsys QuantumATK [Online]. Available https://quantumwise.com
22. Wang J, Ma F, Sun M (2017) Graphene, hexagonal boron nitride, and their heterostructures: properties and applications. RSC Adv 7(27):16801–16822
23. Vorontsov AV, Tretyakov EV (2018) Determination of graphene's edge energy using hexagonal graphene quantum dots and PM7 method. Phys Chem Chem Phys 20(21):14740–14752
24. Aghaei SM, Monshi MM, Torres I, Zeidi SMJ, Calizo I (2018) DFT study of adsorption behavior of NO, CO, NO2, and NH3 molecules on graphene-like BC3: a search for highly sensitive molecular sensor. Appl Surf Sci 427:326–333
25. Hammer BHLB, Hansen LB, Nørskov JK (1999) Improved adsorption energetics within density-functional theory using revised Perdew-Burke-Ernzerhof functionals. Phys Rev B 59(11):7413
26. Kong L, Enders A, Rahman TS, Dowben PA (2014) Molecular adsorption on graphene. J Phys Condens Matter 26(44):443001
27. Tit N, Said K, Mahmoud NM, Kouser S, Yamani ZH (2017) Ab-initio investigation of adsorption of CO and CO2 molecules on graphene: role of intrinsic defects on gas sensing. Appl Surf Sci 394:219–230

# Quantum Fault-Tolerant Implementation of a Majority-Based 4-Bit BCD Adder

**Laxmidhar Biswal, Niladri Pratap Maity, and Hafizur Rahaman**

**Abstract** The greatest irony in the development of viable quantum computers is that decoherence limits scalability and has made quantum information states (qubits) fragile. The recent development of Google's 54-qubit "SYCAMORE"and IBM's 127-qubit "EAGLE"fully programmable quantum information processors were significant breakthroughs in the quantum world that encouraged the research community to contribute to the achievement of quantum supremacy. Quantum computers are based on quantum mechanics, and their Hamiltonians are unitary in nature, which distinguishes them from classical logic. As a result, Boolean logic is no longer useful in quantum computing. To limit the effect of decoherence in this Noisy Intermediate Scale Quantum System(NISQ), a quantum error correcting code, such as surface code with transversal operator-based fault-tolerant architecture, has been used. BCD-Adder is a popular adder logic used in digital computers and calculators to perform arithmetic operations directly in the decimal number system, with major applications in the finance sector (e.g. payroll and tax processing). Again, majority logic transforms the ripple carry adder into the carry-look-ahead adder. Using the Clifford+T-group, We present a 4-bit majority-based BCD adder with a fault-tolerant quantum circuit for 1-digit decimal numbers.

**Keywords** Bcd adder · Clifford+T · Majority

L. Biswal (✉) · N. P. Maity
Department of ECE, Mizoram University, Aizawl, India
e-mail: laxmidhar.cvrce@gmail.com

H. Rahaman
School of VLSI Technology, IIEST Shibpur, Howrah, India
e-mail: hafizur@vlsi.iiests.ac.in



C. Giri et al. (eds.), *Emerging Electronic Devices, Circuits and Systems*,
Lecture Notes in Electrical Engineering 1004,
https://doi.org/10.1007/978-981-99-0055-8_4

# 1 Introduction

The goal of the quantum revolution is to outperform over classical computer in the areas of agriculture, energy, environment, climate, materials science and health [1, 2] in reasonable time, but it is not yet practical due to the decoherence constraint [3]. Quantum computers use the properties of quantum physics at the subatomic level to store, process data and perform computations [4]. The initial power of a quantum computer lies with a quantum bit (qubit), which holds coherence and entanglement between the superposed basis states. The formal definition, and properties on qubit is presented in Sect. 2.

The most significant barrier to quantum processing scalability is the fragile quantum state, which can only be overcome using a quantum error correcting code at the logical level. In this regard, surface code is most promising one with highest error rate threshold value per gate (approximately 0.75%) [5]. As a result, when translating any algorithm into a sequence of unitary operators, the error rate per gate must be kept below the threshold for high fidelity quantum processing, which necessitates the use of a universal transversal primitive gate set [6–9]. In this support, the Clifford+T-group is a well-known collection of basic universal primitive transversal operators. Section 2 discusses the popular Clifford+$T$-group. Indeed, Clifford+T-group with surface becomes *de-facto* platform for designing of fault-tolerant quantum circuit [10].

The quantum computer must solve both classically compliant and interactable problems. In fact, Boolean logic is no longer useful in quantum computing. As a result, finding equivalent high fidelity quantum logic for existing Boolean logic is now an important area of research. It is worth noting here that reversibility is an essential property of quantum logic due to the unitary properties of quantum transformation. Indeed, reversible logic acts as an intermediary logic when mapping any Boolean logic into quantum logic. With this above background, we implement fault-tolerant quantum 4-bit majority-based BCD adder using Clifford+T-group. The BCD-adder is found in computers and calculators that perform arithmetic operations directly in the decimal number system for commercial, financial, and internet-based applications.

The following are the key contributions made in this paper:

- In reversible logic, we proposed an optimised 4-bit majority-based BCD adder for 1 digit decimal number.
- Using the Clifford+T-group, we implemented the proposed 4-bit majority-based BCD adder in fault-tolerant quantum logic.
- We then restructured the fault-tolerant 4-bit BCD adder in a low-circuit-depth environment with additional ancillary cost.

The rest of the paper is structured as follows: Sect. 2, summarizes the reversible and quantum operators, as well as the performance parameters associated with the quantum circuit. Our approach is described in Sect. 3. The experimental results and comparative analysis are described in Sect. 4. Finally, the work is completed in Sect. 5.

## 2 Preliminaries

For a better understanding of the paper, this section provides a detailed background on quantum operators and gate libraries. We also defined the performance parameter associated with the design of fault-tolerant quantum circuits. We also discussed the 4-bit conventional BCD adder.

### 2.1 Quantum Circuits

**Definition 1** ***(Quantum circuit).*** *A quantum circuit is a unitary model for quantum computation that consists of a series of elementary quantum gates.*

**Definition 2** ***(Quantum gate).*** *It is the foundation of a unitary quantum circuit that acts on selected qubits over a fixed time period.*

*All elementary quantum gates are mathematically* $2 \times 2$ *or* $4 \times 4$ *unitary transfer matrices.*

**Definition 3** ***(Clifford+T-group).*** *A set of transversal elementary primitive quantum operators that provide universality which include a CNOT-gate with a single qubit Pauli(X, Y, Z)-gate, S-gate, H-gate, and non-Clifford T-gate [6, 11]. Mathematically, T-gate represents* $4^{th}$*-root of Pauli-Z gate and is represented by a transfer matrix:*

$$T = \begin{bmatrix} 1 & 0 \\ 0 & e^{\frac{i\pi}{4}} \end{bmatrix} \quad (1)$$

*The NCV gate library, like the Clifford+T-group, consists of two qubits CNOT, CV,* $CV^{\dagger}$*, and one qubit V and* $V^{\dagger}$ *gate [12]. It is worth noting that the NCV gate library and the Clifford+T-group are inter-convertible using the identities shown in Fig. 1. Table 1 shows a list of quantum gates and their properties for ease of understanding.*

**Definition 4** *$(T - count)$ It refers to the minimum number of T-gates required to implement a quantum functionality in circuit form.*

**Definition 5** *$(T - depth)$ The minimum number of $T - cycles$ required to execute all of the T-gates of a quantum circuit.*

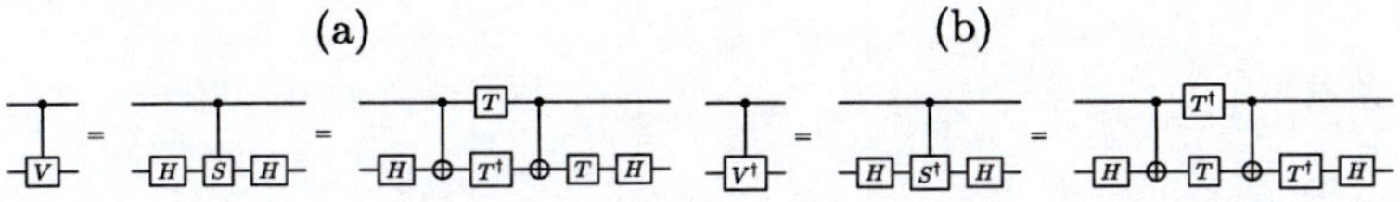


**Fig. 1** Clifford+$T$-based fault-tolerant circuit of $CS$, $CS^{\dagger}$, $CV$, and $CV^{\dagger}$-gates as depicted in [6, Fig. 1b]

**Table 1** Elementary quantum operators and their properties

| Name of elementary quantum gate | Block diagram | Transformation matrix | Properties | Name of elementary quantum gate | Block diagram | Transformation matrix | Properties |
|---|---|---|---|---|---|---|---|
| NOT (X) | $\oplus$ | $\begin{bmatrix} 0 & 1 \\ 1 & 0 \end{bmatrix}$ | $X\|0\rangle = \|1\rangle$<br>$X\|1\rangle = \|0\rangle$ | CNOT(CN) | $\bullet$<br>$\oplus$ | $\begin{bmatrix} 1 & 0 & 0 & 0 \\ 0 & 1 & 0 & 0 \\ 0 & 0 & 0 & 1 \\ 0 & 0 & 1 & 0 \end{bmatrix}$ | $\|00\rangle \rightarrow \|00\rangle$<br>$\|01\rangle \rightarrow \|01\rangle$<br>$\|10\rangle \rightarrow \|11\rangle$<br>$\|11\rangle \rightarrow \|10\rangle$ |
| $Z$ gate | $\boxed{Z}$ | $\begin{bmatrix} 0 & 0 \\ 0 & -1 \end{bmatrix}$ | $X\|0\rangle = \|0\rangle$<br>$X\|1\rangle = -\|1\rangle$ | Hadamard(H) | $\boxed{H}$ | $\frac{1}{\sqrt{2}}\begin{bmatrix} 1 & 1 \\ 1 & -1 \end{bmatrix}$ | $H\|0\rangle = \frac{1}{\sqrt{2}}(\|0\rangle + \|1\rangle)$<br>$H\|1\rangle = \frac{1}{\sqrt{2}}(\|0\rangle - \|1\rangle)$ |
| S gate | $\boxed{S}$ | $\begin{bmatrix} 1 & 0 \\ 0 & i \end{bmatrix}$ | $S\|0\rangle = \|0\rangle$<br>$S\|1\rangle = e^{\frac{i\pi}{2}}\|1\rangle$ | $S^{\dagger}$ gate | $\boxed{S^{\dagger}}$ | $\begin{bmatrix} 1 & 0 \\ 0 & e^{\frac{-i\pi}{2}} \end{bmatrix}$ | $S^{\dagger}\|0\rangle = \|0\rangle$<br>$S^{\dagger}\|1\rangle = e^{\frac{-i\pi}{2}}\|1\rangle$ |
| $T$ gate | $\boxed{T}$ | $\begin{bmatrix} 1 & 0 \\ 0 & e^{\frac{i\pi}{4}} \end{bmatrix}$ | $T\|0\rangle = \|0\rangle$<br>$T\|1\rangle = e^{\frac{i\pi}{4}}\|1\rangle$ | $T^{\dagger}$ gate | $\boxed{T^{\dagger}}$ | $\begin{bmatrix} 1 & 0 \\ 0 & e^{\frac{-i\pi}{4}} \end{bmatrix}$ | $T^{\dagger}\|0\rangle = \|0\rangle$<br>$T^{\dagger}\|1\rangle = e^{\frac{-i\pi}{4}}\|1\rangle$ |

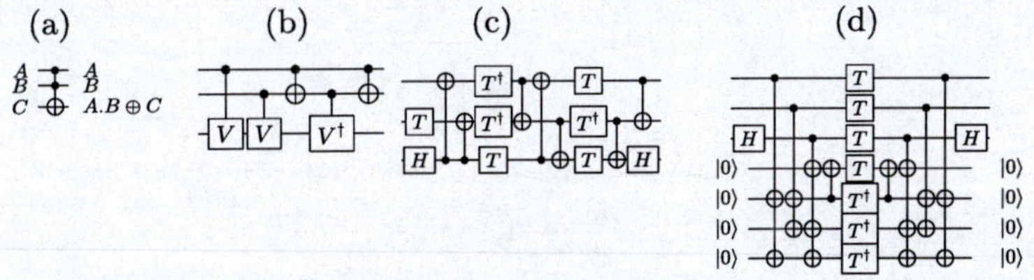


**Fig. 2** **a** Toffoli gate. **b** Fig. 2a represents equivalent NCV-based decomposition. **c** Equivalent Clifford+$T$-based realization of Fig. 2a as depicted in [6, Fig. 1c]. **d** Equivalent unit T-depth based realization of Fig. 2a as depicted in [13, Fig. 1].

**Fig. 3** **a** 3-input majority function. **b** 3-input majority function in reversible logic

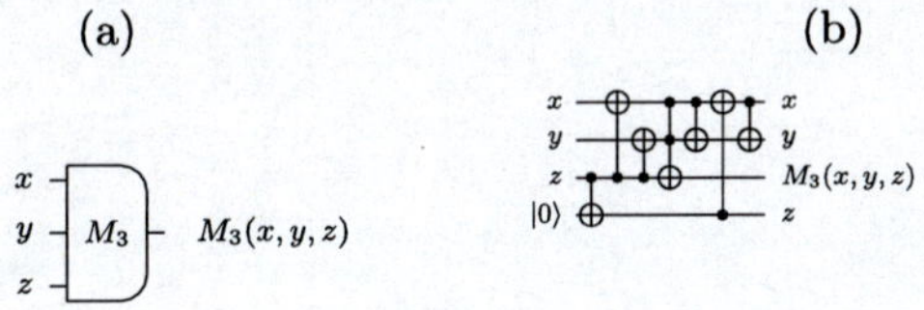


**Definition 6** ***(Toffoli gate)*** *It is a three-qubit reversible gate with two controls and one target that toggles the target bit's state when both control bits are set to* 1. *Fig. 2a and Fig. 2b depict the Toffoli gate and its NCV-based Toffoli gate, respectively [12]. Figure 2c depicts the Clifford+T-based fault-tolerant implementation of Toffoli gate while its unit phase-depth-based design with additional ancillary input is shown in Fig. 2d.*

**Definition 7** ***(Majority function)*** *The majority function is a threshold function that evaluates to logic '1', that is, true if and only if the majority of the inputs are 1. The majority function for n-variables is denoted by* $\langle x_1 x_2 \ldots\ldots x_n \rangle$, *and evaluates to true as '1' if and only if at least* $\frac{n}{2}$ *and* $\frac{n+1}{2}$ *numbers of variables out of n-variables are true for even and odd n, respectively [14]. Figure 3a depicts a schematic of the majority function (*$\langle xyz \rangle$*) for three input variables, namely x, y and z. Its equivalent reversible is shown in Fig. 3b as shown in [15, Fig. 3]. It can also be expressed Mathematically as:*

$$\langle xyz \rangle = M(x, y, z) = xy + yz + zx \tag{2}$$

$$= x \wedge y \vee y \wedge z \vee z \wedge x \tag{3}$$

$$= \begin{cases} x \wedge y \; z = 0 \\ x \vee y \; z = 1 \end{cases} \tag{4}$$

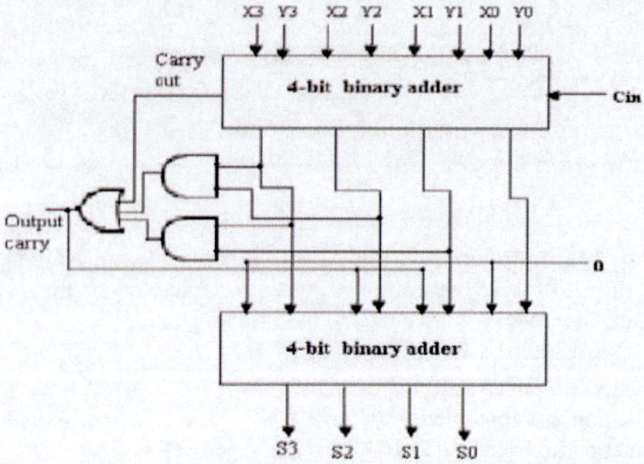


**Fig. 4** Pictorial representation of 4-bit conventional BCD adder

## 2.2 *4-Bit BCD Adder*

A BCD adder is a four-bit binary adder that can add two binary-coded decimal numbers. It appears that this addition operation can be performed using a 4-bit parallel adder, which will provide a 4-bit sum and a 1-bit carry; however, one more parallel adders with correction logic circuit are required. Though BCD code appears to be 4-bit binary, but, it is quite different in practise and is related to the decimal number system. It is worth noting that the BCD adder's input and output are always within the range of '0' to '9' and forbids when the 4-bit sum exceeds 1001, *i.e.*, 1010 to 1111, whose decimal equivalent is '10' to '15' because all of those bits are invalid BCD. For detection and correction of all those invalid sum bit, additional correction logic with a 4-bit parallel adder are used. Correction logic consists two AND and one OR logic, performed detection and add binary 0110 to the sum bits again. A 4-bit BCD adder using Boolean logic is shown in Fig. 4 as depicted in [16, Fig. 1].

# 3 Proposed Methodology

The proposed Boolean logic to quantum logic mapping is a two-tier mapping in which the 4-bit BCD adder's Boolean logic is mapped into reversible logic in the first step, and the equivalent reversible logic is mapped into fault-tolerant quantum logic in the second step by decomposing the universal reversible gate, *i.e.*, the Toffoli gate, into the Clifford+T-group. The entire mapping strategy is presented in two subsections, as follows:

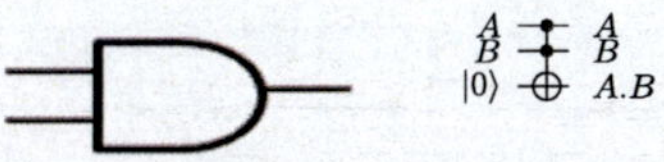


**Fig. 5** **a** 2-input AND gate. **b** Equivalent reversible AND gate.

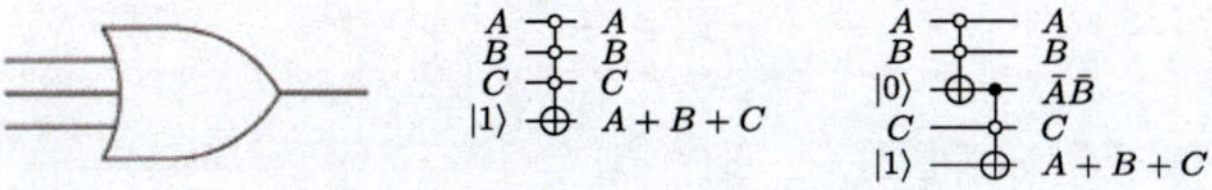


**Fig. 6** **a** 3-input OR gate. **b** Equivalent reversible OR gate using 3-MCT gate. **c** Equivalent reversible OR gate using Toffoli gate.

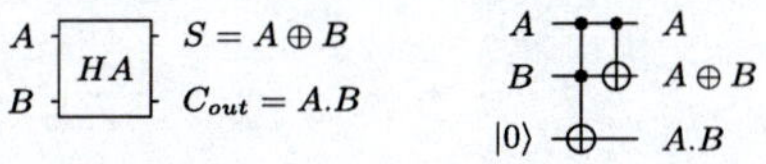


**Fig. 7** **a** Conventional HA **b** Equivalent reversible HA

## *3.1 Reversible 4-Bit BCD Adder*

To obtain an equivalent circuit in reversible logic, each small block of the conventional 4-bit BCD is replaced by its reversible equivalent. Figure 4 depicts a 4-bit BCD adder composed of two parallel adders, two AND logic, and an OR logic. In the case of parallel adders, the digital world discovered that propagation delay is a significant performance disadvantage. Furthermore, the majority-based data structure converts the ripple carry adder into the carry-look-ahead adder [9, 17].

An AND-logic can be implemented by using a Toffoli-gate with constant target bit '0'. Both the 2-input AND gate and its equivalent reversible logic are shown in Fig. 5. An OR-logic, like AND logic, can be implemented using a control-low based Toffoli-gate with a constant target bit of '1'. Both the 3-input OR gate and its equivalent reversible logic is shown in Fig. 6. Throughout the paper, the symbol • represents positive control, whereas; ○ represents negative control Fig. 7.

A 4-bit parallel adder is made up of four full-adders (FA), with the carry of the preceding FA becoming the carry-in for each succeeding FA. To avoid carry propagation delay, we implemented a 4-bit parallel adder using the majority-based FA circuit proposed in [17, Fig. 6] against each conventional FA and its equivalent reversible logic circuit. Figure 8 depicts a majority-based FA and its equivalent reversible circuit for the clarity. It is worth noting that the secondary 4-bit parallel adder does not include input carry, so it only has three FAs after a Half Adder (HA) rather than four. The reversible HA can be implemented using a Peres gate with target bit "0," and the reversible equivalent realization is shown in Fig. 7.

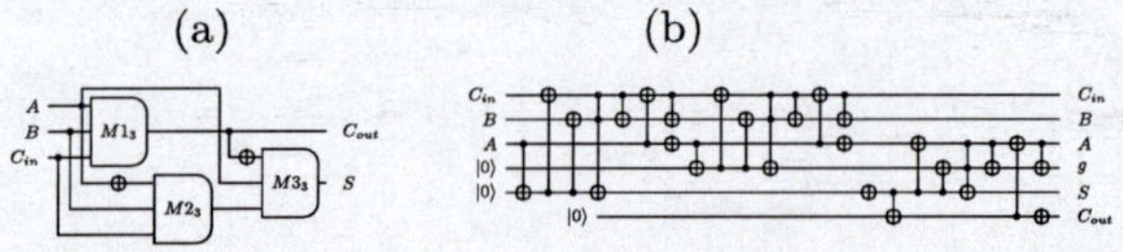


**Fig. 8** **a** Majority-based 1-bit full-adder. **b** Equivalent reversible circuit as depicted in [17, Fig. 5]

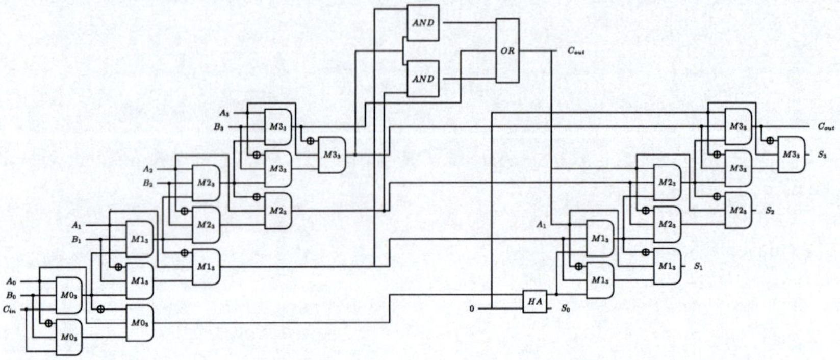


**Fig. 9** 4-bit majority-based BCD adder

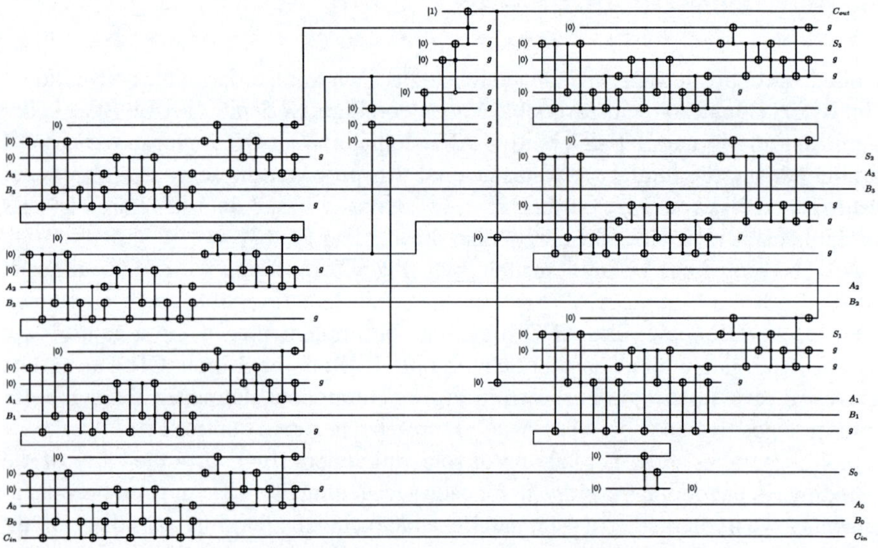


**Fig. 10** 4-bit majority-based reversible BCD adder

Proposed conventional 4-bit BCD adder using majority logic and its equivalent reversible logic are shown in Fig. 9 and Fig. 10, respectively, by considering the reversible realization of conventional AND, OR, HA and 1-bit FA.

## 3.2 Fault-Tolerant Quantum Implementation of a 4-Bit BCD Adder

Compiling the proposed reversible BCD adder into the Clifford+T-group is a significant challenge for the precise synthesis of fault-tolerant quantum circuits. To address this problem, we employ a template matching scheme in which each small unit in Fig. 10 is replaced by its Clifford+T-based quantum circuit equivalent. To implement template matching scheme, Fig. 10 is scanned from left to right and buttom to top approach, revealing that the proposed 4-bit reversible BCD adder is made up of 28 NOT, 132 CNOT, 24 Toffoli gates, 1 Toffoli gate with negative controls, and 1 Toffoli gate with single negative control bit which consider as templates($temp$).

Let us define equivalent circuits against each template before we begin replacing each small unit of Fig. 10 with its equivalent fault-tolerant quantum circuit. While mapping from reversible logic to quantum logic, the NOT and CNOT gates remain intact; however, the Toffoli gate can be replaced by its equivalent Clifford+T-based circuit, as shown in Fig. 3. For the replacement of mixed polarity, i.e., one negative control and two negative control Toffoli gate, we used Clifford+T-based circuit presented in [18, Fig. 8] and Fig. [18, Fig. 7] respectively which is shown in Fig. 11 and Fig. 12 respectively for the better apprehension.

Due to the high latency and hardware cost of the non-Clifford T-gate, the T-count and T-depth become more important in the synthesis of fault-tolerant quantum circuits. $T/T^\dagger$-gate can be executed in parallel by using additional ancillary cost, as shown in Fig. 13. In addition, qubit size detriment the entanglement of quantum state with increasing quibit-count which insists optimization of qubit size. Furthermore the garbage output dissipates more heat [19]. Taking current facts into consideration, two types of fault-tolerant quantum circuits are proposed for the implementation of a BCD adder: one with $T - depth$ and the other is qubit optimized. In this regard, we define two sets of template library, *viz.* $T_Q$ and $T_D$, which consist of Clifford+T-based circuits versus templates optimized for qubits and $T - depth$, respectively.

$$T_Q = \{Fig.\ 2c,\ Fig.\ 11c,\ Fig.\ 12c\} \tag{5}$$

$$T_D = \{Fig.\ 2d,\ Fig.\ 11d,\ Fig.\ 12d\} \tag{6}$$

Now, we resume the replacement approach, replacing each small reversible gate in Fig. 10 with its Clifford+T-based equivalent as stated above, but due to space constraints, we are unable to show the resulting Clifford+T-based fault-tolerant 4-bit BCD adder in quantum logic.

In the qubit-count optimised design approach, each double active high, active low control, and double active low controls Toffoli gate in Fig. 10 is replaced by its equivalent Clifford+T-based fault-tolerant quantum circuit defined in the quantum template library $T_Q$(5). And each has 7 $T - count$ and 3 $T - depth$. Figure 10 shows 24 double active high controlled Toffoli gates, a single active low controlled Toffoli gate, and a double active low controlled Toffoli gate, which resulted in a total T-count

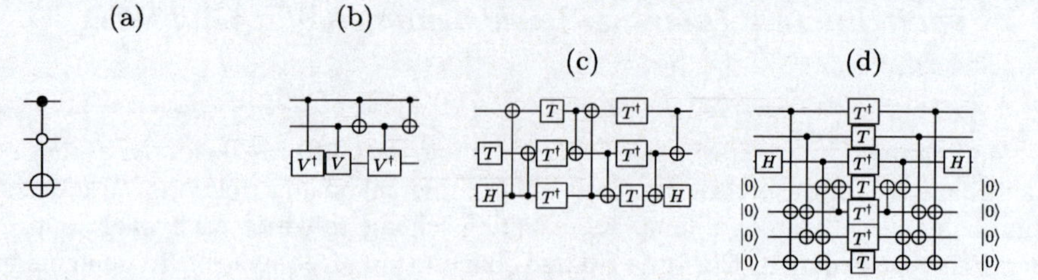


**Fig. 11** **a** Toffoli gate with a negative control. **b** NCV based decomposition of Fig. 11a. **c**.Clifford+T-based quantum implementation of Fig. 11a. **d** Equivalent unit $T$-depth structure of Fig. 11a

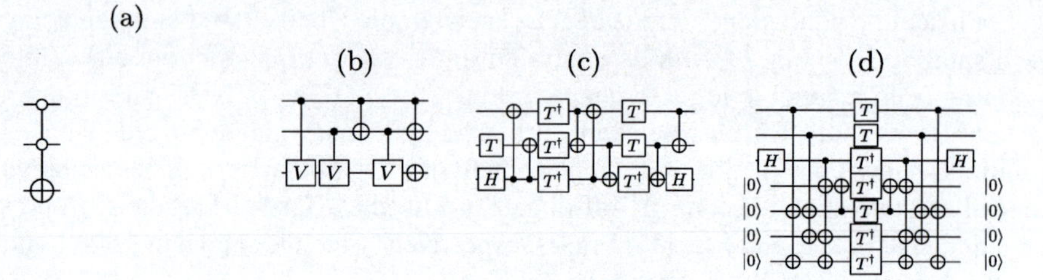


**Fig. 12** **a** Toffoli gate with negative controls. **b** NCV based decomposition of Fig. 12a. **c** Clifford+T-based decomposition of Fig. 12a. **d** Equivalent unit $T$-depth structure of Fig. 12a

**Fig. 13** Equivalent quantum gate identities as depicted in [7, Fig. 3(f)]

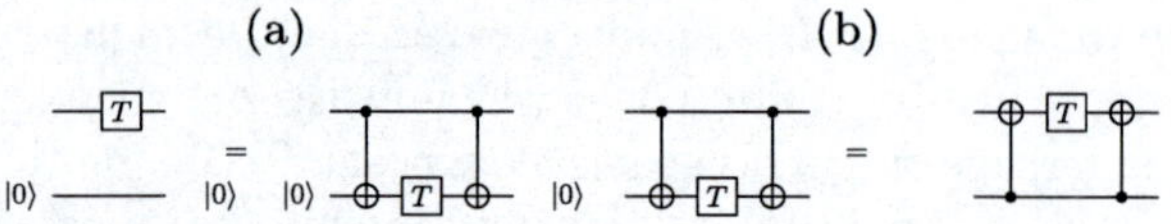


of 182 . Furthermore, 8 pairs of the 24 Toffoli gates can be executed concurrently, implying that 24 $T - cycle$ are required to execute all 16 Toffoli gates. As a result, the total net $T - depth$ of Fig. 10 required to execute all types of Toffoli gates is 54. To put it another way, 54 $T - cycle$ are required to execute all Toffoli gates.

Let's talk about the $T - depth$ optimized design approach, in which Fig. 2d, Fig. 11d and Fig. 12d defined under template library $T_D(6)$, replace the double active high, single active low, and double active low controlled Toffoli gates of Fig. 10, respectively. The out-coming circuit incurred 186 $T - count$ after replacement, which remained equal to the qubit-count optimized design. But, Fig. 2d, Fig. 11d and Fig. 12d provides unit $T - depth$ fault-tolerant quantum circuit by means of 4 additional ancillary cost. It is worth noting that four ancillary inputs are initiated by $|0\rangle$ and provide output that is the same as the input, *i.e.*, $|0\rangle$, allowing the same four lines to be used by all Toffoli gates without distorting their functionality. And the addition of four ancillary lines reduces the $T - depth$ of the overall design by a factor of three, resulting in a $T - depth$ of 18 instead of 54. This design approach incurs additional costs due to the addition of four ancillary qubits.

## 4 Experimental Results

In this paper, a 4-bit BCD adder for a 1-digit decimal adder is implemented in both reversible and fault-tolerant quantum logic. To avoid carry propagation delays, the majority data structure is used in this design approach. A low $T - depth$ fault-tolerant quantum circuit is also introduced to aid in the suppression of the effect of decoherence, which is required to protect the fragile quantum state. Performance parameters such as $T - count$, $T - depth$, and ancillary cost are already calculated and discussed in Sec. 3.2. During reversible implementation, two Toffoli gates are used to implement OR logic instead of a single 3-qubit Toffoli gate as because fault-decomposition of a 3-qubit Toffoli gate incurs more $T - depth$, i.e., 8 [11], whereas two Toffoli gates incurred 6 $T - depth$.

It is worth noting that none of the current state-of-the-arts provide fault-tolerant implementation 4-bit BCD adder quantum logic, so this work cannot be compared to others. However, there have been numerous reports of works relating to the BCD adder in [20–22]. The inclusion of a majority data structure in the design ab-initio of a BCD adder in conventional logic to fault-tolerant quantum logic is the single most significant advantage over the existing BCD adder in conventional and reversible logic, as it converts ripple carry adder to carry look ahead adder.

Furthermore, multiple garbage outputs are shown in Fig. 10 *w.r.t.* sum and carry bit of 4-bit BCD adder, but they are all out-coming the sum and carry bits of the 4-bit parallel adder circuit and can be used for other purposes in practice. The proposed 4-bit reversible BCD adder can serve multiple functions, including four-bit binary adder and BCD adder. In addition, all output corresponds to input lines that have been returned to the initial inputs, allowing inputs to be reused. As a result, the net garbage output is NIL.

## 5 Conclusion

In this paper, we presented a Clifford+$T$-based template matching scheme over majority logic to implement a quantum fault-tolerant 4-bit BCD adder circuit. The matrix decomposition method is used for the mapping from reversible logic to Clifford+$T$-based quantum logic, which is fault-tolerant by default. Our design methodologies are optimized for $T - depth$, $T - count$, garbage outputs, and qubit-count.

## References

1. Peter W Shor. Polynomial-time algorithms for prime factorization and discrete logarithms on a quantum computer. *SIAM review*, 41(2):303–332, 1999
2. Grover Lov K (1997) Quantum mechanics helps in searching for a needle in a haystack. Phys. Rev. Lett. 79(2):325–328
3. David P. DiVincenzo and Ibm. The physical implementation of quantum computation. volume 48. onlinelibrary.wiley.com, 2000
4. R.P.Feynman. Quantum mechanical computers. *Foundations of Physics*, 16(6), 1986
5. Austin G. Fowler, Matteo Mariantoni, John M. Martinis, and Andrew N. Cleland. Surface codes: Towards practical large-scale quantum computation. *Phys. Rev. A*, 86, 2012
6. L. Biswal, C. Bandyopadhyay, A. Chattopadhyay, R. Wille, R. Drechsler, and H. Rahaman. Nearest-neighbor and fault-tolerant quantum circuit implementation. In *2016 IEEE 46th International Symposium on Multiple-Valued Logic (ISMVL)*, pages 156–161, May 2016
7. L. Biswal, R. Das, C. Bandyopadhyay, A. Chattopadhyay, and H. Rahaman. A template-based technique for efficient clifford+t-based quantum circuit implementation. *Microelectronics Journal*, 81, 2018
8. Biswal Laxmidhar, Bhattacharjee Debjyoti, Chattopadhyay Anupam, Rahaman Hafizur (2019) Techniques for fault-tolerant decomposition of a multicontrolled toffoli gate. Phys. Rev. A 100:062326 Dec
9. Laxmidhar Biswal, Habibur Rahaman, and Niladri Pratap Maity. Clifford+t-based implementation of fault-tolerant quantum circuits over xor-majority graph. *Microelectronics Journal*, 116:105212, 2021
10. Michael A. Nielsen and Isaac L. Chuang. *Quantum Computation and Quantum Information: 10th Anniversary Edition*. Cambridge University Press, 2010
11. N. Abdessaied, M. Soeken, M. Kirkedal, Thomsen, and R. Drechsler. Upper bounds for reversible circuits based on young subgroups. *Inf. Process. Lett.*, 114:282–286, 2014
12. Barenco Adriano, Bennett Charles H, Cleve Richard, DiVincenzo David P, Margolus Norman, Shor Peter, Sleator Tycho, Smolin John A, Weinfurter Harald (1995) Elementary gates for quantum computation. Phys. Rev. A 52:3457–3467 Nov
13. P. Selinger. Quantum circuits of $t$-depth one. *Phys. Rev. A*, 87, Apr 2013
14. M. H. Moaiyeri, R. F. Mirzaee, K. Navi, and T. Nikoubin. New high-performance majority function based full adders. In *2009 14th International CSI Computer Conference*, pages 100–104, 2009
15. Anupam Chattopadhyay, Luca Amarú, Mathias Soeken, Pierre-Emmanuel Gaillardon, and Giovanni De Micheli. Notes on majority boolean algebra. In *2016 IEEE 46th International Symposium on Multiple-Valued Logic (ISMVL)*, pages 50–55, 2016
16. H. Thapliyal, S. Kotiyal, and M.B. Srinivas. Novel bcd adders and their reversible logic implementation for ieee 754r format. In *19th International Conference on VLSI Design held jointly with 5th International Conference on Embedded Systems Design (VLSID'06)*, pages 6 pp.–, 2006
17. Laxmidhar Biswal, Bappaditya Mondal, Anindita Chakraborty, and Hafizur Rahaman. Efficient quantum implementation of majority-based full adder circuit using clifford+t-group. In Biswajit Mishra, Jimson Mathew, and Priyadarsan Patra, editors, *Artificial Intelligence Driven Circuits and Systems*, pages 53–63, Singapore, 2022. Springer Singapore
18. Laxmidhar Biswal, Khokan Mondal, Anirban Bhattacharjee, and Hafizur Rahaman. Fault-tolerant quantum implementation of priority encoder circuit using clifford+t-group. In *2020 International Symposium on Devices, Circuits and Systems (ISDCS)*, pages 1–6, 2020
19. C. H. Bennett. Logical reversibility of computation. *IBM Journal of Research and Development*, 17(6), 1973
20. Himanshu Thapliyal and Nagarajan Ranganathan. Design of efficient reversible logic-based binary and bcd adder circuits. *J. Emerg. Technol. Comput. Syst.*, 9(3), oct 2013

21. Low quantum cost realization of reversible binary-coded-decimal adder. *Procedia Computer Science*, 167:1437–1443, 2020. International Conference on Computational Intelligence and Data Science
22. Murugesan Praveena, Keppanagounder Thanushkodi, Natarajan Vijeyakumar (2016) Design of efficient reversible bcd adder-subtractor architecture and its optimization using carry skip logic. Journal of Circuits, Systems and Computers 25(07):1650076

# CNTFET-Based Universal Filter Using DO-CCII

**Mohd Yasir and Naushad Alam**

**Abstract** This paper presents a low voltage low-power dual output second generation current conveyor (DO-CCII)-based universal filter using carbon nanotube FET (CNTFET) technology. The 3 dB BW for voltage gain and current gain is 54.028 GHz and 48.846 GHz, respectively, for DO-CCII. The average power consumed by the DO-CCII is 440.33 μW. It has a DC voltage range from −370 to 410 mV and a current range from −160 to 160 μA. The multi-input single output (MISO) universal filter provides low pass (LP), high pass (HP), bandpass (BP), band reject (BR), and all pass (AP) responses in voltage mode. The filter uses two DO-CCII blocks, two capacitors, and two voltage-controlled resistors, providing resistor behavior. The simulations of all the circuits are performed using HSPICE with a supply voltage of ± 0.5 V. The universal filter operates at 14.79 MHz for controlled voltages of 0.5 V each. The operating frequency can be tuned easily by using the control voltages implementing voltage-controlled resistors. Since the circuit is free from resistors, it can easily be used in analog signal processing ICs for high-frequency applications. The average power consumed by the universal filter is 961.80 μW.

**Keywords** Carbon Nanotube (CNT) · Carbon Nanotube Field Effect Transistor (CNTFET) · Current conveyor · MISO · Universal biquadratic filter · Voltage mode

## 1 Introduction

The dimensions of the current transistors are so low that it has become very challenging to prevent many issues. Some of the issues are scattering effect, decreased gate control over drain current ($I_D$), parasitics, random dopant fluctuation, channel mobility, lithographic limitations, threshold voltage ($V_{TH}$) variability, drain to source

M. Yasir (✉) · N. Alam
Department of Electronics Engineering, Z. H. College of Engineering and Technology, Aligarh Muslim University, Aligarh, UP, India
e-mail: mohdyasir@zhcet.ac.in

N. Alam
e-mail: naushad.el@amu.ac.in



C. Giri et al. (eds.), *Emerging Electronic Devices, Circuits and Systems*, Lecture Notes in Electrical Engineering 1004,
https://doi.org/10.1007/978-981-99-0055-8_5

tunneling, increased heat production, etc., using conventional MOSFET [2]. One of the contenders to replace the conventional MOSFET is carbon nanotube FET (CNTFET) because of its enhanced electrical and material properties like near ballistic charge transport, higher cut off frequency, smaller size, larger drive current, fast switching speed, and low parasitics [26]. The bonds in a graphene sheet between the carbon atoms are very strong while between layers are very weak, which provide unique electrical and material characteristics. The chirality, i.e., how the graphene sheet is rolled, determines the nature of the CNT, whether it is metallic or semiconducting.

Dr. Ijima first invented the CNTs in 1991 [10]. Subsequently, a lot of CNTFET architectures and models have been presented in the literature [4–6, 19]. However, for the design of integrated circuits, the model of CNTFET must be compatible to integrate with other circuit components. This has been done only in SPICE compatible models [4, 5]. In our work, we have used the stanford model presented in [4, 5]. This model includes most of the non-idealities reported in [4–6, 19].

Active filters are used in many integrated circuits (ICs) in communication, biomedical, control, etc., because they can be integrated easily. Active filters, which provide various functions from a single circuit topology, are most popular in signal processing applications. Especially, multi-input single output (MISO) topology is most famous [21]. These biquadratic filters are mostly implemented using various active building blocks like second-generation current conveyor (CCII), voltage differencing buffered amplifier (VDBA), voltage differencing transconductance amplifier (VDTA), operational transconductance amplifier (OTA), voltage differencing differential amplifier (VDDA), differential difference current conveyor (DDCC), current differencing transconductance amplifier (CDTA), fully differential second-generation current conveyor (FDCCII), etc. However, most of the abovementioned filters consume enormous power and operate at lower operating frequency [3, 11, 12, 15–17, 20, 22, 23].

In this paper, a DO-CCII-based universal biquadratic filter using CNTFET technology has been implemented. It provides low voltage, low-power, high-frequency operation. The circuit is suitable for ICs as resistors are implemented using voltage-controlled transistors. At the same time, capacitors are of very low value, i.e., 1 pF, which can easily be fabricated in today's technology. The process variation study uses Monte Carlo simulation using $3\sigma$ and 10% variations.

This paper is divided into five sections. The overview of CNTFET is presented in Sect. 2. Some of the aspects related to DO-CCII are discussed in Sect. 3. The performance of the CNTFET DO-CCII-based MISO universal biquadratic filter is investigated in Sect. 4 using HSPICE. The conclusions are drawn in Sect. 5.

## 2 Overview of CNTFET

The CNTFET technology used to design the circuits in this work is based on the Stanford model [1]. The CNTFET parameters are listed in Table 1. The Stanford model is very accurate and includes several non-idealities of CNTFET like inter

CNT capacitance, charge screening effect, the effect of source-drain extension region, drain to source series resistance, etc. [2].

A CNT can be visualized as a rolled graphene sheet, and its behavior depends on chirality ($C_h$). There are three different methods of rolling graphene sheets. The electrical behavior of CNT depends on $n_1$ and $n_2$. If $n_1 - n_2$ is divisible by 3, its behavior is metallic; otherwise, it is semiconducting [24]. The channel region in CNTFET is made up of CNTs, as shown in Fig. 1. This Stanford model of CNTFET is used because of its similarity with conventional MOSFET [1].

The CNTFET is designed using several parameters like $N$ (number of tubes), $S$ (inter CNT pitch), and $D_{\mathrm{CNT}}$ (diameter of a CNT) as compared to CMOS, where the aspect ratio is the only parameter [2]. The width of the CNTFET depends on three parameters, namely $N$, $S$, and $D_{\mathrm{CNT}}$, by the following relation.

$$W = (N - 1) * S + D_{\mathrm{CNT}} \tag{1}$$

The following equation gives the relation between threshold voltage ($V_{\mathrm{TH}}$) and $D_{\mathrm{CNT}}$ [2].

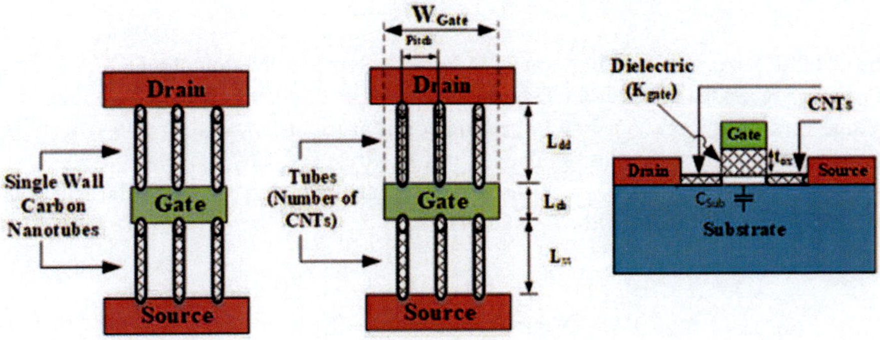


**Fig. 1** Schematic diagram, illustration of modeled CNTFET, and relevant parameters of CNTFET [1]

**Table 1** Technology parameter definitions and default values of a CNTFET [1]

| Device parameter | Default value |
|---|---|
| Oxide thickness of ($T_{\mathrm{OX}}$) | 4.0 nm |
| Chirality of tube ($n_1$, $n_2$) | (19, 0) |
| CNT diameter | 1.49 nm |
| High-k gate oxide material | $HfO_2$ |
| Dielectric constant ($K_{\mathrm{OX}}$) | 16 |
| CNT pitch (S) | 16 nm |
| Physical channel length ($L_{\mathrm{ch}}$) | 32 nm |
| Length of CNT source-side extension region ($L_{ss}$) | 32 nm |
| Length of CNT drain-side extension region ($L_{dd}$) | 32 nm |

$$V_{TH} = \frac{0.436}{D_{CNT}(\text{nm})} \tag{2}$$

where

$$D_{CNT} = \frac{a * (n_1^2 + n_2^2 + n_1 * n_2)^{1/2}}{\pi} \tag{3}$$

where a is the lattice constant for graphene and $n_1$ and $n_2$ are chirality indices specifying the structure of the CNT [18].

The extraordinary electrical properties of CNTFETs for nanotechnology, electronics, optics, and various fields of materials science provide a reliable option compared with conventional CMOS. CNTFETs initially provide a platform for digital circuits like memory components, adders, etc. However, recently, few works are also there on RF and analog applications like voltage-controlled oscillators because CMOS-based circuits are more prone to temperature variations compared to CNTFET-based circuits [7, 8, 13, 14].

## 3 Overview of DO-CCII

The CNTFET-based block diagram and transistor-level realization of a DO-CCII are shown in Fig. 2a, b. The DO-CCII is an active block consisting of a high impedance voltage input terminal Y, a low impedance current input terminal *X*, and two high impedance current output terminals $Z+$ and $Z-$, respectively.

The following equation gives the characteristic equation showing the relation among various terminals of the DO-CCII.

$$\begin{bmatrix} I_Y \\ V_X \\ I_{Z\pm} \end{bmatrix} = \begin{bmatrix} 0 & 0 & 0 \\ 1 & 0 & 0 \\ 0 & \pm 1 & 0 \end{bmatrix} \begin{bmatrix} V_Y \\ I_X \\ V_{Z\pm} \end{bmatrix} \tag{4}$$

where $I_Y$ is the input current at *Y* terminal, $V_X$ is the voltage at *X* terminal, $I_{Z\pm}$ is the output currents at $Z\pm$ terminals, $V_Y$ is the input voltage at *Y* terminal, $I_X$ is input current at *X* terminal, and $V_{Z\pm}$ is the voltages at Z± terminals, respectively.

The number of tubes for CNTFETs used in the DO-CCII is listed in Table 2. The supply voltage used to simulate all the circuits is ± 0.5 V. The chirality of all the transistors used in simulation is (19,0). The value of $I_{REF}$ is selected to be 2 μA. The power consumed by the DO-CCII is 440.33 μW [25].

Figure 3a shows the DC voltage transfer characteristics of DO-CCII. A DC voltage ranging from − 0.5 to 0.5 V is applied at the *Y* terminal, followed by the *X* terminal from −370 to 410 mV. The AC voltage transfer characteristic of DO-CCII is shown in Fig. 3b. The 3 dB BW for voltage gain ($V_X/V_Y$) is 54.028 GHz.

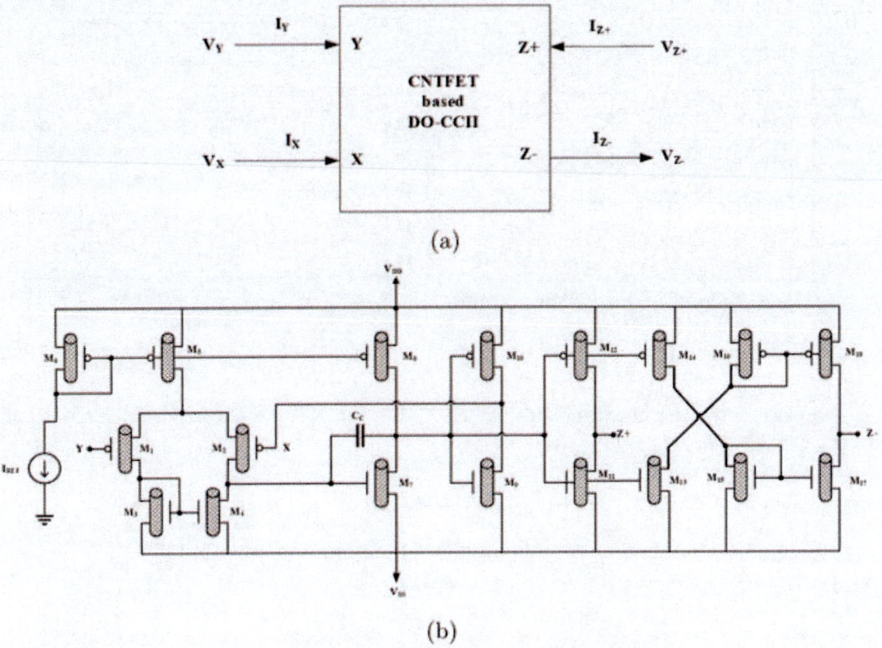


**Fig. 2** **a** Symbolic representation and **b** transistor-level schematic of DO-CCII

**Table 2** Summary of DO-CCII transistor sizing (L = 32 nm for all transistors)

| Transistors | $M_1, M_2$ | $M_3, M_4$ | $M_5, M_6$ | $M_7, M_8$ | $M_9 - M_{18}$ |
|---|---|---|---|---|---|
| Number of tubes | 3 | 2 | 28, 14 | 35, 245 | 20 |

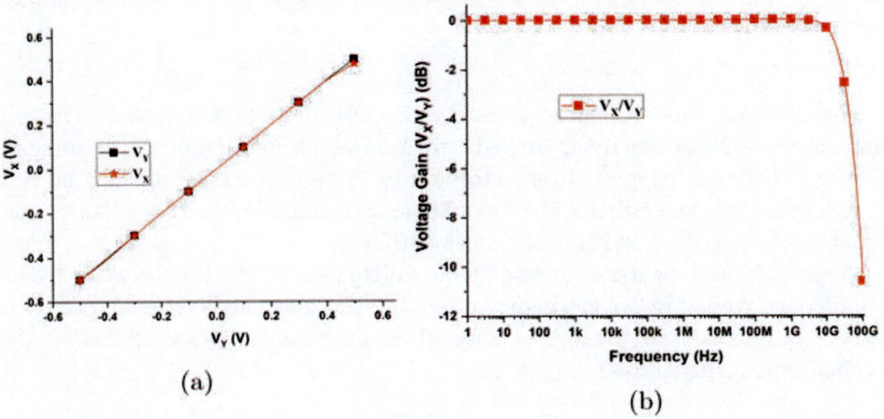


**Fig. 3** **a** Voltage transfer characteristics between $Y$ and $X$ ports **b** frequency response of voltage gain $(V_X/V_Y)$

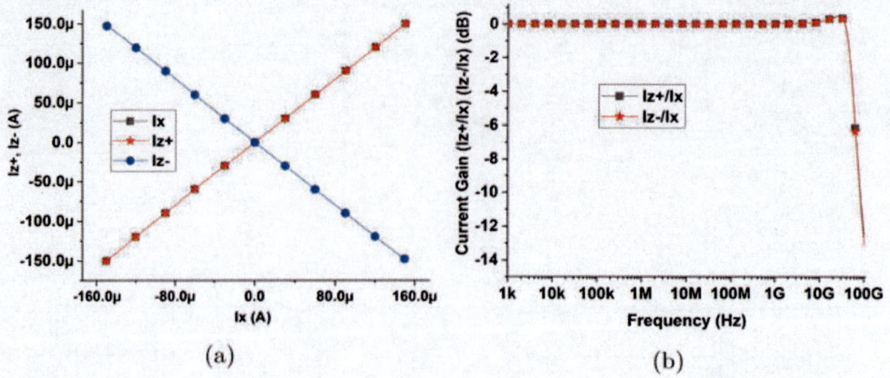


**Fig. 4** **a** Current transfer characteristics among $X$, $Z+$ and $Z-$ ports **b** frequency response of current gains $(I_{Z+}/I_X)$ $(I_{Z-}/I_X)$

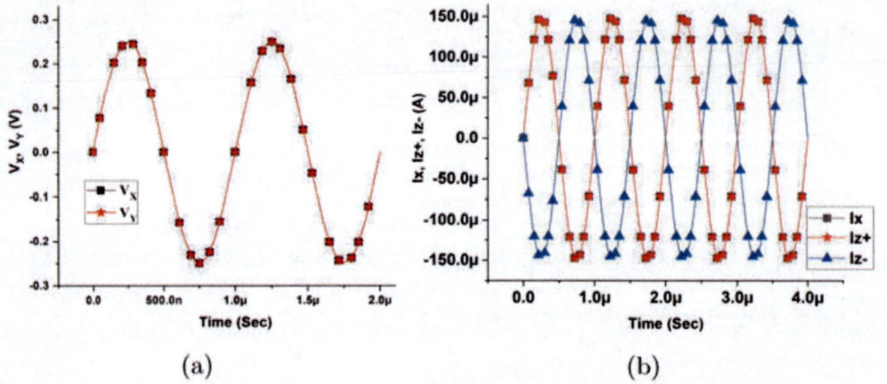


**Fig. 5** Transient response of **a** voltage follower between $Y$ and $X$ ports **b** current transfer among $X$, $Z+$ and $Z-$ ports

Figure 4a shows the DC current transfer characteristics of DO-CCII. A DC current ranging from $-200$ to $200\ \mu A$ is applied at the $X$ terminal, copied at the $Z\pm$ terminals. The range of linear region is from $-160\ \mu A$ to $160\ \mu A$ at $Z\pm$ terminals. The AC current transfer characteristics of DO-CCII are shown in Fig. 4b. The 3 dB BW for current gains $(I_{Z+}/I_X)$ and $(I_{Z-}/I_X)$ is 48.846 GHz.

Figure 5a shows the transient response of voltage follower action between $Y$ and $X$ terminals, while Fig. 5b shows the current transfer characteristics among $X$, $Z+$ and $Z-$ terminals. Both the figures demonstrate the nature of DO-CCII and satisfy its characteristic equation.

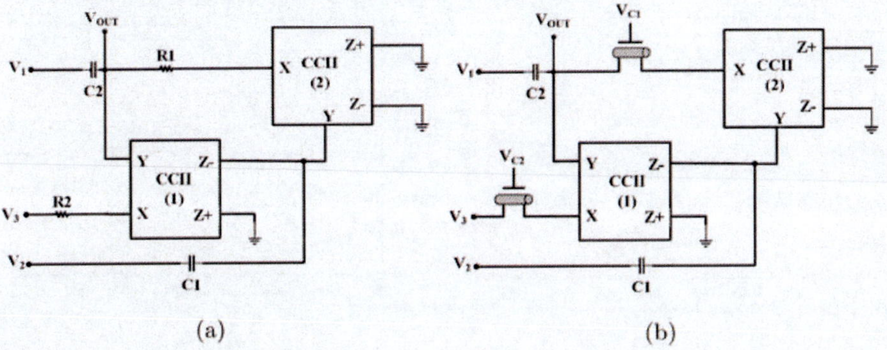


**Fig. 6** **a** Block diagram of filter reported in [9], **b** DO-CCII-based universal filter implementation of Fig. 6a using voltage-controlled resistor

**Table 3** Transfer function of CNTFET DO-CCII-based universal filter

| Filter | LPF | HPF | BPF | BRF | APF |
|---|---|---|---|---|---|
| $V_1$ | 0 | 1 | 0 | 1 | 1 |
| $V_2$ | 0 | 0 | 1 | 0 | −1 |
| $V_3$ | 1 | 0 | 0 | 1 | 1 |

## 4 DO-CCII-Based Universal Biquadratic Filter

A voltage mode (VM) MISO universal filter block diagram using CNTFET-based DO-CCII is shown in Fig. 6a, and its implementation using voltage-controlled resistor is shown in Fig. 6b. The transfer function (TF) of the filter can be obtained using characteristics of the DO-CCII by simple analysis of Fig. 6a in terms of $V_1$, $V_2$ and $V_3$ as given below.

$$V_0 = \frac{s^2 R_1 R_2 C_1 C_2 V_1 + s R_2 C_1 V_2 + V_3}{s^2 R_1 R_2 C_1 C_2 + s R_2 C_1 + 1} \tag{5}$$

All the five TFs of universal filter obtained using Eq. (5) are summarized in Table 3. The equations for natural frequency ($\omega_0$) and quality factor $Q$ from Eq. 5 can easily be obtained and given by the following equations:

$$\omega_0 = \frac{1}{(R_1 R_2 C_1 C_2)^{1/2}} \tag{6}$$

$$Q = \left(\frac{R_1 C_2}{R_2 C_1}\right)^{1/2} \tag{7}$$

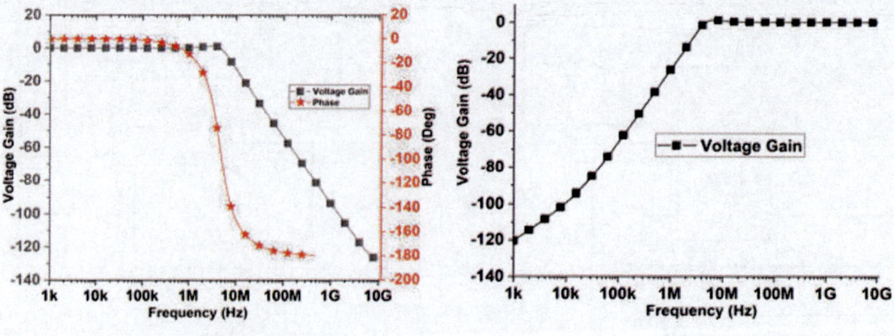


(a) (b)

**Fig. 7** Universal filter frequency response with $V_{C1} = V_{C2} = 0.3$ V. **a** LPF gain and phase **b** HPF gain

Here, $\omega_0$ can be varied using $R_1$ or $R_2$ while $Q$ can be controlled using the ratio of $R_1/R_2$ or $C_2/C_1$. The sensitivity functions for $\omega_0$ and $Q$ can be given as follows:

$$S_{C_1}^{\omega_0} = S_{C_2}^{\omega_0} = S_{R_1}^{\omega_0} = S_{R_2}^{\omega_0} = -\frac{1}{2} \tag{8}$$

$$S_{C_2}^{Q} = S_{R_1}^{Q} = -S_{C_1}^{Q} = -S_{R_2}^{Q} = \frac{1}{2} \tag{9}$$

The above equations show that the universal filter has low passive sensitivity performance.

The simulation of the universal filter is done using the HSPICE tool using the CNTFET model with parameters listed in Table 1. The supply voltages used are $\pm$ 0.5 V, and the average power consumption by the universal filter is 961.80 $\mu$W.

To simulate universal filter gain and phase frequency responses, controlled voltages $V_{C1}$ and $V_{C2}$ are chosen as 0.3 V. The values of $C_1$ and $C_2$ are selected as 1 pF, which provides the values of $Q$ and operating frequency as 1 and 4.57 MHz, respectively. Figures 7a to 8c show the LPF, HPF, BPF, BRF, and APF gain and phase frequency responses.

To simulate the frequency tuning response of universal filter, the controlled voltages $V_{C1}$ and $V_{C2}$ are varied from 0.25 to 0.5 V with fixed values of $C_1$ and $C_2$ to be 1 pF. Table 4 shows the different center frequencies of BPF for different values of $V_{C1}$ and $V_{C2}$. Figure 9a to e shows the gain frequency responses of LPF, HPF, BPF, BRF, and APF, respectively, for varying values of $V_{C2}$. At the same time, $V_{C1}$ is fixed at 0.3 V. Table 4 and Fig. 9a, e show that the operating frequency of the universal filter varies from 2.88 MHz to 14.79 MHz. This significant variation of operating frequency with controlled voltages and lower power consumption makes this universal filter a potential block in analog signal processing ICs.

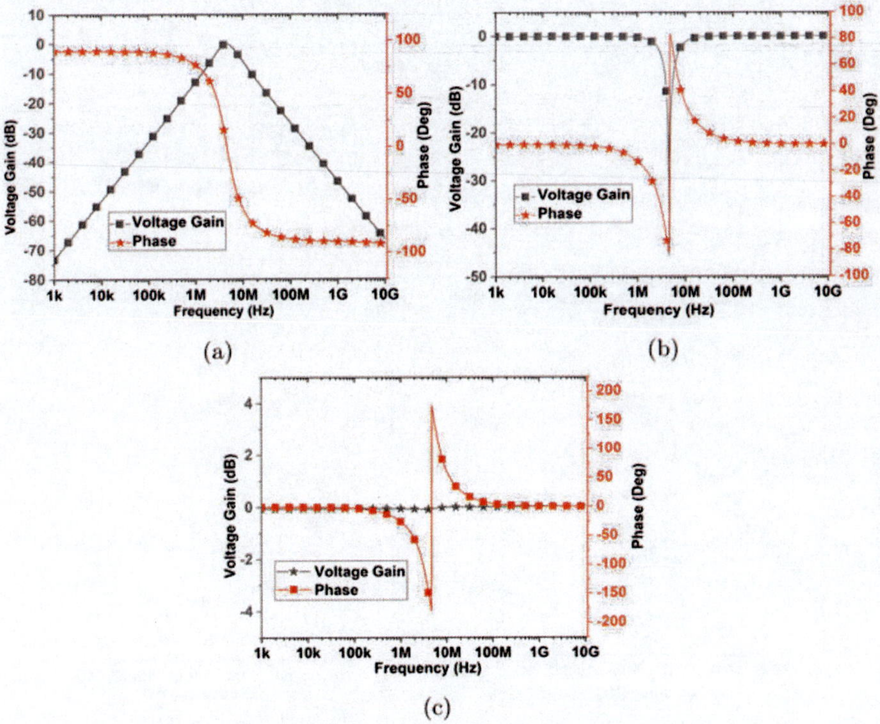


**Fig. 8** Universal filter frequency response with $V_{C1} = V_{C2} = 0.3$ V. **a** BPF gain and phase **b** BRF gain and phase **c** APF gain and phase

**Table 4** Variation of center frequency of BPF with controlled voltages

| | | | | | | | | | | | | |
|---|---|---|---|---|---|---|---|---|---|---|---|---|
| $V_{C1}$ (V) | 0.3 | 0.3 | 0.3 | 0.3 | 0.3 | 0.3 | 0.5 | 0.5 | 0.5 | 0.5 | 0.5 | 0.5 |
| $V_{C2}$ (V) | 0.25 | 0.3 | 0.35 | 0.4 | 0.45 | 0.5 | 0.25 | 0.3 | 0.35 | 0.4 | 0.45 | 0.5 |
| $f_0$ (MHz) | 2.88 | 4.57 | 6.03 | 7.08 | 7.76 | 8.32 | 5.13 | 8.32 | 10.96 | 12.6 | 13.80 | 14.79 |

Figure 10a shows the total harmonic distortion (THD) analysis of the BPF response. A sinusoidal input of different magnitudes (from 10 to 200 $mV_{pp}$) is applied at the input of the universal filter. The amount of distortion increases with the input amplitude but remains within acceptable limits (7.535%) for a significantly large amplitude (200 $mV_{pp}$).

Figure 10b shows the input and output noise with the frequency for the BPF response. It is observed that the maximum value of input and output noise is 704.6 nV/$\sqrt{Hz}$ and 4.536 fV/$\sqrt{Hz}$, respectively, which are significantly smaller.

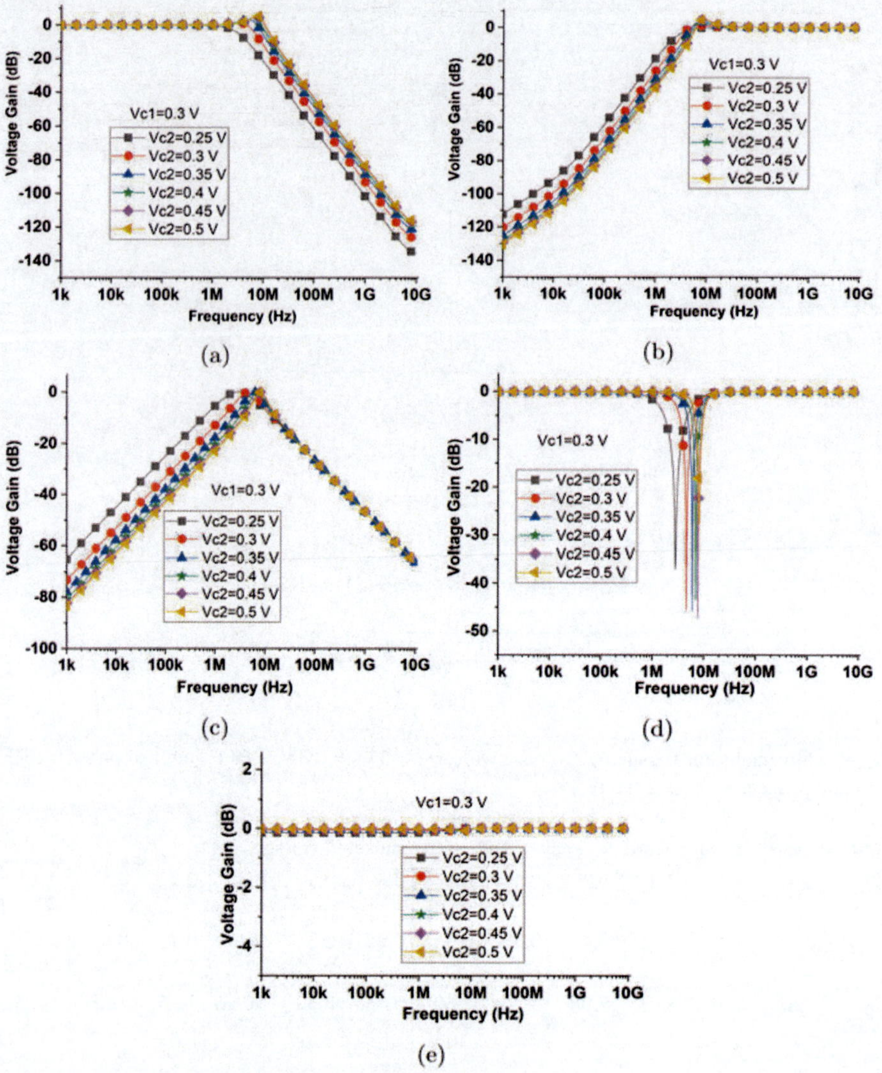


**Fig. 9** Frequency tuning response of universal filter with $V_{C1} = 0.3\,\text{V}$ and different values of $V_{C2}$. **a** LPF gain, **b** HPF gain, **c** BPF gain, **d** BRF gain, **e** APF gain

Figure 11a shows the combined voltage gain frequency responses of all the filters, i.e., LPF, HPF, BPF, BRF, and APF, using the different inputs of the universal filter for the same controlled voltages $V_{C1}$ and $V_{C2}$ of 0.3 V and $C_1$ and $C_2$ of 1 pF. It can easily be observed from Fig. 11a that the frequency of operation for all the filters is 4.57 MHz.

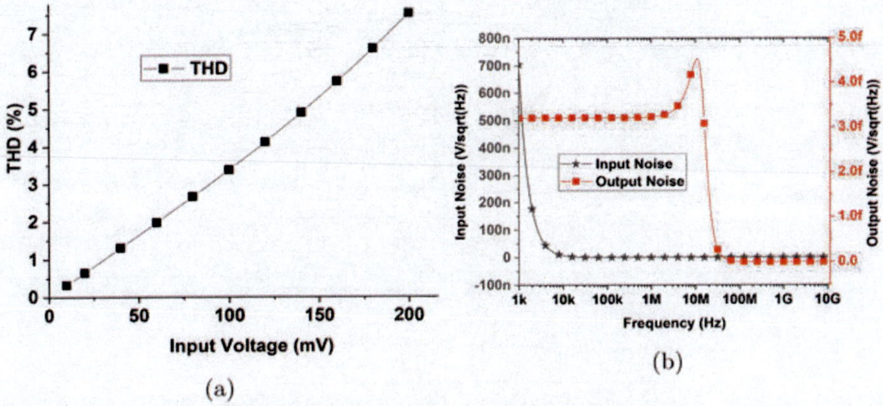


**Fig. 10** **a** THD versus input voltage at 14.79 MHz. **b** Input and output noise versus frequency response of BPF

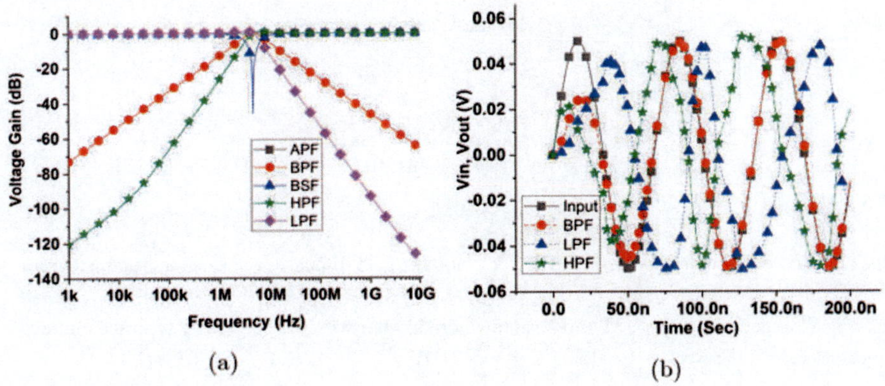


**Fig. 11** **a** Voltage gain frequency responses of universal biquadratic fillter. **b** Transient analysis results of LPF, HPF, and BPF

A 100 $mV_{pp}$ sinusoidal input of 14.79 MHz is applied at different inputs of the universal filters, and output is observed. Figure 11b shows the transient responses of LPF, HPF, and BPF, respectively. This figure shows the signal processing capability of the universal filter. Figure 11b illustrates the combined effect on different filters for the same transient input as above. From Fig. 11b, it can easily be inferred that the phase relationship between the input and output of the different filters is consistent.

The study of process variation and mismatch effect is done using 10% tolerance in capacitor values. The Monte Carlo analysis is done for 100 runs. Figure 12a to c shows the Monte Carlo analysis on BPF voltage gain frequency response, phase frequency response, and histogram of the operating frequency, respectively. All three figures show minimal deviation within acceptable limits with the process variations of capacitor values.

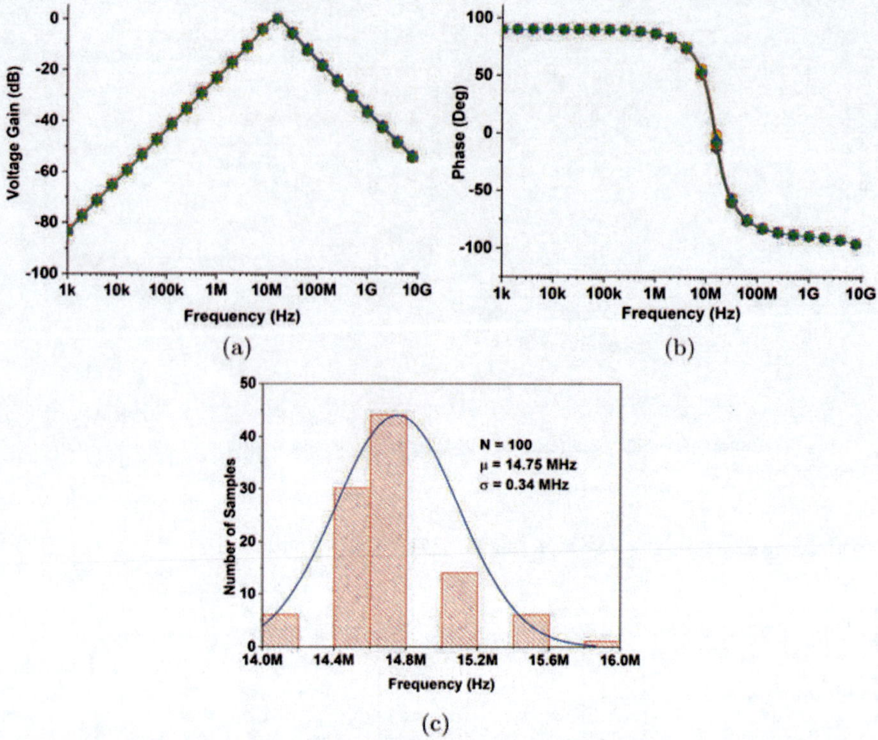


**Fig. 12** Monte Carlo simulations of BPF for **a** voltage gain, **b** phase, **c** corresponding histogram

**Table 5** Comparison of the VM universal biquadratic filter with the previously reported circuits

| Specifications | This work | [2] | [17] | [15] | [23] |
|---|---|---|---|---|---|
| Process (nm) | 32 | 32 | 180 | 450 | – |
| Technology | CNTFET | CNTFET | CMOS | CMOS | BJT |
| Active and passive elements | 2 DO-CCII, 2 C, 2 Switches | 2 OTA, 2 C | 3 DDCC, 2 C, 5 R | 1 VDVTA, 2 C | 3 VDDDA, 2 C, 1 R |
| Resistorless topology | Yes | Yes | No | Yes | No |
| Offer 5 standard Filter functions | Yes | No | Yes | Yes | Yes |
| Supply voltages (V) | ±0.5 | ±0.9 | ±1.8 | ±1 | ±5 |
| Power dissipation (μW) | 961.80 | – | – | 368 | – |
| Tuning range (MHz) | 2.88 – 14.79 | 28.5 – 69.98 | 1.59 | 1318 | 0.109 – 0.202 |

Table 5 shows the comparison of the presented universal filter with relevant universal filters available in the literature. Reference [2] does not offer all the available filter responses. References [17, 23] do not provide resistorless topology, which is why they are not suitable for integration. Also, they have a very low tuning range of operating frequency. Reference [15] provides an excellent tuning range, but it consumes a significant amount of power.

## 5 Conclusion

In this paper, a CNTFET-based DO-CCII and using it a VM MISO universal biquadratic filter has been presented. The DO-CCII has a 3 dB BW of 54.028 GHz for voltage gain and 48.846 GHz for current gain. The DO-CCII consumes an average power of 440.33 μW for its operation. It has a DC voltage range from −370 to 410 mV and a current range from −160 to 160 μA. The VM universal filter provides all the filter responses, LPF, HPF, BPF, BRF, and APF, from a single topology. The operating frequency can be tuned easily by using the control voltages implementing voltage-controlled resistors. Since the circuit is free from any physical resistors, it can easily be used in analog signal processing ICs for high-frequency applications. All the simulations are done using the HSPICE tool with the Stanford model of CNTFET.

## References

1. Carbon nanotube field effect transistors (cnfet) hspice model v. 2.2.1, stanford university. http://nano.stanford.edu/models.php Accessed 06 June 2018
2. Cen M, Song S, Cai C (2017) A high performance cnfet-based operational transconductance amplifier and its applications. Analog Integr Circ Signal Process 91(3):463–472
3. Chen HP (2010) Single ccii-based voltage-mode universal filter. Analog Integr Circ Signal Process 62(2):259–262
4. Deng J, Wong HSP (2007) A compact spice model for carbon-nanotube field-effect transistors including nonidealities and its application-part i: Model of the intrinsic channel region. IEEE Trans Electron Devices 54(12):3186–3194
5. Deng J, Wong HSP (2007) A compact spice model for carbon-nanotube field-effect transistors including nonidealities and its application-part ii: Full device model and circuit performance benchmarking. IEEE Trans Electron Devices 54(12):3195–3205
6. Dokania V, Islam A, Dixit V, Tiwari SP (2016) Analytical modeling of wrap-gate carbon nanotube fet with parasitic capacitances and density of states. IEEE Trans Electron Devices 63(8):3314–3319
7. Garg S, Gupta TK, Pandey AK (2020) A 1-bit full adder using CNFET based dual chirality high speed domino logic. Int J Circ Theory Appl 48(1):115–133
8. Goyal C, Ubhi JS, Raj B (2019) A low leakage TG-CNTFET-based inexact full adder for low power image processing applications. Int J Circ Theory Appl 47(9):1446–1458
9. Horng JW, Lee MH, Cheng HC, Chang CW (1997) New CCNI-based voltage-mode universal biquadratic filter. Int J Electron 82(2):151–156

10. Iijima S (1991) Helical microtubules of graphitic carbon. Nature 354(6348):56–58
11. Kaçar F, Yeşil A (2010) Voltage mode universal filters employing single FDCCII. Analog Integr Circ Signal Process 63(1):137–142
12. Kacar F, Yesil A, Noori A (2012) New CMOS realization of voltage differencing buffered amplifier and its biquad filter applications. Radioengineering 21(1):333–339
13. Khaleqi Qaleh Jooq M, Bozorgmehr A, Mirzakuchaki S (2021) A low-power delay stage ring vco based on wrap-gate CNTFET technology for x-band satellite communication applications. Int J Circ Theory Appl 49(1):142–158
14. Kumar M, Ubhi JS (2019) Design and analysis of CNTFET based 10t SRAM for high performance at nanoscale. Int J Circ Theory Appl 47(11):1775–1785
15. Kumar V, Mehra R, Islam A (2019) Design and analysis of miso bi-quad active filter. Int J Electron 106(2):287–304
16. Kumari S, Gupta M (2018) Design and analysis of tunable voltage differencing inverting buffered amplifier (vdiba) with enhanced performance and its application in filters. Wirel Pers Commun 100(3):877–894
17. Lee CN (2017) Independently tunable plus-type DDCC-based voltage-mode universal biquad filter with miso and SIMO types. Microelectron J 67:71–81
18. Masud M, A'ain A, Khan I, Husin N (2019) Design of voltage mode electronically tunable first order all pass filter in$\pm 0.7$ v 16 nm CNFET technology. Electronics 8(1):95
19. Pacheco-Sanchez A, Fuchs F, Mothes S, Zienert A, Schuster J, Gemming S, Claus M (2017) Feasible device architectures for ultrascaled CNTFETS. IEEE Trans Nanotechnol 17(1):100–107
20. Prasad D, Bhaskar D, Singh A (2009) Multi-function biquad using single current differencing transconductance amplifier. Analog Integr Circ Signal Process 61(3):309–313
21. Sangyaem S, Siripongdee S, Jaikla W, Khateb F (2017) Five-inputs single-output voltage mode universal filter with high input and low output impedance using vdddas. Optik 128:14–25
22. Singh B, Singh AK, Senani R (2013) A new universal biquad filter using differential difference amplifiers and its practical realization. Analog Integr Circ Signal Process 75(2):293–297
23. Supavarasuwat P, Kumngern M, Sangyaem S, Jaikla W, Khateb F (2018) Cascadable independently and electronically tunable voltage-mode universal filter with grounded passive components. AEU-Int J Electron Commun 84:290–299
24. Yanagi K Differentiation of carbon nanotubes with different chirality. In: Carbon nanotubes and graphene. Elsevier, pp 19–38
25. Yasir M, Alam N (2020) Design of CNTFET-based ccii using gm/id technique for low-voltage and low-power applications. J Circ Syst Comput 29(09):2050,143. https://doi.org/10.1142/S0218126620501431
26. Zanjani SMA, Dousti M, Dolatshahi M (2018) A cntfet universal mixed-mode biquad active filter in subthreshold region. Int J RF Microw Comput-Aided Eng 28(9):e21,574

# Designing of Energy-Efficient XOR Gate Implementing DWM Spintronics

**Afreen Khursheed and Kavita Khare**

**Abstract** Scaling dimensions of CMOS integrated circuits bring about a technology slowdown for past decades. Novel spintronics based devices have evolved as promising candidate for future generation integrated circuits. This is because of attractive salient features like superior compatibility with existing MOS process technology, high density of integration, unlimited forbearance and excellent information processing. This paper discusses the implementation of XOR logic gate using the spintronics based on domain wall motion (DWM) phenomenon. This dual input XOR gate is ultra-energy-efficient, robust and also compatible with traditional electrical interconnects. The simulated result shows that the implemented gate shows 56% better energy efficiency in comparison with conventional exclusive OR logic gate. Furthermore, simulating a magnetic impeded full adder using the propound gate shows 28% reduced energy dissipation along with 10% enhanced energy-delay product as compared to conventional one.

**Keywords** Spintronics · Current-induced domain wall motion · Magnetic tunnel junction · Spin transfer torque

## 1 Introduction

Downscaling of CMOS device features, governed by Moore's law, has revolutionaries the technique used to generate process and store the binary information in today's digital era. As we come within reach of nano-CMOS technology, it is extremely knotty to properly handle power consumption issues. Henceforth, as there is surge in value of leakage power dissipation for portable devices, thus there is an imperative

A. Khursheed (✉)
Electronics and Communication Engineering, Indian Institute of Information Technology (IIIT Bhopal), Bhopal, India
e-mail: afreen.khursheed@gmail.com

K. Khare
Electronics and Communication Engineering, Maulana Azad National Institute of Technology, Bhopal, India



C. Giri et al. (eds.), *Emerging Electronic Devices, Circuits and Systems*,
Lecture Notes in Electrical Engineering 1004,
https://doi.org/10.1007/978-981-99-0055-8_6

requirement of low-power alternatives of post-CMOS. With reduction in transistor sizes up to few atomic layers, the quantum effect creeps up [1] showing a sudden rise of leakage current, dynamic power and standby power dissipation. The unrestricted augmentation of leakage and dynamic dissipation makes the integrated circuit crosses its thermal threshold limit and also raises the operating energy requirement of low-power VLSI circuits [2]. For the global interconnects wire, effect of scaling causes a noteworthy amount of dynamics power dissipation along with high signal delay [3]. However, previous research [4–6] carried out in this domain points out that scaling down of supply power and operating the circuit in a subthreshold region reduces power consumption but at the cost of bulky power control monitoring circuit. Another study [7, 8] shows that by adopting power gating strategy and appending sleep transistors in modeling circuit will further suppress standby power consumption. Before applying that strategy the logic information stored in case of memory logic circuit is transferred to a non-volatile memory. The efficacy of the said strategy depends firstly on how long the circuit is in sleep mode and secondly by achieving reduced rate of data transfer in between main memory and non-volatile one. Consequently, fulfilling the ever-increasing demand of state-of-the-art energy-efficient and optimized performance, portable VLSI devices is becoming a bottleneck situation [9]. So to cater these demands invoke researchers to explore alternative variants of conventional semiconductors like double gate MOS field-effective transistor, Si-nano-wire MOS field-effective transistor, fin-type field-effective transistor, carbon nanotube-type and graphene nano-ribbon-type field-effective transistor [10, 11]. In addition to these recently spintronics based devices have also ignited the curiosity of scientists and researchers. Different variants of spintronics such as spin valve, all spin logic device, magnetic-tunneled junction (MTJs), skyrimons as well as domain wall nano-wires (DW) demonstrate better results and hence suitable alternative to MOS technology.

The unearthing of effects like giant magnetoresistance (GMR), spinned transferred torque (STT) and also the effect of tunneled magnetoresistance (TMR) paved way for spintronics field. Hence, post the MOS era, the spintronics is said to be suitable replacement [12, 13]. Spintronics technology makes use of the charge of an electron along with its spin. The spin characteristic is a sort of an intrinsic type of angular momentum associated with elementary particles like protons, electrons and neutrons. The hurdles encountered by conventional charge (q) emanated electronics is resolved by employing the orientation of electron spin and its related magnetic moment as a state variable in case of spintronics technology. The salient feature of this technology is that magnetization as well as spin of electron retains its state for an indefinite time which is contrary to MOS technology where because of leakage current the stored charge is lost. Hence, spintronics technology-based devices are non-volatile in nature.

The exclusive OR logic gate is the elementary logic module for designing various digital arithmetic logic circuits such as subtractor and adder. XOR is widely used in generating parity bits for checking error and also in area of cryptography. The present manuscript models an exclusive OR logic gate using the spintronics based on domain wall motion (DWM) phenomenon. This dual input XOR gate is compatible with traditional electrical interconnects and is proficient to perform high-density integration. Furthermore, the propound device is employed to model a circuit which proves that efficient systems can be constructed using the low-power operation and non-volatility feature of this device.

The organization of the paper is as follows: After a brief introduction in Sect. 1, we start with the Sect. 2 "Background," which consists of the subsections describing the types of spintronics device. Section 3 elaborates on the architecture and working of propound device, while Sect. 4 presents simulation results and discussion. Lastly, Sect. 5 concludes the paper.

# 2 Background

## *2.1 Magnetic Tunnel Junction*

Magnetic tunnel junction has multi-stratum architecture having very fine insulating sheet sandwiched in between two ferromagnetic sheets (FM). Figure 1 shows the architecture of Magnetic tunnel junction [14].

The insulating sheet is termed as *barrier layer,* while the $FM$ sheet which is having preset magnetic orientation is termed as fixed reference layer ($RL$) and another $FM$ sheet having variable magnetic orientation (as $P or AP with reference to RL$) is termed as free layer ($FL$). On either applying current or voltage in a specific direction or under the influence of exterior magnetic field, the magnetic orientation of a free FM layer can be altered. The overall resistance of any magnetic tunnel junction device is ascertained by observing the magnetic orientation of $FL$ apropos to the $RL$. This quantum mechanical effect is because of spin dependent tunneling and is commonly termed as tunnel magneto resistance (TMR) effect in spintronics. In relevance to (TMR) effect, if $FL$ is magnetized in similar direction as the $RL$, then net resistance offered by MTJ device is small and is coined as parallel resistance ($RP$), $w$ hereas if $RL$ and $FL$ are magnetized anti-parallel apropos to each other, then in such a situation overall resistance offered by MTJ device is large and is coined as anti-parallel resistance (RAP). The expression to calculate the TMR effect is

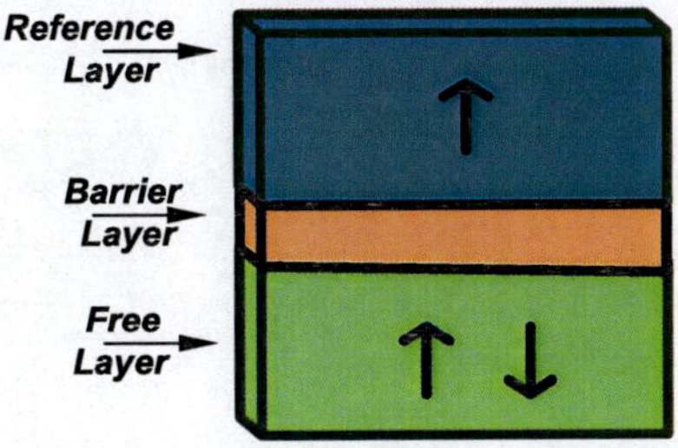


**Fig. 1** MTJ structure

$$\mathrm{TMR} = \frac{G_P - G_{AP}}{G_{AP}} \tag{1}$$

Here, $G_P$ and $G_{AP}$ are the value of conductance in parallel and anti-parallel state, respectively. TMR ratio is the parameter to judge the performance of MTJ device. At the onset in 1975, a TMR ratio of 14% was reported at temperature of 4.2 K by Fe/Ge/Fe architecture. For optimized operation of circuit applications designed using MTJ device at room temperature, it is preferred to have a high value of tunnel magneto resistance (TMR). Initially, in early 90 s highest TMR of 70.4% was achieved at room temperature using amorphous $Al_2O_3$ as barrier layer. Using crystalline $M_gO$ as barrier layer, TMR of 604% is recorded in 2008 at ambient temperature in case of $T_a/C_oF_eB/M_gO/C_oF_eB$ pseudospin valve MTJ device. Researchers vaticinate that TMR of around 1000% can be attained with Heusler + $M_gO$ [15] and spinel $(M_g - Al - O)$ [16] insulating barrier-layered MTJ devices in coming era.

## 2.2 Domain Wall Motion

In addition to MTJ spintronics device, there is yet another promising contender for future VLSI circuits called DWM ferromagnetic nano-wires. Figure 2 shows the architecture of domain wall nano-wire.

The domain wall motion is a well-known mechanism of spintronics technology. Generally, all ferromagnetic substance consists of several minute regions termed as *domains*. Every domain region possesses its own magnetisation and thus behaves as a permanent magnet. A domain wall is primarily an interface which segregates two adjacent magnetic domains of opposite magnetizations [17]. The thickness of domain wall is of the order of nanometers, inside which the rotation of magnetisation takes place from one direction to another. Binary codes can be stored in these minute magnetized regions called domains in the form of a magnetisation vector. Just by sensing trend of magnetic orientation, this binary codes information can be read. As these domain walls are ambulatory in nature, hence it can be pulled by sending

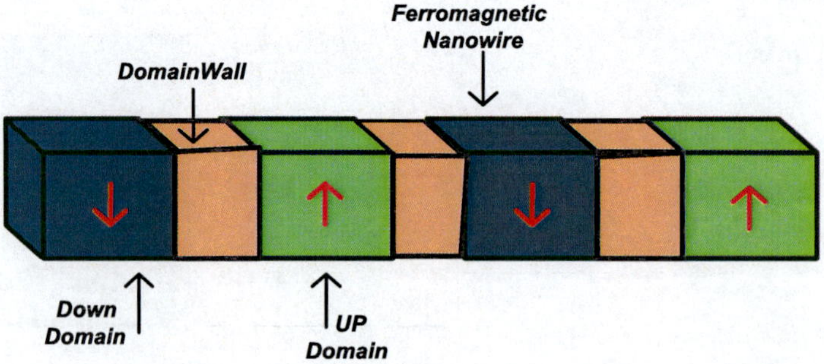


**Fig. 2** Domain wall nano-wire

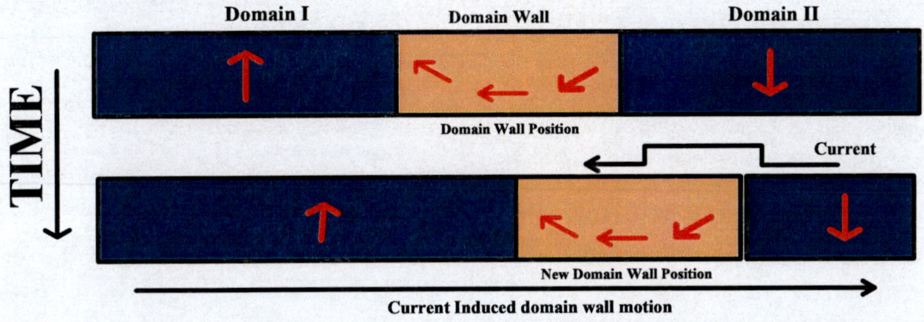


**Fig. 3** Current-induced DWM

pulses of current through nano-wires. This mechanism is termed as current-induced DWM and was first discovered by Berger in 1978 [18]. Figure 3 shows mechanism of current-induced DWM since its discovery numerous applications like logic gates, reconfigurable logic, magnetic memory, full adder and nano-oscillators were designed based on this phenomenon of domain wall motion [19, 20].

As illustrated in Fig. 3, a sequence of domain walls may be present in a long magnetic nano-wire. Notches can be engraved along the wire edges such that it behaves as pinning sites for these minute domain walls region. An electric current of density $(J_{app})$ which is larger than threshold density $(J_{th})$ is made to pass in order to de pin domain walls from these notches and to push them along the wire length. The width of applied current pulse determines the magnitude of $J_{th}$. The velocity of domain wall motion is governed by $J_{app}(> J_{th})$ the direction toward which the domain wall shifts is along that of incoming electrons. The direction can be reversed by changing the direction of applied electric current.

The spin momentum transfer is the phenomenon which is primarily responsible for current-induced motion of domain walls in ferromagnets. The mechanism can be best explained by s-d model. If a spin-polarized current-carrying s-spin electrons pass the domain wall (DW is due to localized d-spin magnetisation), then there is an adiabatic reversal of s-spin electrons. These s-spin electrons which bring about a reaction torque onto the domain walls (localized d-spin magnetisation) ensuring the angular momentum conservation; and the consequence of this is the momentum in domain wall. Hence, it can be stated that exerted spin transfer torque (STT) on magnetized domain wall results in movement of DW along wire length. Past findings shows that STT-driven domain wall motion possessing high speed around 50 m/s and minimum threshold current–density around $10^6$ A/cm$^2$ [21]. Furthermore, on decreasing the thickness of ferromagnetic nano-wire causes mitigation of the threshold current needed to achieve domain wall depinning and also the required current for attaining a specific domain wall velocity. These results motivate researchers to apply DW motion in designing low-power fast logic and memory units.

## 3 Propound Device Architecture and Working

This section discusses about the architecture and working of the proposed device DMW phenomenon.

### 3.1 Architecture

The structure of propound device is illustrated in Fig. 4. The architecture shows a ferromagnetic stratum comprising of magnetization domains divided into five regions, namely $\rho_1, \rho_2, \rho_3, \rho_4, \rho_5$. Here, magnetizations of odd domain regions $\rho_1, \rho_3, \rho_5$ are kept pinned in the fixed direction. Moreover, $\rho_1, \rho_5$ domains are magnetized in similar direction, whereas $\rho_3$ is magnetized in reverse direction. The existence of reversely magnetized domains at the ends of $\rho_2, \rho_4$ nucleates a domain wall in each of these two domains. Since the magnetization of any point in domain $\rho_2$ and $\rho_4$ is governed by relative position of domain wall with respect to the point, it can be maneuvered by shifting domain wall to-or-fro. The threshold current responsible for domain wall motion in the magnetized domains $\rho_2$ and $\rho_4$ is denoted by $I_{\text{th2}}$ and $I_{\text{th4}}$, respectively. It is preferred to have value of $I_{\text{th4}}$ greater than $I_{\text{th2}}$. To attain magnitude of $I_{\text{th4}}$ greater than $I_{\text{th2}}$, either the regions $\rho_3, \rho_4, \rho_5$ are fabricated wider in comparison with regions $\rho_1$ and $\rho_2$ or keeping domains $\rho_1$ and $\rho_2$ thinner than $\rho_3, \rho_4, \rho_5$ will also result in achieving value of $I_{\text{th4}}$ greater than $I_{\text{th2}}$. As the latter condition requires additional steps for patterning a ferromagnetic sheet of diverse thickness, it is generally not adopted for designing.

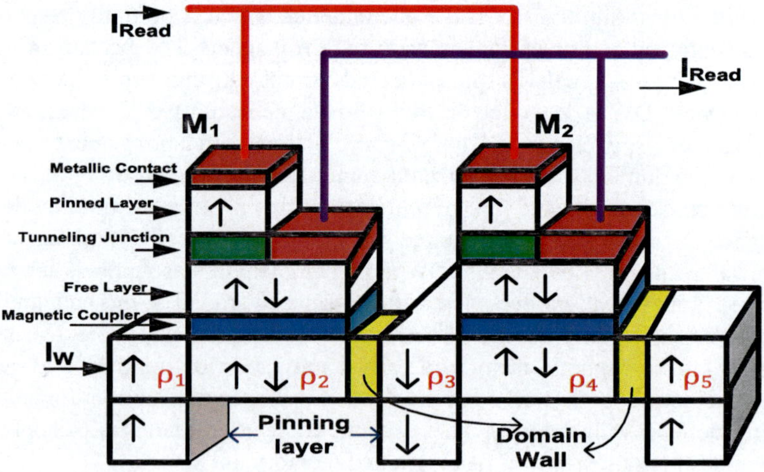


**Fig. 4** Proposed device structure

Figure 4 shows that above regions $\rho_2$ and $\rho_4$ are fabricated above magnetic tunnel junctions $MTJ_1$ and $MTJ_2$, respectively. This configuration assists in reading their magnetic states. Thus, rather than directly having domains $\rho_2$ and $\rho_4$ as free layers of $MTJ_1$ and $MTJ_2$ correspondingly, these magnetic tunnel junctions will consist of free layers having an electrical insulation from domains $\rho_2$ and $\rho_4$. This would result in isolation of read and write current paths from each other and henceforth averts the read current from changing the position of a domain wall. The realization of such choice of design is attained by sandwiching a very fine layer of magnetic oxide in between domain $\rho_2$and the free layerof $MTJ_1$ and also in between domain $\rho_4$and the free layerof $MTJ_2$. The purpose of this oxide layer is to not only aid in creating electrical isolation of regions $\rho_2$ and $\rho_4$ from free layers of corresponding magnetic tunnel junctions but also in creating magnetic coupling between them. Thus, magnetization trend of any particular point in free layer chases the trend of the point in regions $\rho_2$ and $\rho_4$. situated directly beneath it. As aforementioned that in case of magnetic tunnel junctions, the magnitude of resistance depends on position of corresponding domain wall. Since $MTJ_1$ and $MTJ_2$ are parallel connected, the total parallel resistance is denoted by $R_{out}$ and shows the output state of the proposed device.

## 3.2 *Working*

The working of the proposed device shown in Fig. 5 is divided into three phases occurring one after another. These phases are named as *Reset Phase* (RP), *Write Phase* (WP) and *Read Phase* (RdP).

In case of first phase (RP), an electric current ($I_{reset}$) of significant strength is made to pass from domain $\rho_5$to $\rho_1$ of the ferromagnetic sheet. As illustrated in Fig. 5, this reset electric current ($I_{reset}$) is supplied by N-type of MOS transistor

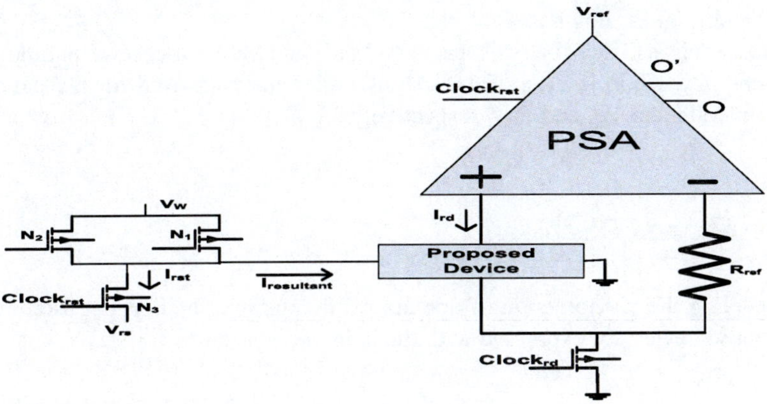


**Fig. 5** Proposed gate structure

**Table 1** Proposed gate truth table

| $I_1$ | $I_2$ | $R_{out}$ | $O$ |
|---|---|---|---|
| 0 | 0 | $\frac{r_p.R_{ap}}{r_p+R_{ap}}(< R_{ref})$ | 0 |
| 0 | 1 | $\frac{r_{ap}.R_{ap}}{r_{ap}+R_{ap}}(> R_{ref})$ | 1 |
| 1 | 0 | $\frac{r_{ap}.R_p}{r_{ap}+R_p}(> R_{ref})$ | 1 |
| 1 | 1 | $\frac{r_{ap}.R_p}{r_{ap}+R_p}(< R_{ref})$ | 0 |

($N_3$). The purpose of $I_{reset}$ is to move both the domain walls toward the right ends of $\rho_2$ and $\rho_4$ irrespective of their current positions. This reset action makes the logic device ready for operating on a fresh new set of inputs. Next in the second phase, (WP) logical inputs $I_1$ and $I_2$ will be applied on the gate node of two N-type of MOSFETs ($N_1$ and $N_2$), respectively. The resultant sum of two currents obtained from $N_1$ and $N_2$ FETs is denoted by $I_{resultant}$. This resultant current will be injected into the device from $\rho_1$to $\rho_5$. Different magnitude of $I_{resultant}$ can be obtained by applying different input combinations. In case when the magnitude of current $I_{resultant} < I_{th2}$, then in such a situation none of the domain walls gets de-pinned. While in case $I_{th2} < I_{resultant} < I_{th4,}$ then in such a situation domain wall in region $\rho_2$ gets de-pinned and because of the impact of spin transfer torque shifts left, while the position of domain wall in region $\rho_4$ is unchanged as it is not affected. Meanwhile in case when $I_{resultant} > I_{th4}$, then both the domain wall moves leftward. Henceforth, the magnitude of $R_{out}$ changes with variation in inputs applied to the gate. Finally, in case of last phase (RdP), the magnitude of output resistance $R_{out}$ is sensed. To sense the value of $R_{out}$ , , a precharged sense amplifier (PSA) is used in the circuit. The purpose of PSA is to undergo comparison of $R_{out}$ with respect to dedicated reference resistance $R_{ref}$ The comparison is done by discharging precharged polarization potential voltage through these resistances during its evaluation phase. Whenever the magnitude of $R_{out} < R_{ref}$, the value of PSA output is logic 0 else logic 1. In order to match the behavior of propound device to the functionality of XOR gate, suitable value of $R_{ref}$ has to be chosen as shown in Table 1.

Here, notations used for resistances in $MTJ_1$ and $MTJ_2$ in case of parallel state are $r_p$ and $R_p$, while in case of anti-parallel state notation used for resistances in $MTJ_1$ and $MTJ_2$ are $r_{ap}$ and $R_{ap}$ , respectively.

## 4 Results and Discussion

For modeling the phenomenon of domain wall, depinning and propagation in case of proposed device are expressed with the help of equation as

$$t_{depin} = 4520|J|^{-2.82} + 0.2285 \tag{2}$$

**Table 2** Physical parameters

| Parameters | Values |
|---|---|
| Length | $40 \times 10^{-9}$ m ($\rho_2, \rho_4$) |
| Width | $20 \times 10^{-9}$ m ($\rho_1, \rho_2$)<br>$50 \times 10^{-9}$ m ($\rho_3, \rho_4, \rho_5$) |
| Thickness | $3 \times 10^{-9}$ m ($\rho_1$ to $\rho_5$) |
| Resistivity | $200\ \Omega \times 10^{-9}$ m |
| Length of MTJ | $12 \times 10^{-9}$ m |
| TMR of MTJ | 125% |
| RA (low) of MTJ | $1.8 \times 10^{-7}\ \Omega$ nm$^2$ |

$$v_{DW} = \frac{1+ \propto \beta}{1+\alpha^2} \frac{g\mu_B P J}{2eM_s} \tag{3}$$

Here, time consumed to depin the domain wall is $t_{\text{depin}}$ (ns), the density of applied current in the direction of intended domain wall movement is $J\left(\frac{\text{MA}}{\text{cm}^2}\right)$, the speed of domain wall movement is $v_{\text{DW}}\left(\frac{\text{nm}}{s}\right)$. The Gilbert damping constant is represented by $\propto$, the coefficient of non-adiabatic spin transfer torque is represented by $\beta$, the land factor is represented by $g$, the electronic charge density is represented by $J$, the spin polarization constant is represented by $P$, the Bohr magnetron constant is represented by $\mu_B$, $M_s$ represents the saturation magnetization, and the charge of electron is represented by $e$.

For running the experimental analysis, CMOS-spin hybrid circuits in Cadence Virtuoso ADE of ST Microelectronics 40 nm PDK is used and the mCell in Verilog-A compact model is used for doing performance evaluation of proposed device. The required parameters for simulation of proposed device are shown in Table 2.

Noteworthy fact about the study done in this manuscript is that to withhold the intention of reducing current needed for domain wall movement low width strips are used. The duration of gate pulse for read, write and read operations is of 1.5 ns. A supply voltage of 60 mV is connected to transistor $N_3$ which is having width size of 0.46 µm. $N_2$ and $N_1$ are designed such that each having width 0.39 µm and connected to supply voltage of 70 mV at their source end. To achieve 1.5 ns duration long reset as well as writes operations, a suitable magnitude of $I_{\text{resultant}}$ and $I_{\text{reset}}$ is needed. To read the output logic state of the device, MOS-type FET-based PSA is used in this manuscript. In order to suppress the total gate-delay, it is advised to overlap the precharge phase of PSA with reset and write stages. The reference resistance $R_{\text{ref}}$ value is set as 18.2 KΩ in PSA of propound gate

Now, a comparative analysis is carried on the proposed exclusive OR gate and the already existing work is presented in [22]. Earlier in [23], Trinh et al provided input operands to the exclusive OR gate after switching unpinned layer of magnetic tunnel junctions. As a proven fact and validated by Venkatesan et al in their research work that spin transfer torque-driven domain wall movement is efficient in performance than the technique of switching magnetic tunnel junctions free layer using STT [23].

**Table 3** Phase-wise energy consumption (Fento Joules) of both the gates

| Input vectors | | Reset | | Write | | Read | |
|---|---|---|---|---|---|---|---|
| $I_1$ | $I_2$ | Prop | Conv | Prop | Conv | Prop | Conv |
| 0 | 0 | 4.6 | 12.9 | 5.8 | 14.5 | 1.27 | 1.22 |
| 0 | 1 | 4.6 | 12.9 | 4.8 | 14.4 | 1.3 | 1.23 |
| 1 | 0 | 4.6 | 12.9 | 4.8 | 14.3 | 1.3 | 1.22 |
| 1 | 1 | 4.6 | 12.9 | 8.2 | 14.0 | 1.17 | 1.23 |

Hence, the proposed design based on domain wall motion described in this paper is superior than already existing works till date.

The concept of using domain wall motion for designing in replacement of programming magnetic junctions using free-layer switching technique improves our proposed gate. For instance moving the domain wall either leftwards or rightwards within mCell device restores binary 1(0) logic in the magnetic tunnel junctions.

Both the types of XOR gates operates in three phases: *Reset Phase* (RP), *Write Phase* (WP) and *Read Phase* (RdP) of 1.5 ns duration. The overall delay of proposed and existing gate is same (= 4.5 ns). Table 3 presents a comparative data describing the phase-wise energy consumption of both the gates.

From the result, it is concluded that proposed gate consumes around 50% of its total energy on reset, 36% on write operation and remaining 14% for read operation. Moreover, scrutinizing data shows that energy consumed for read operation in case of proposed and existing gate is quite equal. But for reset operation, the proposed gate guzzle is 63% less energy, and for write operation, it consumes 69% less energy than existing gate. Thus, it can be stated that proposed gate is 60% energy efficient than conventional one. Total consumption of energy by proposed gate is in the range varying from 6 to 14 fJ approximately. Thus, on an average, it consumes approximately 10.5 fJ of energy under normal conditions. The reason behind the energy-efficient operation of proposed XOR gate is that it requires only two magnetic tunnel junctions in comparison to its counterpart, i.e., existing exclusive OR gate which requires six magnetic tunnel junctions. As a consequence of this, existing XOR gate needs large pairs of input operands, with a rise in number of resets and writes due to its domain wall motion-based operation. Hence, previously designed spintronics based XOR gates were less energy efficient due to their more number of magnetic tunnel junctions switching operations.

Aforementioned paragraphs describe the analysis of proposed gate as an individual logic element. To further validate its supremacy, we analyzed its performance by implementing a full-adder circuit using the proposed logic gate as a circuit element. The methodology used for implementing a full adder is described in below paragraphs.

A pipelined 2-stage adder is taken for the experiment. The set of 2 proposed *exclusive OR gate* is taken for implementing the *SUM* logic of full adder gate: such that during first phase $\text{SUM}_{\text{intermediate}} = I_1 \oplus I_2$ is computed and during second phase $\text{SUM} = \text{SUM}_{\text{intermediate}} \oplus C_{in}$ is computed. These two stages are connected in a cascade manner; the output of first stage is buffered in carry-in $C_{in}$ and provides the input for second stage XOR gate. Finally, a three input majority logic gate during the first phase computes the carry out logic as $C_{out} = \mathrm{AB} + \mathrm{BC_{in}} + C_{in}A$. Output of this gate acts as input to a buffer in secondary phase. Realization of both majority gates and buffers is meant to be in same fashion as the proposed exclusive OR gate. Both are modeled with mCell device and their *Reset Phase* (RP), *Write Phase* (WP) and *Read Phase* (RdP) are of 1.5 ns duration and occurs in succession. As a matter of fact that the *Write Phase* (WP) and *Read Phase* (RdP) of primary stage of full adder coincides with the *Reset Phase* (RP), *Write Phase* (WP) of secondary stage, respectively. Henceforth, the total delay of full adder circuit designed using proposed exclusive OR gates come out to be 6 ns. The mathematical calculation of delay is 1.5 ns (RP stage 1) + 1.5 ns (WP stage 2) + 1.5 ns (RdP stage 1) + 1.5 ns (RdP stage 2). Applying the standard combination of test inputs, we verified the FA with truth table, and hence, the proposed gate can be used to design more sophisticated circuits.

To further justify the superiority of full adder using proposed gate in comparison with conventional XOR gate-based full adder, Table 4 presents a comparative data describing the phase-wise energy consumption of both the adders for different input combinations.

From the result, it is concluded that power consumed by the proposed gates based full adder is froms 35 to 64 fJ and average value is 52 fJ. Thus, the proposed FA dissipate 31% times less energy than conventional FA. The primary cause of this is due to 27% less reset energy, and 57% less write energy is consumed by proposed FA. But this is achieved at the cost of rise in read energy by 73% of proposed FA. For conventional FA, the delay obtained is 4.5 ns and delay in case of the proposed FA is 6 ns. But the overall energy-delay product (EDP) of the proposed FA is 8% time superior to conventional one.

**Table 4** Phase-wise energy consumption of both the full adder

| Input vectors | | | Reset | | Write | | Read | |
|---|---|---|---|---|---|---|---|---|
| a | b | c | Prop | Conv | Prop | Conv | Prop | Conv |
| 0 | 0 | 0 | 25.9 | 34.3 | 0.07 | 39.1 | 9.3 | 2.5 |
| 0 | 0 | 1 | 25.8 | 34.3 | 11.8 | 38.9 | 9.64 | 2.7 |
| 0 | 1 | 0 | 25.8 | 34.3 | 11.9 | 38.7 | 9.63 | 2.7 |
| 0 | 1 | 1 | 25.7 | 34.3 | 26.2 | 38.5 | 10.3 | 2.7 |
| 1 | 0 | 0 | 25.9 | 34.3 | 11.9 | 38.7 | 9.7 | 2.68 |
| 1 | 0 | 1 | 25.4 | 34.3 | 26.8 | 38.5 | 10.1 | 2.68 |
| 1 | 1 | 0 | 25.4 | 34.3 | 17.5 | 38.3 | 9.7 | 2.68 |
| 1 | 1 | 1 | 25.3 | 34.3 | 28.7 | 38.12 | 9.9 | 2.56 |

## 5 Conclusion

This manuscript discusses about the designing of proposed XOR gate using spintronics technology. The proposed gate is then used to model a circuit which proves that efficient systems can be constructed using the low-power operation and non-volatility feature of this device. The experimental result analysis demonstrates that the propound gate shows 56% better energy efficiency in comparison to conventional exclusive OR logic gate. Furthermore, simulating a magnetic impeded full adder using the propound gate shows 28% reduced energy dissipation along with 10% enhanced energy-delay product as compared to conventional one. In the near future, devices based on spintronics technology will evolve as mainstream technology for upcoming era of microelectronics. The initial paragraphs of this paper throws light on various types of spin devices; then historical background of magnetic tunnel junction and domain wall nano-wire is discussed along with its principles of operation. As a well-known fact that traditional MOS technology is based on electronic charge movement is exhausting because of dimension scaling. Consequently, exponential rise in static power due to leakage is reported. This in turn increases total power dissipation in case of energy-efficient portable devices. Henceforth to cater to ever-increasing demand of modern VLSI technology the above-mentioned issue must be resolve at earliest. Spintronics field is an upcoming promising technology. It makes use of spin of an electron along with its charge. Limitation in its fabrication technology is a major hurdle in its rapid development. Spin technology-based devices are better than MOS devices in terms of their scalability, non-volatility, ease to read and write and high endurance. Faithful realization of spin device in electronic circuits is still a challenging task as it calls for a good knowledge of quantum physics, material engineering, fabrication and finally testing methods. Using Verilog-A programming language and then integrating with available CAD tools like Cadence development in some spintronics application for memory and hybrid circuits has taken place.

## References


1. Rabaey JM, Chandrakasan AP, Nikolic B (2005) Digital integrated circuits: a design perspective, 2nd edn. Upper Saddle River, NJ, USA: Pearson Education
2. Khursheed A, Khare K, Haque FZ (2019) Designing of ultra-low-power high-speed repeaters for performance optimization of VLSI interconnects at 32 nm. Int J Numer Model Electron Netw Devices Fields 32(2):1–16. https://doi.org/10.1002/jnm.2516
3. Khursheed A, Khare K (2021) Nano interconnects: device physics, modeling and simulation, 1st edn. CRC Press. https://doi.org/10.1201/9781003104193.
4. Roy K, Mukhopadhyay S, Mahmoodi-Meimand H (2003) Leakage current mechanisms and leakage reduction techniques in deep-submicrometer CMOS circuits. Proc IEEE 91(2):305–327
5. Khursheed A, Khare K (2020) Designing dual-chirality and multi-Vt repeaters for performance optimization of 32 nm interconnects. Circuit World 46(2):71–83
6. Khursheed A, Khare K, Haque FZ (2019) Designing high-performance thermally stable repeaters for nano-interconnects. J Comput Electron 18(1):53–64

7. Weisheng Z et al (2009) High speed, high stability and low power sensing amplifier for MTJ/CMOS hybrid logic circuits. Magn IEEE Trans on 45(10):3784–3787
8. Khursheed A, Khare K, Malik MM, Haque FZ (2018) Performance tuning of very large scale integration interconnects integrated with deep sub Micron repeaters. J Nanoelectron Optoelectron 13(12):1797–1806
9. Moaiyeri MH, Nasiri M, Khastoo N (2016) An ef_cient ternary serial adder based on carbon nanotube FETs. Eng Sci Technol Int J 19(1):271–278
10. Hanyu T, Endoh T, Suzuki D, Koike H, Ma Y, Onizawa N, Natsui M, Ikeda S, Ohno H (2016) Standby-power-free integrated circuits using MTJ-based VLSI computing. In: Proc IEEE 104(10):1844–1863
11. Afreen K, Khare K (2020) Optimized buffer insertion for efficient interconnects designs. Int J Numer Model-Electron Networks Devices Fields 33. https://doi.org/10.1002/jnm.2748
12. Joshi VK (2016) Spintronics: A contemporary review of emerging electronics devices. Eng Sci Technol Int J 19(3):1503–1513
13. Verma S, Kulkarni AA, Kaushik BK (2016) Spintronics-based devices to circuits: Perspectives and challenges. IEEE Nanotechnol Mag 10(4)13–28
14. Hanyu T (2015) Challenge of nonvolatile logic LSI using MTJ-based logicin-memory architecture. In: Zhao W, Prenat G (eds) Spintronics-based computing, Cham, Switzerland: Springer, ch. 5, pp 159–177
15. Yuasa S, Nagahama T, Fukushima A, Suzuki Y, Ando K (2004) Giant room-temperature magnetoresistance in single-crystal Fe/MgO/Fe magnetic tunnel junctions. Nat Mater 3(12):868–871
16. Ikeda S, Hayakawa J, Ashizawa Y, Lee YM, Miura K, Hasegawa H, Tsunoda M, Matsukura F, Ohno H (2008) Tunnel magnetoresistance of 604% at 300K by suppression of Ta diffusion in CoFeB/MgO/CoFeB pseudo-spin-valves annealed at high temperature. Appl Phys Lett 93(8), 082508–1–082508–3
17. Tatara G, Kohno H (2004) Theory of current-driven domain wall motion: spin transfer versus momentum transfer. Phys Rev Lett 92(8):086601
18. Berger L (1996) Emission of spin waves by a magnetic multilayer traversed by a current. Phys Rev B 54(13):9353
19. Matsunaga S, Hayakawa J, Ikeda S, Miura K, Hasegawa H, Endoh T, Ohno H, Hanyu T (2008) Fabrication of a nonvolatile full adder based on logic-in-memory architecture using magnetic tunnel junctions, Appl Phys Express 1(9), Art no 091301
20. Matsunaga S, Hayakawa J, Ikeda S, Miura, T. Endoh K, Ohno H, Hanyu T (2009) MTJ-based nonvolatile logic-in-memory circuit, future prospects and issues. In: Proceedings Design, automation & test in Europe conference & exhibition, Nice, France, pp 433–435
21. Diao Z et al (2005) Spin transfer switching and spin polarization in magnetic tunnel junctions with MgO and AlOx barriers. Appl Phys Lett 87(23):232502
22. Venkatesan R, Kozhikkottu V, Augustine C, Raychowdhury A, Roy K, Raghunathan A (2012) TapeCache: a high density, energy efficient cache based on domain wall memory. In: Proceedings of the 2012 ACM/IEEE international symposium on low power electronics and design, pp 185–190
23. Trinh HP, Zhao W, Klein JO, Zhang Y, Ravelsona D, Chappert C (2013) Magnetic adder based on racetrack memory. IEEE Trans Circuits Syst I Regul Pap 60(6):1469–1477

# Short-Channel Effects in Independently Controlled MG-MOSFET

**Soumajit Ghosh, Mitiko Miura-Mattausch, Hafizur Rahaman, Takahiro Iizuka, and H. J. Mattausch**

**Abstract** Although silicon thickness scaling is the key for reducing short-channel effects, advanced MOSFET structure still suffers from the effect even with multi-gate control. To get a better insight to the origin of the short-channel effects for independently controlled multi-gate MOSFET, investigation is performed here for the very thin SOI layer and BOX thickness. It is found that the microscopic potential distribution within the device is the result of balance among different electric field induced within the device. A new compact model describing the effect is developed by considering the potential distribution at the contact/channel junction explicitly, which is modified by additional induced field by the back-gate voltage. The model was verified with 2D-device simulation results for different channel length devices, and good agreement is verified.

**Keywords** Compact model · Short-channel effect (SCE) · MG-MOSFET · Potential distribution · Sub-threshold slope

## 1 Introduction

Better electrostatic control of MOSFET under subthreshold condition is a key objective for down scaled devices to reduce switching loss. To realize the efficient control, short-channel effects (SCE) must be reduced [1]. Independently controlled multi-gate (MG) MOSFET is widely acknowledged as a potential candidate for reducing SCE [2]. However, it has been verified that the SCE cannot be reduced in spite of aggressive reduction of the device thickness to achieve better gate control (scaling rule for MG). Even though MG-MOSFET is more immune to SCE in comparison with the

S. Ghosh (✉) · M. Miura-Mattausch · T. Iizuka · H. J. Mattausch
Hiroshima University, 1 Chome-3-2, Kagamiyama, Higashihiroshima, Japan
e-mail: ghoshsoumajit@hiroshima-u.ac.jp

H. Rahaman
IIEST Shibpur, Botanical Garden Area, Howrah, India

 
C. Giri et al. (eds.), *Emerging Electronic Devices, Circuits and Systems*,
Lecture Notes in Electrical Engineering 1004,
https://doi.org/10.1007/978-981-99-0055-8_7

bulk counterpart, the increment of lateral electric field under the subthreshold condition still degrades the gate control over channel. The reason has been attribute to an additional electric field induced within the BOX [3]. Our investigation focuses on the origin of specific SCE induced in MG-MOSFET which prevent further device down scaling. It has been demonstrated for symmetrical MG-MOSFET with common gate that the overlap of two potential distributions induced by source and drain/channel junctions are the main origin [4]. Here, we investigate the SCE on independently controlled MG-MOSFET, which has been demonstrated to be suitable even for 0.4 V applications with the independently controlled back-gate voltage [5]. It is shown that the electric field in BOX due to the back-gate control is an additional origin of SCE. Modeling of the effect is also presented. In our present investigation, the drain voltage of 50 mV is considered to extract the simplified SCE origin. 2D-device simulation is applied to study microscopic feature of the effect.

## 2 Origin of SCE in MG-MOSFET

The device structure of the MG-MOSFET for our investigation is depicted in Fig. 1. The channel length is varied, while other device parameter values are fixed. The device is called in different names such as Extremely Thin Silicon on Insulator (ETSOI) and Silicon on Thin Buried Oxide (SOTB), which provide more application possibility for circuit design to achieve required performance. Figure 2 shows simulated $I_{ds} - V_{gs}$ characteristics with a 2D-device simulator at low $V_{ds}$ ($V_{ds} =$ 50 mV) [6] for two channel lengths of $L = 200$ nm and 25 nm. Results for two different bias conditions are compared, (a) the back-gate voltage $V_{bg}$ and (b) the drain voltage $V_{ds}$ variation. It is seen that SCE observed as the threshold voltage ($V_{th}$) shift is remarkable only for the short-channel device in Fig. 2b. However, the $V_{bg}$ variation causes much enhanced SCE as seen in Fig. 2a as the subthreshold-slow degradation even for the long-channel device.

Significant degradation of the subthreshold slope is observed for the short-channel device. As observed, positive biasing enhances SCE for both $L = 200$ nm and 25 nm. This suggests us that $V_{bg}$ plays the similar role as $V_{ds}$ on SCE specific for the thin-layer MG-MOSFET generation. Figure 3 shows a comparison of 2D potential distributions of two channel lengths under the subthreshold condition of $V_{gs} = -0.3$ V. For the long-channel case, the potential distribution along the channel direction shows abrupt reduction at the source/channel junction and increases at the channel/drain junction again. The potential in the channel is less than 0.5 V. In case of short-channel device, however, depletion width at the junctions extended within the SOI layer. The extensions from source and drain overlap each other and increase the channel potential. For the long-channel case, the potential diminishes rapidly along the vertical direction $X$ and it reaches already nearly zero at the SOI/BOX interface. On the contrary, it extends further down to the interface at the BOX/bulk, namely the potential distribution penetrates deep into the bulk. Figure 4 depicts potential distribution of Fig. 3 at different positions and directions in the SOI layer. Figure 4b

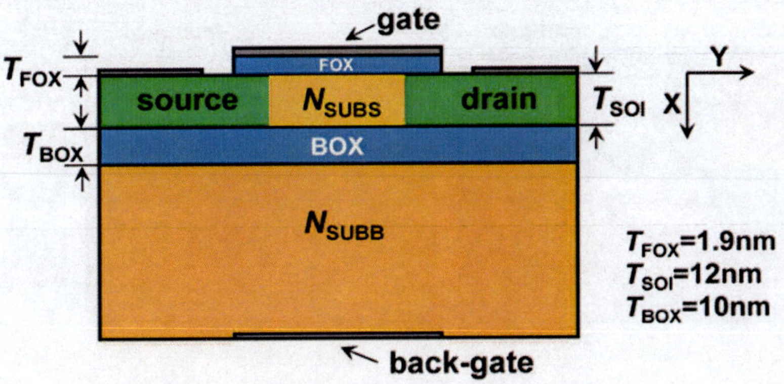


**Fig. 1** Structural description of independently controlled MG-MOSFET

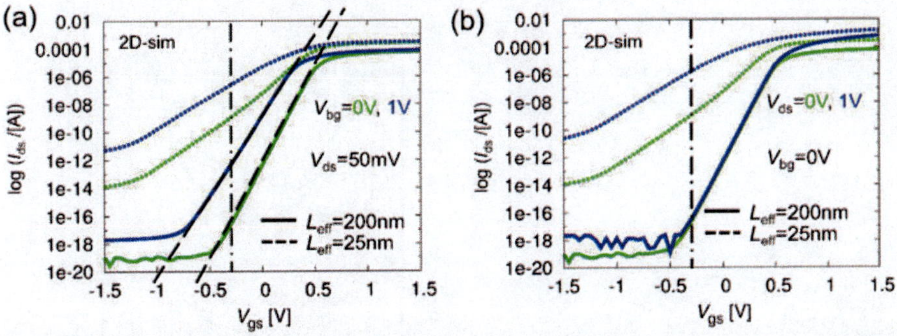


**Fig. 2** log ($I_d$) − $V_{gs}$ plot for long-channel ($L = 200$ nm) and short-channel ($L = 25$ nm) MG-MOSFET for **a** different $V_{bg}$ at $V_{ds} = 50$ mV and **b** different $V_{ds}$ at $V_{bg} = 0$ V. Significant subthreshold slope degradation is observed for both long- and short-$L$ due to positive back-gate biasing for short-channel device indicating stronger SCE. Positive back-gate biasing increases SCE for both $L = 200$ nm and 25 nm

shows those at front-gate surface and back-gate surface along $Y$-direction, and Fig. 4c along $X$-direction at the middle of the channel. It is seen in Fig. 4b that the minimum potential value is quite different for two different channels. With reduced $L$ length, the minimum is much higher, which is valid for both the front- and the back-gate potentials. The potential distribution shown in Fig. 4c is quite smooth as conventionally expected for the long channel. However, that for the short-channel case shows a maximum within the SOI layer near the BOX interface. The difference in two distributions is attributed to the SCE.

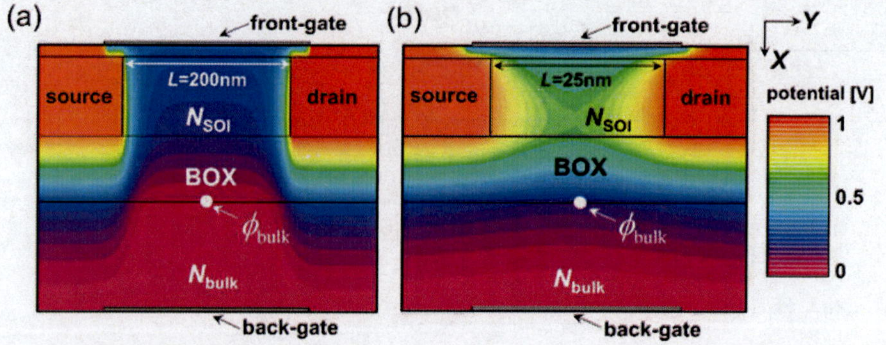


**Fig. 3** Potential distribution for **a** $L = 200$ nm, **b** $L = 25$ nm at $V_{gs} = -0.3$ V, $V_{bg} = 0$ V for $V_{ds} = 50$ mV. It is noteworthy that the vertical potential distribution is mostly located within the SOI layer for long-channel device, whereas it extends further into the bulk for short-channel device

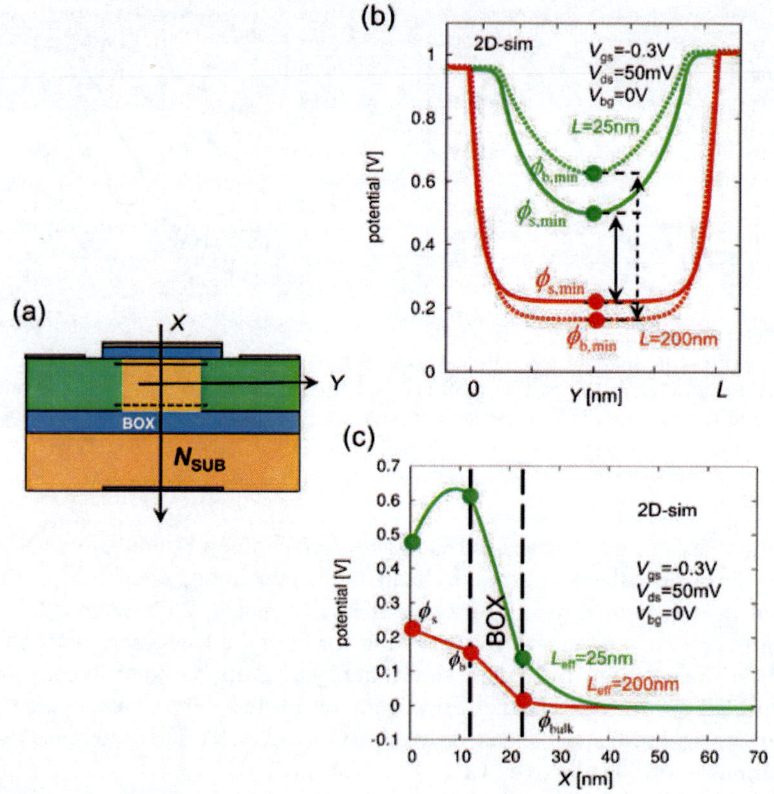


**Fig. 4** **a** Potential distribution profile for $L = 200$ nm and 25 nm along **b** $Y$-direction and **c** $X$-direction at $V_{gs} = -0.3$ V, $V_{bg} = 0$ V for $V_{ds} = 50$ mV, where $\phi_s$ is the front-surface potential and $\phi_b$ is that of the back-gate

## 3 Modeling of SCE in MG-MOSFET

2D-device simulator solves the Poisson equation numerically in 2D or even in 3D. For our investigation, two surface potentials, one at the front-gate and the other at the back gate, are studied here. The surface potentials are distinguished by $\phi_s$ and $\phi_b$ for the front and back, respectively. The Poisson equation includes the quasi-Fermi potential along the channel direction, which is the origin of current flow. Figure 5 compares the potential values of $\phi_s$ and $\phi_b$ as a function of $V_{gs}$ for two different $L$ values. The potential increase is obvious for both $\phi_s$ and $\phi_b$ of the short-channel device. Figure 2 shows that the gate control of the subthreshold current is quite weak for the short-channel device. The reason is seen in Fig. 5, where the back-gate potential $\phi_b$ exceeds the front-gate potential $\phi_s$. Namely, influence of the $V_{bg}$ control is non-negligible for $V_{gs} < V_{th}$.

Figure 6 compares the current–density distribution along the vertical direction $X$, where the contribution of $\phi_b$ for the short-channel device is obvious. For the short-channel device, the most current flows at the back-gate side due to the high $\phi_b$ value. On the contrary, the current flows at the front gate as usual for the long-channel device. Namely, the front-gate control is still dominant. Thus, it is concluded that two potential values ($\phi_s$, $\phi_b$) are keys for modeling the short-channel effect. Figure 7 shows the current-peak position within the SOI layer both for the long- and short-channel lengths. For $V_{gs} > V_{th}$, the gate-controlled inversion charge dominates the total charge, and thus $\phi_s$ exceeds $\phi_b$, resulting the gate-controlled device characteristics.

Modeling of SCE for MG-MOSFET is done by investigating origin of the potential distribution along the channel shown in Fig. 4b. Two different potential minima ($\phi_{min}$) are observed within the SOI layer, at the front- and back-gate sides as shown in Fig. 4b. The potential distribution along the channel is induced by the p/n junction at both source and drain sides due to the built-in potential. The potential minimum is

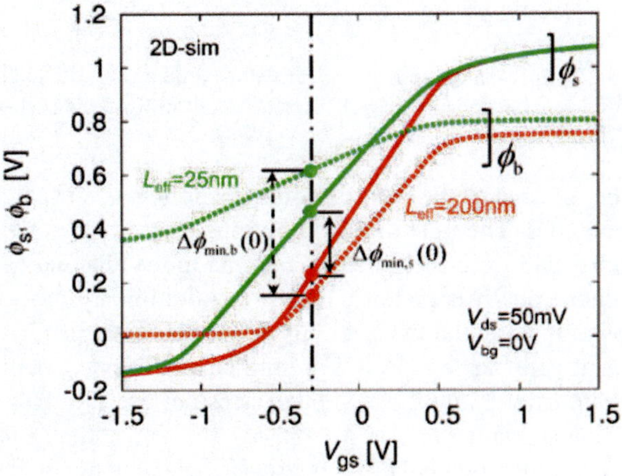


**Fig. 5** Surface potential at the front-gate ($\phi_s$) and back-gate ($\phi_b$) side as a function of $V_{gs}$ for $V_{ds} = 50$ mV and $V_{bg} = 0$ V

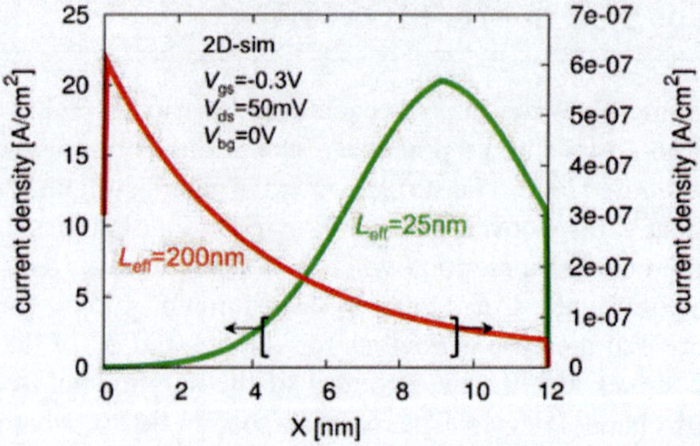


**Fig. 6** Comparison of Current density distribution along *x*-direction for long- and short-channel devices. For long-channel, maximum current flows under profound control of front-gate beneath FOX. For short-channel device, most of the current flows away from FOX, close to BOX

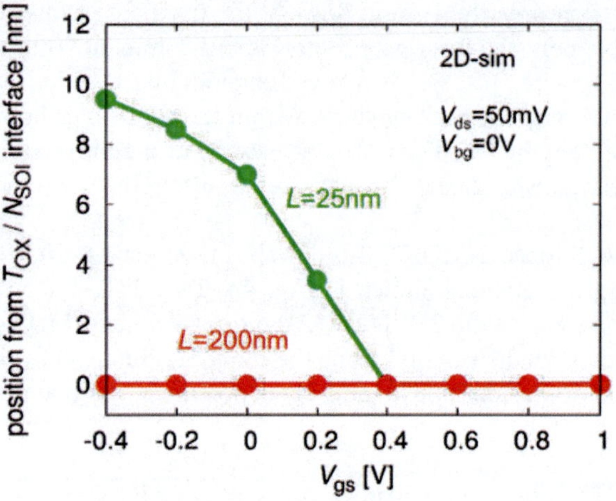


**Fig. 7** Position dependency of current maximum as a function of $V_{gs}$ for $L = 200$ nm and 25 nm. It is seen that the subthreshold current flows mostly at the back-gate side for $L = 25$ nm

the solution of the Poisson equation along the vertical direction at fixed quasi-Fermi potential. The potential distributions from the two *p/n* junctions overlap with each other and raise the potential minima in the channel for the short-channel device as schematically depicted in Fig. 8 [4]. Additionally, the potential minima are influenced by bulk potential ($\phi_{bulk}$ in Fig. 4c) at the same time, depending upon the strength of back-gate biasing ($V_{bg}$). For long-channel device, depletion region from source and drain extend into the SOI layer does not overlap with each other. Potential minima values remain $\phi_{s,\min} = \phi_{s0}$ and $\phi_{b,\min} = \phi_{b0}$ at the front- and back gate, respectively. Therefore, potential difference at each *p/n* junction is assumed not to be influenced by

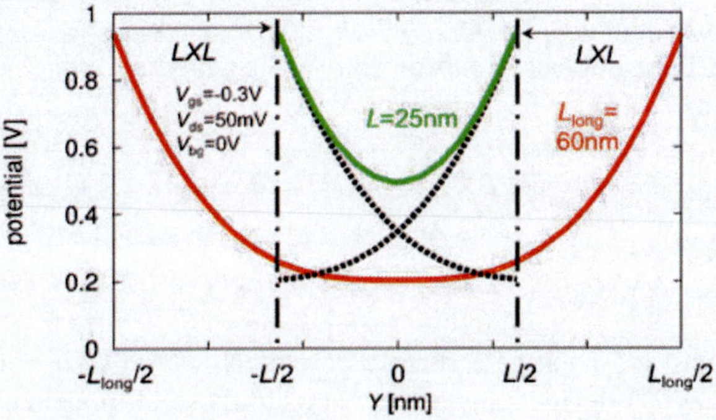


**Fig. 8** Schematic of analytical modeling concept at $V_{gs} = -0.3$ V and $V_{ds} = 50$ mV. The red curve shows a long-channel potential distribution with a cubic function and the black-dashed curves are shifted one of the red curves by *LXL* from two edges. The green curve is sum of the black curves

the carrier injection. For short-channel device, however, depletion region extensions inside the SOI layer overlap with each other, raising the potential minima both along the front- and back-gate side. This is the major origin for the potential increase of $\phi_s$, $\phi_b$ shown in Fig. 4b. Elevation of potential minima increases carrier injection from source.

Analytical model development of the potential minimum change is carried out by assuming the potential distribution along the channel simply with a cubic function

$$\phi(y) = A_0 \times (y + y_{\min})^3 + \phi_{\min} \tag{1}$$

where $y$ indicates the position along channel direction, and $y_{\min}$ indicates the location of $\phi_{\min}$ within the SOI layer. Solving boundary conditions at the source and drain sides, the parameter $A_0$ is obtained for front- and back-gate side, respectively, as

$$A_{0,s} = \frac{(V_{bi} - \phi_{s0})}{(L_{\text{long}}/2)^3} \tag{2}$$

$$A_{0,b} = \frac{(V_{bi} - \phi_{b0})}{(L_{\text{long}}/2)^3} \tag{3}$$

where $\phi_{s0}$ and $\phi_{b0}$ refer to the surface potential in the channel of the long-channel device at front and back gate, respectively. In our investigation, the studied MG_MOSFET maintains its long-channel-subthreshold characteristics approximately down to $L = 60$ nm. For $L < 60$ nm, junction potential distribution starts to superimpose on each other. This phenomenon is replicated by shifting the long-channel potential distribution from each side by $\pm L_{\text{long}}/2 \pm L/2$, assuming that the potential distribution at the junction is kept the same and independent of $L$ as depicted in Fig. 8. Shift in potential minima for the short-channel device at $y = 0$ is attributed to $\Delta\phi_{\min} = \phi_{\text{min\_short}} - \phi_{\text{min\_long}}$ at $y = 0$. $\Delta\phi_{\min}$ (0) has $V_{gs}$ dependency, which is the origin of subthreshold slope degradation. For higher $V_{gs}$, $\Delta\phi_{min}$ (0) diminishes gradually and SCE is suppressed subsequently. The analytical expression for $\Delta\phi_{min}$ ($y$) is expressed as

$$\Delta\phi_{min,s} = (\phi_s(y - LXL) - \phi_{s0}) + (\phi_s(y + LXL) - \phi_{s0}) \tag{4}$$

$$\Delta\phi_{min,b} = (\phi_b(y - LXL) - \phi_{b0}) + (\phi_b(y + LXL) - \phi_{b0}) \tag{5}$$

at respective front- and back-gate surface potentials, where $LXL = (L_{long} - L)/2$. Figure 9 compares 2D-device simulation results to those of model calculation with the cubic function. The comparison is given for $L = 200$ nm and $L = 25$ nm at $V_{gs} = -0.3$ V. At front-gate side, $\Delta\phi_{min,\,s}$ verifies good agreement with 2D-device simulation result for $L = 60$ nm as seen in Fig. 9a. Using the calculated potential minimum $\phi_{s0} + \Delta\phi_{min,\,s}$ at $y = 0$ as boundary condition, the overlapped cubic potential distribution is calculated for $L = 25$ nm as depicted together. However, at back-gate side, model calculation cannot reproduce the shift in potential minima $\phi_{b0} + \Delta\phi_{min,\,b}$ accurately as can be seen in Fig. 9a.

The reason attributes to the disregard of the bulk potential $\phi_{bulk}$. As can be seen in Fig. 4c, $\phi_{bulk}$ for the short-channel device is not equal to zero but keeps certain value, which increases the back-gate potential $\phi_b$ at the same time. Figure 9b shows the result with the non-zero $\phi_{bulk}$. The whole potential distribution shown in Fig. 4c can be calculated by solving the Poisson equation including all possible induced charges. If the front-gate potential $\phi_s$ is accurately calculated, the rest of the potential distribution can be calculated by solving Poisson equation explicitly.

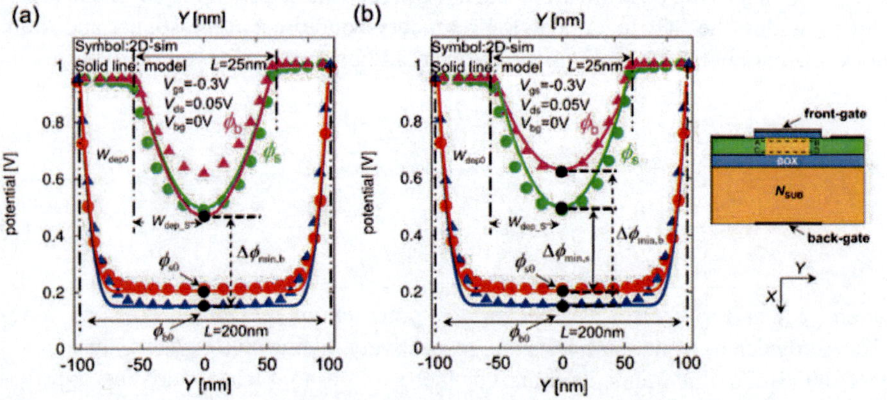


**Fig. 9** Comparison of potential distributions with the developed model to those of 2-D-device simulation along A-B ($\phi_s$: red, green), C-D ($\phi_b$: blue, magenta) for $L = 200$ nm and 25 nm at $V_{gs} = -0.3$ V, $V_{bg} = 0$ V for $V_{ds} = 50$ mV, **a** without taking into account of the $\phi_{bulk}$ contribution and **b** with the contribution

## 4 Conclusion

Thin-layer MOSFETs have been known of their superiority for low-power applications. However, it has been noticed that the short-channel effect cannot be suppressed in spite of aggressive scale down of the layer thickness. Our investigation focused on the subject. We found that the short-channel effect is mostly induced by the depletion extension at the contact/channel junctions, which contributes to the electric field increase along the channel direction in addition to the field induced by the drain voltage. We developed a compact model describing the findings analytically by considering the whole potential distribution within the device along the channel. Good agreement with 2D-device simulation results was achieved. Since device characteristics such as the threshold voltage of the thin-layer MOSFET are strongly influenced not only by the electric field along the channel but also by the back-gate vertical field, solving the Poisson equation with all induced charges is a key for accurate modeling device characteristics. This is our next focus.

## References

1. International Technology Roadmap for Semiconductor (ITRS) (2006)
2. Colinge J-P (ed) (2008) FinFETs and other multi-gate transistors. Springer, NewYork, NY, USA
3. Ghosh S et al (2022) To be published at Proc. EDTM 2022, March, Kokura, Japan
4. Herrera FÁ, Hirano Y, Miura-Mattausch M, Iizuka T, Kikuchihara H, Mattausch HJ, Ito A (2019) Advanced Short-channel-effect modeling with applicability to device optimization—potentials and scaling. IEEE Trans Electron Devices 8;66(9):3726–3733
5. Sugii N et al (2014) Ultralow-power SOTB CMOS technology operating down to 0.4 V. J Low Power Electron Appl 4(2):65–76
6. ATLAS User's Manual, Silvaco, Inc., Santa Clara, CA, USA, Apr 2018

# Reduction of Interconnect Delay and Resistance While Minimizing Grid Area in GNR-Based VLSI Routing Problem

**Subrata Das, Debesh Kumar Das, and Soumya Pandit**

**Abstract** Traditional copper-based interconnect is now facing several challenges. Graphene nanoribbon (GNR)-based interconnect is found to be potential alternative over traditional interconnect due to its outstanding electrical and thermal properties. For special geometric structure, GNR-based interconnect can be bent only in 0°, 60°, and 120° angles. Hence, the routing of GNR is different than that of traditional VLSI routing and may require extra grid area. In this paper, we propose an algorithm for the construction of global routing tree using GNR-based interconnect to reduce the interconnect delay and interconnect resistance with minimized grid area.

**Keywords** Graphene nanoribbon · Hybrid cost · Interconnect resistance · Interconnect delay · Triangular grid

## 1 Introduction

Graphene nanoribbon (GNR) has been proposed to be an alternative material with high potentiality for the use as interconnects due to its excellent thermal and electrical conductivity [1, 2]. Interestingly, compared to copper wire, the power consumption and interconnect delay in GNR can be reduced by up to 50% and 60%, respectively, for global interconnect [3]. Another advantage of GNR is that it is immune to electromigration due to its very strong carbon–carbon bond [4]. Based on the shape of edge, GNR is of two types: zigzag and armchair. Through experiments on band gap with graphene samples, it is observed that GNR with predominantly zigzag edges is metallic, and those with armchair edges are semiconductor [5–7]. Figure 1 shows

S. Das (✉) · S. Pandit
Institute of Radio Physics and Electronics, University of Calcutta, Kolkata, India
e-mail: dsubrata.mt@gmail.com

S. Pandit
e-mail: sprpe@caluniv.ac.in

D. K. Das
Department of Computer Science and Engineering, Jadavpur University, Kolkata, India
e-mail: debeshd@hotmail.com



C. Giri et al. (eds.), *Emerging Electronic Devices, Circuits and Systems*,
Lecture Notes in Electrical Engineering 1004,
https://doi.org/10.1007/978-981-99-0055-8_8

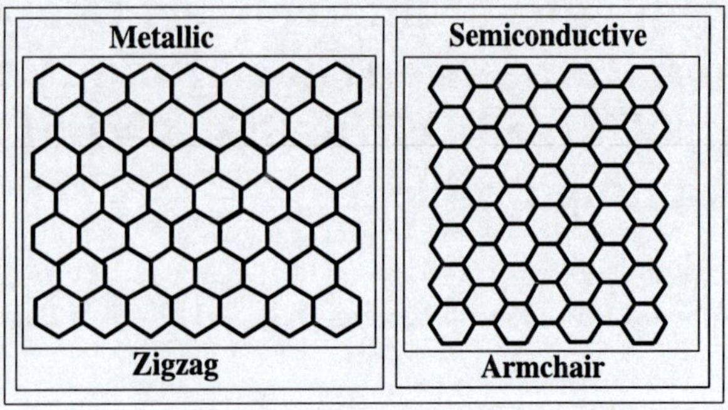


**Fig. 1** Structure of zigzag and armchair GNR

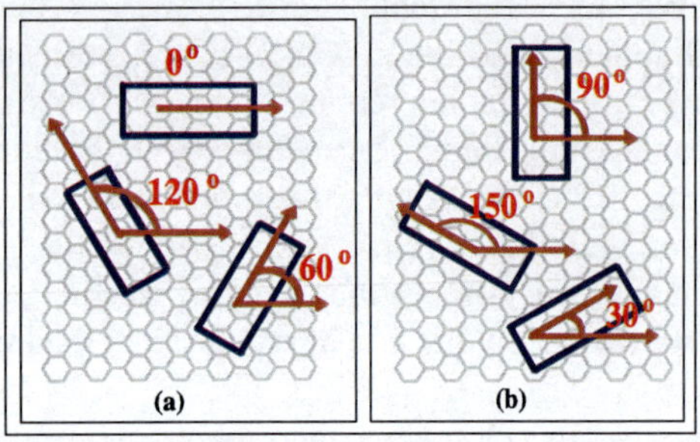


**Fig. 2** Orientation of armchair and zigzag GNR

the structure of the zigzag and the armchair GNR. Figure 2 shows the abstraction of semiconducting and metallic GNR from a graphene sheet as a thin rectangle oriented at (0° , 60°, 120°) and (30°, 90°, 150°) angles [6]. Hence, semiconducting GNR is oriented at 0°, 60°, and 120°, and metallic GNR is oriented at 30°, 90°, and 150° [6]. The bending in 30°, 90°, and 150° angles converts metallic (semiconducting) GNR to semiconducting (metallic). But due to the bending in 0°, 60°, and 120° angles, metallic (semiconducting) GNR remains metallic (semiconducting). Figure 3 shows the bending of armchair and zigzag GNR. As to maintain the same property (metallic/semiconducting), GNR interconnect can be bent only in 0°, 60°, and 120° angles, and the routing grid of GNR-based interconnect is arranged in these specific angles and is known as triangular routing grid. This paper addresses the problem of global routing using GNR-based VLSI interconnect with the goal to reduce the interconnect delay and resistance with minimized grid area.

## *1.1 Literature Review and Motivation*

The use of GNR interconnect in VLSI routing for a pair of source and sink terminals was first proposed in [5] and then improved in [6]. The goal of the research work of these two papers is the optimization of the interconnect resistance that depends on the interconnect length as well as the number and angles of bending. In [5] and [6],

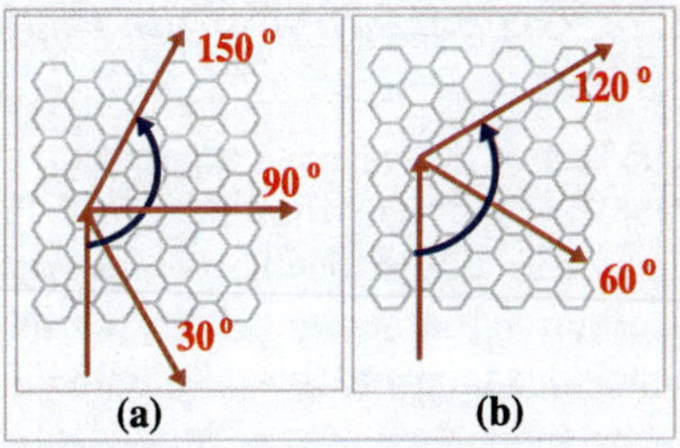


**Fig. 3** Bending of armchair and zigzag GNR

the cost due to interconnect resistance was defined as *hybrid cost*. The routing of GNR-based interconnect for single source and multiple sink terminals for reduction of interconnect resistance and delay was discussed in [8–10]. For a given set of nets in a single-layer GNR routing plane, a routing algorithm to maximize the number of the routed nets with minimizing the bending delay was proposed in [11]. Single-layer delay-driven GNR non-tree routing under resource constraint for yield improvement was presented in [12]. For a given set of GNR nets, a delay-constrained routing algorithm to minimize the number of layers with the constraint that two GNR nets are non-crossing was proposed in [13].

The routing of GNR-based interconnect is different than that of traditional interconnect as this type of interconnect can be bent only in 0°, 60°, and 120° angles. Due to the bending in these special angles, the routing of GNR-based interconnect requires more grid area. Hence, the reduction of grid area may be one of the optimization goals of this kind of routing. The construction of global routing tree for single source and multiple sink terminals to minimize the grid area such that delay from source to each of the each of the sink terminal is within a given limit was proposed in [14].

## *1.2 Contribution and Outline of the Work*

This paper is an extension of the work done in [14]. In this paper, our proposed algorithm can reduce further interconnect delay and resistance with the same grid area as in [14]. We have identified the conditions to reduce further interconnect delay and resistance with same grid area in comparison with that of [14]. This is discussed later in this paper. We discuss the conditions for the reduction of grid area such that the delay of each of the sink terminal is reduced for different positions of the source and sink terminals in triangular grid. We propose algorithm for the construction of global routing tree of GNR-based interconnect for a given source and multiple sink terminals with the optimization goal of reduction of grid area under the condition that delay from source to each of the sink terminal is minimized.

### 1.3 Organization of the Paper

The rest of the paper is organized as follows. Section 2 describes the GNR routing problem. Problem formulation of this research work is described in Sect. 3. The method to reduce the interconnect delay and resistance with minimized the grid area is described in Sect. 4. The experimental result is shown in Sect. 5. Finally, Sect. 6 concludes the paper discussing the future scope.

## 2 GNR Routing Problem

In global routing, a signal net of single source and $n$ sink terminals is given. This set of terminals ($N = N_0, N_1, \ldots N_n$) is connected by a routing tree $T(N)$. The cost of the tree $T(N)$ is the sum of costs of its edges [15]. In case of global routing of GNR-based interconnect, the cost of routing is the total interconnect resistance and total interconnect delay. Total interconnect resistance is the sum resistance of each constituent edges and resistance due to the bending. Interconnect resistance was defined as *hybrid cost* in [5, 6]. Suppose the resistance due to 60° and 120° bending is, respectively, denoted by $\alpha_{60}$ and $\alpha_{120}$. Bending of GNR interconnect in different angles yields different resistances. In [6], it was reported that $\alpha_{120} = 3\alpha_{60}$. Let the total interconnect length is L, the resistance due to unit interconnect length is $W_L$, the number of 60° and 120° bending are, respectively, $n_{60}$ and $n_{120}$, then total interconnect resistance (TR) is given by following equation

$$TR = L \times W_L + n_{60} \times \alpha_{60} + n_{120} \times \alpha_{120} \tag{1}$$

As in [6, 9], here also, we assume $W_L = 1$, $\alpha_{60} = 10$, and $\alpha_{120} = 30$. Figure 4 shows the global routing tree of single source and four sink terminals. Here, $L = 30$, $n_{60} = 3$, $n_{120} = 1$. Hence, total interconnect delay $TR = 30 \times 1 + 3 \times 10 + 1 \times 30 = 90$.

Interconnect capacitance per unit length is denoted by $C_L$. In this paper, *RC delay* is measured from source to each of the sink terminals. Interconnect delay between a pair of terminal is given by the following equation

$$\delta = \frac{1}{2}RC = \frac{1}{2}(L \times W_L + n_{60} \times \alpha_{60} + n_{120} \times \alpha_{120})LC_L. \tag{2}$$

In Fig. 4, the interconnect delay of the sink terminals ($T_1$, $T_2$, $T_3$, and $T_4$) from source terminal is, respectively, $\frac{651}{2}$, $\frac{989}{2}$, $\frac{1121}{2}$, and $\frac{1491}{2}$. The interconnect delay in $SB_1B_2$, $SB_1B_3$ and $SB_1B_3B_4$ is, respectively, $\frac{600}{2}$, $\frac{416}{2}$, and $\frac{1044}{2}$. Hence, total interconnect delay is $\frac{651}{2}+\frac{989}{2}+\frac{1121}{2}+\frac{1491}{2}-\frac{600}{2}-\frac{416}{2}-\frac{1044}{2}=\frac{2192}{2}$.

Now, consider the routing of a single source and three sink terminals as shown in Figs. 5a, b. Here, both these Figs. show the routing of the same source and three sink

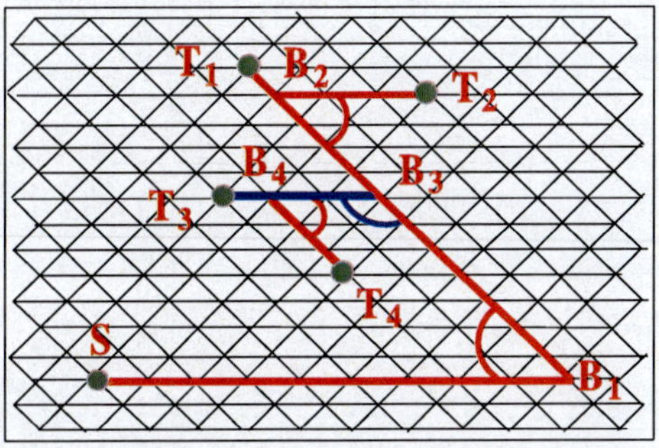


**Fig. 4** Routing tree of single source and four sink terminals

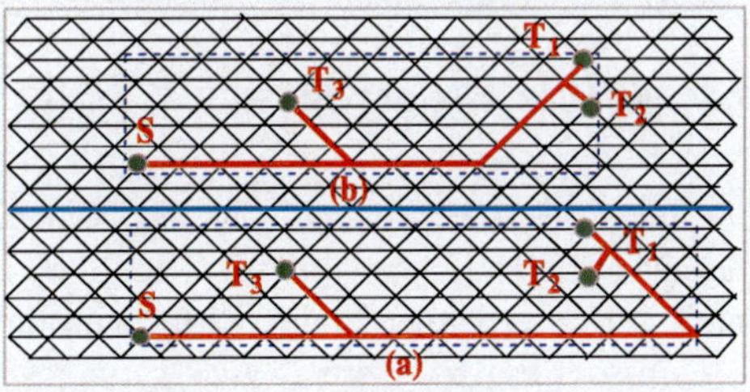


**Fig. 5** Routing tree of single source and three sink terminals

terminals. In Fig. 5a, total interconnect resistance and delay are, respectively, 52 and 424. The grid area required for this routing is $\frac{65\sqrt{3}}{2}$. But for Fig. 5b, the interconnect resistance and delay are, respectively, 67 and $\frac{863}{2}$. The grid area required for this routing is $\frac{55\sqrt{3}}{2}$. Though the interconnect resistance and delay of Fig. 5b are grater than that of 5a, but the area required for the routing in Fig. 5b is less than the routing in Fig. 5a. For GNR-based interconnect, reduction of interconnect delay and resistance with minimized area is considered as an optimization goal.

## 3 Problem Formulation

Given a source (S) and $n$ sink terminals ($T_1, T_2, \ldots, T_n$) in a triangular grid, we have to interconnect them with metallic GNR such that the interconnect delay and resistance are reduced, and area required for the routing is minimized. For a multi-terminal net, VLSI global routing can be described as a steiner tree problem. Here basically, we have to construct a hexagonal steiner tree for the reduction of the interconnect delay and resistance with minimized triangular grid area.

In [14], a similar work was described. In this paper, we have identified few conditions such that interconnect delay can be reduced further than that of [14] though the area of routing is same.

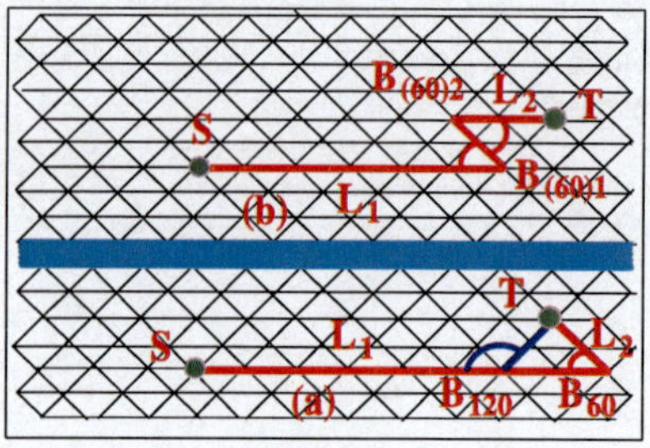


**Fig. 6** Figure for observation 1

# 4 Proposed Method

## 4.1 Initial Observations

Let, for a pair of source and sink terminals, $S(x_s, y_s)$ and $T(x_t, y_t)$ are, respectively, the source and sink coordinates. The following discussions and observations find the minimum interconnect delay with minimum grid area.

Consider Fig. 6a. Here, if $y_t > y_s$ and $(x_t - x_s) > \frac{(y_t - y_s)}{\sqrt{3}}$, then in the routing path $SB_{60}T$, interconnect resistance is minimum. Interconnect resistance, capacitance, and delay in the 60° routing path $SB_{60}T$ are, respectively, given by the following equations.

$$R_{60} = \{(x_t - x_s) + \sqrt{3}(y_t - y_s)\}W_L + \alpha_{60} \tag{3}$$

$$C_{60} = \{(x_t - x_s) + \sqrt{3}(y_t - y_s)\}C_L \tag{4}$$

$$\delta_{60} = \frac{1}{2}\left[\{(x_t - x_s) + \sqrt{3}(y_t - y_s)\}^2 W_L C_L + \alpha_{60}\{(x_t - x_s) + \sqrt{3}(y_t - y_s)\}C_L\right] \tag{5}$$

The required area for the routing of source and sink in this routing path is given by the following equation.

$$\text{area}_{60} = \{(x_t - x_s) + \sqrt{3}(y_t - y_s)\}(y_t - y_s) \tag{6}$$

In the 120° routing path $SB_{120}T$ (Fig. 6a), interconnect resistance, capacitance, delay, and required area are, respectively, given by the following equations.

$$R_{120} = \left\{(x_t - x_s) + \frac{(y_t - y_s)}{\sqrt{3}}\right\} W_L + \alpha_{120} \tag{7}$$

$$C_{120} = \left\{(x_t - x_s) + \frac{(y_t - y_s)}{\sqrt{3}}\right\} C_L \tag{8}$$

$$\delta_{120} = \frac{1}{2}\left[\left\{(x_t - x_s) + \frac{(y_t - y_s)}{\sqrt{3}}\right\}^2 W_L C_L + \alpha_{120}\left\{(x_t - x_s) + \frac{(y_t - y_s)}{\sqrt{3}}\right\} C_L\right] \tag{9}$$

$$\text{area}_{120} = (x_t - x_s)(y_t - y_s) \tag{10}$$

Interconnect resistance, capacitance, delay, and required area in the routing path $SB_{(60)1}B_{(60)2}T$ (Fig. 6b) are, respectively, given by the following equations.

$$R_{2\times 60} = \{(x_t - x_s) + \sqrt{3}(y_t - y_s)\}W_L + 2\alpha_{60} \tag{11}$$

$$C_{2\times 60} = \{(x_t - x_s) + \sqrt{3}(y_t - y_s)\}C_L \tag{12}$$

$$\delta_{2\times 60} = \frac{1}{2}\left[(x_t - x_s) + \sqrt{3}(y_t - y_s)\}^2 W_L C_L + \alpha_{120}\{(x_t - x_s) + \sqrt{3}(y_t - y_s)\}C_L\right] \tag{13}$$

$$\text{area}_{2\times 60} = (x_t - x_s)(y_t - y_s) \tag{14}$$

It is clear to see that the $\text{area}_{2\times 60} = \text{area}_{120} < \text{area}_{60}$.

**Observation 1** *If $y_t > y_s$ and $(x_t - x_s) > \frac{(y_t - y_s)}{\sqrt{3}}$, the routing path $SB_{(60)1}B_{(60)2}T$ (as in Fig. 6) can be can be selected for routing over than that of $SB_{120}T$ if $\delta_{2\times 60} < \delta_{120}$, i.e. if the following constraint is satisfied.*

$$\left[\frac{5}{6}(y_t - y_s)^2 + (x_t - x_s)^2 - 2\sqrt{3}(y_t - y_s)(x_t - x_s)\right]W_L < \alpha_{60}(x_t - x_s) \tag{15}$$

*Otherwise, $SB_{120}T$ path can be preferred for routing.*

Consider Fig. 7a. Here, if $y_t > y_s$ and $x_t < x_s + \frac{(y_t - y_s)}{\sqrt{3}}$ and $x_t \geq x_s + \frac{(y_t - y_s)}{3\sqrt{3}}$, then in the routing path $SB_{60}T$, interconnect resistance is minimum. Interconnect resistance, capacitance, and delay in the 60° routing path $SB_{60}T$ are, respectively, given by the following equations.

$$R_{60} = \{\sqrt{3}(y_t - y_s) - (x_t - x_s)\}W_L + \alpha_{60} \tag{16}$$

$$C_{60} = \{\sqrt{3}(y_t - y_s) - (x_t - x_s)\}C_L \tag{17}$$

$$\delta_{60} = \frac{1}{2}\left[\{\sqrt{3}(y_t - y_s) - (x_t - x_s)\}^2 W_L C_L + \alpha_{60}\{\sqrt{3}(y_t - y_s) - (x_t - x_s)\}\right]C_L \tag{18}$$

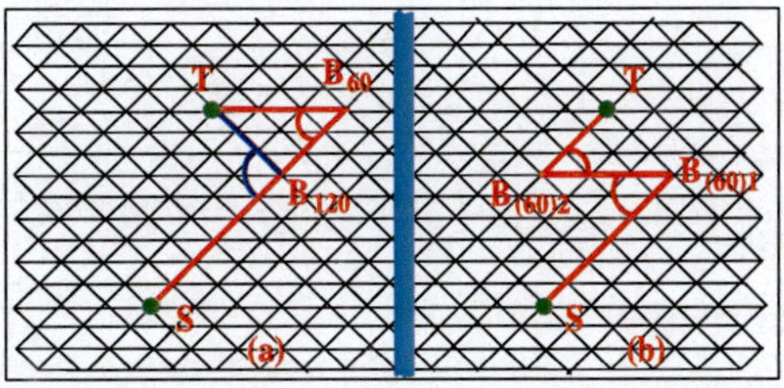


**Fig. 7** Figure for observation 2

The required area for the routing of source and sink in this routing path is given by the following equation.

$$\text{area}_{60} = \frac{(y_t - y_s)^2}{\sqrt{3}} \tag{19}$$

In the 120° routing path $SB_{120}T$ (Fig. 7a), interconnect resistance, capacitance, delay, and required area are, respectively, given by the following equations.

$$R_{120} = \frac{2}{\sqrt{3}}(y_t - y_s)W_L + \alpha_{120} \tag{20}$$

$$C_{120} = \frac{2}{\sqrt{3}}(y_t - y_s)C_L \tag{21}$$

$$\delta_{120} = \left[\frac{2}{3}(y_t - y_s)^2 W_L C_L + \alpha_{120}\frac{1}{\sqrt{3}}(y_t - y_s)C_L\right] \tag{22}$$

$$\text{area}_{120} = \frac{1}{2}\left\{(x_t - x_s) + \frac{(y_t - y_s)}{\sqrt{3}}\right\}(y_t - y_s) \tag{23}$$

Interconnect resistance, capacitance, delay, and required area in the routing path $SB_{(60)1}B_{(60)2}T$ (Fig. 7b), are, respectively, given by the following equations.

$$R_{2\times 60} = \{\sqrt{3}(y_t - y_s) - (x_t - x_s)\}W_L + 2\alpha_{60} \tag{24}$$

$$C_{2\times 60} = \{\sqrt{3}(y_t - y_s) - (x_t - x_s)\}C_L \tag{25}$$

$$\delta_{2\times 60} = \frac{1}{2}\left[\{\sqrt{3}(y_t - y_s) - (x_t - x_s)\}^2 W_L C_L + 2\alpha_{60}\{\sqrt{3}(y_t - y_s) - (x_t - x_s)\}C_L\right] \tag{26}$$

$$\text{area}_{2\times 60} = \frac{1}{2}\left\{(x_t - x_s) + \frac{(y_t - y_s)}{\sqrt{3}}\right\}(y_t - y_s) \tag{27}$$

It is clear to see that the $\text{area}_{2\times 60} = area_{120} < area_{60}$.

**Observation 2** *If $y_t > y_s$ and $x_t < x_s + \frac{(y_t - y_s)}{\sqrt{3}}$ and $x_t \geq x_s + \frac{(y_t - y_s)}{3\sqrt{3}}$, the routing path $SB_{(60)1}B_{(60)2}T$ (as in Fig. 7) can be can be selected for routing over than that of $SB_{120}T$ if $\delta_{2\times 60} < \delta_{120}$, i.e. if the following constraint is satisfied.*

$$\left[\frac{5}{6}(y_t - y_s)^2 - (x_t - x_s)^2\right] < \alpha_{60}(x_t - x_s) \tag{28}$$

*Otherwise, $SB_{120}T$ path can be preferred for routing.*

**Observation 3** *If $y_t > y_s$, $x_t < x_s + \frac{(y_t - y_s)}{\sqrt{3}}$ and $x_t < x_s + \frac{(y_t - y_s)}{3\sqrt{3}}$ (as shown in Fig. 8), the 120° routing path is always preferable for area minimized routing.*

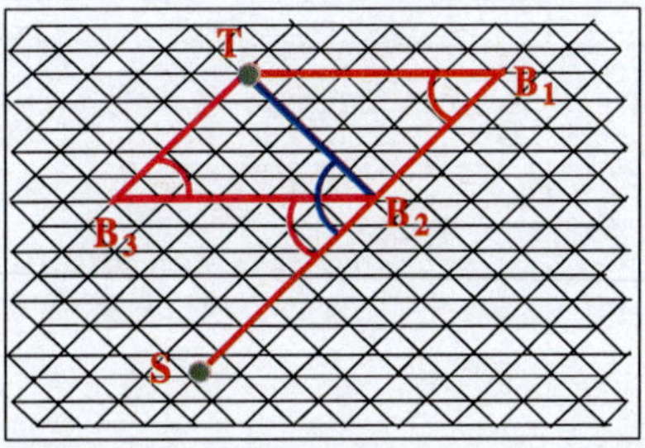


**Fig. 8** Figure for observation 3

From all of the above observations, it is clear to see that if the interconnect delay in the routing path with two 60° bending is less than to that of one 120° routing path, then interconnect resistance is also less in two 60° routing path. But the converse, it not always true.

## *4.2 Algorithm for GNR Routing to Reduce the Interconnect Delay and Resistance with Minimized Grid Area*

For single source (S) and *n* sink terminals ($T_1, T_2, \ldots, T_n$), first we have to find the maximum distant sink terminal as described in [14]. Now based on the positions of source and maximum distant sink terminal, delay may be minimum either in 60° or 120° routing path or in a routing path with two 60° bending as described in observations 1 and 2. Area required for routing for this pair of source and maximum distant sink terminals is minimum in 120° routing path or the routing path with two 60° bending. In [14] 120°, routing path is selected for this routing. But observations 1, 2, and 3 show delay which may be less than 120° routing path with same area. This routing path is basically routing with two 60° bending. Here, we select the routing path with minimum grid area with minimized delay.

The routing path from source to the maximum distant sink is known as the first routing path. As the area is minimum in 120° bending path, 60° routing path is not considered. The area of two 60° routing path is same as the area 120° routing path. Now to select the routing path of maximum distant sink terminal, the minimum delay between 120° or two 60° routing path is considered. Now, there are a pair of 120° and a pair of two 60° routing paths for which delay is minimum. To select the proper bending point, the average distances of the remaining sink terminals from both 120° routing paths are calculated. The bending point is selected so that the average distance is minimum. Now, the remaining sink terminals are divided into two clusters described in [8]. Each of the sink terminals from these two clusters is connected to the existing routing tree in such a way that the grid area required for routing is minimum with minimized interconnect delay and resistance. In order to connect the sink terminals to the existing routing tree, the sink terminals of cluster-1 are considered decreasing distance from the first routing line, and sink terminals of

*Interconnect_sink*$(T_i)$

**Input:** Equation of a $1^{st}$ routing line and Coordinate of a sink terminal
**Output:** Interconnection point of the sink to the existing routing tree

interconnect the sink in such a way so that the area required for routing is minimized
**if** *more than one such paths* **then**
  select the intersecting point for which delay for source to that sink is minimum
**end**

**Fig. 9** Interconnect a sink to the existing routing tree

cluster-2 are considered increasing distance from the first routing path. The function $Interconnect\_sink(T_i)$ as described in Fig. 9 is used for this purpose. The formal description of the algorithm of GNR routing for minimum grid area is described in Algorithm 1.

**Algorithm 1:** GNR routing to minimize interconnect delay and resistance with minimum grid area

**Input**: Single source (S) and n sink terminals ($T_1, T_2, \ldots T_n$) on triangular grid.
**Output**: Global routing tree with reduced interconnect delay and resistance with minimum grid area.

Find the maximum distant sink terminal
Find the average distances $ave_1$ and $ave_2$ of the remaining sink terminals w. r.t. two 120° routing paths.
Calculate the delay in 120° ($\delta_{120}$) and two 60° ($\delta_{2\times60}$) routing paths
**if** $\delta_{120} < \delta_{2\times60}$ **then**
  Select the 120° routing path for which the average distance is minimum for the $1^{st}$ routing path
**else**
  Select the corresponding two 60° routing path for the $1^{st}$ routing path
**end**
$selection\_of\_cluster()$ [as described in [8]]
**for** *each sink* $(T_i)$ *of cluster-1 in decreasing distance order* **do**
  $Interconnect\_sink(T_i)$
**end**
**for** *each sink* $(T_j)$ *of cluster-2 in increasing distance order* **do**
  $Interconnect\_sink(T_j)$
**end**
End of algo.

**Table 1** Experimental result

| Test case | # Sink | Algorithm of [14] | | | Algorithm of this paper | | | % Reduction | |
|---|---|---|---|---|---|---|---|---|---|
| | | Area | Total delay | Total resitance | Area | Total delay | Total resitance | Delay | Resistance |
| 1 | 16 | $280.5\sqrt{3}$ | 3069.5 | 405 | $280.5\sqrt{3}$ | 2975 | 380 | 3.08 | 6.17 |
| 2 | 15 | $308\sqrt{3}$ | 3905 | 381 | $308\sqrt{3}$ | 3850.5 | 366 | 1.39 | 3.94 |
| 3 | 15 | $285\sqrt{3}$ | 2953 | 316 | $285\sqrt{3}$ | 2874.5 | 293 | 2.66 | 7.28 |
| 4 | 15 | $300\sqrt{3}$ | 2357 | 309 | $300\sqrt{3}$ | 2357 | 309 | 0 | 0 |
| 5 | 17 | $308\sqrt{3}$ | 3238.5 | 381 | $308\sqrt{3}$ | 3169.5 | 357 | 2.13 | 6.30 |
| 6 | 18 | $231\sqrt{3}$ | 2991.5 | 355 | $231\sqrt{3}$ | 2927.5 | 339 | 2.14 | 4.51 |
| 7 | 14 | $210\sqrt{3}$ | 3332 | 405 | $210\sqrt{3}$ | 3332 | 405 | 0 | 0 |
| 8 | 14 | $187.5\sqrt{3}$ | 2560.5 | 330 | $187.5\sqrt{3}$ | 2517 | 315 | 1.70 | 4.55 |
| 9 | 16 | $220.5\sqrt{3}$ | 3758 | 317 | $220.5\sqrt{3}$ | 3740.5 | 310 | 0.47 | 2.21 |
| 10 | 16 | $351\sqrt{3}$ | 4304 | 390 | $351\sqrt{3}$ | 4201 | 374 | 2.39 | 4.10 |
| 11 | 17 | $289\sqrt{3}$ | 2708 | 256 | $289\sqrt{3}$ | 2688.5 | 239 | 0.72 | 6.64 |
| 12 | 14 | $400\sqrt{3}$ | 2392 | 262 | $400\sqrt{3}$ | 2310 | 244 | 3.43 | 6.87 |

## 5 Experimental Result

The experimental result is shown in Table 1. In this paper, we have randomly generated 12 data set for our experiment. For each of this data set, there are single source and multiple sink terminals. As in [6, 9], the resistance due to unit length, 60° and 120° bending are, respectively, assumed to be 1, 10, and 30. The capacitance due to unit length interconnect is assumed 1 as in [9]. For each of the data set, we compute the total interconnect resistance, delay, and grid area required for the routing of GNR interconnect with the algorithm proposed in [14] and in this paper. The work of this paper finds the routing path with minimum grid area and minimized delay and resistance in comparison with that of [14]. In this paper, we have discussed that further reduction of interconnect delay and resistance (Observations 1, 2, and 3) may be achieved in comparison with that of [14]. The algorithm in this paper constructs the global routing tree in such a way that not only the area is minimum but also when it connects a sink terminal to the existing routing tree, the interconnect delay and resistance are also minimized. Hence, depending on the positions of source and sink terminals further reduction of interconnect delay and resistance may be obtained in comparison with that of [14]. We observe that except for test case 4 and 7 for all other test cases, interconnect delay and interconnect resistance decrease in the proposed algorithm in comparison with those of [14]. Experimental result shows maximum and average reduction of interconnect delay 3.43%, 1.68%, respectively, and the reduction of interconnect resistance 7.28% and 4.38%, respectively.

## 6 Conclusion

In this paper, we propose an improved algorithm for the construction of global routing tree of GNR-based interconnect so that total interconnect delay and resistance are minimized with minimum grid area. As GNR interconnect can be bent only in some special degrees, the routing using this interconnect requires more grid area than that of traditional interconnect. The reduction of grid area may increase the total interconnect delay and resistance. Our proposed algorithm minimizes interconnection delay and resistance with minimized grid area. For low-power and high-speed VLSI circuit, the routing with minimized delay and resistance is always preferable. The reduction of interconnect resistance may also cause the reduction in the power dissipation due to interconnect. GNR routing with interconnect resistance, delay, and required grid area as joint objective function will be future scope of this work.

## References

1. Naeemi A, Meindl JD (2008) Performance benchmarking for graphene nanoribbon, carbon nanotube, and cu interconnects. In: 2008 International interconnect technology conference. IEEE, pp 183–185
2. Shao Q, Liu G, Teweldebrhan D, Balandin A (2008) High-temperature quenching of electrical resistance in graphene interconnects. Appl Phys Lett 92(20):202108
3. Li H, Xu C, Banerjee K (2010) Carbon nanomaterials: the ideal interconnect technology for next-generation ICS. IEEE Des Test Comput 27(4):20–31
4. Ragheb T, Massoud Y (2008) On the modeling of resistance in graphene nanoribbon (gnr) for future interconnect applications. In: 2008 IEEE/ACM international conference on computer-aided design. IEEE, pp 593–597
5. Yan T, Ma Q, Chilstedt S, Wong MD, Chen D (2011) Routing with graphene nanoribbons. In: 16th Asia and South Pacific design automation conference (ASP-DAC 2011). IEEE, pp 323–329
6. Yan T, Ma Q, Chilstedt S, Wong MD, Chen D (2013) A routing algorithm for graphene nanoribbon circuit. ACM Trans Des Autom Electron Syst (TODAES) 18(4):1–18
7. Han MY, Özyilmaz B, Zhang Y, Kim P (2007) Energy band-gap engineering of graphene nanoribbons. Phys Rev Lett 98(20):206805
8. Das S, Das S, Majumder A, Dasgupta P, Das DK (2016) Delay estimates for graphene nanoribbons: a novel measure of fidelity and experiments with global routing trees. In: 2016 International great lakes symposium on VLSI (GLSVLSI). IEEE, pp 263–268
9. Das S, Das DK, Pandit S (2020) A global routing method for graphene nanoribbons based circuits and interconnects. ACM J Emerg Technol Comput Syst (JETC) 16(3):1–28
10. Sinharay A, Das S, Roy P, Rahaman H (2018) An angular steiner tree based global routing algorithm for graphene nanoribbon circuit. In: International symposium on VLSI design and test. Springer, pp 670–681
11. Yan JT (2018) Single-layer GNR routing for minimization of bending delay. IEEE Trans Comput-Aided Des Integr Circ Syst 38(11):2099–2112
12. Yan JT (2019) Single-layer delay-driven gnr nontree routing under resource constraint for yield improvement. IEEE Trans Very Large Scale Integr (VLSI) Syst 28(3):736–749
13. Yan JT (2020) Delay-constrained GNR routing for layer minimization. IEEE Trans Very Large Scale Integr (VLSI) Syst 28(11):2356–2369

14. Das S, Das DK (2017) A technique to construct global routing trees for graphene nanoribbon (GNR). In: 2017 18th International symposium on quality electronic design (ISQED). IEEE, pp 111–118
15. Samanta T, Ghosal P, Rahaman H, Dasgupta P (2008) Revisiting fidelity: a case of elmore-based y-routing trees. In: Proceedings of the 2008 international workshop on system level interconnect prediction, pp 27–34

# Modeling of Pristine and Intercalation Doped Multilayer Graphene Nanoribbon Conductors with Energy-per-Layer Screening

**Santasri Giri Tunga, Sandip Bhattacharya, Subhajit Das, and Hafizur Rahaman**

**Abstract** The effective resistance of a multilayer graphene nanoribbon (MLGNR) interconnect depends on the number of conducting layers. However, all the layers present in the interconnect do not actively participate in the conduction of current. The number of layers implicitly participating in current conduction is regulated by the interlayer spacing, mean free path, the impact of Fermi level shift, and the number of conducting channels. In this paper, the effective resistance for pristine and different intercalated top-contact MLGNR interconnect has been modeled considering the Fermi level deviation of each layer. Besides, it is also demonstrated that intercalation-doped MLGNR increases the number of effective layers and reduces the effective resistance compared to pristine MLGNR.

**Keywords** Multilayer graphene nanoribbon · Intercalation · Fermi energy deviation · Interlayer screening length

## 1 Introduction

Graphene nanoribbon (GNR) is an extremely slender (less than 50 nm) ribbon obtained from monoatomic structure graphene. The integrated chip industry realizes the prospect of GNR as a potential interconnect material because of its exceptional thermal and electrical conductivity ($5 - 20 \times 10^8$ A/cm$^2$), very large Mean Free Path (MFP), and super mobility [1–4]. A circuit model based on compact physics for GNR interconnect was proposed by Naeemi and Meindl [5, 6]. The number of layers in GNR regulates the electrical properties of GNR. GNR for interconnect applications

S. G. Tunga (✉)
Haldia Institute of Technology, Haldia, India
e-mail: santasri.gt@gmail.com

S. Bhattacharya
SR University, Warangal, Telangana, India

S. Das · H. Rahaman
Indian Institute of Engineering Science and Technology, Howrah, India

 
C. Giri et al. (eds.), *Emerging Electronic Devices, Circuits and Systems*,
Lecture Notes in Electrical Engineering 1004,
https://doi.org/10.1007/978-981-99-0055-8_9

can be used in a single layer (SLGNR) or multilayer (MLGNR). Various studies were performed to analyze the performance of the SLGNR and MLGNR, and it was found that MLGNR performs better than SLGNR [7–10].

An MLGNR interconnect can be connected to the metal contact in the side (SC-MLGNR) or on the top (TC-MLGNR). An analytical MLGNR 2D resistive network model was established by Kumar et al. [9], and based on this model, a performance comparison in time and frequency domain was made for both SC and TC-MLGNR interconnects. The results revealed that the performance of SC-MLGNR is superior to TC-MLGNR [8–12]. Despite that, as the metal contact is connected to all layers of SC-MLGNR, in planar technology, SC-MLGNR fabrication becomes a very complex process. TC-MLGNR fabrication has the advantage of metal contact only on the topmost layer. In our model, we will consider the contact to be the top contact (TC-MLGNR). In [9], a precise and compact model of TC-MLGNR interconnect has been developed considering the impact of the effective number of layers taking part in conduction. Notable improvement has been observed in the Fermi level ($E_f$) and mean free path (MFP) with intercalation doping [13]. Literature [14, 15] experimentally established that in-plane conductivity could be improved by intercalation doped TC-MLGNR and their performance can achieve toward SC-MLGNR. All these analyzes for pristine and intercalated MLGNR were performed considering each layer's $E_f$ to be the same. Literature [9] indicated that for pristine and intercalated MLGNR, $E_f$ is found to change per layer, which affects the effective resistance ($R_{efct}$), yet this effect is not demonstrated and considered in any model.

This work considers the shift in Fermi level ($E_{fs}$) per layer while modeling the $R_{efct}$ of the TC-MLGNR interconnect. For an MLGNR effective number of layers taking part in current conduction is also evaluated. In addition, modeling of $R_{efct}$ with and without energy-per-layer screening, at varied interconnect lengths for pristine and different intercalation doped TC-MLGNR is reflected in this work.

The remaining part of the paper is framed as follows. Section 2 includes determining the number of effective layers taking part in current conduction and the Fermi level corresponding to each of those layers at different widths. Based on these computations modeling of TC-MLGNR conductors is presented in Sect. 3. The conclusion of this work is drafted in Sect. 4.

## 2 Effective Conducting Layer

### 2.1 *Equivalent Single Conductor Model*

The fundamental Equivalent Single Conductor (ESC) model of the MLGNR interconnect is presented in Fig. 1. It consists of lumped elements and distributed elements. For lumped elements, variation in interconnect length does not affect the circuit performance, whereas in distributed elements the effect predominates. The distributed elements comprise the effective resistor, effective inductor, and effective

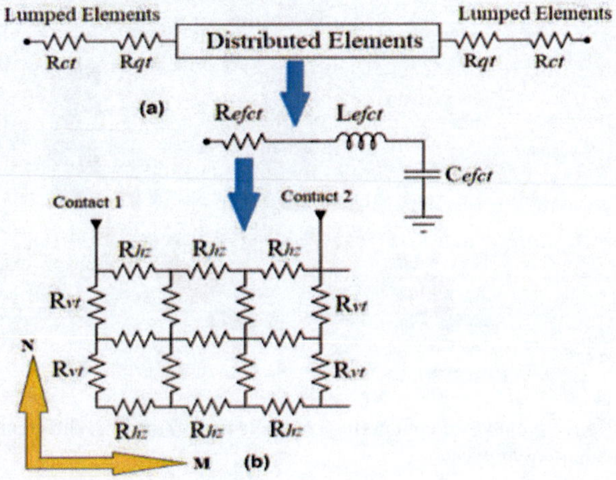


**Fig. 1** **a** Fundamental ESC model. **b** Multiconductor model of MLGNR interconnect

capacitor. The resistor network of the distributed elements is fragmented horizontally and vertically into $M$ segments and $N$ layers, respectively.

The lumped elements of the near and far terminals of the interconnect are presented by conducting channel-dependent quantum resistor $R_{\mathrm{qt}} = h/2\mathrm{e}^2 N_{\mathrm{ch}}$ and contact resistor $R_{\mathrm{ct}} = h/2\mathrm{e}^2 N_{\mathrm{ch}} T_{\mathrm{r}}$ where $T_{\mathrm{r}}$ is the transmission coefficient having a value less than 1 [8] and $N_{\mathrm{ch}}$ is the number of conducting channels. The effective resistor, inductor, and capacitor together build up the distributed elements.

The $R_{\mathrm{efct}}$ is the aggregate of the in-plane or horizontal resistance ($R_{\mathrm{hz}}$) and the perpendicular or vertical resistance ($R_{\mathrm{vt}}$). $R_{\mathrm{hz}}$ and $R_{\mathrm{vt}}$ are stated mathematically as [9]

$$R_{\mathrm{hz}} = \frac{R_{\mathrm{qt}} \Delta l}{N_{\mathrm{ch}} \Lambda_{\mathrm{efct}}} \tag{1}$$

$$R_{\mathrm{vt}} = \frac{\rho_{\mathrm{c}} \delta_{\mathrm{s}}}{W_{\mathrm{d}} \Delta l} \tag{2}$$

where $\rho_c$, $\delta_{\mathrm{s}}$, and $\Lambda_{\mathrm{efct}}$ are the c-axis resistivity, interlayer spacing, and the effective MFP, respectively. $N_{\mathrm{ch}}$ in each GNR layer mainly depends on $E_{\mathrm{f}}$ and width ($W_{\mathrm{d}}$) and is derived from the Fermi–Dirac equation [5]

$$N_{\mathrm{ch}} = \sum_{j=1}^{nc} \left(1 + \mathrm{e}^{\frac{E_f + E_j}{KbT}}\right)^{-1} + \sum_{j=1}^{nv} \left(1 + \mathrm{e}^{\frac{E_j - E_f}{KbT}}\right)^{-1} \tag{3}$$

where $j = 1, 2, 3, \ldots$ a positive integer. $Kb$ and $T$ are the Boltzmann's constant and temperature, respectively.

For an MLGNR interconnect, conductivity can be intensified by intercalation doping [16]. Intercalation doping of MLGNR is an effectual technique to aid the

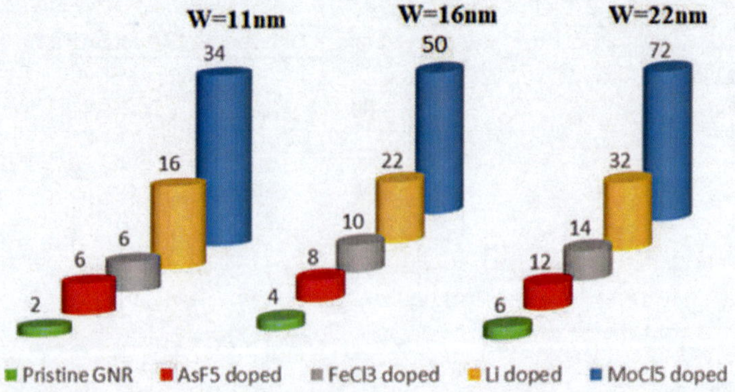


**Fig. 2** Number of conducting channels for pristine and different intercalated MLGNR at different technology nodes

conduction mechanism by reducing the resistivity, interlayer scattering, thereby increasing the effective MFP ($\Lambda_{\mathrm{efct}}$), $E_{\mathrm{f}}$, and $\delta_{\mathrm{s}}$ [16, 17]. With intercalation, there is an increase in $E_{\mathrm{f}}$, which in turn increases $N_{\mathrm{ch}}$ (Fig. 2).

## 2.2 *Computation of Effective Number of Layers*

Ideally, the current conduction in an interconnect is contributed by all the layers ($N_{\mathrm{L}}$) of an MLGNR. For pristine MLGNR, the maximum number of layers that can conduct current has been calculated by Kumar et al. [9] for different $W_{\mathrm{d}}$.

However, with progressive increase in layer number, a deviation in the Fermi energy for each layer is observed, limiting the number of layers implicitly taking part in current conduction. The deviation in Fermi level for each layer ($E_{\mathrm{fs}}$) is expressed as [9]

$$E_{\mathrm{fs}} = E_{\mathrm{f}} e^{\delta s/Ls} \tag{4}$$

The value of $L_{\mathrm{s}}$ is experimentally calculated as 0.54 nm, but it can vary up to 1.2 nm [18]. In our present model, the value of $L_{\mathrm{s}}$ is considered as 1 nm.

The exponential decrease in Fermi level for each layer, limits the number of layers effectively taking part in current conduction, as is observed in Fig. 4. This is true for pristine TC-MLGNR and also for intercalated TC-MLGNR. The present model depicts the variation of the effective number of conducting layers ($N_{L\_\mathrm{Max}}$) at varied $W_{\mathrm{d}}$ for different intercalation doped TC-MLGNR. The same is presented in Fig. 5.

Comparing Figs. 3 and 5, it can be concluded that $N_{\mathrm{L\_Max}}$ for the same $W_{\mathrm{d}}$ is more for intercalated MLGNR than its pristine counterpart.

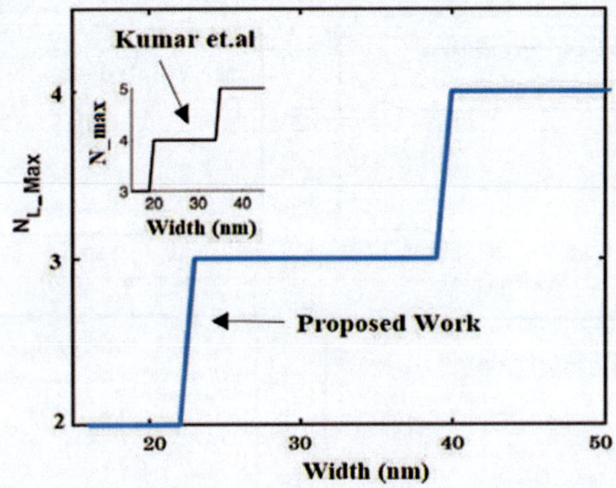


**Fig. 3** Validation of maximum number of layers contributing to current conduction at different widths with analytical data by Kumar et al. [8]

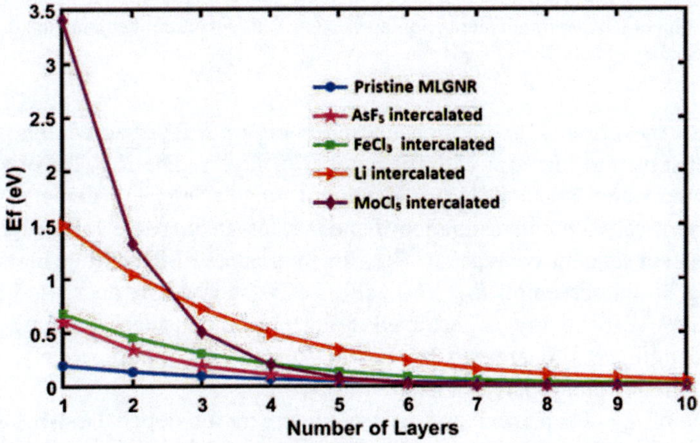


**Fig. 4** Shift of Fermi level in each layer for pristine and different intercalated MLGNR at $Ls$ = 1 nm and $W_d = 16$ nm

## 3 Modeling of TC-MLGNR Conductors Based on Fermi Level Screening

The gradual march of MLGNR-layers toward the substrate causes a subsequent drop of $E_f$ -per layer as shown in Eq. 4. $N_{ch}$ and MFP, which depend on $E_f$, will fall with a decrease in $E_f$ for each layer. Eventually, $R_{vt}$ gradually increases, and $R_{hz}$ gradually decreases toward the substrate. It is demonstrated in [9] that adding more layers to MLGNR saturates $R_{efct}$, provided the $E_f$ for all layers remains constant.

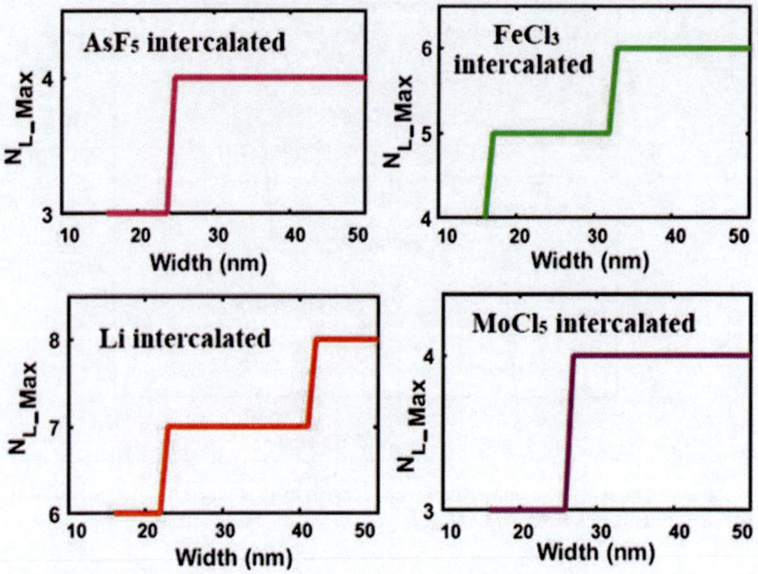


**Fig. 5** Variation of the maximum number of layers contributing to current conduction to width for different intercalated MLGNR at $Ls = 1$ nm

However, in the present model, considering energy-per-layer screening, the $R_{\text{efct}}$ has been observed to increase with the number of layers. The $R_{\text{efct}}$ for pristine and different intercalated MLGNR with and without considering $E_{\text{fs}}$ is depicted in Fig. 6 and 7, respectively. Our investigation demonstrates an increase in the number of layers and a subsequent decrease in $R_{\text{efct}}$ for intercalated MLGNR, where without considering Fermi screening $R_{\text{efct}}$ for $AsF_5$, $FeCl_3$ and $MoCl_5$ decreased by 89%, 93%, and 99% respectively as portrayed in Fig. 7, and considering 1 nm interlayer screening length, $R_{\text{efct}}$ decreased by 16%, 62%, and 44%, respectively, than the pristine MLGNR counterpart at a length of 10 μm.

Variation of $R_{\text{efct}}$ for pristine and different intercalation doped TC-MLGNR with interconnect length is illustrated in Fig. 8. The figure depicts that without considering $L_s$, as interconnect length increases, $R_{\text{efct}}$ drops to a much lower value in different intercalated MLGNR than its pristine counterpart. At shorter interconnects, $R_{\text{efct}}$ increases linearly with interconnect length. It should be noted that higher $E_f$ of intercalation doped MLGNR set off lower $R_{\text{efct}}$. However, considering the Fermi-energy screening per layer, it has been observed that as the length of the interconnect increases, considering the static intercalation, the effective resistance of the intercalated doped MLGNR approaches toward the $R_{\text{efct}}$ of the pristine MLGNR. For shorter interconnect length ($<=$ 10 μm) the increase in $R_{\text{efct}}$ is 48%, 49%, 18%, 95%, and 93%, respectively, for pristine, $AsF_5$, $FeCl_3$, Li, and $MoCl_5$, whereas for longer interconnect ($<=$ 100 μm) the increase is 24%, 23%, 22%, 24%, and 8%, respectively, for pristine MLGNR, $AsF_5$, $FeCl_3$, Li, and $MoCl_5$ intercalated MLGNR.

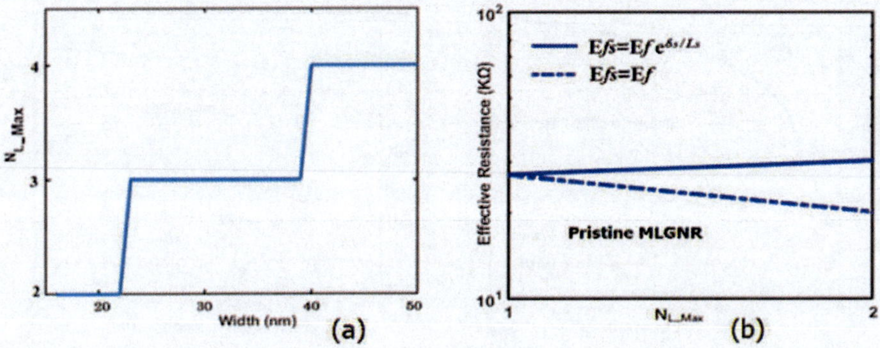


**Fig. 6** **a** The effective number of layers at different width. **b** Effective resistance versus the number of layers with and without considering Fermi-energy screening of Pristine MLGNR at an interconnect length of 10 μm and $W_d$ 16 nm

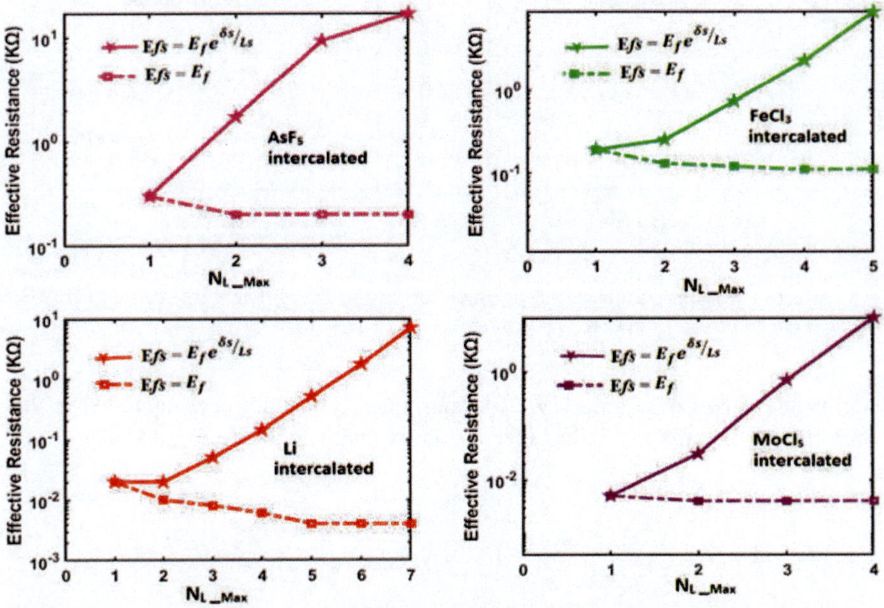


**Fig. 7** Effective resistance versus number of layers with and without Fermi-energy screening for different intercalated MLGNR at an interconnect length of 10 μm and $W_d$ 16 nm

The ESC model presented in Fig. 1a consists of effective inductance as well as capacitance. The magnetic inductance ($L_{Mg}$) and kinetic inductance ($L_{Kn}$) are summed up for getting the effective inductance ($L_{efct}$). The value of $L_{Mg}$ is too

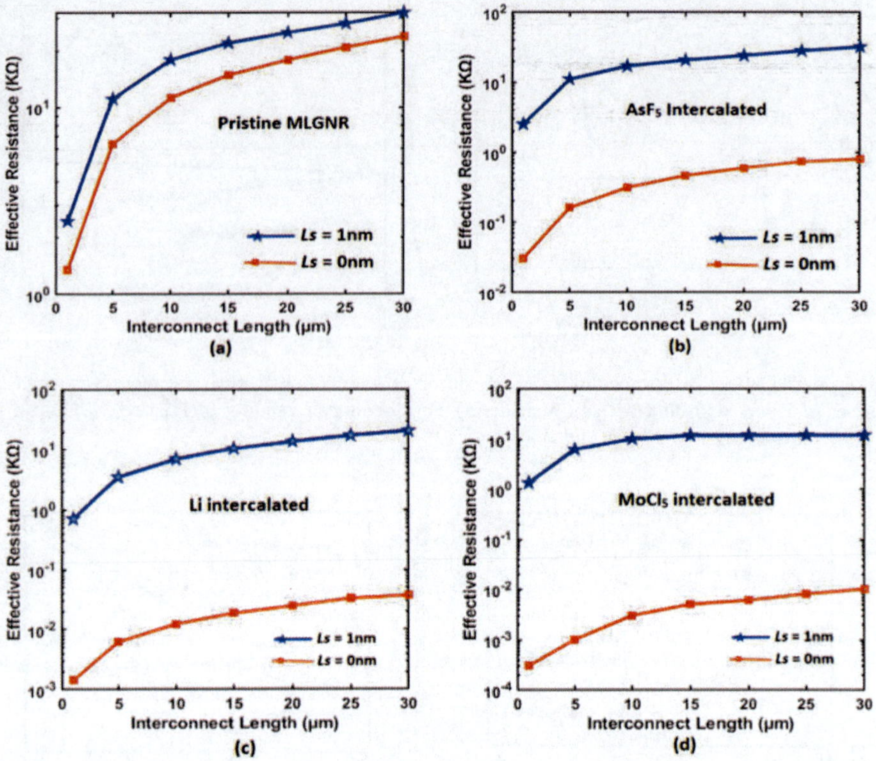


**Fig. 8** Effective resistance to interconnect length for pristine and different intercalated MLGNR for $Ls = 0$ nm and $Ls = 1$ nm at $W_{\rm d} = 16$ nm

insignificant to be considered [19, 20], consequently has not been included in the model; hence, the effective inductance is approximately equal to $L_{\rm Kn}$ and calculated as

$$L_{\rm Kn} = \frac{h}{4e^2 V_f N_{L\_{\rm Max}} N_{\rm ch}} \tag{5}$$

$V_{\rm f}(= 3\pi\gamma a/h)$ stand for the Fermi velocity which depends on the lattice constant $a(\approx 0.246$ nm), the overlap integral $\gamma(\approx 3.16$ eV), and Planks constant $h$ and is determined as [21].

The effective capacitance of the ESC model is a contribution of the quantum capacitance ($C_{\rm qc}$) [21] and electrostatic capacitance ($C_{\rm ec}$) [22, 23].

$$C_{\rm efct} = \left(\frac{1}{C_{qc}} + \frac{1}{C_{\rm ec}}\right)^{-1} \tag{6}$$

## 4 Conclusion

The effective resistance of TC-MLGNR, based on the effective number of layers that are taking part in current conduction considering the Fermi level shift for each layer has been modeled in this paper. The effective resistance is found to be reduced, subsequently saturates if the Fermi level screening per conducting layers is not considered. It is also observed that for the same number of layers with the same length, the effective resistance of intercalated MLGNR is less than pristine MLGNR. However, with deviation in Fermi energy per layer, the effective resistance increases with the gradual rise of layer number. The model also reveals that for longer interconnect, deviation of Fermi energy per layer does not appreciably affect the effective resistance. Consequently, the effective resistances of static intercalation doped MLGNR approaches toward its pristine counterpart, and hence, it could be concluded that the intercalation doped MLGNRs will be a better alternative in the area of local interconnect in VLSI design and applications.

## References

1. Balandin A, Ghosh S, Bao W, Calizo I, Teweldebrhan D, Miao F, Lau CN (2008) Superior thermal conductivity of single-layer graphene. Nano Lett 8(3):902–907
2. Bolotin KI, Sikes KJ, Hone J, Stormer HL, Kim P (2008) Temperature-dependent transport in suspended graphene. Phys Rev Lett 101(9):096–802
3. Murali R, Brenner K, Yang Y, Beck T, Meindl JD (2009) Resistivity of graphene nanoribbon interconnects. IEEE Electron Device Lett 30(6):611–613
4. ITRS International Technology Working Groups (2013) International Technology Roadmap for Semiconductors
5. Naeemi A, Meindl JD (2009) Compact physics-based circuit models for graphene nanoribbon interconnects. IEEE Trans Electron Devices 56(9):1822–1833
6. Naeemi A, Meindl JD (2007) Conductance modeling for graphene nanoribbon (GNR) interconnects. IEEE Electron Device Lett 28(5):428–431
7. Rakheja S, Kumar V, Naeemi A (2013) Evaluation of the potential performance of graphene nanoribbons as on-chip interconnects. Proc IEEE 101(7):1740–1765
8. Kumar V, Rakheja S, Naeemi A (2012) Performance and energy-per-bit modeling of multilayer graphene nanoribbon conductors. IEEE Trans Electron Devices 59:2753–2761
9. Kumar V, Rakheja S, Naeemi A (2011) Modeling and optimization for multilayer graphene nanoribbon conductors. In: IEEE IITC/MAM, pp 1–3
10. Bhattacharya S, Das S, Mukhopadhyay A, Das D, Rahaman H (2018) Analysis of temperature dependent delay optimization model for GNR interconnect using wire sizing method. J Comput Electron 17(4):1536–1548
11. Das S, Bhattacharya S, Das D, Rahaman H (2020) Thermal stability analysis of graphene nano-ribbon interconnect and applicability for terahertz frequency. National Acad Sci Lett
12. Bhattacharya S, Das D, Rahaman H (2017) Stability analysis in top contact and side-contact graphene nanoribbon interconnects. IETE J Res 63(4):588–596
13. Xu C, Li H, Banerjee K (2009) Modeling, analysis, and design of graphene nano-ribbon interconnects. IEEE Trans Electron Devices 56(8):1567–1578
14. Bao W, Wan J, Han X, Cai X, Zhu H, Kim D, Ma D, Xu Y, Munday JN, Dennis Drew H, Fuhrer MS, Hu L (2014) Approaching the limits of transparency and conductivity in graphitic materials through lithium intercalation. Nat Commun 5:4224

15. Jiang J, Kang J, Cao W, Xie X, Zhang H, Chu J, Liu W, Banerjee K (2017) Intercalation doped multilayer-graphene nanoribbons for next-generation interconnects. Nano Lett 17(3):1482–1488
16. Kaur M, Gupta N, Kumar S et al (2020) RF analysis of intercalated graphene nanoribbon-based global-level interconnects. J Comput Electron 19:1002–1013
17. Das S, Das D, Rahaman H (2018) Electro-thermal RF modeling and performance analysis of graphene nanoribbon interconnect. J Computer Electron 17:1695–1708
18. Chen HA, Hsin CL, Huang YT, Tang ML, Dhuey S, Cabrini S, Wu WW, Leone SR (2013) Measurement of interlayer screening length of layered graphene by plasmonic nanostructure resonances. J Phys Chem C 117:22211
19. Nishad A, Sharma R (2014) Analytical time-domain models for performance optimization of multilayer GNR interconnects. IEEE J Sel Topics Quantum Electron 20(1):17–24
20. Li H, Xu C, Banerjee K (2009) Carbon nanomaterials for next-generation interconnects and passives: physics, status and prospects. IEEE Trans Electron Devices 56(9):1799–1821
21. Sarkar D, Xu C, Li H, Banerjee K (2011) High-frequency behavior of graphene-based interconnects—part I: impedance modeling. IEEE Trans Electron Devices 58(3):843–852
22. Stellari F, Lacatia AL (2000) New formulas of interconnect capacitances based on results of conformal mapping method. IEEE Trans Electron Devices 47(1):222–231
23. Zhao W-S, Yin WY (2014) Comparative study on multilayer graphene nanoribbon (MLGNR) interconnects. IEEE Trans Electromagn Compat 56(3)

# Enhancing Lifetime of Non-volatile Memory Caches by Write-Aware Techniques

**S. Sivakumar, Mani Mannampalli, and John Jose**

**Abstract** Traditional memory technologies, such as SRAM, suffer from limited package density and high leakage power. Applications are getting increasingly memory hungry in the age of big data. Non-volatile memories such as STTRAM, PCM, and ReRAM emerged as attractive contenders to replace traditional SRAM-based memories. They have high density and zero leakage power. However, they have a limited write endurance. Non-uniform write patterns in applications can shorten the life of non-volatile memories. Traditional cache block replacement strategy like LRU leads some cache blocks to be accessed more frequently than others, accelerating the wear out of the cache. We present a Write-Aware Last Level Non-Volatile Cache (WALL-NVC), which improves the lifetime up to 5.84× for uni-core, 3.34× for dual-core, and 4.11× for quad-core systems. It reduces intraset write variation in last level caches up to 98.91%, 90.11%, and 94.12% for uni-core, dual-core, and quad-core systems, respectively, using write distribution and NVM-friendly replacement mechanism.

**Keywords** Non-volatile memory · Wear leveling · Lifetime improvement · Write variation

S. Sivakumar (✉) · M. Mannampalli · J. Jose
Department of Computer Science and Engineering, Indian Institute of Technology Guwahati, Guwahati, India
e-mail: sivakumar@iitg.ac.in

J. Jose
e-mail: johnjose@iitg.ac.in

 
C. Giri et al. (eds.), *Emerging Electronic Devices, Circuits and Systems*,
Lecture Notes in Electrical Engineering 1004,
https://doi.org/10.1007/978-981-99-0055-8_10

## 1 Introduction

In the current world, data processing is an inevitable necessity. The amount of data that a processing unit has to handle has expanded significantly in the recent years due to technological advancements, growing popularity of IoT-based devices, social media, and video streaming platforms. We expect the trend of data-intensive applications to continue for the years to come. Applications that run from handheld devices to supercomputers require greater processing power and memory than before. Because of their poor package density and high leakage power, traditional memory technologies such as SRAM are inadequate to handle this demand for large on-chip memory.

Spin Transfer Torque RAM (STT-RAM) [1], Phase Change RAM (PCRAM) [2], and Resistive RAM (ReRAM) [3] are all promising non-volatile technologies that can replace conventional memories. They feature a high package density and little leakage power, making them ideal for realizing large memories [4, 5]. However, these emerging non-volatile technologies have a significant drawback in terms of write latency and write endurance. The maximum number of writes a memory cell can withstand before it permanently wears out is termed as write endurance. When non-volatile memories are used in applications with non-uniform write patterns, some memory cells wear out faster than others. Absence of write-aware cache replacement policies can result in frequent writes to some cache blocks can also force memory cells to wear out soon. These circumstances highlight the need for a system that reduces the amount of writes or distributes them evenly when using non-volatile memories at various levels of the memory hierarchy. Our work aims to extend the life of non-volatile memory when it is employed as a last level cache. EqualWrites [6], a state-of-the-art wear-leveling techniques which reduce the intraset write variation and improve lifetime, compares LRU and random replacement policy for cache blocks. Both of these policies, however, are not customized for NVMs. Replacement plans like Refresh-Aware Replacement Policy (RFR) [7] help to extend the life of NVMs. However, they are difficult to implement because they are designed for write-optimized cache memory and do not work with traditional wear-leveling schemes. We believe a more NVM-friendly replacement policy combined with a good wear-leveling method can greatly boost lifetime.

In this work, we make the following major contributions:

- We analyze write variations in the last level NVM cache and draws meaningful conclusions.
- We propose Write-Aware Last level Non-Volatile Caches (WALL-NVC), which can reduce the intraset variation, thereby increasing its lifetime.
- For WALL-NVC, we use an NVM-friendly replacement policy called Least Recently Used Cold Block (LRU-CB), which also contributes to increase the lifetime.
- We test WALL-NVC using SPEC 2006 [8] benchmarks on the gem5 cycle-accurate simulator [9], and find that our proposed method outperforms other state-of-the-art solutions.

## 2 Related Work

As previously stated, write endurance refers to the maximum number of writes that a non-volatile memory may withstand before failing. High write variation applications, as well as malicious apps that target NVMs' limited write endurance, can significantly reduce their lifetime. Several lifetime enhancement strategies were proposed in the past which can be broadly divided into two categories: write avoidance and write distribution techniques. In write avoidance techniques, we reduce the number of writes to the NVM using write avoidance strategies such as early termination, inversion, and encoding. We try to evenly spread writes across the memory in write distribution techniques so that a few memory cells do not wear out at a faster pace than others due to frequent writing to them. Write variation can occur within a set (intraset write variation) or across sets (interset write variation). EqualWrites [6] is a strategy for reducing intraset variation by swapping frequently written hot blocks with less frequently written cold blocks within a set. EqualChance [10] is a wear-leveling technique that involves changing the physical location of a frequently written data block on a regular basis. Another solution is $i^2$WAP [11], which reduces intraset and interset write variations through probablistic flushing of frequently written blocks and set swapping. Unlike traditional replacement policies like LRU, which focus on increasing write pressure on write-intensive blocks, replacement policies like RCR, RFR, FCR, [7], and others prioritize endurance.

## 3 Motivation

To study the write variations of different applications, we analyze maximum and average writes to LLC for uni-core architecture. To facilitate this, we model these architectures in gem5 [9] with two levels of cache and main memory. For L1-I and L1-D caches, we use 32 KB, 4-way set associative configuration. The 512 KB unified L2 cache is 8-way set associative and we use 8 GB main memory. The block size is 64 bytes. The number of kilo writes per 1 billion instruction window for selected benchmarks from the SPEC CPU2006 suite [8] is shown in Fig. 1. We plot the maximum writes per way as well as average writes across ways. Our study shows that write variations can occur within a set and also across the set. This reinforces the need for good wear-leveling policy for NVM-based LLC.

The write variation that occurs within a set is is quantified using the coefficient of intraset variation (IntraV). Write variation that occurs across sets is quantified using the coefficient of interset variation (InterV). IntraV and InterV are given below.

$$\text{IntraV} = \frac{100}{N.\text{Write}_{\text{avg}}} \sum_{k=1}^{N} \sqrt{\frac{\sum_{l=1}^{M} (W_{k,l} - \sum_{m=1}^{M} \frac{W_{k,m}}{M})^2}{M-1}} \tag{1}$$

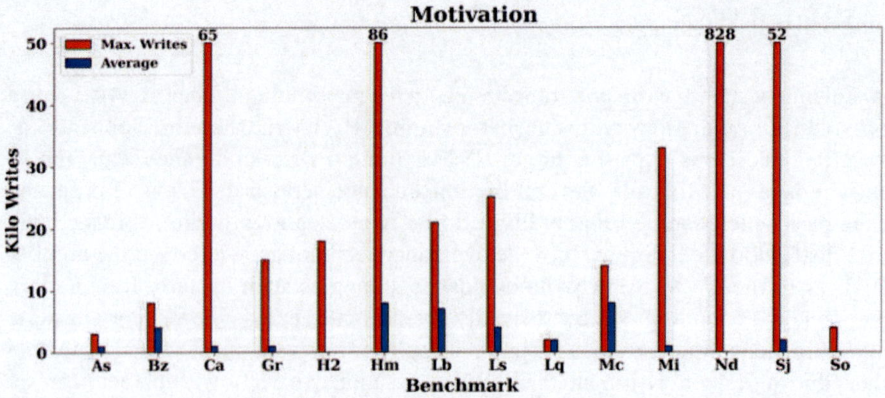


**Fig. 1** Average and maximum writes per way (Kilo writes per 1 billion instructions) of various SPEC CPU 2006 benchmarks: height difference between bars of a given benchmark indicates intensity of write level variations

$$\text{InterV} = \frac{100}{\text{Write}_{\text{avg}}}\sqrt{\frac{\sum_{k=1}^{N}\left(\sum_{l=1}^{M}\frac{W_{k,l}}{M} - W_{\text{avg}}\right)^2}{N-1}} \tag{2}$$

where $N$ is number of sets in cache.
$M$ is the number of ways in a set.
$W_{k,l}$ is the write count in set $k$ and way $l$.
$\text{Write}_{\text{avg}}$ is average write count given by

$$\text{Write}_{\text{avg}} = \frac{\sum_{k=1}^{N}\sum_{l=1}^{M} W_{k,l}}{N.M} \tag{3}$$

Low IntraV and InterV values suggest more evenly distributed writes within and across the cache sets, respectively. The $\text{Write}_{\text{avg}}$ shows the average number of writes in the cache memory. Popular cache replacement policies, such as least recently used (LRU), pseudo-LRU, and others, consider the most recent use of a cache block into account while choosing a victim block for replacement. However, in non-volatile memories where write endurance is a major concern, the number of writes to the victim block can impact the cache memory's lifetime. To the best of our knowledge, state-of-the-art wear-leveling techniques do not study the role of replacement policy in improving the lifetime. EqualWrites, which reduces the intraset write variation, compares between LRU and random replacement policy. However, both of these policies are not custom made for NVMs. This inspires us to investigate how to effectively combine a wear-leveling technique and a better replacement policy tailor made to enhance the endurance of NVM caches.

## 4 Write-Aware Last Level Non-volatile Cache

To improve the lifetime of NVM while running applications having non-uniform writes and protecting them against targeted malicious attacks by repeated writes to specific blocks, we propose Write-Aware Last Level Non-Volatile Cache (WALL-NVC). Unlike most of the state-of-the-art wear-leveling techniques, WALL-NVC is a dual-stage wear-leveling technique. The first stage is a new least recently used cold block (LRU-CB) replacement policy, that takes care of selecting a better victim block for cache replacement in NVMs. The second stage employs a traditional write distribution strategy that works in tandem with LRU-CB to increase lifetime. The following sections discuss these stages in detail.

### *4.1 LRU-CB Replacement Policy*

To the best of our knowledge, the impact of cache block replacement policy on the write endurance of NVM-based caches is not explored so far by any wear-leveling mechanisms. A good cache replacement policy for NVMs should enhance write endurance and reduce intraset write variation. Ideally, it should preserve the hottest blocks and prevent frequent evictions while reducing the write variation across blocks. When the cache hit rate is high, the number of replacements is less; hence, the impact of replacement policies is minimal. Traditional replacement policies like LRU and pseudo-LRU do not consider the write count of the block while selecting a victim block. To meet these objectives, we propose a simple NVM-friendly cache block replacement policy called as least recently used cold block (LRU-CB).

The basic concept of LRU-CB policy is to choose a block from the set that is less frequently written as the victim block, thereby making writes to the set more uniform. To ensure that the blocks are not evicted frequently, a weighted aggregate average of each block's LRU age and write index is calculated. The block with the lowest aggregate average is picked as the victim block. To facilitate this, we attach a write counter to each block. When one of the write counters in a set reaches its saturation value, the write counters of blocks of that set are bitwise right-shifted. This downgrading of the counter value forces the least significant bit (LSB) of the counter to be lost, resulting in a minor loss of precision. This ensure that the respective values of write counters are downgraded (with a small error margin) before getting wrap around. We study the impact of LRU-CB by running different benchmarks and find that LRU-CB marginally improves the lifetime of NVM caches. This marginal improvement demonstrates the necessity for a complementary technique to LRU-CB in order to increase its performance.

## 4.2 *Impact of LRU-CB with Write Distribution*

Write-aware replacement policies have a limited impact on the endurance of NVM caches while running applications with high L1 cache hit rates as they trigger fewer evictions. Write distribution policies increase lifetime by distributing writes evenly. Lifetime of NVM caches can be extended by combining a good wear-leveling policy with a write-aware replacement policy rather than doing so separately.

We compare the effectiveness of EqualWrites technique with the pseudo-LRU policy and LRU-CB in order to verify the impact of LRU-CB when used in conjunction with a standard state-of-the-art wear-leveling technique. Figures 2, 3, and 4 show the relative lifetime, intraset variation, and hit rate of NVM caches, respectively while running different benchmarks in SPEC CPU 2006 suite. From the graphs, we can observe that EqualWrites with LRU-CB increases the lifetime of L2 cache upto 1.39× than the combination of EqualWrites with pseudo-LRU. LRU-CB reduces the intraset variation up to 83.08% without affecting the hit rate. This improvement is visible across all benchmarks thereby ascertaining that LRU-CB is a better cache block replacement algorithm for NVM caches.

## 4.3 *Write Distribution in WALL-NVC*

LRU-CB policy improves the performance of EqualWrites technique. But this comes with a high overhead. This is because LRU-CB needs extra counters apart from ones used for EqualWrites. This motivated us for the need for wear-leveling policy that reduces the intraset variation and improves lifetime and synergizes with

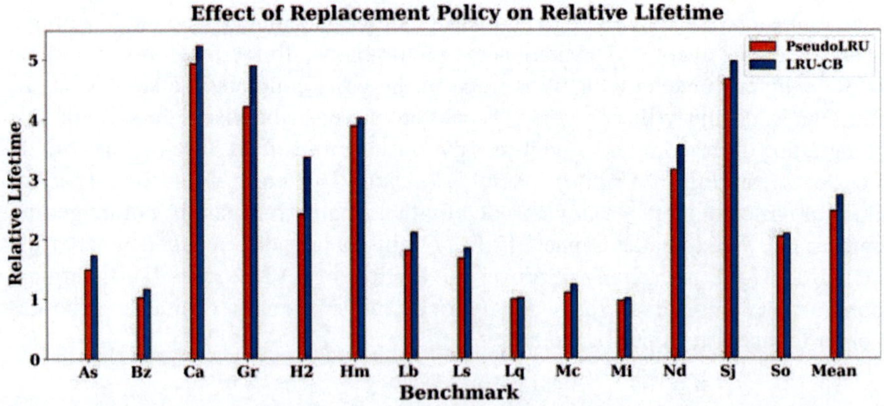


**Fig. 2** Comparison of relative lifetime of NVM-based L2 cache using EqualWrites with pseudo-LRU and LRU-CB replacement policies

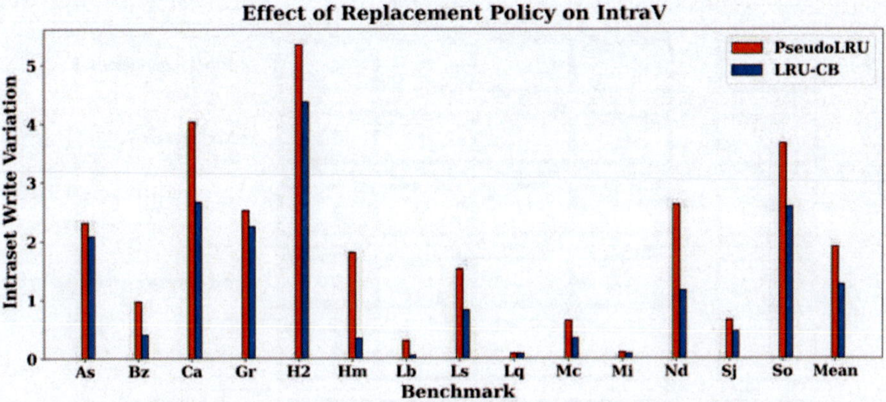


**Fig. 3** Comparison of intraset variation of NVM-based L2 cache using EqualWrites with pseudo-LRU and LRU-CB replacement policies

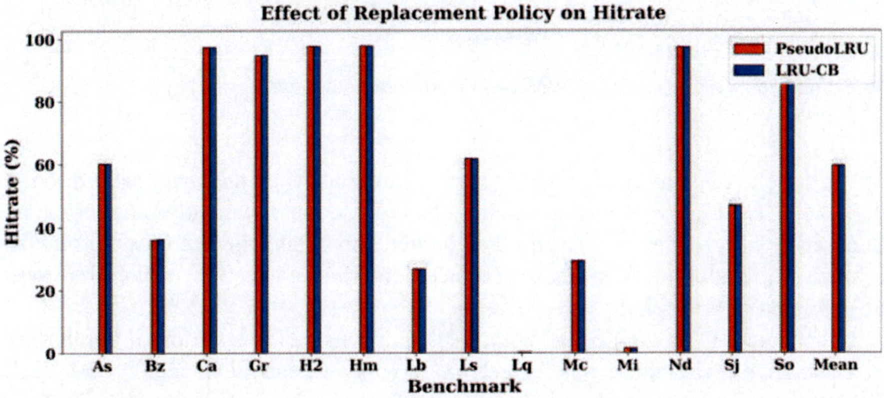


**Fig. 4** Comparison of hit rate of NVM-based L2 cache using EqualWrites with pseudo-LRU and LRU-CB replacement policies

LRU-CB. Like other popular wear-leveling techniques, WALL-NVC also works on the principle of redirection of the writes from hot blocks to cold blocks.

Each set of an $n$-way set associative WALL-NVC has $(n + 1)$ counters: one set counter and $n$ block counters. For each write hit to WALL-NVC, the corresponding set and block counter are updated. Once the set counter reaches a prefixed threshold $T$, it selects a write redirection target among blocks of that set. Block with least writes is preferred as the redirection target, and hence it selects the block with zero write count. Swapping is initiated between the accessed block and redirection target blocks, if such block is available. If target block is invalid, instead of swapping, it writes to the target block and invalidates hot line. If a block with zero write count is unavailable, if the target is not found, all counters (including set counter) are decreased by the value of the least written block, which delays write redirection for a few more writes.

**Fig. 5** Sample counter updating of WALL-NVC for threshold value, $T = 50$

This is done to prevent unnecessary write redirections when the write pattern to the set is more uniform. Note that decreasing the block counter value does not affect its functionality. Moreover, decreasing the counter also delays the need for the bitwise right shift operation of counters for replacement victim selection, which improves the precision of the technique.

To understand the working of WALL-NVC, we use an illustration of a four-way set associative cache block of WALL-NVC having a threshold value, $T = 50$. Each cache set is associated with a set counter and four block counters which keep track of the write count of each set and block, respectively. Let us consider a specific set A whose four blocks are $B0$, $B1$, $B2$, and $B3$. Consider an instance where the values of set counter and block counters of A are shown as in the first row in Fig. 5. Write hit in a block increments block counter and A's set counter. Once the set counter reaches the threshold value (50), it searches for a write redirection target for heavily written block ($B2$). Since there is no target block with zero counts, the values of all block counters and the set counter are decremented by the value of least count (here it is 2 for $B3$) to create a block with zero counts. After the decrement operation, the cache is operated normally by incrementing the counters on write hits. When the set counter reaches the threshold again and write redirection is initiated, the redirection takes place by swapping the contents of most written block ($B2$) with least written block ($B3$) by swapping their contents using swap module. As $B2$ and $B3$ are valid blocks, write redirection results in an extra write in $B2$ and $B3$ and set counter is reset.

**Table 1** System configuration

| | |
|---|---|
| CPU | 1 GHz, uni-core, dual-core, quad-core, ALPHA |
| L1 cache 64 B block | Private, 32 KB, SRAM-based split cache, 4-way set associative |
| L2 cache | Shared 512 KB, NVM-based unified cache 64 B block, 8-way set associative |
| Main memory | 8 GB |

**Table 2** Benchmark classification based on WPKI

| Category | Benchmark |
|---|---|
| Low | Namd (Nd), soplex (So), calculix (Ca), astar (As), gromacs (Gr) |
| Mid | Milc (Mi), libquantum (Lq), sjeng (Sj), bzip2 (Bz) |
| High | Leslie3d (Ls), lbm (Lb), mcf (Mc), hmmer (Hm) |

## 5 Experimental Setup and Result Analysis

We use gem5 [9], a cycle-accurate event-based open source architectural simulator, to model and evaluate the performance of the last level WALL-NVC cache on uni-core, dual-core, and quad-core system architectures. Ruby module is used for simulating memory module, and MESI protocol is used for maintaining cache coherence. Details of the system configuration are given in Table 1.

We use the SPEC CPU 2006 benchmarks [8] to evaluate the performance of state-of-the-art techniques as well as WALL-NVC. Based on number of Writes Per Kilo Instruction (WPKI) to last level cache, these benchmarks are classified into three categories: Low (WPKI $\leq 9$), mid ($10 \leq$ WPKI $\leq 29$), and high (WPKI $\geq 30$) as given in Table 2. This classification helps us to understand the impact of the existing and proposed techniques for applications having different write characteristics. For architectural modeling in uni-core system, we assign one benchmark instance to the core. For dual-core and quad-core architectures, we create workloads by mixing instances of two and four benchmarks, respectively. We run these benchmarks and workload categories on an unoptimized NVM LLC (baseline), two state-of-the-art write balancing approaches: EqualWrites technique and EqualChance technique and the proposed WALL-NVC with a threshold value $T = 50$ (WALL-NC50).

### *5.1 Performance Analysis*

Figure 6 shows the comparison of intraset variation (IntraV) for various architectures under study in uni-core systems. We can clearly see that for benchmarks *leslie3d*, *lbm*, *mcf*, *milc*, and *bzip2*, the writes are happening more or less uniformly across different ways of a set there by having lower IntraV for all architectures. This behavior is more

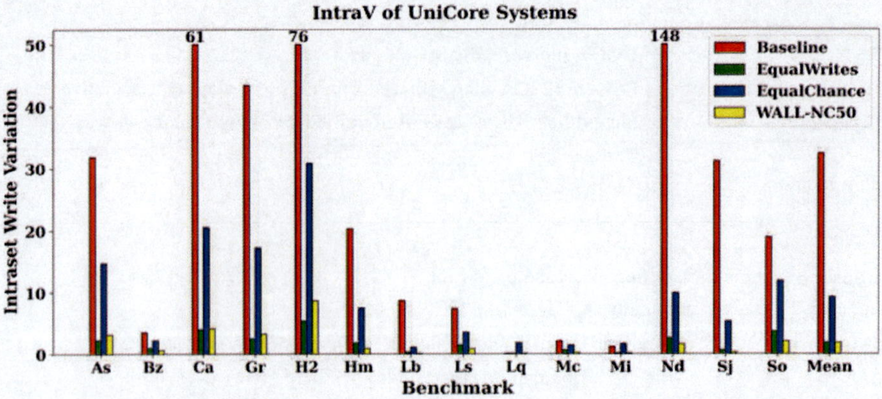


**Fig. 6** Comparison of IntraV for various NVM architectures in uni-core system (shorter the bar, the better)

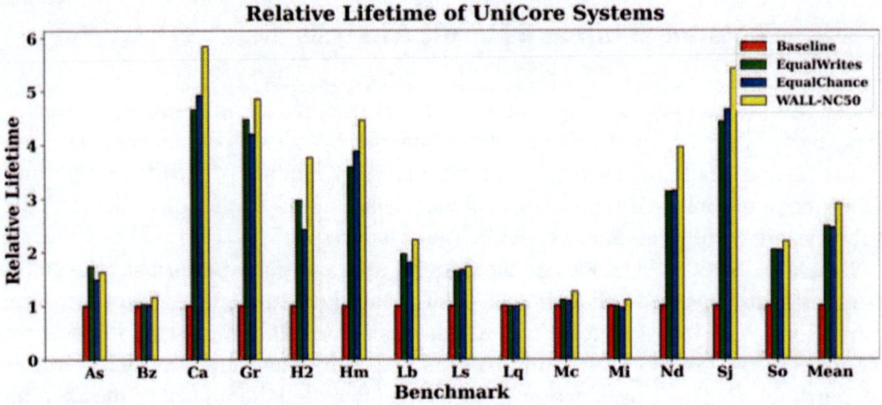


**Fig. 7** Comparison of lifetime for various NVM architectures normalized to base configuration in uni-core system (taller the bar, the better)

prominent in benchmarks with mid and high WPKI. In some low WPKI benchmarks like *namd, calculix* and *gromacs*, there is a great improvement in IntraV that we could achieve. So we conclude that the write variance depends a lot on pattern of write hits on a given set than number of writes over a period of time. However, we can see that WALL-NC50 gives a low write variance irrespective of benchmarks classification that makes it suitable add-on to NVM caches.

Figure 7 shows comparison of lifetime for various architectures under study normalized to baseline in uni-core systems. The inverse of the maximum write count to an LLC block is used to calculate lifetime. As expected we observe that benchmarks with high WPKI have limited improvement in lifetime due to the heavy writes to LLC. Compared to the baseline architecture, on an average, WALL-NVC ($T = 50$)

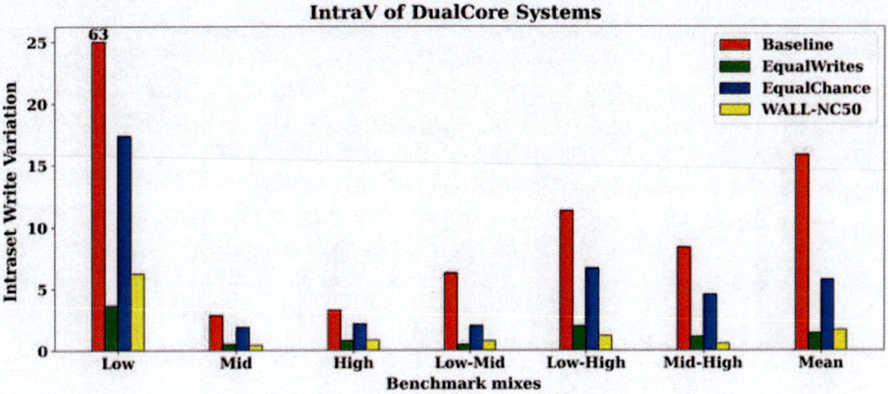


**Fig. 8** Comparison of IntraV for various NVM architectures in dual-core system (shorter the bar, the better)

improves lifetime by 2.90× and shows 1.16× and 1.18× improvement compared to EqualWrites and EqualChance, respectively.

We also analyze the performance of our technique on dual-core and quad-core systems as well. WALL-NC50 shows an average lifetime improvement of NVM by 2.25× and 1.63× compared to baseline systems for dual-core and quad-core, respectively, as shown in Figs. 10 and 11. It improves 1.07× on dual-core systems when compared to EqualWrites and 1.02× when compared to Equalchance. Lifetime improvement of 1.10× and 1.02× is achieved for quad-core systems, respectively. Similarly, we plot the intraset variation in dual-core and quad-core systems in Figs. 8 and 9, respectively. Since multicore framework needs more than one benchmark to run depending on number of core, we create workload consisting of benchmark mixes. Depending on WPKI values of the constituent benchmarks, we create workloads marked as low, mid, low-high, mid-high, etc. On careful analysis, we find that due to the multiple applications accessing the shared NVM-based LLC, the write variation created by writes of one core gets reduced by writes from other core. Hence, we could achieve only minor improvement in the lifetime improvement using mid-high workloads in dual-core and quad-core systems.

## 5.2 *Sensitivity Analysis*

We study the impact of the threshold value ($T$) by experimenting with five different values, $T = \{10, 30, 50, 70, 100\}$. Based on the endurance improvement and associated overhead, we fix the default value of $T$ as 50. We also conduct a detailed sensitivity analysis on various threshold values. Table 3 gives the mean lifetime improvement with respect to baseline and intraset write variation, using different threshold values in WALL-NVC.

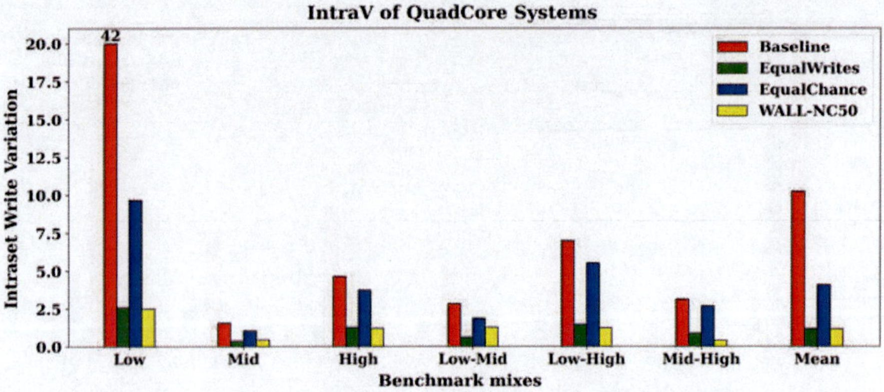


**Fig. 9** Comparison of IntraV for various NVM architectures in quad-core system (shorter the bar, the better)

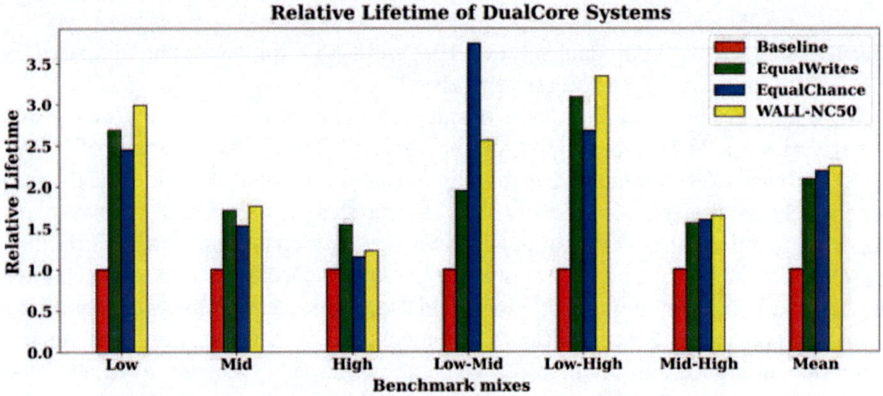


**Fig. 10** Comparison of lifetime for various NVM architectures normalized to base configuration in dual-core system (taller the bar, the better)

Another parameter that can impact the performance of the proposed architecture is the weightage given to LRU-CB while selecting a victim block for cache replacement. As discussed earlier, we compute the weighted aggregate average of each cache block using its LRU age and write count index. We explore two variants: (a) 80% for LRU age and 20% for write count index-0.2W and (b) 60% for LRU age and 40% for write count index-0.4W. We compare these two variants in a uni-core system. The results for IntraV, relative lifetime wit respect to baseline, and LLC hit rate are given in Figs. 12, 13, and 14, respectively. We can see that 0.2W gives 1.13 time lifetime improvement and 2.16% improvement in IntraV than 0.4W. We could also observe that there is not much impact on hit rate for these two variants in any benchmarks. Since 0.2W and 0.4W are better than baseline, we propose that a minimum weightage

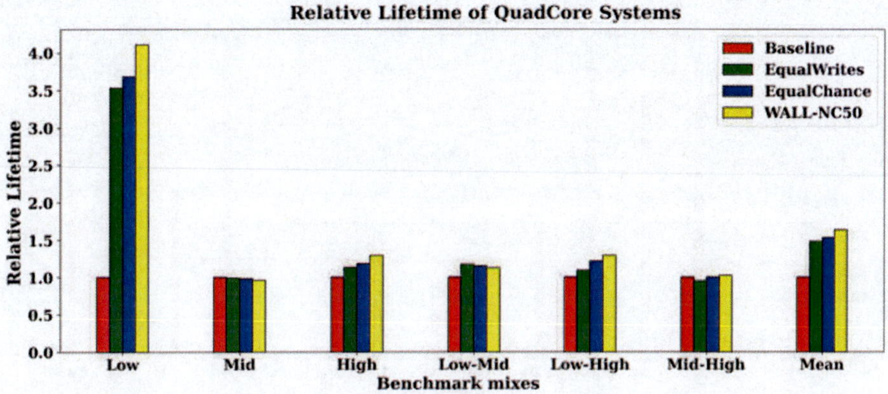


**Fig. 11** Comparison of lifetime for various NVM architectures normalized to base configuration in quad-core system (taller the bar, the better)

**Table 3** Relative lifetime improvement (LT) and IntraV of WALL-NVC for different threshold values

| | Uni-core | | Dual-core | | Quad-core | |
|---|---|---|---|---|---|---|
| | LT | IntraV | LT | IntraV | LT | IntraV |
| Baseline | 1 | 32.41 | 1 | 15.85 | 1 | 10.22 |
| 10 | 2.32 | 6.55 | 2.08 | 7.28 | 1.31 | 6.58 |
| 30 | 2.54 | 2.46 | 2.63 | 1.48 | 1.68 | 0.34 |
| 50 | 2.90 | 1.85 | 2.25 | 1.65 | 1.63 | 1.17 |
| 70 | 2.56 | 4.08 | 2.23 | 2.59 | 1.59 | 1.43 |
| 100 | 2.57 | 4.20 | 2.35 | 2.26 | 1.53 | 1.52 |

to LRU-CB (0.2W) should be given to get a suitable victim block. At the same time, overemphasize to write count index (0.4W) diminishes the role of LRU age.

## 5.3 *Overhead Analysis*

WALL-NVC employs two types of counters: a set counter for each set and a block counter for each block. It also necessitates the use of a swapping module to swap the contents of hot and cold data blocks. The swapping module has 64 buffers, each of which is 64 bytes in size that incurs a total storage overhead of 2%. The SRAM-based counters and swap buffers incur a maximum power and area overhead of 0.47% and 1.47%, respectively, when compared with baseline configuration. The cache blocks replacement policy, LRU-CB uses the same counters for victim selection; hence, it does not incur any additional overhead.

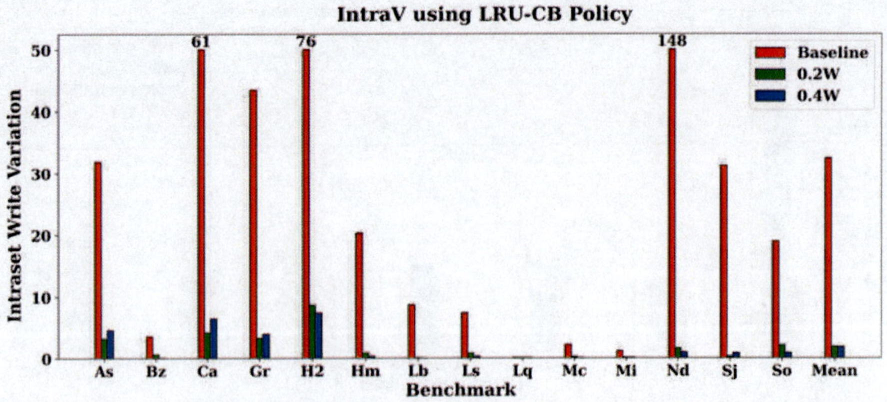


**Fig. 12** Comparison of IntraV for WALL-NC50 variants in uni-core system (shorter the bar, the better)

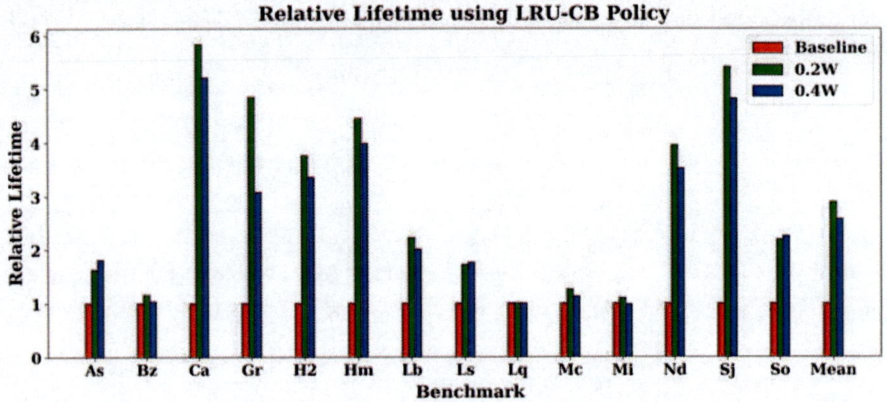


**Fig. 13** Comparison of relative lifetime for WALL-NC50 variants in uni-core system (taller the bar, the better)

# 6 Conclusion

Limited write endurance of NVM is always a critical challenge. In this paper, we proposed a new architecture called as WALL-NVC that used a write distribution policy and an NVM-friendly least recently used cold block cache replacement policy to improve lifetime of NVM caches. We observed that both the write distribution policy as well as the write-aware replacement policy contributed equally to the improved performance. Experimental results showed that with minimal area and power overhead, our technique improved the lifetime for uni-core, dual-core, and quad-core systems. To achieve further lifetime improvement, we look forward to incorporate a dynamic adaptive replacement policy based on run time inputs. Dynamic power-gating can be also be applied to adapt diverse write patterns of applications.

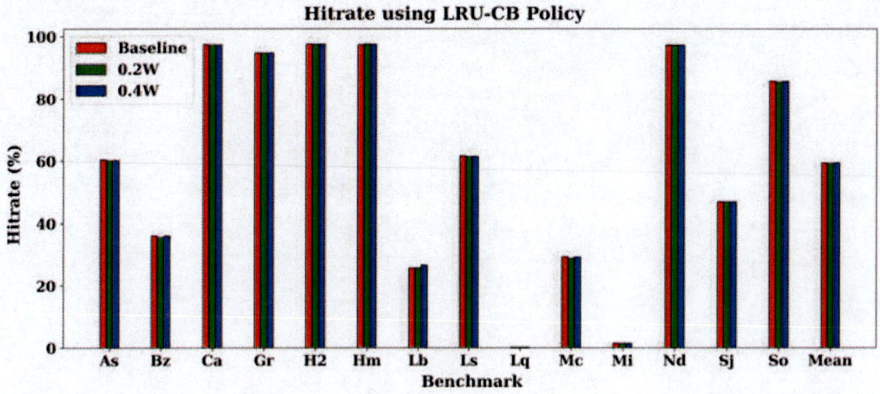


**Fig. 14** Comparison of LLC hit rate for WALL-NC50 variants in uni-core system (taller the bar, the better)

# References


1. Chi P, Li S, Cheng Y, Lu Y, Kang SH, Xie Y (2016) Architecture design with STT-RAM: opportunities and challenges. In: 2016 21st Asia and South Pacific design automation conference (ASP-DAC), pp 109–114. https://doi.org/10.1109/ASPDAC.2016.7427997
2. Wong HP et al (2010) Phase change memory. In: Proceedings of the IEEE, vol 98, no 12, pp 2201–2227. https://doi.org/10.1109/JPROC.2010.2070050
3. Akinaga H, Shima H (2010) Resistive Random Access Memory (ReRAM) based on metal oxides. Proc IEEE 98(12):2237–2251. https://doi.org/10.1109/JPROC.2010.2070830
4. Mittal S, Vetter JS (2016) A survey of software techniques for using non-volatile memories for storage and main memory systems. IEEE Trans Parallel Distrib Syst 27(5):1537–1550. https://doi.org/10.1109/TPDS.2015.2442980
5. Dong X, Xu C, Xie Y, Jouppi NP (2012) NVSim: a circuit-level performance, energy, and area model for emerging nonvolatile memory. IEEE Trans Comput Aided Des Integr Circ Syst 31(7):994–1007. https://doi.org/10.1109/TCAD.2012.2185930
6. Mittal S, Vetter JS (2016) EqualWrites: reducing intra-set write variations for enhancing lifetime of non-volatile caches. IEEE Trans Very Large Scale Integr (VLSI) Syst 24(1):103–114. https://doi.org/10.1109/TVLSI.2015.2389113
7. Saraf P, Mutyam M (2019) Endurance enhancement of write-optimized STT-RAM caches. In: Proceedings of the international symposium on memory systems (MEMSYS '19). Association for Computing Machinery, New York, NY, USA, pp 101–113. https://doi.org/10.1145/3357526.3357538
8. Henning John L (2006) SPEC CPU2006 benchmark descriptions. SIGARCH Comput Archit News 34(4):1–17. https://doi.org/10.1145/1186736.1186737
9. Binkert N et al (2011) The gem5 simulator. SIGARCH Comput Archit News 39(2):1–7. https://doi.org/10.1145/2024716.2024718
10. Mittal S, Vetter JS (2014) EqualChance: addressing intra-set write variation to increase lifetime of non-volatile caches. In: 2nd workshop on interactions of NVM/flash with operating systems and workloads (INFLOW 14)
11. Wang J, Dong X, Xie Y, Jouppi NP (2013) i2WAP: improving non-volatile cache lifetime by reducing inter- and intra-set write variations. In: 2013 IEEE 19th international symposium on high performance computer architecture (HPCA), pp 234–245. https://doi.org/10.1109/HPCA.2013.6522322

# Microfluidic Dilution by Recycling Arbitrary Stock Solutions Using Various Mixing Models

**Abhishek Ghosh, Debraj Kundu, Sudip Poddar, Shigeru Yamashita, Robert Wille, and Sudip Roy**

**Abstract** Microfluidic biochips are being widely used for automating biochemical laboratory protocols, and several algorithms for automated sample preparation (dilution and mixing of reagent fluids) were reported in the literature. Almost all the sample preparation algorithms assumed the availability of pure sample fluid (i.e., with 100% concentration) ignoring the fact that pure samples may not always be readily available in stock. In fact, in many practical situations, a number of arbitrary concentrations of the sample fluid are discarded as wastes, which can be recycled to reduce the sample preparation cost (usage of pure sample, etc.) and generate the desired target concentration of the fluid required elsewhere. The traditional microfluidic biochips support (1:1) mixing model, for which there exists only a few old algorithms in literature, (namely generalized dilution algorithm, i.e., *GDA* and dilution/mixing with reduced wastage, i.e., *DMRW*) which were solely proposed for solving dilution problem by recycling arbitrary stock solutions (*RASS*) with traditional biochips. Although, a variety of microfluidic biochips have been developed over the years, no sample preparation algorithm is proposed for solving *RASS* problem for such modern biochips—which may provide a cost-effective solution for *RASS*. In order to fill this gap, in this paper, we propose a "cost-effective" heuristic solution (called *hRASS*) for dilution of a sample fluid from its arbitrary stock solu-

A. Ghosh
Government College of Engineering and Leather Technology, Kolkata, India

D. Kundu · S. Roy
Indian Institute of Technology Roorkee, Roorkee, India

S. Poddar (✉)
Johannes Kepler University Linz, Linz, Austria
e-mail: sudippoddar2006@gmail.com

S. Yamashita
Ritsumeikan University, Kyoto, Japan

R. Wille
Software Competence Center Hagenberg GmbH (SCCH), Technical University of Munich, Germany, Hagenberg im Mühlkreis, Austria

 
C. Giri et al. (eds.), *Emerging Electronic Devices, Circuits and Systems*,
Lecture Notes in Electrical Engineering 1004,
https://doi.org/10.1007/978-981-99-0055-8_11

tions catering various mixing models supported by modern microfluidic biochips. Simulation results confirm the superiority of the proposed method and show that *hRASS* can improve the solution quality by 36.8% and 21% on average for a large number of random testcases over state-of-the-art methods (e.g., *DMRW* and *GDA*, respectively).

**Keywords** Biochips · Microfluidics · Sample preparation · Dilution · Mixing models · Stock solutions

## 1 Introduction

Microfluidic biochips revolutionized the healthcare industry by integrating several fluidic functionalities (e.g., sample preparation, point-of-care diagnostics, drug discovery, *DNA* analysis) on a single chip [4, 5, 11, 22, 24]. It is expected that microfluidic biochips will replace the bulky biochemical equipment (which are operated manually in labs or hospitals) in the foreseeable future, by automatically performing microfluidic operations on a tiny devices. In practice, two types of biochip technologies are used: (i) continuous flow microfluidic (*CMF*) biochips [13], where fluids can flow through the micro-channels using micropumps and microvalves, and (ii) digital microfluidic (*DMF*) biochips [20], where fluid droplets are manipulated on a 2D array of electrodes applying electrical actuations. Recent advancement of fabrication technology also led to the development of micro-electrode-dot-array (*MEDA*) [25] and a fully programmable valve array (*FPVA*) [6] biochips.

Sample preparation is an essential part of any biochemical analysis, and in general, 90% cost and 95% time are spent on this task during molecular diagnosis [7]. Many algorithms are reported with microfluidic biochips for preparing a desired target concentration factor (*CF*) of a sample fluid starting from pure sample (100% concentration) and buffer (0% concentration) [1, 2, 8, 14–19, 23, 24, 26]. So far, almost all these sample preparation methods assumed the availability of pure sample and based on it determined the sequence of mix-split steps (as a directed graph) for producing the desired *CF* of the sample. However, the availability of pure samples may be limited in practical scenario. For example, physiological samples, such as infant's blood and *DNA* evidence (collected from a crime place) [10]. Note that the DNA fingerprinting method, called restriction fragment length polymorphism (*RFLP*), typically requires a large amount of *DNA* samples for performing a single biochemical test repeatedly [3]. The overall analysis may become useless if the sample preparation method does not consider the minimization of usages of reactants seriously enough. Interestingly, the waste fluids discarded in the previous tests may be utilized in the successive tests. Thus, the waste fluids, if recycled, can reduce the usage of expensive reactant fluids when multiple tests are carried out simultaneously.

Almost all prior works [1, 8, 23, 24] on sample preparation with microfluidic biochips suffer from a key limitation that they assumed an unlimited availability of pure samples as input. Also, due to the architectural constraints, existing dilution methods with *DMF* biochips use only (1:1) mixing model.[1] *CMF* biochips overcome this limitation to some extent by allowing the use of various mixing models. Few dilution algorithms, namely *VOSPA* [9], *FloSPA-D* [1] were also proposed for *CMF* biochips. Recently, a dilution algorithm called *DPMD* is proposed for dilution using *FPVA* biochips, allowing multiple mixing models [8]. Similarly, for *MEDA* biochips, dilution algorithms, namely *WSPM* [12] and *FacDA* [23] were proposed that utilizes the advanced various mixing models supported by a *MEDA* biochip. However, all of these methods fail to produce the desired *CF*, when only the arbitrary stock solutions are available as inputs along with the buffer solution. Although, the *DMF*-based dilution algorithms, namely *DMRW* (dilution and mixing with reduced wastage) [22] and *GDA* (generalized dilution algorithm) [21] can provide solution for *RASS* problem, they fail to utilize the full power (in terms of available mixing models) of modern biochips for constructing the dilution tree/graph. Thus, they deliver mediocre solutions for *RASS* even though cost of the solutions can be further improved for modern biochips.

In this paper, we propose a cost-effective heuristic method called *hRASS* for solving the problem of recycling arbitrary stock solutions (*RASS*) considering all available mixing models of modern microfluidic biochips. The proposed methods can produce a target *CF* of the sample fluid form its stock solutions having arbitrary *CF*s. They can utilize the full power of allowable mixing models supported by those modern biochips (e.g., *MEDA* or *FPVA* biochips). The proposed algorithms produce any desired target *CF* of the sample fluid within the tolerable error limit, while minimizing an user-defined objective function. Simulation results show that proposed methods *hRASS* outperform the prior works *DMRW* and *GDA* in terms of the number of mixing steps, the discarded amount of waste fluids, and the usage of arbitrary stock solutions.

The remainder of the paper is organized as follows. Section 2 describes the preliminaries of sample preparation and Sect. 3 presents the motivation and problem formulation. The proposed methodology is discussed in Sect. 4 and Sect. 5 shows the simulation results. Finally, Sect. 6 concludes the paper.

## 2 Preliminaries of Sample Preparation

In dilution, two or more different concentrations of a sample fluid are mixed followed by a split operation as indicated by the sequence of mix-split steps,[2] represented by a directed graph (known as dilution tree or graph) to achieve the target *CF*. More

[1] In the (1:1) mixing model, two unit-volume droplets are mixed.

[2] In one mix-split step, two or more equal/unequal volumes of fluids are mixed and split into two or more equal/unequal volume of fluids after mixing.

precisely, dilution is the addition of more solvent (e.g., buffer) to produce a lower *CF* of the sample or input fluid. Note that *CF* of a pure sample (buffer) fluid is assumed to be 1 (0). Generally, each *CF* is represented as $x : y$, where $x, y \in \mathbb{N}$. Therefore, the target *CF* of the sample is expressed as $\frac{x}{N}$, where $N = x + y$. Based on the error-tolerance limit $\epsilon$, the target $CF = x : y$ is converted into another $CF = x' : y'$, where $N' = x' + y'$. Depth $d$ of the dilution graph is determined by $N' = M^d$, where $M$ is the mixer size.

More often, a diluted fluid is prepared from a set of higher concentrated stock solutions,[3] which were frequently obtained as the byproduct in other biochemical tests using the same sample. The waste fluids discarded in one dilution process can be utilized as stock solutions to prepare another target *CF* of the same sample fluid. Hence, if we mix $k$ number of arbitrary stock solutions $A_1, A_2, \ldots, A_k$ having *CF*s as $C_1, C_2, \ldots, C_k$ at a volumetric ratio $V_1 : V_2 : \ldots : V_k$ using a Mixer-$M$ (an on-chip mixer of size $M$), where $C_i \in (0, 1)$, $V_i \in \mathbb{Z}^+$, $0 \leq V_i < M$, and $\sum_i V_i \leq M$, then the resultant *CF* $C_t$ $(0 < C_t < C_k)$ can be written as $C_t = \dfrac{C_1 \times V_1 + C_2 \times V_2 + \cdots, C_k \times V_k}{V_1 + V_2 + \cdots, V_k}$. Moreover, the cost of any sample preparation algorithm can be estimated after scanning the obtained dilution tree, where (i) $n_m$ is the total number of mix-split steps, (ii) $n_i$ is the total number of unit-volumes of the stock solutions used, (iii) $n_b$ is the total number of unit-volumes of buffer solution used, and (iv) $n_w$ is the total number of unit-volumes of waste fluid generated.

For a Mixer-$M$, we denote the set of all mixing models as $X_M$ having total volume strictly equal to $M$ units, where for any $x \in X_M$, $x$ is $(V_1 : V_2 : \ldots : V_R)$ and $\sum_i V_i = M$. For example, $X_4 = \{(2{:}2), (1{:}3), (1{:}2{:}1), (1{:}1{:}1{:}1)\}$ denotes the set of all mixing models available in a Mixer-4. Figure 1 shows various mixing models available for different types of on-chip mixers in modern microfluidic biochips. Note that *DMF* biochips can use a $r \times c$ $(r, c \in \mathbb{Z}^+)$ array mixer (Mixer-2) to realize the (1:1) mixing model supported by it as shown in Fig. 1a, whereas, *CMF* biochips can use a ring shaped rotary mixer (Mixer-$M$ for fixed $M$), which can realize various mixing models (see Fig. 1b). *FPVA* biochips can use $w \times h$ $(w, h \in \mathbb{Z}^+)$ ring shaped rotary mixer to implement all mixing models of Mixer-$M$, where $M = 2h + 2w - 4$ (see Fig 1c). Interestingly, *MEDA* biochips can support more powerful mixing models by the virtue of its inherent granularity. In *MEDA* biochips, the same mixer Mixer-$M$ can be reconfigured for different values from 2 to $M$, i.e., $X_M = \{X_M \bigcup X_{M-1} \bigcup \cdots \bigcup X_2\}$ (see Fig 1d).

[3] The term stock solution is frequently used in biochemistry to refer to the concentrated solutions, from which a working target concentration is achieved.

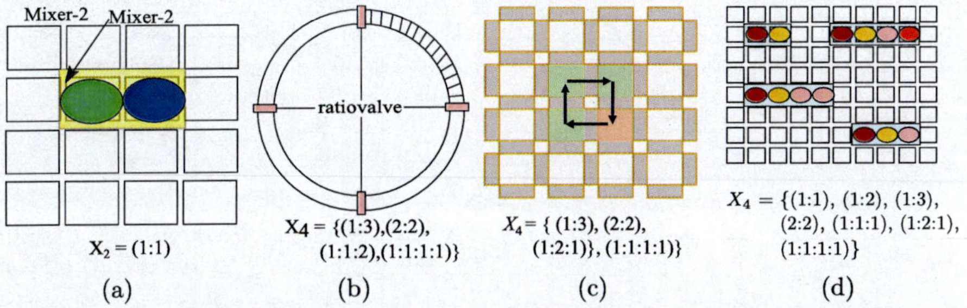


**Fig. 1** Mixing models supported by **a** *DMF* biochips using Mixer-2, **b** *CMF* biochips using Mixer-4, **c** *FPVA* biochips using Mixer-4, and **d** *MEDA* biochips using Mixer-4

## 3 Motivation and Problem Formulation

### *3.1 Motivation*

The *DMRW* algorithm [22] was proposed for dilution of a sample fluid from the supply of two arbitrary stock solutions and aims to minimize waste droplets during sample preparation. As it is designed for *DMF* biochips, it determines a dilution graph based on the (1:1) mixing model using two stock solutions ($R = 2$), one lower than target *CF* and another greater than target *CF*. Later on, another *DMF*-based dilution algorithm *GDA* was reported for minimizing the number of mix-split steps while preparing a target *CF* from two or more stock solutions. An example for producing the target CF = $\frac{23}{64}$ using *DMRW* and *GDA* is shown in Fig. 2a, b, where the set of arbitrary stock solutions are $\{\frac{1}{8}, \frac{7}{8}\}$ and Mixer-2 is used (since *DMRW* and *GDA* developed only for *DMF* biochips). But the cost of the solution can be reduced if we utilize all mixing models (e.g., as shown in Fig. 1d) available in modern microfluidic biochips (e.g., *FPVA*, *MEDA*) for constructing the dilution tree. A dilution tree for the same example target *CF* obtained by the proposed method (*hRASS*) is shown in Fig. 2c. It can be seen that the proposed method reduces the cost of noticeably compared to *DMRW* and *GDA*. This motivates us to propose the dilution algorithms to solve the problems of *hRASS*, which can produce a target *CF* by taking the full advantages of all the mixing models supported by a Mixer-*M* in modern biochips.

### *3.2 Problem Formulation*

The problem of automated dilution of a sample fluid from two or more arbitrary stock solutions of the same fluid is called as recycling arbitrary stock solutions (*RASS*) and it can be described as follows.

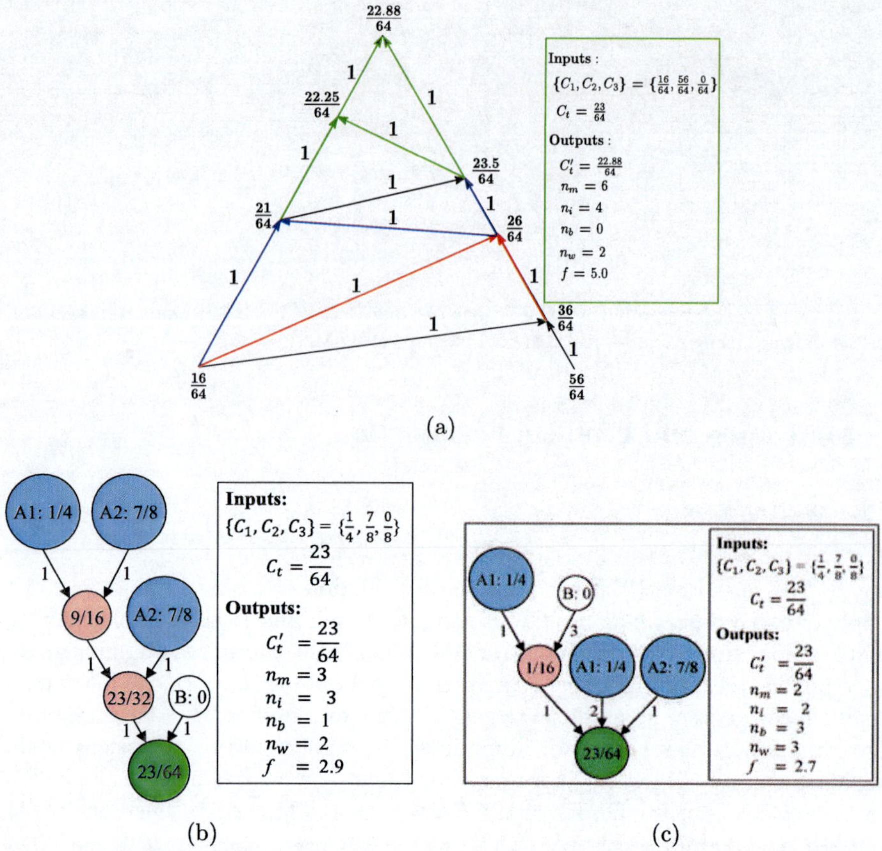


**Fig. 2** Dilution trees generated by **a** *DMRW* ($M = 2$), **b** *GDA* ($M = 2$), **c** proposed method (*hRASS*) for $M = 4$

**Inputs**:

- A set of $R$ different stock solutions $\{A_1, A_2, \ldots, A_R\}$ of the sample fluid, where *CF* of any $A_i$ is $C_i \in \mathbb{R}$ and $C_i \in (0, 1)$.
- A pure buffer solution $B$, where *CF* of $B$ is 0.
- Target *CF* $C_t = x : y = \frac{x}{x+y}$, where $x, y \in \mathbb{Z}^+$ and $C_t \in (0, 1)$.
- Mixer size $M$.
- Error-tolerance $\epsilon$ ($0 \leq \epsilon < 1$) of the *CF* $C_t'$ obtained in the process, where *CF*-error $\Delta = C_t - C_t'$ and $C_t' \in (0, 1)$.
- Weights $\alpha$, $\beta$, and $\gamma$ of $n_m$, $n_i$, and $n_w$, respectively, where $0 < (\alpha, \beta, \gamma) \leq 1$ such that $\alpha + \beta + \gamma = 1$.

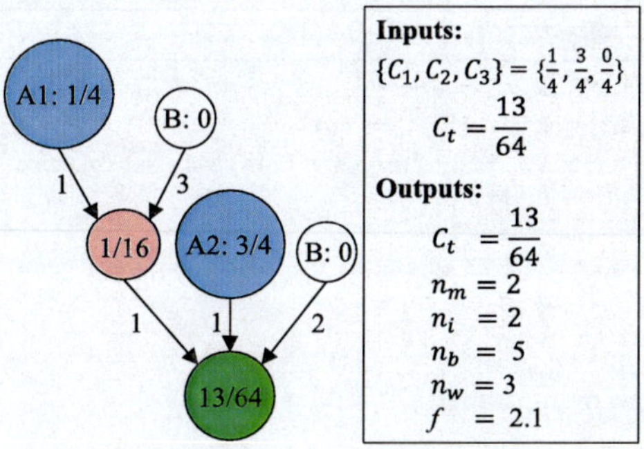


**Fig. 3** Dilution tree ($iTREE$) for the target CF = $\frac{13}{64}$ generated by Algorithm 1 ($vGDA$) for $M = 4, d = 2, R = 2$ (excluding buffer $B$)

**Output**: $M$-ary directed dilution tree[4] $D_g$ representing the sequence of mix-split steps each of which can be implemented by Mixer-$M$ to produce the target *CF* $C_t$ within the error-tolerance limit $\epsilon$, i.e., $\Delta < \epsilon$.
**Objective**: Minimize $f$, where $f$ is the objective function defined as: $f = \alpha \times n_m + \beta \times n_i + \gamma \times n_w$.

## 4 Proposed Methodology

We propose a heuristic based-solution *hRASS* to solve the problem of recycling arbitrary stock solution *RASS*. Note that the proposed method works in two phases. In the first phase, it determines a best solution $S_g(X_M)$ over a large solution space, by considering the algorithm *GDA*, a set of mixing models $X_M$, and a Mixer-$M$. Whereas, in the second phase, the proposed method (*hRASS*) works on the solution tree (generated in the first phase) by considering the set of mixing models $X_M, X_{M-1}, \ldots, X_2$ for a Mixer-$M$ and provides a final solution $S_h$ by choosing the best among the set of all solutions in $\{S_g(X_M) \bigcup S_g(X_{M-1}) \bigcup \cdots \bigcup S_g(X_2)\}$.

More precisely, the proposed method works on a base mixing tree generated by *GDA* (which is based on (1:1) mixing model) and generates a multiple mixing model based mixing tree as a solution tree. Note that *GDA* considers only (1:1) mix-split steps while constructing the mixing tree, which is supported by *DMF* biochips. So, to serve the purpose, the proposed method first converts the *GDA*-based mixing tree to another intermediate mixing tree ($iTREE$) which is based on $(n : m)$ mix-split steps (where $m + n = M$ for mixer size $M$). We describe the detailed steps of creating such a tree ($iTREE$) from the base mixing tree (provided by *GDA*) in Algorithm 1 and named it as $vGDA$. We also show the dilution tree ($iTREE$) obtained by the $vGDA$ Algorithm 1 for the target CF = $\frac{13}{64}$ in Fig. 3 for demonstration purpose.

---

[4] An $M$-ary tree is a rooted tree in which each node has no more that $M$ children.

**Algorithm 1:** *vGDA*$(\{C_1, C_2, \ldots, C_R\}, B, C_t, M, \epsilon, d, \alpha, \beta, \gamma)$

Set $C_{R+1} = 0$ (for buffer) and dilution tree $\mathcal{T} = \phi$;
Approximate $C_i$ as $\frac{B_i}{M^d}$ and $C_t$ as $\frac{T}{M^d}$, where $B_i, T \in \mathbb{Z}^+$;
Create $\mathcal{W}$ as a matrix of $R + 1$ rows and $d + 1$ columns;
**for** $i = 1$ *to* $R + 1$ **do**
    $\mathcal{W}[i][0] = B_i$;
/* append $(d + 1)$ G.P. terms in each row of $\mathcal{W}$, such that the common ratio is $\frac{1}{M}$ */
**for** $i = 1$ *to* $R + 1$ **do**
    **for** $j = 1$ *to* $d + 1$ **do**
        $\mathcal{W}[i][j] = B_i \times \frac{1}{M^j}$;
Create a set $Q$ with all numbers of $\mathcal{W}$ excluding the $(R + 1)$th row of $\mathcal{W}$;
For each member $x \in Q$, associate a weight $w_x = \frac{1}{M^j}$, if $x$ is $jth$ G.P. term in $\mathcal{W}$;
Find a collection of elements (duplicates allowed) as a set $P$ from set $Q$, where $P \subset Q$ by using *SMT* solver such that $|T' - T| < \epsilon * M^d$, where $T' = \sum_{\forall x, x \in P} x$;
A solution to the problem obtained by the *SMT* solver is an entry matrix $\chi$ (i.e., $P$) denoting the selection of numbers from $\mathcal{W}$ (i.e., $Q$);
/* More than one solutions may be obtained. */
**for** *each of the solutions obtained* **do**
    Compute $V = \sum_{\forall x, x \in P} w_x$;
    **if** $V \leq 1$ **then**
        The solution is valid;
        $S_1$ = radix-$M$ summation of the first R rows of $\chi$ matrix
        $S_2$ = radix-$M$ equivalent of $M^d - S_1$;
        Calculate radix-$M$ $d + 1$ digits of the $(R + 1)$th row of $\chi$ using $S_1$ and $S_2$;
        Determine $d + 1$ digit number for $R + 1$th row of $\chi$; /* for buffer */
        Obtain the dilution tree $\mathcal{T}$ by scanning $\chi$ entry matrix from right to left;
**for** *all the valid solutions* **do**
    Estimate the cost parameters $n_m, n_i, n_w$;
    $f = \alpha \times n_m + \beta \times n_i + \gamma \times n_w$; /* Store $f$-value of each valid solution */
/* return the dilution tree $\mathcal{T}$ having minimum $f$-value from all the valid solutions */
**return** $\mathcal{T}, f$;

**Algorithm 2:** *hRASS*$\{(C_1, C_2, \ldots, C_R\}, B, C_t, M, \epsilon, d, \alpha, \beta, \gamma)$

Set $C_{R+1} = 0$ (for buffer) and dilution tree $\mathcal{T} = \phi$ and $f_{\min}$ as a large integer;
Approximate $C_i$ as $\frac{B_i}{M^d}$ and $C_t$ as $\frac{T}{M^d}$, where $B_i, T \in \mathbb{Z}^+$;
/* checking all possible $\mathcal{T}$ for different $M$ */
**for** $m = 2$ *to* $M$ **do**
    *vGDA*$(\{C_1, C_2, \ldots, C_R\}, B, C_t, m, \epsilon, d, \alpha, \beta, \gamma)$;
    **if** $f < f_{\min}$ **then**
        $f_{\min} = f$;
        $D_g = \mathcal{T}$;
/* returns the dilution tree having minimum $f$-value from all the valid solutions */
**return** $D_g$;

Finally, the proposed method considers the dilution tree $iTREE$ (returned by Algorithm 1) for generating a mixing tree which is based on $(n : m)$ mix-split steps (where $m + n \leq M$ for mixer size $M$). The rationale for choosing such a mixing

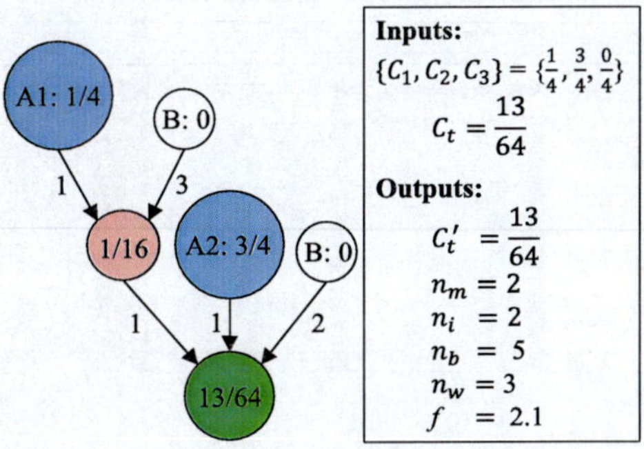


**Fig. 4** Dilution tree for the target CF = $\frac{13}{64}$ determined by the proposed method (*hRASS*) for $M = 4, d = 2, R = 2$ (excluding buffer $B$)

model lies in the fact that multiple mixing models are supported by various modern microfluidic biochips. The overall procedure of this algorithm ($hRASS$) is also described in Algorithm 2. In *hRASS*, for a mixer size of $M$, we define the weight matrices corresponding to each $m$ in the range 2 to $M$, i.e., for each $m$, we make weight matrices having geometric progression (G.P.) with common ratio as $\frac{1}{m}$ instead of always being $\frac{1}{M}$. Moreover, the entry matrix is a radix-$m$ entry matrix for each $m$ in the range 2 to $M$. We also form a sub-collection $P$ from the set $Q$ (i.e., $P \subset Q$) for each $m$, where $2 \leq m \leq M$, such that $|T - T_m'| < \epsilon * M^d$. Hence, for each $m$, we will get an $m$-ary tree. The dilution tree obtained by the proposed method (*hRASS*) for the target CF = $\frac{13}{64}$ is shown in Fig. 4 for demonstration purpose.

## 5 Simulation Results

We implemented *vDMRW*[5] (a variant of *DMRW* [22]), *GDA* [21], *vGDA* (a variant of *GDA*), and the proposed algorithm *hRASS* in Python programming language (version 3.6.8) and compared their performances for evaluating the effectiveness of the proposed method. Note that *DMRW* was proposed only for two arbitrary stock solutions, whereas *vDMRW* works for any number of stock solutions. In order to do that, *vDMRW* chooses $C_{\max}$ as higher value $C_h$ and sets lower value $C_l$ to 0 ($C_{\min}$) when $C_t < C_{\min}$ ($C_t \geq C_{\min}$), where $C_{\min}$, $C_{\max}$, and $C_t$ denote the minimum, maximum value of the set of stock solutions, and the target *CF*, respectively. We preformed all simulations with a 2.526 GHz Intel Core 2 Duo E7200(2) computer having 3GB RAM running 32-bit Ubuntu 18.10 OS. We considered the weights of $\alpha$, $\beta$, $\gamma$ as 0.6, 0.3, 0.1, respectively, for comparing the performance of different algorithms.

We first consider Mixer-2, i.e., we assume that only (1:1) mixing model is available for generating the solution. By considering such a scenario, we randomly generated 60 target *CF*s and performed experiments on different methods (*vDMRW*, *GDA*, and *hRASS*. Note that, while performing experiments, we varied the depth of dilution trees ($d$) from 3 to 5 and assume that only two stock solutions ($R$) are available. We

[5] The method *vDMRW* is designed from *DMRW* for fair comparison here.

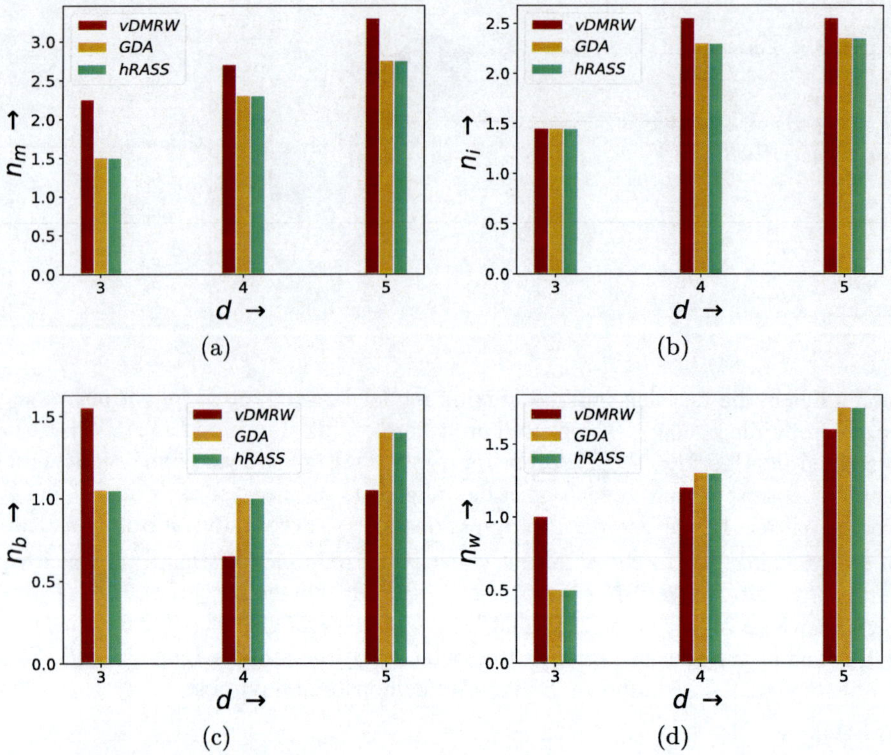


**Fig. 5** Distributions of **a** avg. #mix-split steps ($n_m$), **b** avg. #unit-volumes of stock solutions ($n_i$), **c** avg. #unit-volumes of buffer ($n_b$), and **d** avg. #unit-volumes of waste ($n_w$) varying the depth of dilution trees as $d = 3, 4, 5$ for *vDMRW*, *GDA*, and *hRASS*, when two stock solutions ($R$) are available and $M = 2$ that allows only (1:1) mixing model

then calculated average values of $n_m$, $n_i$, $n_b$, and $n_w$ and plot the results in Fig. 5 for evaluating the performance of these methods. We observed that the *GDA* and *hRASS* provide similar performance in all parameters. However, both of them provide better quality solution than *vDMRW* most of the times.

We again performed experiments by considering a different scenario to draw a conclusion on the performance of these methods. During the experiment, we randomly generated a dataset (having 80 target *CF*s), we varied the number of available stock solution from 2 to 5, and we assume that mixer 2 is available. We again calculated the results and report them in in Fig. 6. It can be easily concluded from the figure that the performance of *GDA* and *hRASS* becomes equal when mixer 2 is available, whereas both of them outperform *vDMRW* in most of the cases.

We evaluated the performance of *vDMRW*, *vGDA*, and *hRASS* when size of the mixer is greater than 2. Here, we did not consider *GDA* algorithm since it is applicable only for mixer 2. So, for the sake of completeness, here we consider *vGDA* algorithm which can provide solution for mixer size greater than 2. During experiments, we

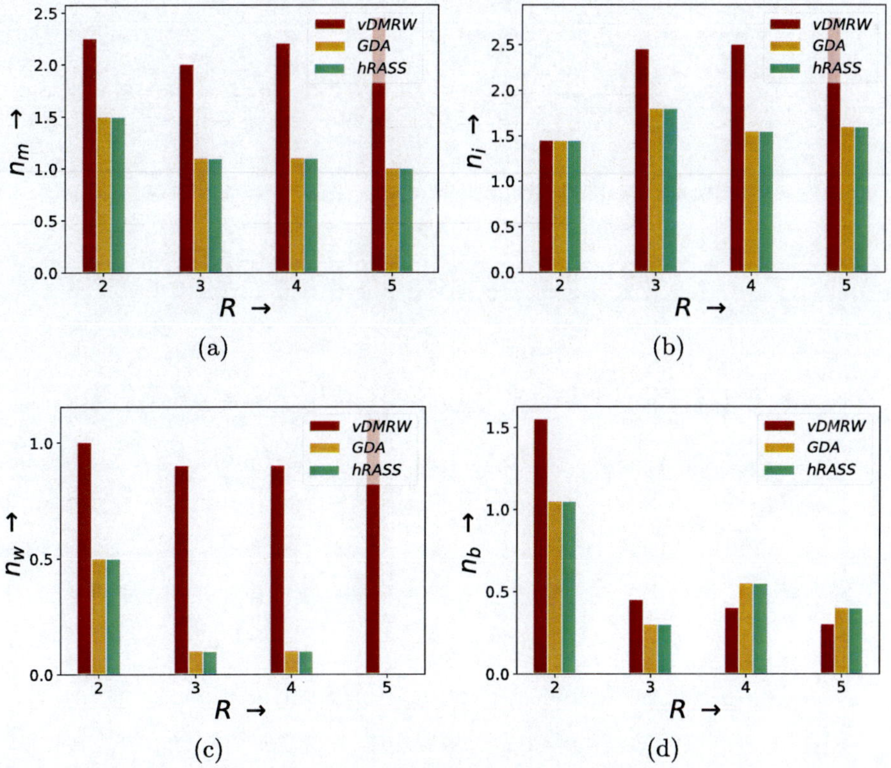


**Fig. 6** Distributions of **a** avg. #mix-split steps ($n_m$), **b** avg. #unit-volumes of stock solutions ($n_i$), **c** avg. #unit-volumes of waste ($n_w$) and **d** avg. #unit-volumes of buffer ($n_b$) varying the number of available stock solutions as $R = 2, 3, 4, 5$ for *vDMRW*, *GDA* and *hRASS*, when $M = 2$ that allows only (1:1) mixing model and depth $d$ of the dilution trees is 3

randomly generated a large dataset with 540 target *CFs* and compare the performance of different methods using the objective function $f$ defined earlier in Sect. 3.2. Here, we assume the values of $\alpha$, $\beta$, and $gamma$ as 0.6, 0.3, and 0.1, respectively. However, these values can be changed depending on the user-requirements. While calculating the values of $f$ for these methods, we first varied mixer size from 3 to 5. Then, for each mixer size, we varied the depth of the trees from 3 to 5. We varied the number of stock solutions from 3 to 5 for each depth of the tree. Finally, the results are reported in Table 1. In order to evaluate the quality of the solutions, we also reported the overall improvement of *hRASS* compared to *vDMRW* and *vGDA* in the table. It can be easily observed that the overall performance of *hRASS* is improved compared to *vDMRW* (36.8%) and *vGDA* (21%), respectively—which confirms superiority of the proposed method.

**Table 1** Comparison of performance parameters (i.e., $f = 0.6n_m + 0.3n_i + 0.1n_w$, and $\Delta = C_t - C_t'$) for three algorithms *vDMRW*, *vGDA*, and *hRASS* over 540 testcases [where $I_1 = \frac{(f1-f3)}{f1} \times 100\%$ and $I_2 = \frac{(f2-f3)}{f2} \times 100\%$]

| Mixer size $M$ | Accuracy $d$ | #Stock solutions $R$ | vDMRW | | | vGDA | | | hRASS | | | %Improvement | |
|---|---|---|---|---|---|---|---|---|---|---|---|---|---|
| | | | $f1$ | $\Delta_1$ | $n_{sol}$ | $f2$ | $\Delta_2$ | $n_{sol}$ | $f3$ | $\Delta_3$ | $n_{sol}$ | $I_1$ | $I_2$ |
| 3 | 3 | 3 | 3.4 | 0.1 | 1.0 | 2.0 | 0.2 | 122.8 | 1.9 | 0.2 | 127.8 | 44.1 | 5.0 |
| | | 4 | 3.5 | 0.2 | 1.0 | 1.8 | 0.2 | 215.8 | 1.8 | 0.1 | 238.1 | 48.6 | 0.0 |
| | | 5 | 3.6 | 0.2 | 1.0 | 2.5 | 0.2 | 174.7 | 1.9 | 0.1 | 218.0 | 47.2 | 24.0 |
| | 4 | 3 | 4.6 | 0.2 | 1.0 | 2.4 | 0.1 | 194.7 | 2.4 | 0.1 | 209.0 | 47.8 | 0.0 |
| | | 4 | 4.6 | 0.2 | 1.0 | 3.7 | 0.3 | 187.9 | 2.2 | 0.2 | 254.9 | 52.2 | 40.5 |
| | | 5 | 4.4 | 0.2 | 1.0 | 3.8 | 0.2 | 149.6 | 2.6 | 0.1 | 266.9 | 40.9 | 31.6 |
| | 5 | 3 | 5.7 | 0.2 | 1.0 | 4.7 | 0.3 | 193.9 | 3.1 | 0.3 | 226.7 | 45.6 | 34.0 |
| | | 4 | 5.2 | 0.2 | 1.0 | 5.5 | 0.2 | 147.3 | 3.4 | 0.2 | 263.2 | 34.6 | 38.2 |
| | | 5 | 5.9 | 0.3 | 1.0 | 5.4 | 0.1 | 118.2 | 4.1 | 0.2 | 225.2 | 30.5 | 24.1 |
| 4 | 3 | 3 | 3.8 | 0.2 | 1.0 | 2.5 | 0.2 | 259.4 | 1.8 | 0.2 | 347.0 | 52.6 | 28.0 |
| | | 4 | 4.3 | 0.2 | 1.0 | 3.1 | 0.2 | 241.3 | 2.3 | 0.2 | 419.6 | 46.5 | 25.8 |
| | | 5 | 4.4 | 0.2 | 1.0 | 3.2 | 0.3 | 194.7 | 2.3 | 0.2 | 342.5 | 47.7 | 28.1 |
| | 4 | 3 | 5.6 | 0.2 | 1.0 | 3.9 | 0.2 | 212.2 | 3.0 | 0.3 | 363.2 | 46.4 | 23.1 |
| | | 4 | 6.1 | 0.2 | 1.0 | 4.8 | 0.3 | 174.6 | 3.6 | 0.3 | 332.1 | 41.0 | 25.0 |
| | | 5 | 5.8 | 0.2 | 1.0 | 5.2 | 0.2 | 144.8 | 3.8 | 0.2 | 288.4 | 34.5 | 26.9 |
| | 5 | 3 | 7.3 | 0.2 | 1.0 | 6.1 | 0.2 | 190.6 | 4.6 | 0.3 | 361.1 | 37.0 | 24.6 |
| | | 4 | 7.2 | 0.2 | 1.0 | 5.9 | 0.2 | 139.9 | 5.6 | 0.2 | 270.1 | 22.2 | 5.1 |
| | | 5 | 7.0 | 0.2 | 1.0 | 7.4 | 0.3 | 112.8 | 6.5 | 0.2 | 224.5 | 7.1 | 12.2 |
| 5 | 3 | 3 | 5.2 | 0.2 | 1.0 | 3.0 | 0.3 | 239.3 | 2.5 | 0.2 | 439.1 | 51.9 | 16.7 |
| | | 4 | 4.7 | 0.2 | 1.0 | 3.7 | 0.2 | 204.3 | 2.7 | 0.2 | 413.8 | 42.6 | 27.0 |
| | | 5 | 5.1 | 0.2 | 1.0 | 4.3 | 0.3 | 183.2 | 3.3 | 0.2 | 358.8 | 35.3 | 23.3 |
| | 4 | 3 | 6.8 | 0.2 | 1.0 | 5.5 | 0.2 | 223.1 | 4.2 | 0.3 | 443.8 | 38.2 | 23.6 |
| | | 4 | 6.6 | 0.2 | 1.0 | 5.6 | 0.3 | 174.1 | 4.7 | 0.2 | 335.0 | 28.8 | 16.1 |
| | | 5 | 6.9 | 0.2 | 1.0 | 6.2 | 0.2 | 136.0 | 5.6 | 0.2 | 268.9 | 18.8 | 9.7 |
| | 5 | 3 | 8.1 | 0.2 | 1.0 | 7.8 | 0.3 | 179.8 | 6.0 | 0.2 | 354.9 | 25.9 | 23.1 |
| | | 4 | 7.5 | 0.2 | 1.0 | 8.3 | 0.2 | 135.7 | 6.6 | 0.2 | 270.7 | 12.0 | 20.5 |
| | | 5 | 8.7 | 0.2 | 1.0 | 8.4 | 0.3 | 102.7 | 7.5 | 0.2 | 210.2 | 13.8 | 10.7 |
| Avg. improvement (in %) | | | | | | | | | | | | 36.8 | 21.0 |

The corresponding $n_{sol}$ values (total number of various solution trees obtained by any method) are also shown

# 6 Conclusions

In this work, we propose a dilution algorithm for recycling arbitrary stock solutions of a sample fluid leveraging various mixing models. We observed that the performance of the proposed method is similar to the state-of-the-art method *GDA* in traditional biochips, i.e., when mixer 2 is only available. However, experimental results on a wide-range of dataset confirm the applicability of the proposed method in modern microfluidic biochips (i.e., which allow multiple mixing models). Obtaining a more generalized and hybrid kind of $M$-ary trees as the dilution trees for such problems is kept as an open problem in this paper.

# References

1. Bhattacharjee S et al (2017) Dilution and mixing algorithms for flow-based microfluidic biochips. IEEE Trans CAD 36(4):614–627
2. Bhattacharjee S, Wille R, Huang JD, Bhattacharya BB (2018) Storage-aware sample preparation using flow-based microfluidic labs-on-chip. In: Proceedings of DATE. IEEE, pp 1399–1404
3. Butler JM (2009) Fundamentals of forensic DNA typing, 2nd edn. Academic Press, Cambridge (2009)
4. Chin C et al (2008) Microfluidics-based diagnostics of infectious diseases in the developing world. Nat Med 17(8):1015–1019
5. Einav S et al (2008) Discovery of a hepatitis c target and its pharmacological inhibitors by microfluidic affinity analysis. Nat Biotechnol 26(9):1019–1027
6. Fidalgo LM et al (2011) A software-programmable microfluidic device for automated biology. Lab Chip 11:1612–1619
7. Gascoyne PRC et al (2004) Dielectrophoresis-based sample handling in general-purpose programmable diagnostic instruments. Proc IEEE 92:22–42
8. Gupta A et al (2019) Design automation for dilution of a fluid using programmable microfluidic device-based biochips. ACM TODAES 24(2)
9. Huang C et al (2015) Volume-oriented sample preparation for reactant minimization on flow-based microfluidic biochips with multi-segment mixers. In: Proceedings of DATE, pp 1114–1119
10. Kalanick KA (2011) Phlebotomy technician specialist, 2nd edn. Cengage Learning, Boston
11. Karns K et al (2011) Human tear protein analysis enabled by an alkaline microfluidic homogeneous immunoassay. Anal Chem 83(21):8115–8122
12. Li Z et al (2017) Droplet size-aware and error-correcting sample preparation using micro-electrode-dot-array digital microfluidic biochips. IEEE TBCAS 11(6):1380–1391
13. Mark D et al (2010) Microfluidic lab-on-a-chip platforms: requirements, characteristics and applications. Chem Soc Rev 39:1153–1182
14. Poddar S, Wille R, Rahaman H, Bhattacharya BB (2018) Error-oblivious sample preparation with digital microfluidic lab-on-chip. IEEE Trans CAD 38(10):1886–1899
15. Poddar S et al (2019) Optimization of multi-target sample preparation on-demand with digital microfluidic biochips. IEEE Trans CAD 38(2):253–266
16. Poddar S, Banerjee T, Wille R, Bhattacharya BB (2020) Robust multi-target sample preparation on MEDA-biochips obviating waste production. ACM Trans DAES 26(1):7:1–7:29
17. Poddar S, Bhattacharjee S, Fang SY, Ho TY, Bhattacharya BB (2021) Demand-driven multi-target sample preparation on resource-constrained digital microfluidic biochips. ACM Trans DAES 27(1):7:1–7:21

18. Poddar S, Bhattacharya BB (2022) Error-tolerant biochemical sample preparation with microfluidic lab-on-chip. CRC Press, Boca Raton (in publication stage)
19. Poddar S, Fink G, Haselmayr W, Wille R (2021) A generic sample preparation approach for different microfluidic labs-on-chips. IEEE Trans CAD 41(11):4612–4625
20. Pollack MG et al (2002) Electrowetting-based actuation of droplets for integrated microfluidics. Lab Chip 2:96–101
21. Roy S et al (2014) Theory and analysis of generalized mixing and dilution of biochemical fluids using digital microfluidic biochips. ACM JETC 11(1):2:1–2:33
22. Roy S et al (2010) Optimization of dilution and mixing of biochemical samples using digital microfluidic biochips. IEEE Trans CAD 29(11):1696–1708
23. Saha S et al (2019) Factorization based dilution of biochemical fluids with micro-electrode-dot-array biochips. In: Proceedings of ASPDAC, pp 462–467
24. Thies W et al (2008) Abstraction layers for scalable microfluidic biocomputing. Nat Comput 7(2):255–275
25. Wang G et al (2011) Digital microfluidic operations on micro-electrode dot array architecture. IET Nanobiotechnol 5:152–160
26. Zhong Z, Wille R, Chakrabarty K (2019) Robust sample preparation on digital microfluidic biochips. In: Proceedings of ASPDAC, pp 474–480

# Design and Analysis of Posit Processing Engine with Embedded Activation Functions for Neural Network Applications

**Pranose J. Edavoor, Aneesh Raveendran, Vivian Desalphine, and David Selvakumar**

**Abstract** This paper presents a novel design of Posit-based Processing Engine with embedded activation functions for neural network applications. The Posit Processing Engine (PPE) is split into a Posit-based Multiply Accumulate Unit (MAC) and a Posit-based activation unit. The proposed design computes the dot product with lower precision in a single unit (iterative) and computes the activation output of the dot product result. The presented approach achieves reduced area and less delay compared to IEEE-754 based Floating-Point Processing Engine. Furthermore, Posit MAC with Quire support is proposed, which improves the system's overall accuracy. The designs are modelled using Verilog HDL, functionally verified, synthesized, and implemented targeting Xilinx UltraScale VCU118 FPGA board. The proposed Posit-based Processing Engine is validated on LeNet Handwritten digit recognition system and CIFAR-10 image classification system. Both the applications show comparable accuracy compared to traditional Floating-Point (single precision) Processing Engines with reduced hardware utilization.

**Keywords** Posit · Neural networks · Processing engine · Handwritten digit recognition system · Image classification system

P. J. Edavoor (✉) · A. Raveendran · V. Desalphine · D. Selvakumar
Secure Hardware and VLSI Design Group, Centre for Development of Advanced Computing, Bangalore, India
e-mail: pranoseje@cdac.in

A. Raveendran
e-mail: raneesh@cdac.in

V. Desalphine
e-mail: viviand@cdac.in

D. Selvakumar
e-mail: david@cdac.in



C. Giri et al. (eds.), *Emerging Electronic Devices, Circuits and Systems*,
Lecture Notes in Electrical Engineering 1004,
https://doi.org/10.1007/978-981-99-0055-8_12

# 1 Introduction

Machine Learning, particularly Deep Learning, has recently been used in a wide variety of applications such as Natural Language Processing, Speech Recognition, Computer Vision, and Prediction. Deep Learning, on the other hand, necessitates lot of computations during training and inference. Computational complexity could be reduced if optimizations are made in both hardware and algorithm. Lower precision Floating-Point representations for training and inferencing have been the subject of recent studies [1–3]. The use of Fixed-Point model parameters to replace Floating-Point model parameters with lesser precision has been studied and widely adopted. The accuracy of inference using Fixed-Point numbers is considerably degraded when precision is less than or equal to 8 bits. In 2017, Posit number format, a type-III variant of Unum, was introduced by Gustafson [4]. Posit number system is found to be a suitable drop-in replacement for the IEEE-754 standard for Floating-Point arithmetic and Fixed-Point arithmetic for deep learning inference application [5]. Posits provide better dynamic range, tapered accuracy, and parameterized precision which are suitable for Deep Learning applications. The use of Posit number system is efficient in Deep Learning network, as most of the network weights fall in the range of $-1$ to 1 and the coding space of Posit also falls in that range [6, 7].

A quantization technique, when done with a Posit unit in the deep neural network, significantly reduces the memory requirement and computational cost, and it can also be used to improve power efficiency. Mostly, 8-bit or 16-bit Posit format is used in the deep learning system, and 32-bit Posit format is used in scientific computation. 32-bit Posit format could be used to replace the standard 64-bit Floating-Point format [7].

This work presents a new approach to replace IEEE-754 Floating-Point (Single Precision) model parameters with Posit (8, 0) and Posit (16, 1) parameters for LeNet-5 handwritten digit recognition (HDRS) and CIFAR-10 image classification applications. FPGA implementation of the Posit Processing Engine along with existing Processing Engines is also discussed in this work. Using this approach, accuracy can be maintained even with reduced precision. In addition, a Posit Processing Engine (PPE) with embedded activation function is also discussed in this paper.

The rest of the paper is organized as follows: Section 2 discusses the related works. Section 3 reviews the Unum number system and Posit number representation in detail. Section 4 discusses the Processing Engine and the proposed Posit-based Processing Engine. This section also discusses the implementation of approximate activation functions using Posit. Section 5 discusses the improvements of PPE against previous Floating-Point implementations and presents accuracy evaluations. Finally, Section 6 concludes this paper.

## 2 Related Works

In the previous decade, for training and inferencing model parameters in Neural Network applications, IEEE-754 standard [8] for Floating-Point arithmetic was used. Lately, Fixed-Point, bfloat, and half Precision Floating-Point format are being used for training and inference. The work in [2] presents an analysis of Posit, Float, and Bfloat. The results showcase Posits to have higher accuracy, wider dynamic range and could be used in systems with lower performance requirements and less memory usage. The work concludes that Posits and Bfloats could be used for machine learning algorithms. Posit has spawned many new hardware implementations due to its superiority in dynamic range and tapered accuracy. R. Chaurasiya et al. introduced a parameterized adder/subtractor and a parameterized multiplier [9]. Leading bits were determined using the leading zeros detector (LZD) and the leading one detector (LOD) in these two studies. These two efforts transformed negative Posit numbers to the corresponding opposite numbers to process them. The implementation has two limitations: both LOD and LZD produce redundant area when converting the two's complement. Alouani et al. [6] present a sturdiness study between 32-bit IEEE-754-2008 and 32-bit Posit representations. The article has proposed a speculative analysis for both number representations for double bit flips and single bit flips, and experiments were carried out on machine learning applications. The article concluded that Posit demonstrates higher robustness compared to IEEE-754 representations. M. K. Jaiswal et al. proposed an algorithm flow for hardware architecture of Floating-Point to Posit convertor, Posit to Floating-Point converter, Posit multiplier, and Posit adder unit, as well as to demonstrate the implementation detail on FPGA platform with 8, 16, 24, 32, and 64-bit implementations with varying exponent sizes (es) [10]. J. Johnson et al. proposed an approximate logarithmic multiplier in combination with the Posit number systems, which optimized area, power, and delay [11]. In this work, the authors have optimized the multiplication operation using Posit number system. Fatemi Langroudi et al. [5] proposed a Deep Convolutional Neural Network (DCNN) by using Posit number system for digit recognition and image classification. The author also compares the Fixed-Point and Posit number system that is used to represent the model parameters. The article concludes that the normalized Posit number system outclassed the Fixed-Point number system in terms of memory utilization and accuracy if the two number systems are compared in the equal dynamic range, i.e. $-1$ to 1. Cococcioni et al. [12] designed an energy-efficient Posit processing unit (PPU) as an alternative to Floating-Point processing unit (FPU), which was used in the autonomous driving assistant system (ADAS). In this, 16-bit Posits were used to replace the conventional FPU as ADAS require 16-bit Floating-Point representations for safety-critical application. Zhang et al. [13] investigated and arrived at an efficient architecture of Posit-based multiply accumulate (MAC) unit generator for deep learning applications. The generator generates Verilog code for user-defined total bit-width and exponent bit-width. The work has been analyzed for delay, area, and power. The work in [14] presents a mixed-precision hardware and software co-design framework for DNNs targeting edge devices. The framework supports DNN training and inference using Posit and other number formats. The results demonstrate that 16-bit Posits perform better than 16-bit Floating-Point in DNN training. Wan et al. [15] provide a study of the impact of various bit precisions on speech-to-text neural applications. The study shows Posit numerical representation (8-bit precision) is best suited for speech recognition inference. The study also shows that the Posit-based hardware (in terms of power and area) would be less expensive compared to Float. The work in [16] describes an algorithm for multiplying two Posit numbers and integration into FloPoCo framework. Posit is analyzed for inference and training stages of neural networks, and the results while working with the MNIST data set are promising when compared to Floating-Point. Zhou et al. [17] proposed a high-speed area-efficient Multiply accumulator for FPGAs. In the work presented by Nambi et al. [18], a modified and novel representation of Posit number system is used to represent the trained parameters of DNNs. The work proposes a Posit to Fixed-Point converter for enabling energy-efficient, high-performance hardware implementations for ANNs with minimum effect to the output accuracy. After conversion, all arithmetic operations are performed using Fixed-Point operators. The work presented in [19] by Raul Murillo et al. describes a framework for performing DNN training and inference based on Posit. The work showcases the whole training phase performed with Posit, obtaining better results than training with Floating-Point. Langroudi et al. [20] investigate ultra-low precision DNN training. The work presents an adaptive Posit numerical format in the context of DNNs, particularly targeting edge computing. The evaluation is performed on Fashion-MNIST, CIFAR-10, and ImageNet data sets, showing promising results. The work presented in [7] proposes a Posit implementation of deep neural network for breast tumour diagnosis using data set from [21]. In our proposed work, application of Posit and Posit Processing Engine (PPE) with Quire for Deep Neural Network has been investigated. Furthermore, hardware implementation of Posit Processing Engine with embedded activation function along with mathematical modelling and representation of activation functions in terms of Posit is also discussed.

## 3 Posit Number System and Processing Engine

### *3.1 Posit Number System*

John Gustafson introduced Unums (Universal Number), a different way to represent real numbers. It uses a finite number of bits to represent the real number. Unums have three versions, and Posit is the latest one that is introduced as an alternative to the IEEE-754 standard Floating-Point number system. IEEE-754 Floating-Point number system has a sign bit, a set of bits to denote the exponent term and a set of bits called mantissa, whereas Posits add extra bits known as regime. Therefore, a Posit number has four parts: sign bit, regime, exponent, and fraction. The sign and regime bits have high priority compared to others [4]. The exponent and fraction part

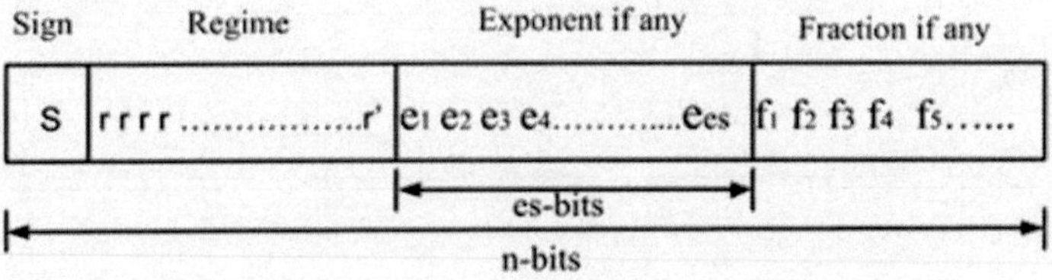


**Fig. 1** Posit number representation

of Posit do not have a fixed length. The two numbers mainly specify a Posit number system; the first is the total number of bits denoted as '$n$', and the second is the number of bits devoted to the exponent, which is typically represented as 'es'. Posit is expressed as Posit $P(n, \text{es})$. The first bit of the Posit number system represents the sign bit, which represents a negative number if set to '1'. The sign bit is followed by regime bits which can be varied, and for an '$n$' bit, Posit number regime can be of length 2 to $n - 1$. It can be either running ones followed by a zero or running zeros followed by a one. The exponent bits and fraction bits occupy the rest of the bit positions. Max value of exponent bits is specified by the number 'es', and any bit left after the exponent belongs to the fraction part. The binary representation of the Posit number system is shown in Fig. 1.

### *3.2 Generic Processing Engine*

Deep learning requires a large number of operations and large storage for the weights and activation. Domain-specific hardware accelerators are used to handle the huge computational challenges associated with Deep Learning algorithms and processes. Processing Engines (PE) are used to handle the computational process by using a parallel computation mechanism. Block diagram of a generic Processing Engine is given in Fig. 2. Processing Engine consists of a multiplier and adder, which are required to solve the computational process of deep learning. It has a specific multiplier element (ME) array and adder tree for multiplication and addition. Multiplier elements have multiplexers that exchange the input data between adjacent ME or from the register file to various multiplier elements groups, which is used for the computation. The ME and adder tree are connected together with the help of a switch network to perform multiplication and accumulation (MAC) operations. The results of MAC are stored in the data registers. The ME array and adder tree are required to handle different parallel computations based on kernel size and other parameters. As shown in Processing Engine, the multiplication and accumulation (MAC) operations require input features map along with weights values in each cycle and to increase parallelism, and a large number of PE are required. Parallelization of the workload is done based on the number of PE arrays (row and column) and the tiling loop.

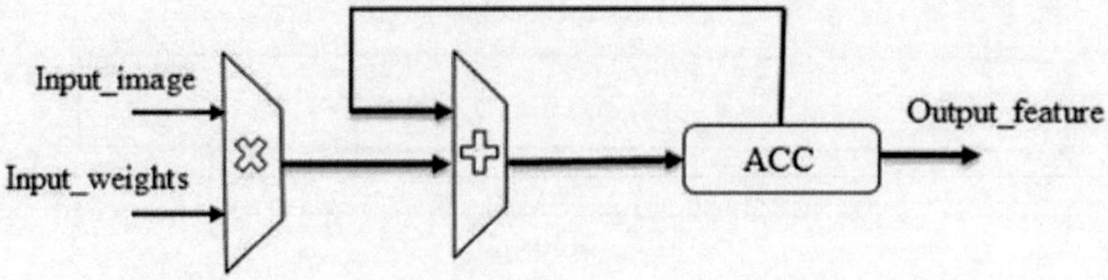


Fig. 2 Generic processing engine

Fig. 3 Posit processing engine

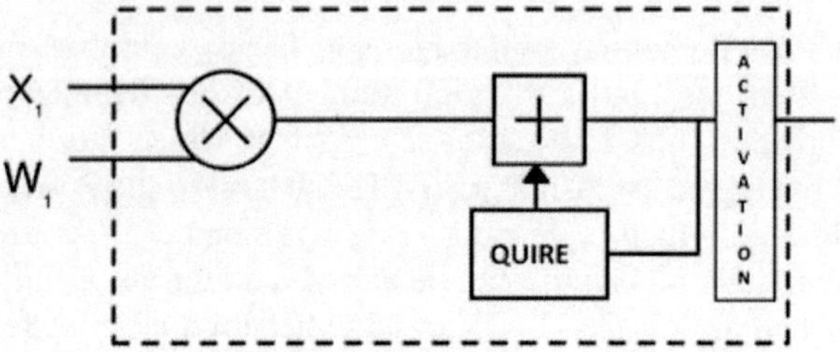


# 4 Proposed Posit-Based Processing Engine

The proposed Posit Processing Engine (PPE) comprises a Posit-based MAC unit that accumulates the result and a Posit-based activation unit that computes the activation function of the accumulator output. The detailed block diagram of PPE is depicted in Fig. 3.

## 4.1 Posit-Based Multiply Accumulate Unit with Quire

The MAC module consists of a Posit multiplier, adder, and a Quire accumulator. The multiplier is a Posit multiplier that multiplies two Posit numbers with es = 0 and handles special cases like 0, extreme values like $\pm\infty$. The product is normalized and converted into a Fixed-Point number. The normalized output is then accumulated in a Quire register. In this module, the output from the multiplier is accumulated as Fixed-Point values, which are shifted by an exponential parameter in a 64-bit wide register called Quire. The adder is a 64-bit integer adder that adds the Quire output with the normalized Posit multiplier output. The width of the Quire accumulator for $r$ iterative multiplications can be calculated using Eq. 1.

$$\text{quire\_size} = [\log_2(r)] + 2 \times \left\lceil \log_2\left(\frac{\text{maxpos}}{\text{minpos}}\right)\right\rceil + 2 \tag{1}$$

where maxpos is the maximum value and minpos is the minimum value that can be represented using the number format.

## 4.2 *Posit-Based Activation Module*

Activation functions are a quintessential part of Deep Neural Networks to introduce nonlinearity in the system, enabling the model to learn complex patterns in the data. Therefore, designing a Processing Engine that efficiently processes the activation functions is important. This paper proposes a novel methodology to compute activation functions like sigmoid, Tanh, ReLU, and Leaky ReLU along with the MAC unit within the Processing Engine. All the activation functions are approximated using bit operations, which can be performed in a single cycle.

**Sigmoid Activation**

Sigmoid activation functions are most commonly used in any Deep Neural Network applications. The expression for Sigmoid function is given in Eq. 2

$$\text{Sigmoid}(x) = \frac{1}{1 + \mathrm{e}^{-x}} \tag{2}$$

It is well known that sigmoid expressions are computed using Taylor series. The Taylor series expansion of $\mathrm{e}^x$ is given by

$$F(x) = 1 + \frac{x}{1!} + \frac{x^2}{2!} + \frac{x^3}{3!} \tag{3}$$

On substituting Eq. 3 in Eq. 2 we get:

$$\text{Sigmoid}(x) = \frac{1}{1 + 1 + \frac{(-x)}{1!} + \frac{(-x)^2}{2!} + \frac{(-x)^3}{3!}} \tag{4}$$

Here in order to approximate sigmoid using Posit, we analyse the above sigmoid equation in two parts, i.e. $0 < x < 1$ and $1 < x < infinity$. Considering the input values in $0 < x < 1$ range the expression for $\mathrm{e}^x$ can be modified as

$$F(x) = 1 + \frac{-x}{1!} \tag{5}$$

as the larger powers of $x$ can be negligible as the value of $x$ is $0 < x < 1$ range. On substituting this in Eq. 2 and approximating higher powers of $x$, Eq. 2 gets modified as

$$\text{Sigmoid}(x) = \frac{1}{2} + \frac{x}{4} \tag{6}$$

Considering a Posit number with es = 0 in [0, 1] range, the resolution of Posit number behaves similar to Fixed-Point number. This can be proved by considering a Posit number, and the real number $z$ can be represented as Eq. 7.

$$z = 2^k.(1 + \phi.2^{-F}) \tag{7}$$

where $\phi$ represents the fraction value and $F$ represents the fraction bit length. $k$ value depends on the regime length of the Posit numbers. On rearranging Eq. 7 by imposing conditions of stop bit and sign bit on values ranging [0, 1], we get Eq. 8. The resolution of Posit number [0, 1] behaves similar to Fixed-Point numbers (Z).

$$z = Z.2^{-(N-2)} \tag{8}$$

A Posit with zero exponent bits can also be understood as a Fixed-Point number with a shift of $(N - 2)$ bits, according to Eq. 8. Thus, for a Posit with zero exponent bits in range [0 1], multiplication and division by a dyadic number can be executed by right shift and left shift operations similar to Fixed-Point numbers which are bit operations.

Inversion operation is the basic operation required in any activation function realization. In order to realize inversion operation, if value of $z > 1$, it is just the reduction of length of regime by one bit, which is again a bit operation.

The ones complement operator $y = 1 - z$ is another handy bit manipulation function to develop using bit operations for activation functions. It can be noted that when $z \subset [0, 1]$ then $y$ also belongs to $y \subset [0, 1]$.

On applying the conditions to the above equations, we get Eq. 9

$$y = 1 - 2^{N-2}.(2^{-F} + \phi) \tag{9}$$

which can be further reduced to Eq. 10

$$Y = 2^{N-2} - Z \tag{10}$$

Equation 10 can be easily achieved as an integer subtraction which is again a bit wise operation.

All these bit operations can be combined to implement sigmoid function. Equation 6 can be implemented easily using shift and invert operations. A right shift operations will divide the corresponding value by 2, and invert will add value 1, then again, one right shift will divide the existing real value by 2 using the properties of bit operations. Thus, sigmoid can be implemented by just shifting of two bits and flipping of the sign bit, which are simple bit operations.

The sigmoid values lies between 0 and 1, and while implementing other activation functions using sigmoid function, all the above properties are still valid.

#### Tanh Activation

Tanh activation can be implemented by modification of the Tanh function derived from the approximated sigmoid. Tanh function expressed using sigmoid can be expressed using Eqs. 11 to 13.

Tanh activation can be expressed using Eq. 11

$$\tanh(x) = \frac{e^x - e^{-x}}{e^x + e^{-x}} \tag{11}$$

Modifying Eq. 2, we get equation for sigmoid represented by Eq. 12

$$\text{sigmoid}(x) = \frac{e^x}{1 + e^x} \tag{12}$$

Substituting Eq. 12 in Eq. 11, tanh can be expressed in terms of sigmoid with basic bit operations.

$$\tanh(x) = 2(\text{sigmoid}(x)) - 1 \tag{13}$$

**ReLU Activation**

Similarly, ReLU can be implemented by comparison with the sign bit and selecting input or zero accordingly.

**Leaky ReLU Activation**

Leaky ReLU is usually not used in hardware implementation because of complex hardware, and therefore, models are not developed with leaky ReLU even though results are better. Replacing ReLU with leaky ReLU during the training process can increase the accuracy of the model. Leaky ReLU using Posit can also be implemented using simple bit operations on the sigmoid function.

Equation 14 represents the equation for leaky ReLU

$$\text{Leaky\_relu}(x) = e^x - 1 \tag{14}$$

Sigmoid($-x$) can be represented using Eq. 15

$$\text{Sigmoid}(-x) = \frac{1}{1 + e^x} \tag{15}$$

On modifying Eq. 15, we derive at Eq. 16

$$1/(2.\text{Sigmoid}(-x)) = \frac{1 + e^x}{2} \tag{16}$$

On further rearranging of Eq. 16 and substituting in 14, we get

$$\text{Leaky\_relu}(x) = 2.[1/(2.\text{Sigmoid}(-x)) - 1] \tag{17}$$

Using Eq. 17, Leaky ReLU can be implemented using sigmoid and simple bit operations.

## 5 Experimental Results

In order to validate the proposed design, we have chosen two different Convolution Neural Network-based models, namely MNIST-based Handwritten digit recognition system and CIFAR-10-based image classification system.

The LeNet-based HDRS consists of six different layers: two convolutional layers, two max-pooling layers, and two dense layers. The training data is collected from the MNIST database, an open-source database for handwritten digits recognition. It consists of 60,000 training data sets and 10,000 test data sets. Training data (training data + validation data) is used to train the model, and testing data is used for the testing of the model for test accuracy. MNIST data set consists of greyscale image data of dimension $(28 * 28 * 1)$, which is fed as input to the model. The first layer of the neural network is the convolutional layer having 30 kernels of size $(5 * 5 * 1)$, and the output of the first layer (convolution layer) having dimension $(24 * 24 * 30)$ is passed through ReLU activation and is fed as input to the second layer (max-pooling layer) having a kernel of size $(2 * 2)$. The output of max-pooling operation of dimension $(12 * 12 * 30)$ is fed as input to the third layer (convolutional layer), having 15 kernels of size $(3 * 3 * 30)$. The output of the third layer having dimension $(10 * 10 * 15)$ is fed through ReLU activation and is fed as input to the fourth layer (max-pooling layer) having a kernel of size $(2 * 2)$. Max-pooling layer output of dimension $(5 * 5 * 15)$ is flattened for classification task as it needs to be fed to fully connected (dense layer) having a dimension of $(375 * 500)$. Output from dense layer $(1 * 500)$ after ReLU activation is fed to final layer (Dense layer) of dimension $(500 * 10)$ through softmax layer that gives the probabilities of output as one dimension vector having dimension $(1 * 10)$. A detailed diagram of the LeNet model is given in Fig. 4.

CIFAR-10 image classification model consists of eleven layers that include six convolutional layers, three max-pooling layers, and two dense layers. CIFAR-10 is a subset of 80 million tiny images data set, an open-source image recognition database. This is a data set of 50,000 $32 \times 32$ colour training images (out of which 10,000 images are used for validation) and 10,000 test images, labelled over 10 categories. Image with dimension $(32 * 32 * 3)$ is fed as input to the convolutional layer having 32 kernels of size $(3 * 3 * 3)$ with a padding value of 1. The output of the first layer (convolution layer) having dimension $(32 * 32 * 32)$ is fed as input to

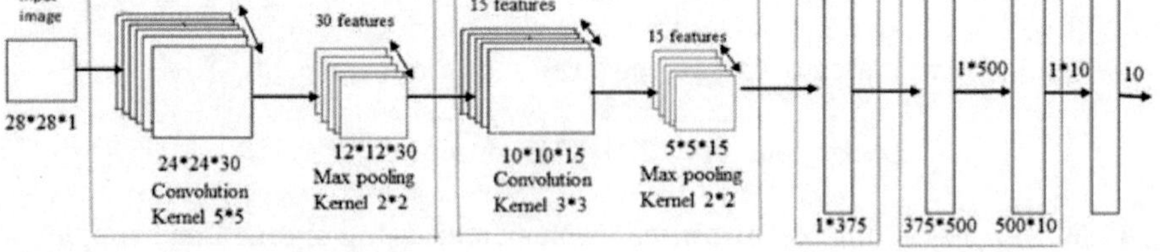


**Fig. 4** LeNet model network

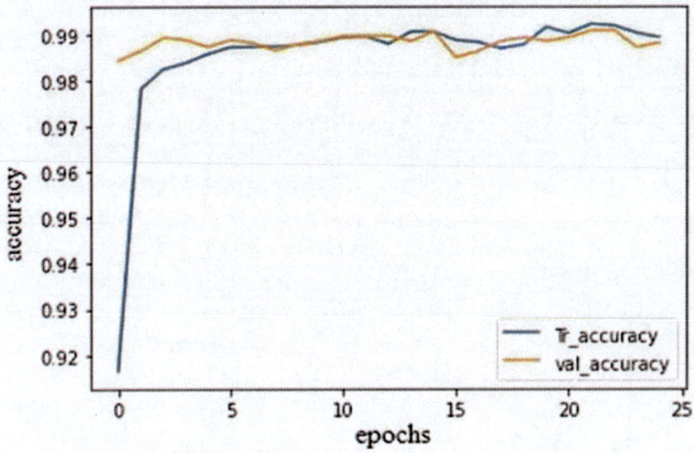


**Fig. 5** LeNet accuracy versus epoch

the second layer (convolution layer) having 32 kernels of size (3 ∗ 3 ∗ 32) with a padding value of 1. The output of the second layer having dimension (32 ∗ 32 ∗ 32) is fed as input to the third layer (max-pooling layer) having a kernel of size (2 ∗ 2). The output of max-pooling operation (16 ∗ 16 ∗ 32) is fed as input to the fourth layer (convolutional layer) having 64 kernels of size (3 ∗ 3 ∗ 32), and the fourth layer output having dimension (14 ∗ 14 ∗ 64) is fed to the convolutional layer with 64 kernels of size (3 ∗ 3 ∗ 64). The output having dimension (12 ∗ 12 ∗ 64) is passed through a max-pooling layer having a kernel of size (2 ∗ 2). Further, the output of the sixth layer having dimension (6 ∗ 6 ∗ 64) is fed as input to the seventh layer (convolution layers) having 128 kernels of size (3 ∗ 3 ∗ 64) and output of seventh layer having dimension (4 ∗ 4 ∗ 128) is fed as input to eighth layer (convolutional layer) having 128 kernels of size (3 ∗ 3 ∗ 128). The output of the eighth layer with dimension (2 ∗ 2 ∗ 128) is fed to the ninth layer (max-pooling layer) having a kernel of size (2 ∗ 2). The output of max-pooling layer having dimension (1 ∗ 1 ∗ 128) is fed as input to the tenth layer (flatten layer). The output (1 ∗ 128) is fed to two fully connected layers (Dense layers) of dimension (128 ∗ 128) and (128 ∗ 10). The final output is passed through a softmax layer which gives the probabilities of 10 classification outputs. The diagram of the CIFAR model is given in Fig. 6.

Training of LeNet-based model is done for 25 epochs and it uses ReLU as the activation function for all the layers, and last layer has softmax activation. Learning rate is set to 0.01, and it uses the categorical cross-entropy function for loss calculation. Training is carried out in single precision Floating-Point format using TensorFlow framework, and all the model parameters are in fp32 format. The model obtained training accuracy of 99.08%, validation accuracy of 98.82%, and test accuracy of 98.98%. Training accuracy and validation accuracy variations based on epoch are shown in Fig. 5. Trained model is saved in HDF5 format and is used to extract the parameters of the neural networks (Fig. 6).

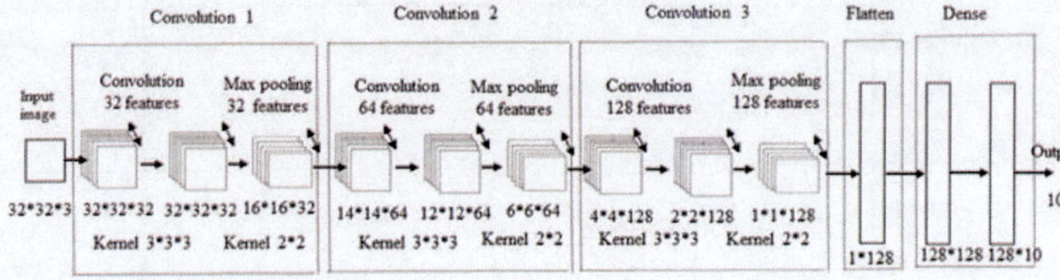


**Fig. 6** CIFAR model network

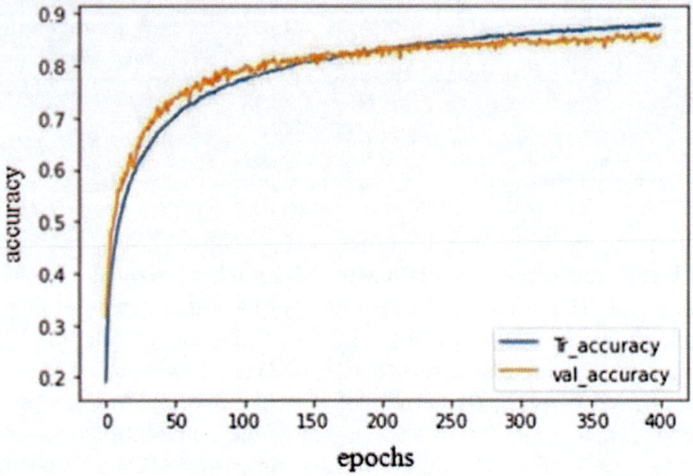


**Fig. 7** CIFAR accuracy versus epoch

Training of CIFAR-based image classification model is carried for 400 epochs, and it uses ReLU as the activation function for all the layers, and last layer has softmax as activation. Initial learning rate is set for 0.01 using SGD as an optimizer and categorical cross-entropy as a loss function. The model obtained training accuracy of 88.24, validation accuracy of 85.43, and test accuracy of 85.0. Training and validation accuracy based on epoch are shown in Fig. 7.

Different layer parameters, including model weights and biases, are extracted using a separate Python script and is stored in CSV format using a 32-bit Floating-Point number. The model parameters are then loaded from the CSV file and converted to Posit format using NumPy and Posit packages. All the layer and activation operations are converted to corresponding iterative MAC operations using Python script with Posit and Quire support using NumPy and softposit. The models are replicated using these Posit-based MAC operations, and the testing of 10,000 images from the MNIST data set and CIFAR data set is carried out to compare the accuracy with the 32-bit Floating-Point model.

**Table 1** Accuracy comparison of CNN models

| Number format | MNIST (%) | CIFAR-10 (%) | Breast cancer [21] (%) |
|---|---|---|---|
| This work (32-bit floating-point) | 98.98 | 85.0 | 90.1 |
| This work (16-bit floating-point) | 96.42 | 82.10 | – |
| This work (16-bit fixed-point) | 98.02 | 80.18 | – |
| This work (8-bit fixed-point) | 91.53 | 48.1 | 57.8 |
| Deep positron posit (8, 0) [7] | – | – | 85.89 |
| Deep PeNSieve posit (8, 0) [19] | 98.77 | 68.88 | – |
| Adaptive posit posit (8, 0) [20] | 92.60 | 91.1 | – |
| This work [Posit (16, 1)] | 98.98 | 85.0 | – |
| This work [Posit (8, 0)] | 98.18 | 81.01 | – |

**Table 2** Hardware comparison of posit processing engine with existing processing engines

| Designs | Look-up tables (LUT) | Registers | DSP slices |
|---|---|---|---|
| 32-bit floating-point [17] | 2226 | 1062 | 0 |
| This work (16-bit fixed-point) | 256 | 289 | 1 |
| This work (8-bit fixed-point) | 326 | 273 | 0 |
| This work [Posit (16, 1)] | 2124 | 361 | 0 |
| This work [Posit (8, 0)] | 684 | 88 | 0 |

The accuracy comparison of the proposed Posi-based LeNet HDRS system and CIFAR-10 image classification system with Floating and Fixed-Point system is depicted in Table 1. It can be seen that the Posit-based system gives comparable accuracy with ideal fp32 system with reduced bit precision from 32 to 8, 16 bit and better accuracy compared to Fixed-Point systems. It is to be noted that the models used in comparison with Deep PenSieve [19] and Adaptive Posit [20] are different, and fp32 implementation of the models gives an accuracy of 99.17% and 69.32%, 92.54%, and 91.54% for MNIST and CIFAR data sets. In addition, Posit-based deep learning implementation of Wisconsin Breast Cancer [21] model is also projected in Table 1. It can be seen that compared to same precision Fixed-Point, Posit implementations give better accuracy.

The hardware comparison of the proposed Posit-based processing engine with the existing state-of-the-art floating, fixed, Posit Processing Engine is depicted in Table 2. Our proposed design was modelled using Verilog HDL, functionally verified with test vectors (exhaustive) using Questasim, synthesized and implemented targeting Xilinx UltraScale VCU118 FPGA board. Single stage Posit (8, 0) achieved maximum operating frequency of 148 MHz. It can be observed that Posit (8, 0) processing engine utilizes least hardware keeping 8-bit Fixed-Point Processing engine as an exception. Even though 8-bit Fixed-Point processing engine utilizes least hardware, and accuracy for both models is least which makes it unsuitable for Convolution Neural Network-based applications.

## 6 Conclusion

The proposed work presents an approach to design a novel Posit-based processing engine with embedded activation functions for neural network applications. A Posit-based MAC with Quire support was proposed along with mathematical modelling and representation of activation functions in terms of Posit. Hardware implementation for Posit Processing Engine with embedded activation was also discussed in this work. Simulations for handwritten digit recognition using CNN-based LeNet model and CIFAR-10 image classification model using the proposed Posit Processing Engine were carried out. It was observed that Posit-based implementation of models gave better accuracy with reduced precision. Reduced precision lowers the memory overhead in terms of hardware implementations. Furthermore, the hardware utilization of Posit Processing Engine and existing processing engines is also discussed. It was found that the proposed design achieves comparable accuracy with 32-bit Floating-Point models with reduced bit precision. The model achieves better accuracy when compared to same precision with reduced hardware overhead, making Posit a suitable alternative for Neural Network applications.

## References

1. WRPN and apprentice: methods for training and inference using low-precision numerics (2018). arxiv:1803.00227v1
2. Romanov AY et al (2021) Analysis of posit and bfloat arithmetic of real numbers for machine learning. IEEE Access 9
3. DiCecco R et al (2017) Fpga-based training of convolutional neural networks with a reduced precision floating-point library. In: International conference on field programmable technology (ICFPT)
4. Gustafson Y (2017) Beating floating-point at its own game: posit arithmetic. Supercomput Front Innov Int J 4(2)
5. Fatemi Langroudi SH, Pandit T, Kudithipudi D (2018) Deep learning inference on embedded devices: fixed-point vs posit. In: 1st workshop on energy efficient machine learning and cognitive computing for embedded applications
6. Buoncristiani N, Shah S, Donofrio D, Shalf J (2020) Evaluating the numerical stability of posit arithmetic. In: 2020 IEEE international parallel and distributed processing symposium (IPDPS), LA, USA, New Orleans
7. Carmichael Z, Langroudi HF, Khazanov C, Lillie J, Gustafson JL, Kudithipudi D (2019) Deep positron: a deep neural network using the posit number system. In: 2019 design, automation and test in Europe conference and exhibition
8. IEEE standards Board and ANSI (2008) IEEE standards for binary floating-point arithmetic. IEEE Std., 754-2008
9. Chaurasiya R et al (2018) Parameterized posit arithmetic hardware generator. In: 2018 IEEE 36th international conference on computer design (ICCD), pp 334–341. https://doi.org/10.1109/ICCD.2018.00057
10. Jaiswal MK, So HKH (2018) Universal number posit arithmetic generator on FPGA. In: Proceedings of the design, automation and test in Europe conference and exhibition, Dresden, Germany
11. Johnson J (2018) Rethinking floating-point for deep learning. arXiv e-prints [Online]. Available: arxiv:1811.01721

12. Cococcioni M, Ruffaldi E, Saponara S (2018) Exploiting posit arithmetic for deep neural networks in autonomous driving applications. In: 2018 international conference of electrical and electronic technologies for automotive. IEEE
13. Zhang H, He J, Ko S (2019) Efficient posit multiply-accumulate unit generator for deep learning applications. In: 2019 IEEE international symposium on circuits and systems (ISCAS)
14. Cheetah: mixed low-precision hardware and software co-design framework for DNNs on the edge. arxiv:1908.02386v1
15. Wan Z, Mibuari E, Yang E-Y, Tambe T (2018) Study of posit numeric in speech recognition neural inference. Harvard University, Cambridge, MA, USA, Technical report CS247r
16. Montero RM, Del Barrio AA, Botella G (2019) Template-based posit multiplication for training and inferring in neural networks. arXiv:1907.04091v1
17. Zhou B, Wang G, Jie G, Liu Q, Wang Z (2021) A high-speed floating-point multiply-accumulator based on FPGAs. IEEE Trans Very Large Scale Integr (VLSI) Syst 29(10):1782–1789. https://doi.org/10.1109/TVLSI.2021.3105268
18. Nambi S, Ullah S, Lohana A, Sahoo SS, Merchant F, Kumar A (2021) ExPAN(N)D: exploring posits for efficient artificial neural network design in FPGA-based systems. IEEE Access 9
19. Murillo R, Del Barrio AA, Botella G (2020) Deep PeNSieve: a deep learning framework based on the posit number system. Digital Signal Process 102:102762
20. Langroudi HF, Karia V, Gustafson JL, Kudithipudi D (2020) Adaptive posit: parameter aware numerical format for deep learning inference on the edge. In: 2020 IEEE/CVF conference on computer vision and pattern recognition workshops (CVPRW)
21. Street WN, Wolberg WH, Mangasarian OL (1993) Nuclear feature extraction for breast tumor diagnosis. Biomed Image Process Biomed Vis Int Soc Opt Photonics 1905:861–871

# Multinet Global Routing Algorithm for On-Chip Optical Interconnects to Minimize Optical Signal Loss

**Anik Saha, Subhajit Chatterjee, Supriyo Srimani, Tuhina Samanta, and Hafizur Rahaman**

**Abstract** Advancement in the field of integrated optics in recent years increases the need for the development of automated routing techniques for on-chip optical interconnect to reduce the design cycle. The integrated optical system is a planar technology that lacks the intrinsic signal restoration capabilities of static CMOS. Therefore, minimization of signal degradation in terms of waveguide crossings and optical waveguide curvature is considered to be the most critical constraint for on-chip optical interconnect. In this work, an automated, planar global routing approach for integrated optical waveguides is presented. We propose a heuristic multinet multi-pin global routing algorithm for on-chip optical interconnect. This is achieved by creating a minimal area bounded zone (MABZ) to cluster all pins of a net, which will be followed by spine Manhattan routing (SMR) to carry out rectilinear routing along with horizontal and vertical directions. The proposed techniques implemented on ISPD 2008 benchmark circuits have shown promising results.

**Keywords** Optical Interconnect · Multinet Global Routing Algorithm · Minimal Area Bounded Zone · Spine Manhattan Routing · Waveguide Loss Minimization

## 1 Introduction

The design necessities of power, delay, noise, and bandwidth due to metal interconnect have become more stringent as CMOS technology has been scaled aggressively for the past decade. Today, most of the ultra-fast processors need high-speed inter-

A. Saha · S. Chatterjee · S. Srimani · T. Samanta · H. Rahaman
School of VLSI Technology, IIEST, Shibpur, Howrah, India
e-mail: t_samanta@it.iiests.ac.in

H. Rahaman
e-mail: rahaman_h@it.iiests.ac.in

S. Srimani (✉)
Department of Electronics and Communication Engineering, DSCSITSC, Kolkata, India
e-mail: supriyo.srimani@dsec.ac.in



C. Giri et al. (eds.), *Emerging Electronic Devices, Circuits and Systems*, Lecture Notes in Electrical Engineering 1004,
https://doi.org/10.1007/978-981-99-0055-8_13

connect that allow individual processors fast access to memory, logical cores, and external I/O devices. It has become extremely difficult for conventional copper interconnect to satisfy these design requirements at a frequency of approximately above 1 GHz due to increased crosstalk and frequency-dependent signal attenuation [1, 2]. One promising candidate to solve the issue of high-frequency and high-performance computing is on-chip optical interconnect-based routing to connect multiple chip modules. To be precise, optical interconnect is a way of communication through optical waveguides placed on silicon wafers [3, 4]. It is considered a potential solution to overcome the performance constraints since fiber optic cables have a negligible frequency-dependent loss, low crosstalk interference, low latency, and high bandwidth. Recent progress in silicon photonics has provided more flexibility to the fabrication laboratories to apply existing infrastructure and fabrication processes for fabricating electro-optical devices [5, 6].

However, despite all these advantages, radiative losses that cause signal attenuation occurs due to a finite radius of curvature. This type of loss, commonly referred to as *waveguide bending loss* [7, 8], is one of the major concerns for optical interconnect routing as it may lead to signal loss in the order of 0.001–0.35 dB per bend. Apart from that, it is evident that multiple waveguides can cross during routing, which may lead to signal losses in the order of 0.1–0.4 dB per crossing [9, 10]. So, signal loss minimization in terms of waveguide crossings and waveguide bends are the most critical constraints for on-chip optical interconnect routing.

Generally, the routing problem is solved by the use of a two-stage approach—*(a) Global routing* and *(b) Detailed routing*. Objectives in the global routing phase include wire-length minimization, signal integrity, congestion-aware routing. The next stage is detailed routing, where actual net tracks and via's for nets are assigned, followed by the paths guided by global routing. Finding a realistic congestion-aware global routing of a two-pin net is an *NP-Hard* problem. Thus, several heuristic approaches have emerged over the past decade to obtain approximate solutions for routing. Conventional metal routing primarily focuses on wire-length minimization to reduce delay and meet the timing and power constraints. However, high-performance optical interconnect does not emphasize fiber-length minimization as data are transferred in the form of an electromagnetic wave. Therefore, waveguide crossings and fiber bends are considered to be the most critical constraints of fiber optic communication.

In this work, a methodology to obtain multinet global routing is proposed for on-chip optical interconnects such that overall waveguide bend loss and crossing loss are minimized. The remainder of this work is structured as follows: Sect. 2 discusses a few similar works. Section 3 represents the formulation of the problem. Section 4 describes the proposed method of constructing MABZ and SMR for each type of net, thereby computing the waveguide bend and crosstalk loss. Section 5 describes the empirical observations. Section 6 concludes the work.

# 2 Literature Survey

An in depth survey on global rectilinear Steiner tree routing followed by existing optical interconnect routing is provided in this section.

## 2.1 *Global Routing*

Generally, two kinds of methodologies solve global routing problems: (i) sequential routing and (ii) concurrent routing.

In a sequential routing problem, the *Maze Runner* heuristic sorts the nets based on specific parameters. In the presence of wire blockages, the maze runner algorithm provides an effective routing solution connecting the shortest path between the linked pins. However, sequential routing inherently can consider one net at a time, and the nets must be ordered in some specific way. Thus, an extension to multinet domain necessitates the consideration of several nets at the same time and in a particular order, i.e., based on their importance, bounding-box areas, or numbers of terminals.

In concurrent routing, integer programming approaches are often utilized, attempting to route all nets concurrently. Generally, the problem is modeled as a *Zero–One Integer Linear Programming* (0–1 ILP) problem [11] where one tree is selected to route each net. An effective global router based on progressive ILP and adaptive maze routing is studied in [12].

In the last few years, various researchers optimized a single net without including wire congestion issues. When the edge cost is defined according to congestion, the minimum Steiner tree algorithm proved to be an effective solution. An effective algorithm based on particle swarm optimization has been applied to construct a multilayer obstacle-avoiding X-architecture Steiner minimal tree in [13]. The timing issue of global routing in the VLSI design cycle has been addressed by [14]. In contrast to wire-length reduction, the author in [15] attempts to solve the challenge of adding RC delay restrictions into global routing.

## 2.2 *Optical Interconnect Routing*

One of the primary area of this current investigations is directed toward architectural explorations for photonic interconnection networks in multi-core processor system [16, 17]. There has been literature's at the logic or functional design level over the use of optical components as building blocks, coupled by waveguides, to design optical computing systems [18, 19]. At the physical level, [20] illustrates a full-custom layout of photonic structures using Cadence Design Systems Virtuoso, a commercial CMOS-based layout tool. During physical synthesis, automated design strategies such as automated floor planning, placement, and waveguide routing while optimizing for physical characteristics like bend loss, insertion loss, and crossing loss issues have yet to be investigated.

A formal system-level approach to analyze the crosstalk noise and signal-to-noise (SNR) in arbitrary fat-tree-based optical network on-chip (ONoC) has been studied in [21]. Recently, a detailed study on coherent and incoherent crosstalk noise in inter-chip/intra-chip optical interconnection networks has been presented in [22]. A crossing-aware channel routing algorithm has been proposed in [23] where crossing minimization was considered as the primary metric for signal degradation.

# 3 Problem Formulation

The implementation of global routing using optical interconnect is mainly hindered due to the following two phenomena, resulting in significant signal degradation, i.e., routing performance.

## 3.1 Waveguide Curvature Loss

Radiative optical signal losses occur when a fiber bends with a limited radius of curvature [24, 25]. When the bend radius is reduced to a value less than a specific critical radius of curvature, the loss becomes more pronounced. This critical radius of curvature $R_c$ for multi-mode fiber can be approximated as [26]:

$$R_c = \frac{3n_1^2\lambda}{4\pi\left[n_1^2 - n_2^2\right]^{3/2}} \tag{1}$$

where $\lambda$ is the transmission wavelength, $n_1$ is the core refractive index, $n_2$ is the cladding refractive index. Optical waveguides can be subjected to two types of bends: (a) macroscopic bend and (b) random microscopic bends.

*Micro-bending* (bends too small to observe) occur when surface pressure causes bending of the fiber core at the core-cladding interface. Figure 1a depicts how surface generated micro-bends lead the transmitted mode to surpass the critical angle at the fiber micro-bend. *Macro-bending* (fiber bends large enough to observe) losses (shown in Fig. 1b) occur when fibers are physically bent beyond the point where the critical angle is surpassed. As illustrated in Fig. 2a, a bend of radius $R$ in a step-index single-mode fiber with a core cross-sectional area of $w \times b$ and refractive index $n_1$ is wrapped by a cladding of refractive index $n_2$.

As illustrated in Fig. 2b, the waveguide bend loss when the bend length $L$ is high may be spatially separated into transition losses from regions $I$ and $III$, and a pure bend loss from area $II$. For the fundamental mode, the ratio of the output power to the input power, i.e., $P_o/P_i$ can be expressed as:

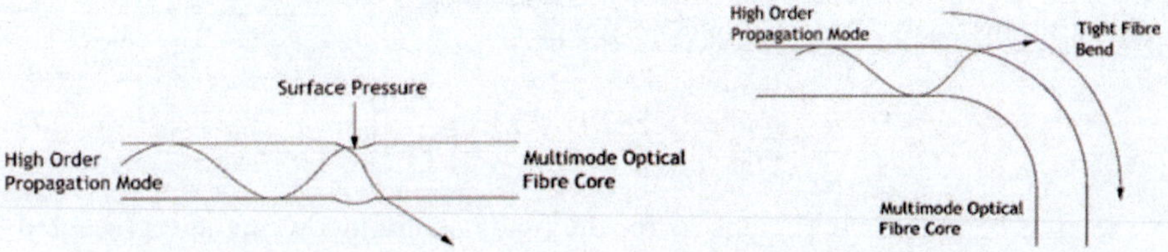


(a) Example of Microscopic Curvature Loss (b) Example of Macroscopic Curvature Loss

**Fig. 1** Curvature loss

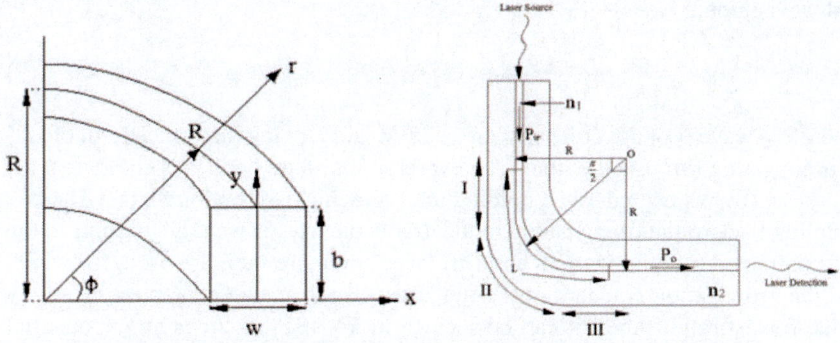


(a) Structure of Planar Rectangular Curved Waveguide (b) 90° Curved Rectangular Waveguide

**Fig. 2** Rectangular curved waveguide

$$\frac{P_o}{P_i} = \left.\frac{P_o}{P_i}\right|_{I} \times \left.\frac{P_o}{P_i}\right|_{II} \times \left.\frac{P_o}{P_i}\right|_{III} \tag{2}$$

where,

$$P_i = P_i|_I, \quad P_o|_I = P_i|_{II}, \quad P_o|_{II} = P_i|_{III} \tag{3}$$

and

$$P_o|_{III} = P_o \tag{4}$$

From symmetry, the transition components can be considered to be equal. So,

$$\left.\frac{P_o}{P_i}\right|_{I} = \left.\frac{P_o}{P_i}\right|_{III} \tag{5}$$

Again, pure bend loss component can be represented by [24]

$$\left.\frac{P_o}{P_i}\right|_{II} = e^{-2\alpha L}, \quad \left.\frac{P_o}{P_i}\right|_{II} = e^{-2\alpha R\phi} \tag{6}$$

Substituting Eqs. (5) and (6) into Eq. (2) gives:

$$\ln\left[\frac{P_o}{P_i}\right] = -2\alpha R\phi + 2\ln\left[\frac{P_o}{P_i}\right]\Big|_I \tag{7}$$

The pure bend loss coefficient $2\alpha$ for the fundamental mode can be represented as [24]:

$$2\alpha = Ke^{-\frac{2}{3}\frac{\gamma^3}{\beta^2}R} \tag{8}$$

where $K$, $\gamma$, $\beta$ are some constants whose value depends on physical parameters of optical waveguide.

$$\gamma = \left[\beta^2 - n_{2^2}k^2\right]^{1/2}, \quad \beta = n_2 k \tag{9}$$

where $k$ = wave propagation constant $= 2\pi/\lambda$ and $\lambda$ being the wavelength of EM wave propagating through the fiber. The expression of pure bend loss coefficient $2\alpha$ given by Eq. (8) has been derived holding the following assumption that (i) There is no coupling back to the fundamental mode; (ii) as the fiber is weakly guiding, so the loss is unaffected by polarization, and (iii) large curvature radii are used, for which $\beta$, and the propagation constant of the fundamental mode, can be approximated by its value for a straight fiber. It can be seen from Eq. (8) that $2\alpha$ is an exponential function of $\frac{\gamma^3}{\beta^2}R$ which on substituting for $\gamma$ and $\beta$ is found to be proportional to $2\pi b^{\frac{3}{2}}\left(\frac{R}{\lambda}\right)$ where $b$ is a dimensionless physical quantity and is basically the ratio of the integrated EM field inside the core to the total integrated field. Thus, we see that pure bend loss coefficient $2\alpha$ will be an exponential function that depends strongly on transmission wavelength as well as on the bend radius. By eliminating the transition loss, the pure bend loss component can be written as

$$\frac{P_o}{P_i} = e^{-2\alpha R\phi} \tag{10}$$

In two-dimensional rectilinear Manhattan routing, only, 90° bends are possible. For simplicity, we considered the radius of curvature to be 10 μm, and the wavelength of the light used to be 1152 nm. Bend loss of −0.378 dB is computed for each waveguide bend using Eq. (10) as a function of the radius of curvature and wavelength.

### 3.2 Waveguide Crossings

Current silicon-on-insulator (SOI) photonic circuits are usually constrained to a single layer, unlike traditional CMOS technology, which allows for flexible routing due to many-layered interconnects. Therefore, routing of more complex photonic circuits

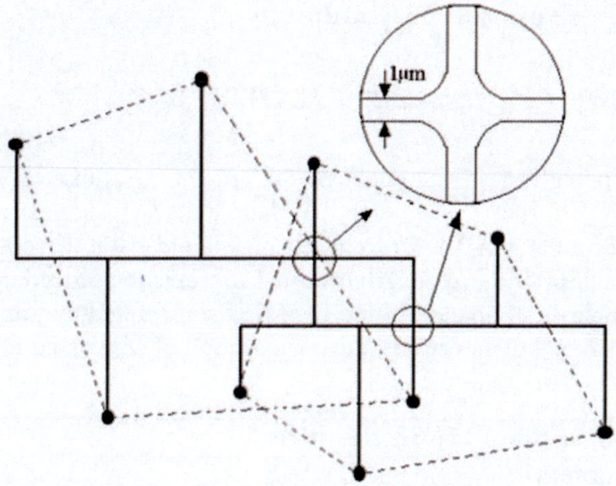


**Fig. 3** Waveguide crossing

is difficult without low-waveguide crossings and low loss. When a direct waveguide crossing occurs between two fibers, lateral confinement in the photonic waveguide is lost at the crossing, allowing light diffraction to occur more easily, resulting in considerable signal loss. Even if the crossing is many micrometers long, these direct waveguide crossings represent just a tiny disturbance of the straight waveguide in low-contrast waveguides, and the beam is scarcely diffracted. This isn't the case with high-contrast systems like SOI, which include wide-angle spatial components in the guided mode of the photonic wires. Calculations and experiments show losses between 0.1–0.4 dB per crossing for silicon waveguide [9, 10] of 1 μm width and for a transmission wavelength range of 1520–1580 nm. Figure 3 represents waveguide crossing between multiple nets.

The above two issues proved to be the most critical constraints for on-chip optical interconnect routing. Since there hasn't been much work on global routing with optical waveguides, we set out to answer the following question in this work:

*Given a set of pins of different nets, is it possible to obtain a global Manhattan routing of all these nets using optical waveguides as interconnect, such that the total number of waveguide crossings and bends are reduced, leading to a considerable reduction in signal losses?*

In the multinet multi-pin global routing problem, we consider a number $\kappa$, say, of multi-pin nets $\Pi_i$, $i = 1$ to $\kappa$ where each net consisting of a set of fixed terminals or pins $\{p_0, p_1, ..., p_n\}$. For a given set of such multi-pin net, the primary objective of this work is to construct a minimal area bounding zone followed by spine structure Manhattan routing to achieve (i) minimal bend count and (ii) a minimal number of waveguide crossings. It may consequently lead to a significant reduction in signal degradation due to waveguide bends and crosstalk. It is also assumed that the signal going from one source pin to the sink pin of any net has the same wavelength.

# 4 Proposed Methodology

## 4.1 Construction of MABZ

For a given a set of points $S$={$p_0$, $p_1$, ....., $p_n$} of each $\Pi_i$ multi-pin net, we attempt to construct a zone, such that there exits $\Omega \subseteq S$, where $\Omega$ is a subset of points that form the MABZ. Every point in $S$ is either on the boundary of the zone or in the interior of the zone. Algorithm 1 presents the procedure to construct MABZ where the overall bounded zone $\Omega$ of $S$ is constructed by joining $\Omega_{\text{min}}$ and $\Omega_{\text{max}}$ together. Figure 4 demonstrates the construction of convex hull.

---

**Algorithm 1** MABZ algorithm

---

**procedure** MABZ($S = \{p_0, p_1, ..., p_n\}$)<br>
  Sort $S$ by increasing the $x$-coordinate first, then the $y$-coordinate.<br>
  Let $p[]$ be an $n$-point sorted array.<br>
  $p_{minmin}$→index of $p$ with min $x$ first and min $y$ second.<br>
  $p_{minmax}$→index of $p$ with min $x$ first and max $y$ second.<br>
  $p_{maxmin}$→index of $p$ with max $x$ first and min $y$ second.<br>
  $p_{maxmax}$→index of $p$ with max $x$ first and max $y$ second.<br>
  Computation of the lower vertex chain stack as given:<br>
  Let $L_{min}$ be the lower line joining $p_{minmin}$ with $p_{maxmin}$.<br>
   Push $p_{minmin}$ onto the stack.<br>
   **for** $i \leftarrow p_{minmax}$+1 to $p_{maxmin}$-1 **do**<br>
    **if** $p[i]$ is above or on $L_{min}$ **then**<br>
     Ignore it and continue.<br>
    **while** there are at least 2 points on the stack **do**<br>
     $p_{t_1} \leftarrow$ the top point on the stack.<br>
     $p_{t_2} \leftarrow$ the second point on the stack.<br>
     Use *isLeft()* comparisons.<br>
     **if** $p[i]$ is strictly left of the line from $p_{t_1}$ to $p_{t_2}$ **then**<br>
      break;<br>
     Pop the top point $p_{t_1}$ off the stack.<br>
    Push $p[i]$ onto the stack.<br>
   Push $p_{maxmin}$ onto the stack.<br>
   Similarly, compute the upper vertex chain stack.<br>
   $\Omega \leftarrow$ the join of the lower and upper vertex chain stack.<br>
   **return** $\Omega$ ▷ $\Omega = \{x, y\}_{MABZ}$

---

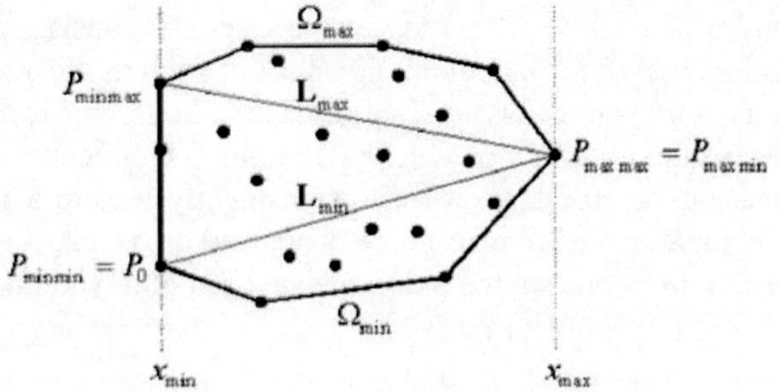


**Fig. 4** Construction of MABZ

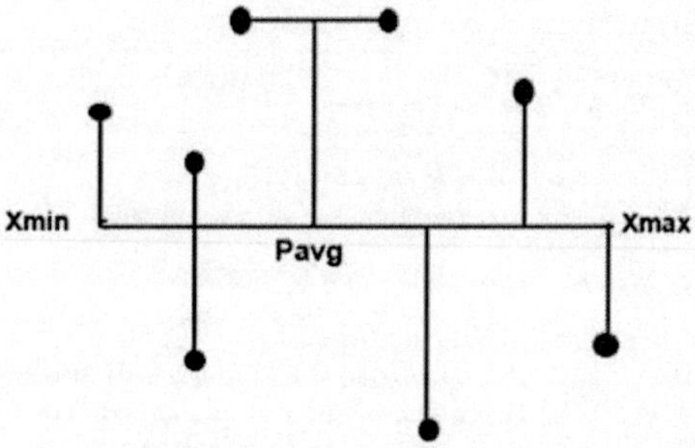


**Fig. 5** Spine manhattan routing

## 4.2 Construction of SMR and Computation of Waveguide Bend Loss

Before evaluating waveguide bend count, it is necessary to construct Manhattan routing. In this work, our primary objective is to minimize bending loss by simply reducing the fiber bend count of each net such that a transmitted optical signal encounters minimal bend from its source terminal to the destination pin. Steiner tree-based routing is very popular for metal interconnects in present-day integrated circuits where wire-length minimization is the major concern in system performance. But, in optical interconnect-based high-performance computing systems, polymer fiber length does not affect the system performance [27], so we propose a unique method to construct Manhattan routing for each $\Pi_j$ nets, $j = 1$ to $\kappa$ within the corresponding MABZ. We name it as *Spine Manhattan Routing*. Figure 5 shows SMR-based routing. Algorithm 2 describes the construction of SMR and evaluation of waveguide bend count.

## 4.3 Computation of Waveguide Crossings

Once the construction of *MABZ* is complete for $\prod_i$ nets, where $i = 1$ to $\kappa$, we figure out the number of waveguide crossings between $\prod_i$ nets after SMR stage. If $(MABZ)_i$ and $(MABZ)_j$ represent the corresponding zones for *i*th and *j*th net; then, for waveguide crossing, the following two scenarios have been analyzed in this work:

**Internet crossing**: If $(MABZ)_i$ and $(MABZ)_j$ overlap, then there exist a probability of waveguide crossing between *i*th and *j*th net. So, our objective is to locate the intersection between each net with respect to the remaining nets recursively, followed by the computation of the number of crossings within the overlapping zone. This type of crossing loss occurs due to wavelength mismatch between multiple nets and also on the variation of the width of the optical waveguide at the crossing region.

**Intranet crossing**: After the *SMR* stage, there exits crossing between the pins of a particular net when the projection of multiple pins is at the same point on the spine.

**Algorithm 2** Waveguide Bend Loss

```
procedure SMR- BEND(S = {p0, p1, ..., pn})
  for ∏i nets i ← 1 to κ do
    pavg_i ← zoneAvg(Ωi);
    Let avgLine → line through pavg_i
    avgLine extends from x_min to x_max of Ωi
    ψ ← 0;
    upList[p.x, p.y] ← divideNet(S);
    lowList[p.x, p.y] ← divideNet(S);
    for elements in upList[] do
      if p.xi ∈ upList[i] = p.xj ∈ upList[i] then
        if p.yi ∈ upList[i] ≠ p.yj ∈ upList[i] then
          Project {x, y}max over avgLine ;
          Creation of SMR;
          ψ ++;
      if p.yi ∈ upList[i] = p.yj ∈ upList[i] then
        if p.xi ∈ upList[i] ≠ p.xj ∈ upList[i] then
          Connect {x, y}min with {x, y}max;
          Project {x, y}min over avgLine;
          Creation of SMR;
          ψ ← ψ + 2;
    for elements in lowList[] do
      Apply similar logic to create SMR and bend count;
  return ψ                                        ▷ ψ ← Total Bend Count
```

**Table 1** Types of operator to minimize waveguide crossings

| Operator | Activity | Shape of interconnect | Penalty |
|---|---|---|---|
| $\phi_1$ | Flipping | 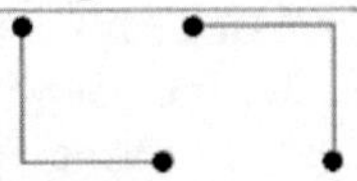 | Losses incurred will not increase |
| $\phi_2$ | Sliding | 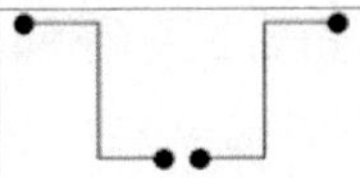 | Increase in unit bend loss in each operation |

This type of crossing loss occurs due to the variation in the width of the fiber near the crossing point. The intra-net crossing is considered to be somewhat less than the former one.

For convenience, in this work, we consider the crossing loss to be 0.4 dB for both inter- and intra-waveguide crossings. The process of evaluating waveguide crossings arising from SMR of multiple nets has been explained in Algorithm 3.

**Crossing Minimization Scheme** Once the SMR is constructed for $\Pi_j$ nets, $j = 1$ to $\kappa$, the pairs of SMR with intersecting edges are identified. In this work, we propose a pair of operator $\phi_1$ and $\phi_2$ which, when applied to these SMR, leads to a possible transformation of different sets of SMR, such that the number of intersections is reduced and new SMR is created. The function of the operators that we implemented is summarized in Table 1.

**Algorithm 3** To evaluate Waveguide Crossings

```
procedure CROSS(S = {p0, p1, ..., pn})
    χ ← 0;
    for ∏i nets, i ← 1 to κ do
        Let pavg_i ← zoneAvg(Ω);
        for ∏j nets, j ← 1 to κ do
            if ∏i ≠ ∏j then
                pavg_j ← zoneAvg(Ω);
                if pavg_i.y < pavg_j.y then
                    upList[p.x, p.y] ∈ i^th net ← divideNet(S);
                    if avgLine of j^th net covers upList[].x then
                        if upList[].y ∈ i^th > pavg_j.y then
                            cros[p.x, p.y] ← {upList[].x, pavg_j.y}
                            χ++;
                else
                    lowList[p.x, p.y] ∈ i^th net ← divideNet(S);
                    if avgLine of j^th net covers lowList[].x then
                        if lowList[].y ∈ i^th < pavg_j.y then
                            cros[p.x, p.y] ← {lowList[].x, pavg_j.y}
                            χ++;
            j++;
        i++;
    return χ                                    ▷ χ ← Total Waveguide crossings
```

**Algorithm 4** Crossing Minimization Scheme

```
procedure FLIP(E_SMR, cros[])
    for each edge e ∈ E_SMR do
        if (p.x, p.y) ∈ cros[] exits in e then
            if isMovable(e) ⇒ true then
                createMovable(e);
        for each movable set do
            if noOverlap(E_s) with adjacent edge then
                Flip the edge using φ1 operator;
                Slide the edge using φ2 operator;
                Create new route;
    return NewSMR;
```

The main motivation to build these operators is to maximize the flexibility of the SMR without changing the topology. Since the terminals of a net are fixed, crossing multiple edges can be minimized by flipping the edge of SMR instance whenever intersecting edges are identified. We define a set of movable segments and adjacent edges to $m, E_m$, as a movable set $(m, E_m)$. Flexible candidate is an edge that has the potential to become flexible by moving the movable segment. If $cros[]$ array contains all the crossing coordinate obtained as a result of routing of $\prod_i$ nets, $i = 1$ to $\kappa$ and $E_{SMR}$ represents the edge set of SMR of a multi-pin net then function *isMovable(e)* checks whether the edge *e* can be flipped or not. Function $createMovable(e)$ generates the movable set associated with edge $e$. Algorithm 4 demonstrates the flipping operation.

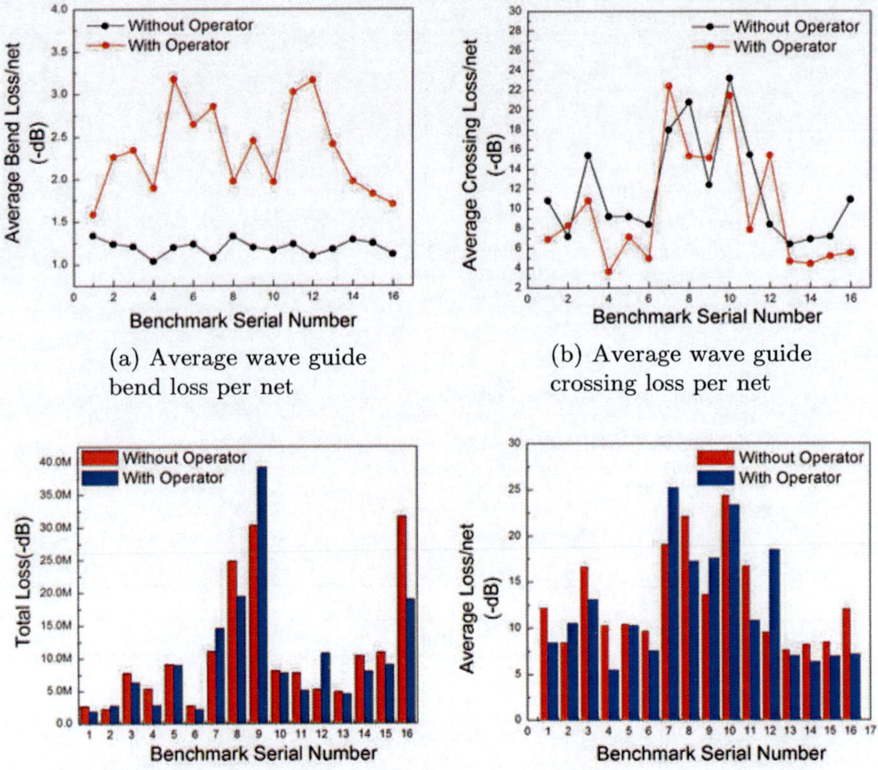


(a) Average wave guide bend loss per net

(b) Average wave guide crossing loss per net

(c) Total loss

(d) Average loss per net

**Fig. 6** Estimation of loss for ISPD 2008 global routing benchmarks

# 5 Results and Discussion

In this work, our proposed technique is applied in C++ on a system running at 2.8 GHz Intel Core-i3 processor on a Linux environment with 2 GB RAM. The proposed algorithm is validated for benchmark circuit available at [28]. Routing solutions have been evaluated, taking waveguide bending and crossings as a metric. Table 2 summarizes the outcomes of an algorithm, which indicate that the proposed method can handle a large number of multi-pin nets without compromising CPU execution time. Analysis of results summarized in Table 2 reveals that our global optical router provides an efficient routing solution with minimal optical signal degradation in terms of waveguide bends without applying flipping and sliding operators during SMR. However, on applying crossing minimization operators, a significant reduction of crossing loss has been achieved on compromising bend loss. This resulted in a trade-off between the number of fiber bends and crossings for routing.

**Table 2** Benchmark simulation result summary

| Benchmark serial # | Benchmark name | # of nets | Waveguide bend count | | Average bend loss/net (−dB) | | Crossing count | | Average crossing loss/net (−dB) | |
|---|---|---|---|---|---|---|---|---|---|---|
| | | | Without operators | With operators | Without operators | With operators | Without operators | With operators | Without operators | With operators |
| 1 | adaptec1 | 219794 | 779781 | 923134 | 1.34 | 1.58 | 5934438 | 3772814 | 10.80 | 6.86 |
| 2 | adaptec2 | 260159 | 856149 | 1560954 | 1.24 | 2.26 | 4682862 | 5372441 | 7.20 | 8.26 |
| 3 | adaptec3 | 466295 | 1502534 | 3264052 | 1.21 | 2.34 | 17983275 | 12564757 | 15.42 | 10.77 |
| 4 | adaptec4 | 515304 | 1430691 | 2576520 | 1.04 | 1.89 | 11851992 | 4628355 | 9.20 | 3.59 |
| 5 | adaptec5 | 867441 | 2764890 | 7286504 | 1.20 | 3.17 | 19951143 | 15422731 | 9.20 | 7.11 |
| 6 | bigblue1 | 282974 | 929604 | 1980818 | 1.24 | 2.64 | 5942454 | 3485776 | 8.40 | 4.92 |
| 7 | bigblue2 | 576816 | 1651899 | 4360617 | 1.08 | 2.85 | 25956720 | 32269407 | 18 | 22.37 |
| 8 | bigblue3 | 1122340 | 3950636 | 5869838 | 1.33 | 1.97 | 58361680 | 42948820 | 20.80 | 15.30 |
| 9 | bigblue4 | 2228903 | 7087911 | 14487853 | 1.20 | 2.45 | 69092583 | 84278041 | 12.39 | 15.12 |
| 10 | newblue1 | 331663 | 1028155 | 1724647 | 1.17 | 1.96 | 19236454 | 17755283 | 23.20 | 21.41 |
| 11 | newblue2 | 463213 | 1528602 | 3705704 | 1.24 | 3.02 | 17918159 | 9052655 | 15.47 | 7.81 |
| 12 | newblue3 | 551667 | 1450806 | 6068337 | 1.10 | 3.15 | 11585007 | 21224824 | 8.40 | 15.38 |
| 13 | newblue4 | 636195 | 1997065 | 4071648 | 1.18 | 2.41 | 10179126 | 7365513 | 6.40 | 4.63 |
| 14 | newblue5 | 1257555 | 4297065 | 6557041 | 1.29 | 1.97 | 21755701 | 13821247 | 6.91 | 4.39 |
| 15 | newblue6 | 1286452 | 4275900 | 6207118 | 1.25 | 1.82 | 23115864 | 16588351 | 7.18 | 5.15 |
| 16 | newblue7 | 2635625 | 7847413 | 11860312 | 1.12 | 1.70 | 71768068 | 36135776 | 10.89 | 5.48 |

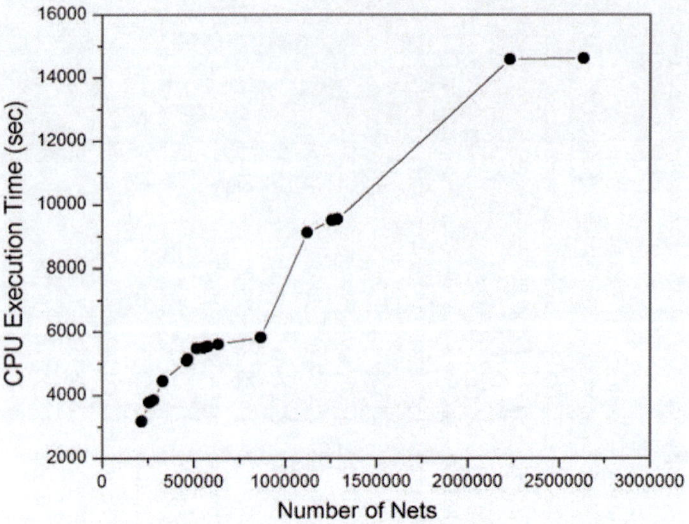


**Fig. 7** CPU execution time for different benchmarks

It is visible from the graphical plot explained in Fig. 6d that even though bend loss increases on the application of crossing minimization operators, the average loss per net for most of the benchmarks has improved significantly on applying operators and produces better results meeting our objective. However, Fig. 6b indicates that flipping and sliding operation could not create an impact on waveguide crossing for some benchmark instances that include −2, 7, 9, and 12. Figure 6a and b present the comparison of average bend loss and crossing loss per net, respectively, for all benchmarks before and after using operators. Comparison between the total signal loss for each benchmark instance before and after using operator is demonstrated in Fig. 6c. The runtime limit for each benchmark is set to 24 h. If a router needs over than 24 h to route a benchmark, it is regarded to have failed in routing. Figure 7 represents the CPU execution time for running ISPD benchmarks, and the maximum time of execution taken among all benchmarks is found to be 14460 sec.

## 6 Conclusion

This work presents a hierarchical partitioning-based global router for on-chip optical interconnect, suitable for Manhattan-based routing grid, that provides flexibility for integrated optics. The presented experimental result reveals that optical interconnect routing in photonic integrated circuits may be automated using global routing approaches, optimizing for waveguide crossing constraints and bend loss as a primary objective. In our work, MABZ and SMR techniques combined provide an excellent method for automated optical waveguide routing with a significant reduction in optical signal degradation due to fiber bends and crossings. We further attempt to address the waveguide crossing issues by the application of two operators $\phi_1$ and $\phi_2$, which in turn results in a significant reduction of waveguide crossing loss. In addition, the proposed work can be extended by: (i) further minimization of waveguide crossings and curvature loss by the use of non-Manhattan 'X' routing (ii) the congestion and crosstalk-aware global routing for optical systems can also be explored.

## References

1. Kodi A, Louri A (2007) Performance adaptive power-aware reconfigurable optical interconnects for high-performance computing (hpc) systems. In: SC'07: Proceedings of the 2007 ACM/IEEE conference on Supercomputing. IEEE, pp 1–12
2. Bashir J, Peter E, Sarangi SR (2019) Bigbus: a scalable optical interconnect. ACM J Emerg Technol Comput Syst (JETC) 15(1):1–24
3. Jalali B, Fathpour S (2006) Silicon photonics. J Lightwave Technol 24(12):4600–4615
4. Zhang Y, Xiao X, Zhang K, Li S, Samanta A, Zhang Y, Shang K, Proietti R, Okamoto K, Yoo SB (2019) Foundry-enabled scalable all-to-all optical interconnects using silicon nitride arrayed waveguide router interposers and silicon photonic transceivers. IEEE J Sel Top Quantum Electron 25(5):1–9
5. Peach M (2012) Silicon photonics forges ahead. Optics org
6. Yang S, Yang L, Luo F, You B, Ni Y, Chen D (2019) Multi-node all-optical interconnect network routing for data-center parallel computers. IEEE Photonics J 11(2):1–8
7. Schermer RT, Cole JH (2007) Improved bend loss formula verified for optical fiber by simulation and experiment. IEEE J Quantum Electron 43(10):899–909
8. Qian Y, Kim S, Song J, Nordin GP, Jiang J (2006) Compact and low loss silicon-on-insulator rib waveguide 90 bend. Opt Express 14(13):6020–6028
9. Bogaerts W, Dumon P, Van Thourhout D, Baets R (2007) Low-loss, low-cross-talk crossings for silicon-on-insulator nanophotonic waveguides. Opt Lett 32(19):2801–2803
10. Chen H, Poon AW (2006) Low-loss multimode-interference-based crossings for silicon wire waveguides. IEEE Photonics Technol Lett 18(21):2260–2262
11. Sherwani NA (2012) Algorithms for VLSI physical design automation. Springer Science & Business Media
12. Cho M, Pan DZ (2007) Boxrouter: a new global router based on box expansion and progressive ilp. IEEE Trans Comput-Aided Des Integr Circuits Syst 26(12):2130–2143
13. Liu G, Huang X, Guo W, Niu Y, Chen G (2014) Multilayer obstacle-avoiding x-architecture steiner minimal tree construction based on particle swarm optimization. IEEE Trans Cybern 45(5):1003–1016
14. Grewal G, Xu M (2006) An efficient graph-based steiner tree heuristic for the global routing of macro cells. Can J Electr Comput Eng 31(4):211
15. Scheifele R (2016) Rc-aware global routing. In: Proceedings of the 35th international conference on computer-aided design, pp 1–8
16. Ding D, Zhang Y, Huang H, Chen RT, Pan DZ (2009) O-router: an optical routing framework for low power on-chip silicon nano-photonic integration. In: Proceedings of the 46th annual design automation conference, pp 264–269
17. Petracca M, Lee BG, Bergman K, Carloni LP (2008) Design exploration of optical interconnection networks for chip multiprocessors. In: 2008 16th IEEE Symposium on high performance interconnects. IEEE, pp 31–40
18. Condrat C, Kalla P, Blair S (2011) Logic synthesis for integrated optics. In: Proceedings of the 21st edition of the great lakes symposium on Great lakes symposium on VLSI, pp 13–18
19. Kotiyal S, Thapliyal H, Ranganathan N (2012) Mach-zehnder interferometer based design of all optical reversible binary adder. In: 2012 Design, automation and test in Europe conference and exhibition (DATE). IEEE, pp 721–726

20. Orcutt JS, Ram RJ (2010) Photonic device layout within the foundry cmos design environment. IEEE Photonics Technol Lett 22(8):544–546
21. Nikdast M, Xu J, Duong LH, Wu X, Wang Z, Wang X, Wang Z (2014) Fat-tree-based optical interconnection networks under crosstalk noise constraint. IEEE Trans Very Large Scale Integr (VLSI) Syst 23(1): 156–169
22. Duong LH, Wang Z, Nikdast M, Xu J, Yang P, Wang Z, Wang Z, Maeda RK, Li H, Wang X et al. (2016) Coherent and incoherent crosstalk noise analyses in interchip/intrachip optical interconnection networks. IEEE Trans Very Large Scale Integr (VLSI) Syst 24(7):2475–2487
23. Condrat C, Kalla P, Blair S (2014) Crossing-aware channel routing for integrated optics. IEEE Trans Comput-Aided Des Integr Circuits Syst 33(6):814–825
24. Marcuse D (1976) Curvature loss formula for optical fibers. JOSA 66(3):216–220
25. Gloge D (1972) Bending loss in multimode fibers with graded and ungraded core index. Appl Opt 11(11):2506–2513
26. Keiser G (2000) Optical fiber communications, vol. 2. McGraw-Hill New York
27. Haurylau M, Chen G, Chen H, Zhang J, Nelson NA, Albonesi DH, Friedman EG, Fauchet PM (2006) On-chip optical interconnect roadmap: challenges and critical directions. IEEE J Sel Top Quantum Electr 12(6):1699–1705
28. ISPD 2008 Global routing benchmaks. http://www.ispd.cc/contests/08/ispd08rc.html#head-benc

# Performance Enhancement of Dielectric Engineered Doping Less InGaN Tunnel FET for Low Power Analog/Radio Frequency Applications

**Arnab Som and Sanjay Kumar Jana**

**Abstract** This literature represents the design and analysis of III-Nitrite-based Doping-less Tunnel Field Effect Transistors (TFET). On-state current improvement at low supply voltage at the device level is very much essential for low-power circuit-level assessments. Proper bandgap and lower electron-effective mass-based materials are required for the development of high-performance TFET devices. $In_{0.75}Ga_{0.25}N$ is a suitable material for TFET due to its low effective mass and high electron density of states without doping. Homojunction TFETs have been designed based on $In_{0.75}Ga_{0.25}N$ and Silicon as channel material, respectively. The dielectric engineering has been performed. The proposed $In_{0.75}Ga_{0.25}N$ doping less TFET with Heterogate dielectric ($HfO_2/Al_2O_3$) at $HfO_2$ length under the gate ($L_{HfO_2}$) of 20 nm exhibits better immunity to short channel effects. The analog/RF performances have been investigated and compared with the other devices. The proposed device provides an average sub-threshold slope of 16 mV/Dec, which is 67.3% lower than silicon-based doping-less TFET. At drain bias of 0.3 V, the proposed device offers the maximum cut-off frequency ($f_T$) of 45.7 GHz and the maximum gain bandwidth product (GBP) of 21.8 GHz, which is much larger than other reported III-V material-based TFET devices. The improved device can be regarded as a suitable device for radio frequency applications with lower power dissipation.

**Keywords** $In_{0.75}Ga_{0.25}N$ · Tunnel FET (TFET) · Dielectric engineering · Hetero-gate dielectric

A. Som (✉) · S. K. Jana
National Institute of Technology, Sikkim, Ravangla 737139, India
e-mail: phec190004@nitsikkim.ac.in

S. K. Jana
e-mail: skjnit@nitsikkim.ac.in



C. Giri et al. (eds.), *Emerging Electronic Devices, Circuits and Systems*,
Lecture Notes in Electrical Engineering 1004,
https://doi.org/10.1007/978-981-99-0055-8_14

# 1 Introduction

Nowaday, the integrated circuit technology is scaled down to nano-size to decrease the power dissipation and fabrication cost. In the nanoscale regime, MOSFET suffers from various obstacles like several short channel effects, larger leakage current, lower value of Ion/Ioff, limited sub-threshold slope [1–4]. To overcome these difficulties an alternative device, i.e., Tunnel FET (TFET) has been invented which can also continue the moor's law. Due to several advantages over MOSFETs like lower subthreshold slope value, lesser leakage current, high Ion/Ioff current ratio, lower threshold voltage, TFETs are getting enormous attention [5–9]. Instead of this, the main drawback with silicon-based TFET is lower ON-state current. To minimize the drawback, several researchers have reported many solutions such as gate dielectric engineering, gate material work-function engineering, and bandgap engineering [10–16]. Recently, the doping-less or charge plasma-based structure of TFET becomes very popular. For TFET devices, the source region and the drain region can be realized by using appropriate metals with specific work functions [17–20]. The Silicon-based doping less TFET still provides a lower ON current ($I_{ON}$) due to the reason of the larger electron-effective mass of silicon material and smaller lateral electric field across the source-channel junction. Therefore, to minimize that above-mentioned problem, a gate engineering concept is to form an abrupt $n+$ pocket without dopant diffusion, which can enhance the $I_{ON}$ of the device [21, 22]. $In_xGa_{1-x}N$ materials have shown better performance in physically doped TFET devices than silicon-based TFETs [23–26]. For Dopingless TFET, $In_xGa_{1-x}N$ material with an "In" fraction of 0.75 has shown superior ON-state performance [14]. The entire or full compositional range of indium can be fabricated by the plasma-assisted molecular beam epitaxy [27, 28]. $In_{0.75}Ga_{0.25}N$ can be realized and can become the best suitable material for high-performance doping-less TFET devices. By utilizing the excellent material properties of $In_{0.75}Ga_{0.25}N$, more investigations are required to achieve higher ON state performances at lower supply voltages.

In this work, a dielectric engineered heterodielectric ($HfO_2$ + $Al_2O_3$) $In_{0.75}Ga_{0.25}N$ DL-TFET with a tunnel gate is studied and compared with a silicon-based DL-TFET. The length of gate oxide dielectric ($HfO_2$) above the channel is varied from 0 to 25 nm. The effect of dielectric engineering on analog/RF performances has been investigated. This article is organized as follows. Different device architectures and different simulation input parameters are discussed in Sect. 2. Several results are analyzed in Sect. 3, where dc and RF analyses are mainly investigated. The conclusion part is described in Sect. 4.

## 2 Structures of Devices and the Parameters Used for Simulations

The schematic views of Silicon Doping-Less TFET, $In_{0.75}Ga_{0.25}N$ Doping-Less TFET, and proposed $In_{0.75}Ga_{0.25}N$ Dielectric Engineered Doping-Less TFET ($In_{0.75}Ga_{0.25}N$ DE DL-TFET) are presented at Fig. 1.

The design parameters of devices considered in simulations [14, 29] are given in Table 1.

Platinum (Pt) is chosen as the source metal, and Hafnium (Hf) is chosen as the drain metal. The thickness of the silicon film is considered as 10 nm to induce uniform carrier distribution, which is lesser than the Debye length. Debye length = $((\varepsilon V_T)/(q N))^{1/2}$, where $N$ and $\varepsilon$ denote the carrier concentration and the dielectric constant of body material respectively, $V_T$ is considered as thermal voltage [19]. To reduce the ambipolar current, the gap between the gate and drain (Lgd) and the gap between source and gate (Lgs) have been chosen as 15 nm and 2 nm, respectively [14]. The

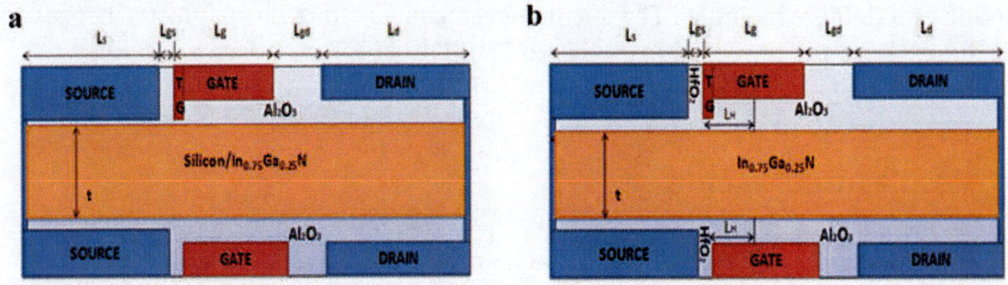


**Fig. 1** Schematic views of **a** Silicon/$In_{0.75}Ga_{0.25}N$ DL-TFET, **b** Proposed $In_{0.75}Ga_{0.25}N$ DE DL-TFET

**Table 1** The device design parameters

| Parameters | Values |
|---|---|
| Body thickness (nm), t | 10 |
| Gate length (nm), Lg | 50 |
| Source length (nm), Ls | 100 |
| Drain length (nm), Ld | 100 |
| Drain/gate space (nm), Lgd | 15 |
| Source/gate space (nm), Lgs | 2 |
| Gate oxide material | $Al_2O_3$ and $HfO_2$ |
| Source work function (eV) | 5.93 |
| Gate material work function (eV) | 4.5 |
| Drain material work function (eV) | 3.9 |
| TG work function (eV) | 3.5 |
| $HfO_2$ length above channel region ($L_H$) (nm) | 0–25 |

**Table 2** The property parameters of materials considered in this work

| Parameters | Silicon | $In_{0.75}Ga_{0.25}N$ |
|---|---|---|
| Band gap | 1.12 | 1.1125 |
| Light hole effective mass | 0.56 | 0.295 |
| Effective mass of electron | 0.26 | 0.1025 |
| Hole mobility ($cm^2$/V.s) | 500 | 20 |
| Electron mobility ($cm^2$/V.s) | 1350 | 1050 |
| Static dielectric constant | 11.7 | 12.75 |

material property parameters of silicon and $In_{0.75}Ga_{0.25}N$ [14, 30, 31] are given in Table 2.

Silvaco Atlas device simulation tool is used here for device simulations [32]. To include the tunneling phenomena accurately, the nonlocal band-To-band tunneling model is included in this simulation work. To capture the effects of carrier recombinations, the Shockley Read Hall and the recombination models (Auger) are included. Two different mobility models, i.e., field-dependent and concentration-dependent mobility models are included. The quantum mechanical effects are neglected because the thickness of the proposed $In_{0.75}Ga_{0.25}N$ DE DL-TFET is chosen as 10 nm.

## 3 Results and Discussions

For comparative analysis, device performances of Silicon-based Doping less TFET, dielectric engineered InGaN ($In_{0.75}Ga_{0.25}N$) DL-TFET and, InGaN ($In_{0.75}Ga_{0.25}N$) DL-TFET has been simulated under the consideration of the same device parameters. Indium fraction is chosen as 0.75 for all InGaN-based devices.

The transfer characteristics of InGaN and silicon-based DL-TFET devices are shown in Fig. 2a, b in both the linear and log scale. The result shows that $I_{ON}$ increases for higher $L_{HfO_2}$. Key short channel effects, average SS($SS_{avg}$), and $I_{ON}/I_{OFF}$ have been investigated for different $L_{HfO_2}$. The simulation result shows that after 20 nm, the $I_{ON}$ and $I_{OFF}$ change. After 20 nm of $L_{HfO_2}$, due to higher $I_{OFF}$, $SS_{avg}$ increases, and $I_{ON}/I_{OFF}$ decreases. For further analysis, we have considered $L_{HfO_2}$ as 20 nm due to its better short channel effect performances. Figure 3a, b present the energy band (EB) diagrams of all devices. The band diagram at the channel region shows change due to the hetero-gate dielectric and the Tunnel Gate (TG) metal-induced $n+$ pocket. The change in the energy band diagram can be credited to the maximum electric field induced by the hetero dielectric near the source-channel interface region as shown in Fig. 3c. The transfer characteristics show that for $V_{ds} = 0.3$ V and $V_{gs} = 0.6$ V, the proposed device exhibits $6.05 \times 10^{-5}$A/μm. The proposed device has a $2.83 \times 10^2$ times improvement in ON-state current ($I_{ON}$) than the silicon-based device. Improved $I_{ON}$ in the proposed device is credited to the hetero-dielectric induced electric field and also credited to the excellent material properties, such as the smaller electron and hole effective mass and the direct-band gap feature in $In_{0.75}Ga_{0.25}N$ material, which directly affects the tunneling probability $P_{\text{tun}}$, at Kane's formula [30].

$$P_{\text{tun}} \sim \frac{E^2 m_r^{1/2}}{E_g^{1/2}} \exp\left(-\frac{C_2 m_r^{1/2} E_g^{3/2}}{E}\right) \tag{1}$$

where the electric field is denoted as $E$, $m_r$ represents the effective mass and $C_2$ is considered as the constant. The improved ON current (3 times) in the proposed InGaN DE DL-TFET in comparison with the single dielectric InGaN-based device is mainly due to hetero-dielectric induced electric field (Fig. 3c), which also increases the tunneling probability. The average SS ($SS_{avg}$) is given by

$$SS_{avg} = (V_t - V_{OFF})/(\log I_{V_t} - \log I_{OFF}) \tag{2}$$

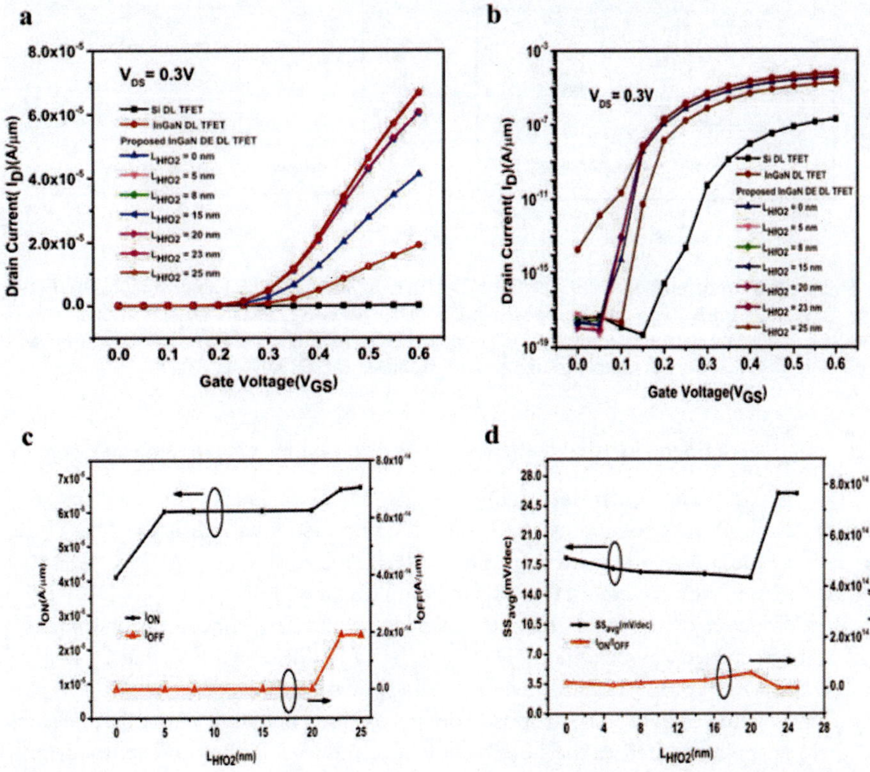


**Fig. 2** **a** Transfer characteristics (linear scale) of proposed InGaN DE DL-TFET, InGaN DL-TFET and Silicon DL-TFET. **b** Transfer characteristics (log scale) of proposed InGaN DE DL-TFET, InGaN DL-TFET and Silicon DL-TFET. **c** $I_{ON}$ and $I_{OFF}$ of Proposed InGaN DE DL-TFET for different $L_{HfO_2}$ (0–25 nm). **d** $SS_{avg}$ and $I_{ON}/I_{OFF}$ of Proposed InGaN DE DL-TFET for different $L_{HfO_2}$ (0–25 nm)

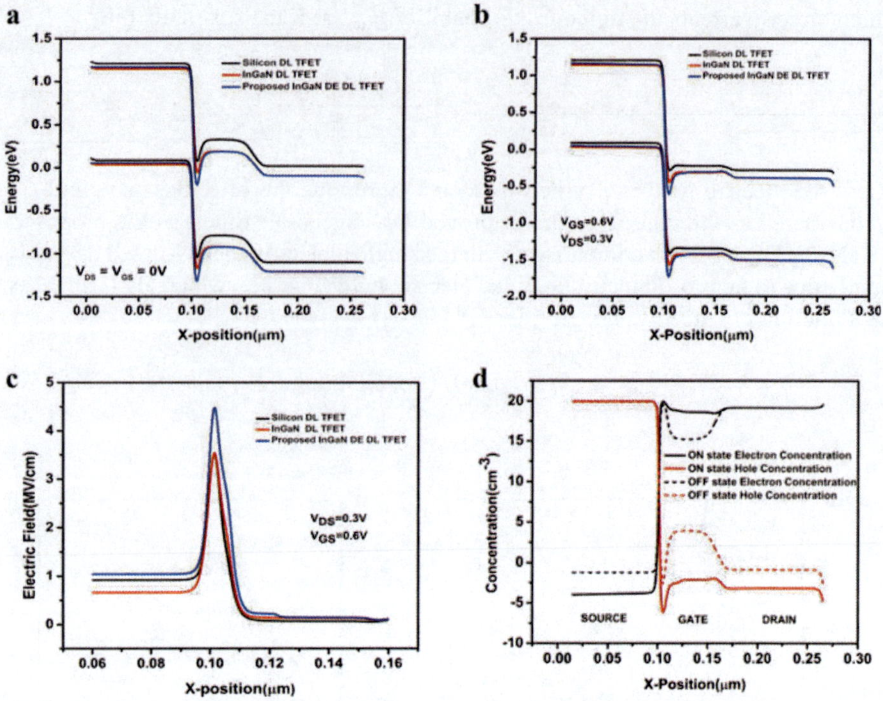


**Fig.3** EB diagram of proposed InGaN DE DL-TFET, InGaN DL-TFET and Silicon DL-TFET under OFF state. **b** EB diagram of proposed InGaN DE DL-TFET, InGaN DL-TFET and Silicon DL-TFET under ON state. **c** Electric field of proposed InGaN DE DL-TFET, InGaN DL-TFET and Silicon DL-TFET. **d** Carrier concentration plot of proposed InGaN DE DL-TFET

where $V_t$ is defined as the voltage when the value of current achieves $1 \times 10^{-7}$ A/μm. The proposed $In_{0.75}Ga_{0.25}N$ DE DL-TFET (16 mV/decade) exhibits a 67.3% improvement in $SS_{avg}$ in comparison with Silicon DL-TFET (49 mV/decade). The proposed $In_{0.75}Ga_{0.25}N$ DE DL-TFET (16 mV/decade) also provides a 26.1% improvement in $SS_{avg}$ in comparison with $In_{0.75}Ga_{0.25}N$ DL-TFET (21.66 mV/decade). Figure 3d presents the electron and hole concentration distributions of the proposed device under ON ($V_{DS} = 0.3$ V and $V_{GS} = 0.6$ V) and OFF ($V_{DS} = 0.0$ V and $V_{GS} = 0.0$ V) conditions. Figure 4a shows the transconductance ($g_m$) plot for different gate voltages. The proposed device exhibits better $g_m$ than other devices and that can be attributed to the hetero-dielectric induced higher ON-state current. The parasitic capacitances like gate-source capacitance ($C_{gs}$), gate-drain capacitance ($C_{gd}$), and gate-gate capacitance ($C_{gs}$) have been investigated. In TFET devices, the band to band tunneling phenomena across the source-channel interface region causes the conduction current. Under ON conditions the thickness of the tunneling barrier reduces and that allows the charges to tunnel across the source and channel interface. For the above reason, $I_{ON}$ is lower in TFET in comparison with the MOSFET devices.

The smaller $I_{ON}$ in TFET causes reduced $C_{gs}$ [33]. In OFF state conditions, the lower potential drop across the drain-channel interface causes larger $C_{gd}$. In MOSFET, $C_{gd}$ is smaller due to large reverse-bias [34]. So, for TFET the overall gate capacitance is mainly due to $C_{gd}$, whereas in MOSFET devices the overall gate capacitance is due to both $C_{gs}$ and $C_{gd}$. Figure 4b shows that the proposed device exhibits larger $C_{gg}(C_{gg} = C_{gs} + C_{gd})$. The hetero-dielectric of the proposed device induces a larger amount of electron charges, enhances the $C_{gs}$ value, and also improves $C_{gg}$. The key parameters for Analog and Radio Frequency applications, the Cut-off frequency, and the gain-bandwidth product has been investigate. The cutoff frequency is defined as

$$f_T = g_m/\left(2\pi\left(C_{gs} + C_{gd}\right)\right) \tag{3}$$

The GBP is

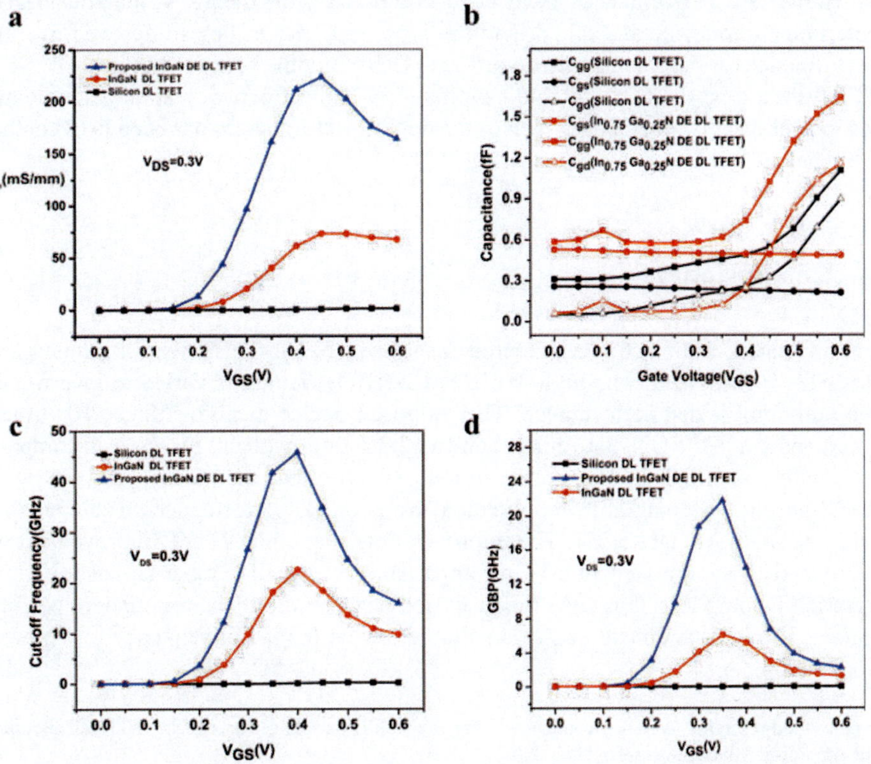


**Fig. 4** **a** Transconductance ($g_m$) of proposed InGaN DE DL-TFET, InGaN DL-TFET and Silicon DL-TFET. **b** Capacitance of proposed InGaN DE DL-TFET and Silicon DL-TFET. **c** Cut-off frequency of proposed InGaN DE DL-TFET, InGaN DL -TFET and Silicon DL-TFET. **d** Gain bandwidth product (GBP) of proposed InGaN DE DL-TFET, InGaN DL-TFET and Silicon DL-TFET

**Table 3** Comparison of the device performances at $V_{DS} = 0.3$ V and $V_{GS} = 0.6$ V

| Device Name | $I_{ON}$(A/μm) | $I_{OFF}$(A/μm) | $g_m$(mS/mm) | $f_T$(GHz) | GBP(GHz) |
|---|---|---|---|---|---|
| Heterojunction DL TFET(29) | $1.67 \times 10^{-5}$ | $8.5 \times 10^{-14}$ | 69.3 | 13 | 5.23 |
| $In_{0.75}Ga_{0.25}N$ DE–DL-TFET (presentwork) | $6.054 \times 10^{-5}$ | $1.021 \times 10^{-18}$ | 229 | 45.7 | 21.8 |

$$GBP = g_m/(2\pi 10C_{gd}) \quad (4)$$

Figure 4c, d show that the improved device exhibits $f_T$ and GBP of 45.7 GHz and 21.8 GHz at a $V_{DS}$ of 0.3 V. Improved transconductance due to higher $I_{ON}$ plays the main role in the enhancement of ($f_T$) and GBP.

The device performances have been compared with the III-V material-based heterojunction device at Table 3, and the proposed device shows improved overall performances at the same applied voltages. Therefore, the proposed InGaN DE DL-TFET should be the best suitable candidate for radio frequency applications with low power dissipation. Comparison of the device performances has been listed in the table below.

## 4 Conclusion

In conclusion, a dielectric engineered $In_{0.75}Ga_{0.25}N$ DL-TFET with a tunnel gate has been investigated. The high–k dielectric ($HfO_2$) length is varied to investigate the optimum device performance. The proposed device exhibits $2.83 \times 10^2$ times improvement in the ON-state current and a 67.3% improvement for $SS_{avg}$ in comparison with silicon DL-TFET. Furthermore, dielectric engineered $In_{0.75}Ga_{0.25}N$ DL-TFET has shown remarkable improvement in analog/RF performances in comparison with $In_{0.75}Ga_{0.25}N$ DL-TFET. The improved device exhibits 71.5% improvement in GBW and 76% improvement in $f_T$ in comparison with the III-V material-based hetero junction Doping less TFET. As the proposed device exhibits better efficient performance, it might be considered as a suitable device for radio frequency low-power applications.

**Acknowledgements** We should acknowledge the Solid State Electronics Lab, NIT Jamshedpur for providing Silvaco Atlas Tool.

## References


1. Masuda H, Nakai M, Kubo M (1979) Characteristics and limitation of scaled-down MOSFET's due to two-dimensional field effect. IEEE Trans Electron Devices 26(6):980–986
2. Nirschl T, Fischer J, Fulde M, Bargagli-Stoffi A, Sterkel M, Sedlmeir J, Weber C, Heinrich R, Schaper U, Einfeld J, Neubert R (2006) Scaling properties of the tunneling field effect transistor (TFET): device and circuit. Solid-State Electron 50(1):44–51
3. Colinge JP (2008) FinFETs and other multi-gate transistors. Springer, New York
4. Bangsaruntip S, Cohen GM, Majumdar A, Sleight JW (2010) Universality of short-channel effects in undoped-body silicon nanowire MOSFETs. IEEE Electron Device Lett 31(9):903–905
5. Wang PF, Hilsenbeck K, Nirschl T, Oswald M, Stepper C, Weis M, Schmitt-Landsiedel D, Hansch W (2004) Complementary tunneling transistor for low power application. Solid-State Electron 48(12):2281–2286
6. Choi WY, Park BG, Lee JD, Liu TJ (2007) Tunneling field-effect transistors (TFETs) with subthreshold swing (SS) less than 60 mV/dec. IEEE Electron Device Lett 28(8):743–745
7. Koswatta SO, Lundstrom MS, Nikonov DE (2009) Performance comparison between pin tunneling transistors and conventional MOSFETs. IEEE Trans Electron Devices 56(3):456–465
8. Seabaugh AC, Zhang Q (2010) Low-voltage tunnel transistors for beyond CMOS logic. In: Proceedings of the IEEE. IEEE, pp 2095–2110
9. Anand S, Amin SI, Sarin RK (2016) Analog performance investigation of dual electrode based doping-less tunnel FET. J Comput Electron 15(1):94–103
10. Boucart K, Ionescu AM (2007) Double-gate tunnel FET with high-k gate dielectric. IEEE Trans Electron Devices 54(7):1725–1733
11. Ram MS, Abdi DB (2015) Dopingless PNPN tunnel FET with improved performance: design and analysis. Superlattices Microstruct 82:430–437
12. Wang Y, Wang YF, Xue W, Cao F (2016) Asymmetric dual-gate tunneling FET with improved performance. Superlattices Microstruct 91:216–224
13. Kwon RH, Lee SH, Yoon YJ, Seo JH, Jang YI, Cho MS, Kim BG, Lee JH, Kang IM (2017) InGaAs-based tunneling field-effect transistor with stacked dual-metal gate with PNPN structure for high performance. J Semicond Technol Sci 17(2):230–238
14. Duan X, Zhang J, Wang S, Li Y, Xu S, Hao Y (2018) A high-performance gate engineered InGaN dopingless tunnel FET. IEEE Trans Electron Devices 65(3):1223–1229
15. Han G, Zhao B, Liu Y, Wang H, Liu M, Zhang C, Hu S, Hao Y (2015) Investigation of performance enhancement in InAs/InGaAs heterojunction-enhanced N-channel tunneling field-effect transistor. Superlattices Microstruct 88:90–98
16. Ahish S, Sharma D, Vasantha MH, Kumar YBN (2016) Device and circuit level performance analysis of novel InAs/Si heterojunction double gate tunnel field effect transistor. Superlattices Microstruct 94:119–130
17. Kondekar PN, Nigam K, Pandey S, Sharma D (2016) Design and analysis of polarity controlled electrically doped tunnel FET with bandgap engineering for analog/RF applications. IEEE Trans Electron Devices 64(2):412–418
18. Hueting RJ, Rajasekharan B, Salm C, Schmitz J (2008) The charge plasma PN diode. IEEE Electron Device Lett 29(12):1367–1369
19. Kumar MJ, Janardhanan S (2013) Doping-less tunnel field effect transistor: design and investigation. IEEE Trans Electron Devices 60(10):3285–3290
20. Sahu C, Singh J (2014) Charge-plasma based process variation immune junctionless transistor. IEEE Electron Device Lett 35(3):411–413
21. Bashir F, Loan SA, Rafat M, Alamoud ARM, Abbasi SA (2015) A high performance gate engineered charge plasma based tunnel field effect transistor. J Comput Electron 14(2):477–485
22. Abbassi SA, Bashir F, Loan SA, Alamoud ARM, Nizamuddin M, Rafat M (2016) Hetero gate material and dual oxide dopingless tunnel FET. In: Proceedings of IMECS, pp. 1.
23. Ghosh K, Singisetti U (2014) RF performance and avalanche breakdown analysis of InN tunnel FETs. IEEE Trans Electron Devices 61(10):3405–3410

24. Li W, Sharmin S, Ilatikhameneh H, Rahman R, Lu Y, Wang J, Yan X, Seabaugh A, Klimeck G, Jena D, Fay P (2015) Polarization-engineered III-nitride heterojunction tunnel field-effect transistors. IEEE J Exploratory Solid-State Comput Devices Circ 1:28–34
25. Peng Y, Han G, Wang H, Zhang C, Liu Y, Wang Y, Zhao S, Zhang J, Hao Y (2016) InN/InGaN complementary heterojunction-enhanced tunneling field-effect transistor with enhanced subthreshold swing and tunneling current. Superlattices Microstruct 93:144–152
26. Li W, Cao L, Lund C, Keller S, Fay P (2016) Performance projection of III-nitride heterojunction nanowire tunneling field-effect transistors. Phys Status Solidi (a) 213(4):905–908
27. Aseev P, Rodriguez PES, Gómez VJ, Alvi NUH, Mánuel JM, Morales FM, Jiménez JJ, García R, Senichev A, Lienau C, Calleja E (2015) Near-infrared emitting In-rich InGaN layers grown directly on Si: towards the whole composition range. Appl Phys Lett 106(7):072102
28. Fabien CA, Gunning BP, Doolittle WA, Fischer AM, Wei YO, Xie H, Ponce FA (2015) Low-temperature growth of InGaN films over the entire composition range by MBE. J Cryst Growth 425:115–118
29. Liu H, Yang LA, Jin Z, Hao Y (2019) An In 0.53 Ga 0.47 As/In 0.52 Al 0.48 As heterojunction dopingless tunnel FET with a heterogate dielectric for high performance. IEEE Trans Electron Devices 66(7):3229–3235
30. Verhulst AS, Vandenberghe WG, Maex K, Groeseneken G (2008) Boosting the on-current of an-channel nanowire tunnel field-effect transistor by source material optimization. J Appl Phys 104(6):064514
31. Kane EO (1960) Zener tunneling in semiconductors. J Phys Chem Solids 12(2):181–188
32. ATLAS device simulation software. Silvaco int, Santa Clara (2016)
33. Yang Y, Tong X, Yang LT, Guo PF, Fan L, Yeo YC (2010) Tunneling field-effect transistor: capacitance components and modeling. IEEE Electron Device Lett 31(7):752–754
34. Mookerjea S, Krishnan R, Datta S, Narayanan V (2009) Effective capacitance and drive current for tunnel FET (TFET) CV/I estimation. IEEE Trans Electron Devices 56(9):2092–2098

# Voltammetric Detection and Controlled Inhibition of Decarboxylation of Gallic Acid (GA) in Green Tea Using Eugenol

**Aindrila Roy, Debopam Bhattacharya, Chirantan Das, Basudev Nag Chowdhury, Anupam Karmakar, and Sanatan Chattopadhyay**

**Abstract** The present work is focused on investigating the effect of eugenol on oxidative decarboxylation of gallic acid in green tea by employing cyclic voltammetry (CV) technique. Such technique has been employed to examine the nature of variation of voltammograms obtained from green tea solutions subjected to different settling and boiling times, with particular emphasis on the oxidation of gallic acid (GA). The oxidation peak at −0.35 V in the voltammograms is observed to originate during decarboxylation process of GA to produce pyrogallol and carbon dioxide ($CO_2$) which have several detrimental effects on human health. The controlled addition of eugenol during the preparation of tea has been observed to inhibit such decarboxylation.

**Keywords** Cyclic voltammetry · Green tea · Gallic acid · Pyrogallol · Eugenol

## 1 Introduction

Green tea is globally recommended to be regularly consumed for the improvement of a wide variety of health-related issues. It exhibits several beneficial properties such as anti-inflammatory [1], antioxidant [2], antiviral [3], antibacterial [4], antiarthritic [5], and cholesterol-lowering effects [6]. It also has valuable applications in the prevention of cancer and cardiovascular diseases [7]. All such properties of green tea originate from its components such as polyphenols like catechins, tannins and flavonoids, vitamins, and minerals [8] providing significant positive effects on human health. However, green tea can cause adverse effects on human health due to chemical change of some of its components depending on boiling time, temperature, repeated boiling and also even if kept for a longer period of time after preparation. It is worthy

A. Roy · D. Bhattacharya · C. Das · B. Nag Chowdhury · A. Karmakar · S. Chattopadhyay (✉)
Department of Electronic Science, University of Calcutta, 92 A.P.C. Road, Kolkata 700009, India
e-mail: scelc@caluniv.ac.in

S. Chattopadhyay
Centre for Research in Nanoscience and Nanotechnology (CRNN), University of Calcutta, JD Block, Sector III, Bidhannagar, Kolkata 700098, India

 
C. Giri et al. (eds.), *Emerging Electronic Devices, Circuits and Systems*,
Lecture Notes in Electrical Engineering 1004,
https://doi.org/10.1007/978-981-99-0055-8_15

to mention that several redox reactions occur simultaneously in tea after and during its preparation, and therefore, the boiling time, temperature, and settling time are crucial parameters to modulate both aromatic and anti-oxidant properties of such green tea. Therefore, a comprehensive understanding and relevant analysis of such redox reactions will lead to successful monitoring of the preparation of green tea.

Numerous research works suggest that several spectroscopic and chromatographic techniques including IR [9], UV–Vis [9, 10], thin layer chromatography [11], and high-performance liquid chromatography [11] are able to detect the presence of catechins and its derivatives in green tea. However, such techniques are incapable for real-time study of the redox reactions occurring in tea prepared under several conditions. In this perspective, electrical or electrochemical characterization systems offer great advantages in terms of their rapid, sensitive, precise, point-of-care, and real-time diagnosis [12]. Moreover, the antioxidant properties of polyphenols in tea present in their oxidized forms involves the exchange of electrons, which makes the electrochemical methods of detection to be the most appropriate technique for such studies [8].

Among many components, tannic acid constitutes almost 40 mg/g of tea [13], and its natural oxidation produces gallic acid (GA), which is a phenolic acid that undergoes oxidative decarboxylation to form pyrogallol [14–16]. Such pyrogallol has been reported to cause several health hazards such as irritation on ingestion [17], severe gastrointestinal illness [18] and thrombocytopenia [19]. Further, the decarboxylation of GA produces in-vivo $CO_2$ which is also very harmful for human health [18]. In this context, the present work focuses on the prevention of decarboxylation of GA to pyrogallol by incorporating controlled amount of eugenol in tea during its preparation. Eugenol also finds great medicinal applications for its inherent antioxidant and anti-carcinogenic properties [20]. Interestingly, it is imperative to mention that such eugenol is the most naturally abundant phenolic compound in cloves (*Syzygium aromaticum*), tulsi or holy basil (*Ocimum tenuiflorum*), and cinnamon (*Cinnamomum verum*). Therefore, the addition of an appropriate amount of such natural phenolic compounds during the preparation of tea will be beneficial both for the health as well as taste.

The current research sought to investigate the effect of addition of eugenol in green tea solutions by studying its impact on the variations in oxidative decarboxylation of GA from cyclic voltammograms. The effect of GA decarboxylation in green tea samples has been identified and verified with pure GA solution through voltametric study. Accordingly, cyclic voltammograms of tea samples with different settling and boiling times have been recorded and analyzed. The relevant voltammograms display noticeable contribution of GA decarboxylation which can be observed from their anodic peaks. Such peaks have been observed to diminish with the controlled incorporation of eugenol.

# 2 Materials and Methods

## 2.1 *Sample Procurement and Preparation*

Prior to the measurements, fresh green tea leaves were procured from local markets. GA was procured from SRL CHEM, India. Eugenol was procured from Sigma Aldrich, India. Green tea solution was prepared by mixing 1 g of tea leaves in 30 ml de-ionized (DI) water, and this proportion was maintained for all the remaining sets of test samples. GA solution was prepared by adding 1.6 mg GA in 30 ml DI water. The volume of eugenol in both tea and GA solutions is maintained to be 0.02 ml throughout the experiments. All the relevant samples have been weighed using Contech Analytical Balance (Model CAS-54) having a precision of 0.0001 g. The heating time of the test samples in every step of the measurements has been set to 3 min except mentioned otherwise.

## 2.2 *CV Technique*

In this work, all the measurements have been performed employing CV technique consisting of platinum 'working' and 'counter' electrodes and an Ag/AgCl 'reference' electrode. Each of the three electrodes have been carefully immersed in the samples under test and connected to the potentiostat. The potentiostat applies a potential over the working electrode with respect to the reference electrode potential and records the current flowing between the counter and working electrodes. The current–potential data has been recorded for every sample to be analyzed and displayed as current versus potential plots known as voltammograms. The electrodes are responsible for sensing the redox properties of the compounds in each test sample where anodic positive convention is maintained. The sensing mechanism is based on the redox reactions occurring at the surface of the working electrode which is accompanied by the motions of the ions in the sample solution to balance the charges during the potential sweep. Therefore, results obtained from cyclic voltammetry reveal the redox properties of green tea under different test conditions and provide a comparative study in presence and absence of eugenol.

# 3 Experimental Details

## 3.1 *CV Technique and Its Equivalent Circuit*

CV measurements of each sample have been performed employing an electrochemical workstation (CH Instruments, Model 660E) as shown in Fig. 1a; and

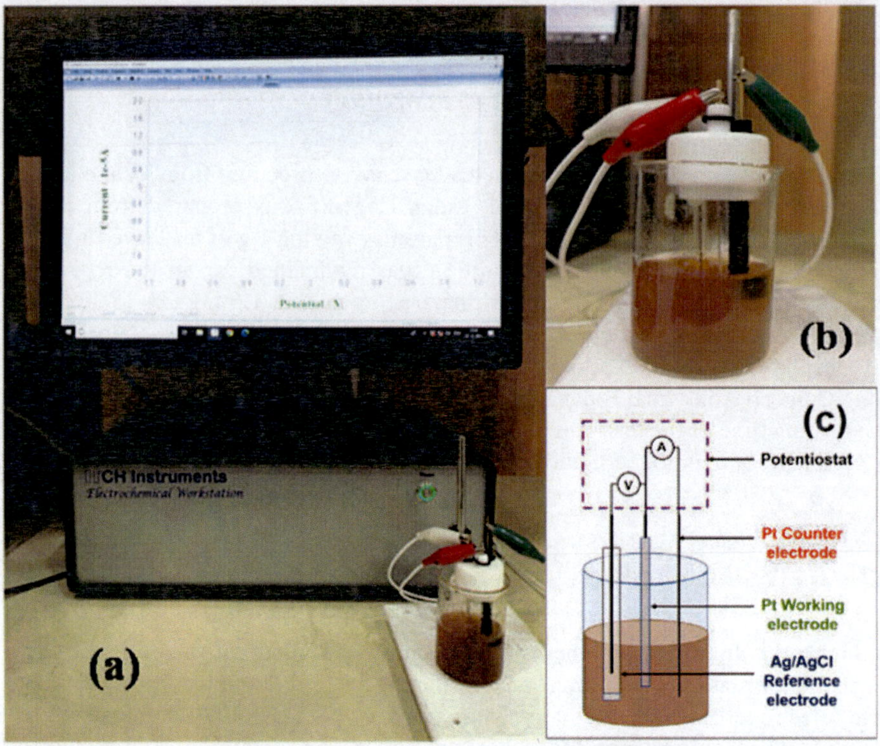


**Fig. 1** **a** Real-time cyclic voltammetry measurements of green tea samples, **b** three-electrodes immersed in a test sample and **c** schematic of the measurement setup

Fig. 1b depicts a real-time image of the three-electrode setup immersed inside a test tea sample solution. Figure 1c represents the schematic of the experimental setup consisting of the test solution and three electrodes connected to the potentiostat. All the measurements were conducted in a potential range of $-1$ to 1 V (versus Ag/AgCl reference electrode) at a scan rate of 0.1 V/s and sensitivity 1 $\mu$A/V. Each plot consists of two segments: oxidation and reduction, thereby completing a redox loop in an anodic positive configuration.

# 4 Results and Discussion

## *4.1 Voltammetric Analysis*

The redox properties of green tea samples have been explored in this section employing CV technique, focusing at the detectable changes in oxidation of GA in tea with variation in both settling and boiling time. Further, the nature of variation of such voltammograms with the incorporation of eugenol in green tea has been systematically analyzed to develop a comprehensive understanding on the oxidation–reduction processes occurring in such samples.

It is worthy to mention at this point that green tea consists of an array of components including catechins, caffeine, tannins, flavonoids, vitamins, and minerals [8]. Catechins are a type of polyphenol and a component that imparts astringency in tea, and one of its major kind, epigallocatechin gallate (EGCG), is found abundantly in green tea. Other major components include epigallocatechin, epicatechin gallate, theaflavins, epicatechin, and GA [21]. All these components are responsible for the antioxidant properties of tea and show distinct electrochemical behavior. Previous works have reported consistent oxidation potentials of such components of tea with respect to saturated calomel electrodes (SCE) [22] and glassy carbon electrodes (GCE) [21, 23] which led to the oxidation potential of GA to be 0.32 V for SCE and 0.2 V for GCE. However, platinum working electrode is used in the present work and cyclic voltammograms have been recorded for both tea and GA separately for analysis. Such voltammograms depict an oxidation peak at −0.35 V for GA which is observed to be consistent for all other measurements.

## 4.2 *Effect of Settling Time*

The natures of variation of cyclic voltammograms of green tea solutions for different settling times have been represented as solid lines in Fig. 2. Initially, the tea solution was heated at 100 °C for 3 min followed by cooling down to room temperature before its measurement for excluding the combined impact of temperature. It must be noted that before measurements to be performed, the tea leaves were filtered out after boiling. This ensures a fixed amount of constituents extracted from tea on boiling to be maintained. For studying the effect of time variation on voltametric properties, measurements have been performed after 5, 10, 15, 20, 25, and 30 min from the time of initial measurement just after boiling followed by cooling to room temperature, as shown in Fig. 2.

From the voltammograms of green tea, several anodic/oxidation peaks have been observed at −0.5 V and −0.35 V along with cathodic/reduction peaks at 0.1 V and 0.5 V. On comparison of the same with the voltammogram of GA (represented as dotted lines in Fig. 2), the anodic peak at −0.35 V is attributed to the oxidation of GA. Although GA is produced in tea during boiling, it has been observed that the initial measurement (0 min) does not show any peak resembling its oxidation. An oxidation peak at −0.35 V is observed to appear only after a time duration of 5 min and gets prominent for all successive measurements till 30 min. This suggests that the oxidative decarboxylation of GA becomes significant after 5 min and occurs continuously in tea after its extraction during boiling.

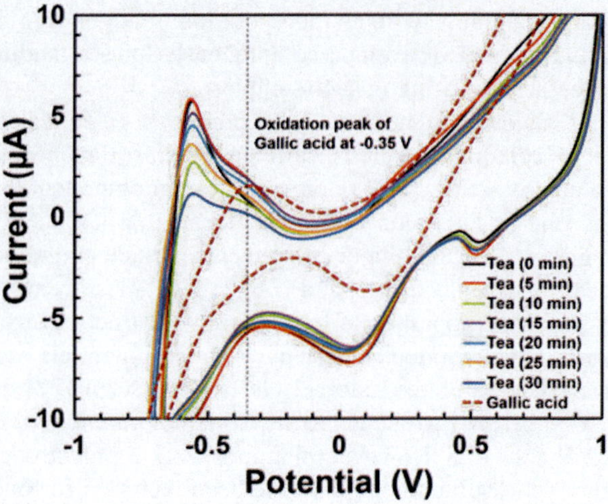


**Fig. 2** Cyclic voltammograms of green tea measured at different settling times in the potential range −1 to 1 V with pure GA showing an oxidation peak at −0.35 V

## *4.3 Effect of Boiling Time*

The variation of the voltametric properties of green tea for different boiling times has been investigated, where four different tea solutions have been boiled at 100 °C for 3, 5, 7, and 9 min, respectively, and have been plotted in Fig. 3. It is apparent from Fig. 3 that for a boiling period of 3 min, there is no contribution of GA oxidation to the anodic peak of green tea.

It suggests that although GA is extracted while boiling, its oxidation does not occur during the same period of time. However, for longer periods of boiling green tea at 100 °C, the presence of GA oxidation peak at −0.35 V starts appearing and gets more prominent in the voltammograms of the test tea solution that has been boiled for 9 min. Thus, it can be concluded that for a boiling time of 9 min, GA is

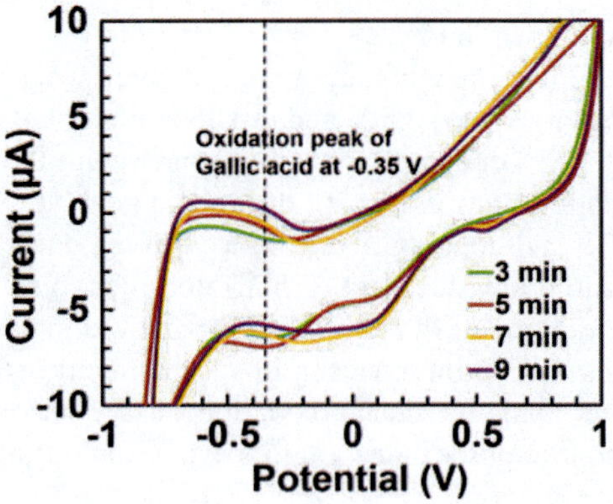


**Fig. 3** Plot showing cyclic voltammograms of green tea for four different boiling times in the potential range −1 to 1 V

extracted as well as oxidized to form pyrogallol in the solution, which is suggested not to be carried out to avoid the associated health hazards [17–19].

It is also apparent from Fig. 3 that the relevant oxidation current increases with the increment in boiling time thereby suggesting continuous oxidation of GA inside the solution. Therefore, the decarboxylation of GA depends on both the settling and boiling times, and it has been observed to become a considerable oxidation reaction occurring in green tea for settling time of 5 min and more and a minimum boiling time of 9 min.

## 4.4 *Effect of Eugenol on the Oxidation of Green Tea*

As mentioned earlier, the decarboxylation reaction of GA, in reality, is a harmful process for simultaneously releasing $CO_2$ and forming pyrogallol [14–16]. Such decarboxylation reaction, as depicted in Fig. 4a, is catalyzed by gallate decarboxylase naturally present in tea or at moderately high temperature. It is imperative that such reaction should be prevented to improve the health benefits of tea. Hence, the present work proposes the esterification of GA to be a possible method of averting the formation of pyrogallol. Figure 4a also illustrates the basic mechanism of esterification reaction. GA, being a phenolic acid undergoes esterification reaction with phenols, like eugenol [24, 25], which is the major component present in cloves, tulsi, cinnamon, and others. In general, cloves are often added to tea to enhance its aroma and flavor, and eugenol is responsible for such astringency and aroma. However, alongside providing such flavor to tea, a controlled addition of eugenol in tea leads to the esterification of GA in tea that can successfully prevent the production of pyrogallol and $CO_2$.

The esterification of GA by eugenol is verified by the absence of the GA oxidation peak on controlled addition of eugenol, as shown in Fig. 4b. During stirred heating of eugenol (0.02 ml) with GA (1.6 mg) in 30 ml DI water at 100 °C, instead of undergoing oxidation to form pyrogallol, GA participates in its esterification reaction with eugenol. For concentrations of eugenol below 0.02 ml significant change was not observed. Further, when incorporated in tea, eugenol shows absolutely similar effects which suggests its reaction with GA is undisturbed by other components of tea. From the respective plots of Fig. 4c, it is apparent that the peak at −0.35 V, which appears due to the decarboxylation of GA in tea, is diminished significantly on addition of the same amount of eugenol.

It can therefore be concluded that in the presence of eugenol, which may be obtained naturally from cloves, tulsi, and other spices like cinnamon, the production of pyrogallol, i.e., the harmful oxidized product of GA, can be significantly suppressed. Thus, observations from the present research suggest that controlled addition of cloves, tulsi, or cinnamon in tea can be highly beneficial for human health.

**Fig. 4** **a** Decarboxylation (i) and esterification (ii) reactions of GA and cyclic voltammograms showing the effect of incorporation of eugenol in **b** GA solution and **c** green tea

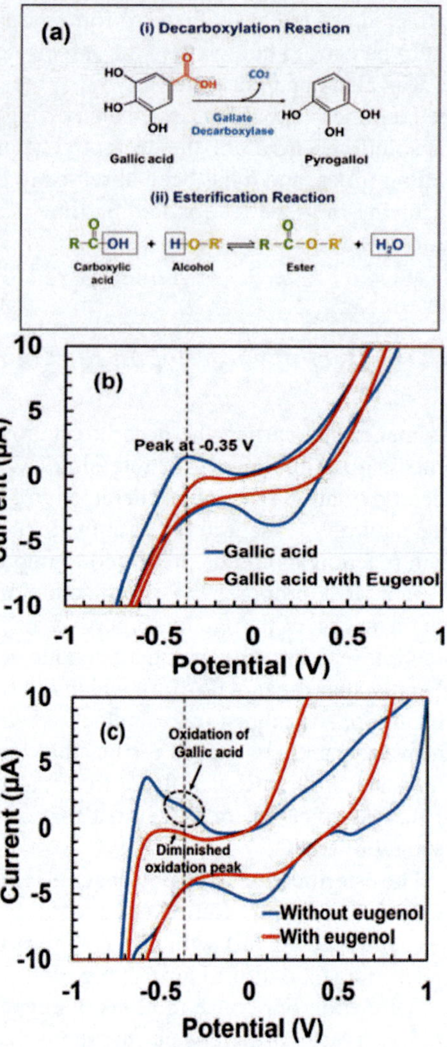


## 5 Conclusion

In the present work, the oxidation properties of GA in green tea along with its suppression using eugenol by employing cyclic voltammetry technique has been investigated. The voltammogram results suggested significant impact of settling and boiling time on the redox reactions occurring in green tea. A prominent GA oxidation peak at −0.35 V has been obtained due to the use of platinum working electrode. The initiation of GA oxidation process in tea has been observed to be at settling and boiling times of 5 min and 9 min, respectively. The controlled addition of eugenol suppresses the conversion of GA to pyrogallol and the formation of $CO_2$. This phenomenon has been attributed to the esterification reaction occurring between GA and eugenol and thus averting the formation of pyrogallol and $CO_2$. Therefore, the current research illustrates a systematic approach to enhance the beneficial effects of green tea by methodically investigating and modifying the redox reactions with the controlled incorporation of eugenol. Therefore, the consumption of tea by adding spices and herbs containing eugenol such as cloves and tulsi can be beneficial for health.

**Acknowledgements** The authors would like to acknowledge the Department of Electronic Science, University of Calcutta and WBDITE for providing infrastructural support.

## References

1. Donà M, Dell'Aica I, Calabrese F, Benelli R, Morini M, Albini A, Garbisa S (2003) Neutrophil restraint by green tea: inhibition of inflammation, associated angiogenesis, and pulmonary fibrosis. J Immunol 170(8):4335–4341
2. Osada K, Takahashi M, Hoshina S, Nakamura M, Nakamura S, Sugano M (2001) Tea catechins inhibit cholesterol oxidation accompanying oxidation of low density lipoprotein in vitro. Comp Biochem Physiol C Toxicol Pharmacol 128(2):153–164
3. Weber JM, Ruzindana-Umunyana A, Imbeault L, Sircar S (2003) Inhibition of adenovirus infection and adenain by green tea catechins. Antiviral Res 58(2):167–173
4. Sudano RA, Blanco AR, Giuliano F, Rusciano D, Enea V (2004) Epigallocatechin-gallate enhances the activity of tetracycline in staphylococci by inhibiting its efflux from bacterial cells. Antimicrob Agents Chemother 48(6):1968–1973
5. Haqqi TM, Anthony DD, Gupta S, Ahmad N, Lee MS, Kumar, GK, Mukhtar H (1999) Prevention of collagen-induced arthritis in mice by a polyphenolic fraction from green tea. Proc Nat Acad Sci 96(8):4524–4529
6. Raederstorff DG, Schlachter MF, Elste V, Weber P (2003) Effect of EGCG on lipid absorption and plasma lipid levels in rats. J Nutr Biochem 14(6):326–332
7. Chacko SM, Thambi PT, Kuttan R, Nishigaki I (2010) Beneficial effects of green tea: a literature review. Chin Med 5(1):1–9
8. Ziyatdinova GK, Nizamova AM, Aytuganova II, Budnikov HC (2013) Voltammetric evaluation of the antioxidant capacity of tea on electrodes modified with multi-walled carbon nanotubes. J Anal Chem 68(2):132–139
9. Lin H, Gan T, Wu K (2009) Sensitive and rapid determination of catechol in tea samples using mesoporous Al-doped silica modified electrode. Food Chem 113(2):701–704
10. Atomssa T, Gholap AV (2015) Characterization and determination of catechins in green tea leaves using UV-visible spectrometer. J Eng Technol Res 7(1):22–31
11. Kiani A, Raoof JB, Nematollahi D, Ojani R (2005) Electrochemical study of catechol in the presence of dibuthylamine and diethylamine in aqueous media: part 1. Electrochemical investigation. Electroanalysis Int J Devoted Fundam Pract Aspects Electroanalysis 17(19):1755–1760
12. Sun YG, Cui H, Li YH, Lin XQ (2000) Determination of some catechol derivatives by a flow injection electrochemiluminescent inhibition method. Talanta 53(3):661–666
13. Savolainen H (1992) Tannin content of tea and coffee. J Appl Toxicol 12(3):191–192
14. Li W, Wang C (2015) Biodegradation of gallic acid to prepare pyrogallol by Enterobacter aerogenes through substrate induction. BioResources 10(2):3027–3044
15. Guzmán-López O, Loera O, Parada JL, Castillo-Morales A, Martínez-Ramírez C, Augur C, Gaime-Perraud I, Saucedo-Castaneda G (2009) Microcultures of lactic acid bacteria: characterization and selection of strains, optimization of nutrients and gallic acid concentration. J Ind Microbiol Biotechnol 36(1):11–20

16. Goldberg I, Rokem JS (2019) Organic and fatty acid production, microbial, pp 358–382
17. Fiege H, Voges HW, Hamamoto T, Umemura S, Iwata T, Miki H, Fujita Y, Buysch HJ, Garbe D, Paulus W (2000) Phenol derivatives. In: Ullmann's encyclopedia of industrial chemistry
18. Gupta YK, Sharma M (2001) Reversal of pyrogallol-induced delay in gastric emptying in rats by ginger (Zingiber officinale). Methods Find Exp Clin Pharmacol 23(9):501–504
19. Bruges G, Venturini W, Crespo G, Zambrano ML (2018) Pyrogallol induces apoptosis in human platelets. Folia Biol 64(1):23–30
20. Adeyemi JA, Arowolo OK, Olawuyi ST, Alegbeleye D, Ogunleye A, Bamidele OS, Adedire CO (2018) Effect of co-administration of green tea (Camellia sinensis) on clove-(Syzygiumaromaticum) induced hepatotoxicity and oxidative stress in Wistar rats
21. Kilmartin PA, Hsu CF (2003) Characterisation of polyphenols in green, oolong, and black teas, and in coffee, using cyclic voltammetry. Food Chem 82(4):501–512
22. Furuno K, Akasako T, Sugihara N (2002) The contribution of the pyrogallol moiety to the superoxide radical scavenging activity of flavonoids. Biol Pharm Bull 25(1):19–23
23. Kondo K, Kurihara M, Fukuhara K (2001) Mechanism of antioxidant effect of catechins. Methods Enzymol 335:203–217
24. Makuch E, Nowak A, Günther A, Peech R, Kucharski Ł, Duchnik W, Klimowicz A (2020) Enhancement of the antioxidant and skin permeation properties of eugenol by the esterification of eugenol to new derivatives. AMB Express 10(1):1–5
25. Fadilah AY, Yanuar A, Arsianti A, Andrajati R, Indah Paramita R (2017) Silico study, synthesis, and cytotoxic activity of esterification of eugenol and gallic acid against HT-29 cell line. Orient J Chem 33(6):3009–3014

# Impact of a Tubular Dielectric Medium on Peak Noise and Crosstalk Delay in a Coaxial TSV

**Maya Chandrakar and Manoj Kumar Majumder**

**Abstract** In the era of deep sub-micron technology, the coupling capacitance plays a major role due to the higher integration density that degrades the performance of 3D IC-based through silicon vias (TSVs). These TSVs with an oxide liner ($SiO_2$) experiences major reliability issues, such as increased coupling capacitance and electromagnetic interference. In order to address these concerns, this paper examines the electrical performance of coaxial TSV for the first time that offers improved electrical performance compared to conventional cylindrical and tapered shapes. At 32 nm technology, an equivalent *RLGC T*-type network is proposed for a coaxial TSV for different via heights and inner shell radius by considering the impact of a tubular dielectric medium with different polymer dielectrics. Besides, the impact of peak noise and crosstalk induced delay on via performance are also demonstrated. A significant improvement in crosstalk induced delay and peak noise using a coaxial TSV with Benzocyclobutene (BCB) as a polymer dielectric are observed as 2.5% and 38.8%, respectively, at a via height of 20 μm.

**Keywords** Crosstalk · Peak noise · Tubular dielectric medium · Metal oxide semiconductor (MOS) effect · Through silicon via (TSV)

## 1 Introduction

The heart and vital element of 3D integration, through silicon via (TSV), is a vertical interconnect passing through the silicon (Si) substrate, enables the stacking of heterogeneous technologies [1, 2]. Thus it effectively reduces the physical density of the ICs, allowing Moore's law to be still maintained or even exceeded [2, 3]. It offers several benefits such as increased integration density, shorter interconnection length,

M. Chandrakar (✉) · M. K. Majumder
Department of Electronics and Communication Engineering, Dr. SPM International Institute of Information Technology, Naya Raipur, Chhattisgarh, India
e-mail: mayac721@gmail.com

M. K. Majumder
e-mail: manojk@iiitnr.edu.in

 
C. Giri et al. (eds.), *Emerging Electronic Devices, Circuits and Systems*,
Lecture Notes in Electrical Engineering 1004,
https://doi.org/10.1007/978-981-99-0055-8_16

improved noise tolerance, and electrical performance, etc. [4]. Despite the indisputable potential advantages, TSVs encounter difficulties in real-world applications when the operating frequency increases to improve bandwidth per channel. As a matter of fact, the inherent frequency-dependent loss of silicon substrate, crosstalk, inter- and intra-substrate coupling, and electromagnetic interference (EMI), etc., impose restrictions on high-speed data transmission and channel bandwidth [5, 6]. At present, a large variety of TSV configurations with various physical attributes have been fabricated and explored, including cylindrical, tapered, annular, and coaxial TSVs, etc. In order to resolve the concerns related to the substrate loss, crosstalk, etc., the coaxial TSVs (CTSVs) with self-shielding characteristics have been proposed [7]. Depending on its structural characteristics, the coaxial TSV, which employs an inner shell as a central signal-carrying conductor and is surrounded by a concentric outer shell as a ground return path, could be the best TSV structure in electrical performance [8]. Since coaxial TSVs provide an inherent shielding and EMI suppression, additional ground TSVs are no longer required. Also, because the separation between the inner and outer conductors of coaxial TSVs can be filled with the dielectric insulating layers, the electric and magnetic fields carrying the signal are confined to the dielectrics with slight leakage outside the shield. As a result, coaxial TSVs provide high noise immunity and good thermal and electrical performance. Besides the electrical performance of the TSVs such as crosstalk, propagation delay, and power consumption, an essential concern on 3D integration is reliability that is associated with the thermal and mechanical aspects of TSVs. Moreover, the performance of the TSVs depends on the type of liner layer used. The most commonly used liner layer in TSVs is silicon dioxide ($SiO_2$). The massive discrepancy in the coefficients of thermal expansion (CTE) of copper (Cu) (16.5 ppm/K), Si substrate (2.6 ppm/K), and $SiO_2$ liner (0.5 ppm/K) introduces significant stresses in the Cu, $SiO_2$ liner, and Si substrate when the TSVs are subjected to high temperatures during the fabrication process [9]. As a result, the Cu protrusion and interfacial deformation have an impact on the overall performance of a TSV [10]. Also, due to the large relative permittivity value of $SiO_2$, high coupling capacitance occurs between the TSVs that introduces more crosstalk delay between the neighboring TSVs [11]. As a consequence, dealing with the aforementioned difficulties in TSVs is a challenging task. In this regard, several polymer liners (such as Benzocyclobutene (BCB), Polypropylene Copolymer (PPC), and Polymide (PI)) have been proposed as either part of TSVs or a replacement to standard $SiO_2$ insulation layers to address these issues [12]. To circumvent the reliability concerns encountered in Cu filled TSVs, the polymer liners with less CTE values, Young's modulus, relative permittivity, and more consistent thicknesses can be used in TSVs. Therefore, to further enhance its thermal–mechanical characteristics, a coaxial TSV configuration with a suitable polymer liner is needed.

Previously, Lee et al. [13] presented high-frequency temperature-dependent *RLGC* models for two neighboring TSVs and TSV channels. They also evaluated the effect of temperature on TSV-TSV channel noise coupling. However, the derived *RLGC* equations were primarily focused on $SiO_2$ liner-based cylindrical-shaped TSVs. Thereafter, Su et al. [14] derived closed-form formulae based on the

Maxwell's equations to evaluate the insulator and substrate capacitances of a tapered TSV. The observations indicated that the tapered TSVs had lesser crosstalk than cylindrical TSVs. Also, the equations were particularly suited for TSVs with a high aspect ratio (AR), with only a 4% difference between the analytical and simulation results. Later, by incorporating the advantages of the coaxial TSV and differential signaling approach, Fu et al. [15] established the equivalent circuit model of a shielded-differential annular through silicon via (SD-ATSV). Although the authors have considered the modeling of different TSV shapes in the suggested electrical models but the impact of the coupling capacitance was constrained only to the $SiO_2$ liner. Afterwards, Su et al. [16] presented an equivalent *RLGC* model of a novel TSV structure named partial coaxial TSV (PC-TSV) to suppress TSV-induced substrate noise. In this structure, the TSV filler was surrounded by a BCB layer to reduce signal leakage to the Si substrate. Based on the aforementioned state-of-the-art research [13–16], it is evident that a detailed analysis of modeling of a coaxial shaped TSV with the influence of various liner materials is required in order to achieve the improved thermal and electrical performance. Furthermore, a comprehensive analysis is essential to apostrophize the peak noise and worst-case crosstalk induced delay of TSVs by considering all physical geometries.

In this paper, an electrical modeling of the coaxial TSV is demonstrated for the first time, including the influence of a tubular dielectric medium (combination of $SiO_2$ and different polymer liner materials such as BCB, PI, and PPC) on peak noise and crosstalk induced delay at high frequencies. The crosstalk interference occurs due to the Miller's effect, which states that when all the signals given to the aggressor and victim lines are switched in the opposite directions simultaneously, then the victim line experiences an increased coupling capacitance. Hence, a voltage peak can be seen on the victim line that has a severe effect on the TSV performance. In order to demonstrate these effects, the closed-form *RLGC* expressions for the coaxial TSV with a tubular dielectric medium are derived. In the presented TSV configuration, a thick low-*k* with a small loss tangent uniform tubular dielectric medium is desired that is placed between the two shells of the coaxial TSV in order to provide both electrical isolation and mechanical support. Also, this medium incorporates $SiO_2$ with a high resistivity value that aids in suppressing eddy current generated inside the substrate when frequency increases. Additionally, the via is isolated from the Si substrate by the oxide liner and depletion layers that effectively prevent the crosstalk interference between the neighboring TSVs. A micro-bump is employed as a contact to connect the TSV to the functional block of the dies. The underfill and Inter Metal Dielectric (IMD) layers are provided to separate the bump from the Si substrate to avoid cross-coupling between these two conducting regions. This paper analyzes the peak noise and crosstalk performance using a Driver-Via-Load (DVL) at 32 nm technology node, wherein the via line is modeled with Cu as filler material, and inner and outer shells are separated by the tubular dielectric medium. The primary reason for choosing the aforementioned technology parameters is that using a liner beyond 32 nm enhances leakage making it unsuitable for low-power and high-performance applications. Moreover, a specific *T*-type network is used to construct the electrical circuit model. With this approach, the capacitive charging current flows over half of the TSV line that minimizes the crosstalk interference.

This paper is organized in the following sections: Sect. 1 sheds light on the existing state-of-the-art research scenario and discusses the modeling of coaxial shaped TSV, taking into account the impact of the tubular dielectric medium on TSV performance. Section 2 presents the physical configuration of a coaxial TSV structure, its important design parameters and an equivalent electrical model. This section also includes the detailed analytical *RLGC* equations for modeling a coaxial TSV employing a tubular dielectric medium with several polymer liner materials. The impact of various liner materials in TSV is investigated in Sect. 3, that examines the effect of a tubular dielectric medium for different via height and inner radius of coaxial TSV. Finally, Sect. 4 provides a brief summary of this work.

## 2 TSV Configuration and Equivalent Circuit Model

This section describes the modeling of a coaxial TSV using an accurate geometrical shape at 32 nm technology node. The equivalent circuit model of a coaxial via shape at high frequency includes the effects of the tubular dielectric medium, liner, underfill layer, and bump is also included in this section. The analytical closed-form *RLGC* expressions are also employed to develop the *T*-type network-based electrical model. The model is proposed using accurate physical dimensions and properties of the materials as variables in the *RLGC* equations. Accordingly, a novel electrical modeling of a coaxial TSV is introduced with the consideration of a tubular dielectric medium in the following sub-section.

### 2.1 TSV Structure and Physical Parameters

This sub-section provides a detailed representation of the coaxial TSV shape along with assessable values of structural features [17–19] and material properties that will eventually be used in the modeling of comparable *RLGC* model. The structural configuration and top cross-sectional view of coaxial TSV are shown in Fig. 1a and b, respectively. As shown in Fig. 1a, $d_{TSV}$, $d_{inner}$, $h_{TSV}$, $t_{SiO_2}$, $t_{polymerliner}$, and $t_{dep}$ represent outer shell diameter, inner shell diameter, via height, oxide layer thickness, polymer liner thickness, and depletion layer thickness, respectively. Wherein *a*, *b*, and *c* represent inner shell radius, outer shell radius, and ring thickness, respectively (shown in Fig. 1b).

Considering the coaxial shaped TSV, Tables 1, 2, and 3 illustrate the via structural dimensions and material characteristics of liners, copper and Si substrate, and TSV, respectively. The presented coaxial TSV configuration consists of a micro-bump that is used to provide a contact to the functional block of the dies and an IMD and underfill layers are used to isolate the bumps from the Si substrate. Cu is employed

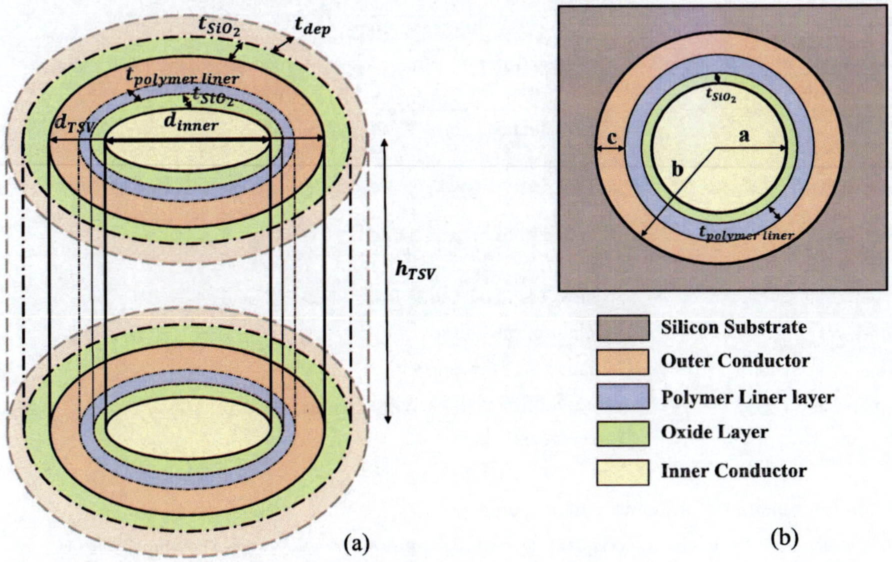


**Fig. 1** **a** The structural configuration and **b** the top cross-sectional view of a coaxial TSV

as a conductive filler material in the TSV and bump. The substrate is composed of lossy silicon, although the depletion region is formed with lossless silicon to prevent the leakage. It is required to isolate the via filler of the outer shell from the Si substrate by surrounding the TSV with an oxide layer in order to analyze the MOS effect. In addition, to provide both electrical isolation and mechanical support, a uniform tubular dielectric medium (combination of $SiO_2$ and different polymer liner materials such as BCB, PPC, and PI) is provided between the two shells of the coaxial TSV. The ratio of thicknesses of polymer liner to oxide liner is considered as 8:1, where another ratio of total dielectric thickness ($t_{\text{Total}} = t_{\text{SiO}_2} + t_{\text{Polymerliner}}$) to inner radius is considered as 1:5. Therefore, it provides different values of inner radius to be obtained by varying the thickness of the oxide layer. From Fig. 1b, the equation of outer radius can be obtained as $b = a + t_{\text{SiO2}} + t_{\text{polymerliner}} + t_{\text{SiO2}} + c$ and the value of ring thickness can be calculated as: $c = (b - 55t_{\text{SiO2}})$. Considering the aforementioned expressions, Table 4 summarizes the quantitative values of ring thickness, inner radius, and polymer liner thickness at different oxide layer thickness.

## 2.2 *Equivalent Electrical Model of a Coaxial TSV*

An equivalent *RLGC* modeling of Cu-based coaxial TSV is demonstrated in this sub-section by considering the various polymer materials in combination with an oxide layer that is employed to provide separation between the two shells of the

**Table 1** Structural dimensions of a via at 32 nm technology

| Symbolic notation of via dimension | Specification | Technology dependent value ($\mu$m) |
|---|---|---|
| $d_{\mathrm{TSV}}$ | Via diameter | 2–10 |
| $h_{\mathrm{TSV}}$ | Via height | 20–120 |
| $t_{\mathrm{siO_2}}$ | Thickness of oxide layer | 0.03–0.1 |
| $t_{\mathrm{SiO_2,bot}}$ | Thickness of bottom oxide layer | 0.03–0.1 |
| $t_{\mathrm{dep}}$ | Thickness of depletion layer | 0.757 |
| $h_{\mathrm{Bump}}$ | Height of the bump | 5 |
| $d_{\mathrm{Bump}}$ | Diameter of the bump | 15 |
| $h_{\mathrm{IMD}}$ | Height of the IMD layer | 5 |
| $p_{\mathrm{TSV}}$ | Pitch of the via | 20 |

**Table 2** Properties of different materials used in TSV

| Material | Dielectric constant | CTE (ppm/K) | Young's Modulus (GPa) |
|---|---|---|---|
| Copper | – | 16.5 | 120 |
| Si substrate | 11.9 | 2.6 | 169 |
| $SiO_2$ | 3.9 | 0.5 | 73 |
| PBO | 3.0 | 35–55 | 2.9 |
| PPC | 2.9 | 58–100 | 1.5–3 |
| PI | 3.5 | 20 | 3 |
| BCB | 2.65 | 42 | 2.9 |

**Table 3** Material characteristics of different materials used in TSV

| Symbolic notation | Specification | Value |
|---|---|---|
| $\sigma_{\mathrm{Si}}$ | Si substrate conductivity | 10 [S/m] |
| $\sigma_{\mathrm{TSV}}$ | TSV filler conductivity | $5.9524 \times 10^7$ [S/m] |
| $\sigma_{\mathrm{Bump}}$ | Bump filler conductivity | $5.9524 \times 10^7$ [S/m] |
| $\rho_{\mathrm{TSV}}$ | TSV filler resistivity | $1.68 \times 10^{-8}$ [Ω-m] |
| $\rho_{\mathrm{Bump}}$ | Bump filler resistivity | $1.68 \times 10^{-8}$ [Ω-m] |
| $\varepsilon_{r,\mathrm{Si}}$ | Dielectric constant of Si substrate | 11.9 |
| $\varepsilon_{r,\mathrm{SiO_2}}$ | Dielectric constant of $SiO_2$ | 3.9 |
| $\varepsilon_{r,\mathrm{IMD}}$ | Dielectric constant of IMD | 4 |
| $\varepsilon_{r,\mathrm{Underfill}}$ | Dielectric constant of underfill | 7 |
| $\mu_{r,\mathrm{TSV}}$ | TSV filler relative permeability | 1 |
| $\mu_{r,\mathrm{Bump}}$ | Bump filler relative permeability | 1 |

**Table 4** Quantitative values of ring thickness, inner radius, and polymer liner thickness

| Thickness of oxide layer ($\mu$m) | $t_{SiO2} = 0.03$ | $t_{SiO2} = 0.04$ | $t_{SiO2} = 0.05$ | $t_{SiO2} = 0.06$ |
|---|---|---|---|---|
| Ring thickness (c) ($\mu$m) | 3.35 | 2.8 | 2.25 | 1.7 |
| Inner radius (a) ($\mu$m) | 1.35 | 1.8 | 2.25 | 2.7 |
| Polymer liner thickness ($t_{polymerliner}$) ($\mu$m) | 0.24 | 0.32 | 0.4 | 0.48 |

coaxial TSV. Figure 2 depicts the structure, various parasitics, and an equivalent *RLGC* circuit model of a coaxial TSV with the connected bumps.

The structural dimensions and various material properties (provided in Tables 1, 2, and 3) are used to develop closed-form *RLGC* equations expressions for a coaxial TSV, as demonstrated in the subsequent sub-sections.

### Via Resistance

Via resistance occurs as a result of the space and void present in the bump and TSV filler that signifies the heat dissipation in these conducting regions of the TSV. Thus, the equivalent via resistances of inner and outer shells ($R_{via,in}$ and $R_{via,out}$) can be defined as

$R_{via,out} = r'_{via,out} \times h_{TSV}$ and $R_{via,in} = r'_{via,in} \times h_{TSV}$; where $r'_{via,out}$ and $r'_{via,in}$ are the *p.u.l.* equivalent via resistances of outer and inner shells as shown in Fig. 2.

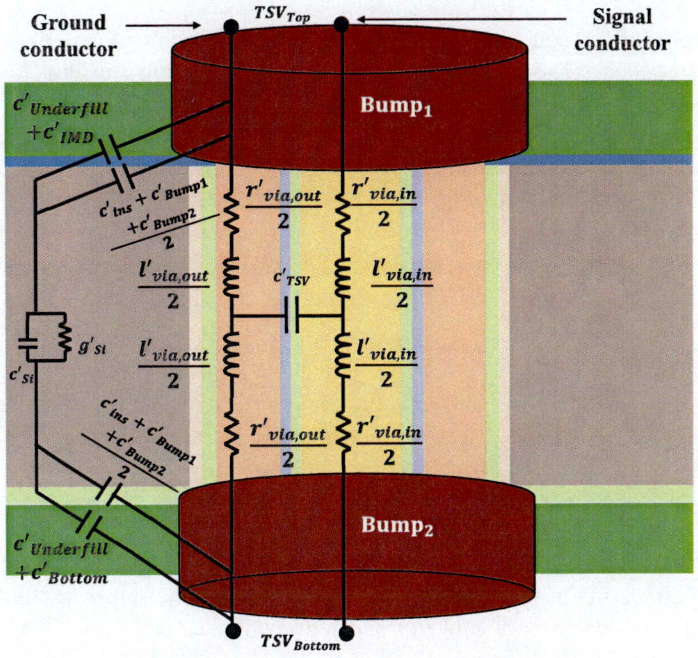


**Fig. 2** An equivalent *RLGC* circuit model of the Cu-based coaxial TSV

The equations of $R_{\text{via,out}}$ and $R_{\text{via,in}}$ of Fig. 2 can be expressed as

$$R_{\text{via,out}} = R_{\text{TSV,out}} + R_{\text{Bump,out}} \tag{1}$$

$$R_{\text{via,in}} = R_{\text{TSV,in}} + R_{\text{Bump,in}} \tag{2}$$

The expression for $R_{\text{TSV,out}}$ can be obtained as:

$$R_{\text{TSV,out}} = \sqrt{\left(R_{\text{dc,TSV,out}}\right)^2 + \left(R_{\text{ac,TSV,out}}\right)^2} \tag{1a}$$

The expressions of $R_{\text{dc,TSV,out}}$ and $R_{\text{ac,TSV,out}}$ are provided in [20].

Similarly, the expression for $R_{\text{TSV,in}}$ can also be calculated using Eq. (1a).by considering $d_{\text{TSV}} = d_{\text{inner}}$.

Likewise, the expression for $R_{\text{Bump,out}}$ can be obtained as:

$$R_{\text{Bump,out}} = \sqrt{\left(R_{\text{dc,Bump,out}}\right)^2 + \left(R_{\text{ac,Bump,out}}\right)^2} \tag{1b}$$

However, the expressions of $R_{\text{dc,Bump,out}}$ and $R_{\text{ac,Bump,out}}$ are provided in [20].

In a similar manner, the expression for $R_{\text{Bump,in}}$ can also be obtained using Eq. (1b) by considering $d_{\text{Bump}} = d_{\text{Bump,inner}}$.

**Via Inductance**

The influence of an equivalent via inductance is important to be considered at increased frequency. The equivalent via inductances (as shown in Fig. 2) of inner and outer shells $\left(L_{\text{via,out}} \text{ and } L_{\text{via,in}}\right)$ can be expressed as

$L_{\text{via,out}} = l'_{\text{via,out}} \times h_{\text{TSV}}$ and $L_{\text{via,in}} = l'_{\text{via,in}} \times h_{\text{TSV}}$; where $l'_{\text{via,out}}$ and $l'_{\text{via,in}}$ are the *p.u.l.* equivalent via inductances of outer and inner shells as shown in Fig. 2.

where $L_{\text{via,in}}$ and $L_{\text{via,out}}$ of Fig. 2 can be expressed as:

$$L_{\text{via,out}} = L_{\text{TSV,out}} + L_{\text{Bump,out}} \tag{3}$$

$$L_{\text{via,in}} = L_{\text{TSV,in}} + L_{\text{Bump,in}} \tag{4}$$

The expressions of $L_{\text{TSV,out}}$ and $L_{\text{Bump,out}}$ are provided in [20], wherein the expressions of $L_{\text{TSV,in}}$ and $L_{\text{Bump,in}}$ can be calculated considering $d_{\text{TSV}} = d_{\text{inner}}$ and $d_{\text{Bump}} = d_{\text{Bump,inner}}$ respectively.

**Conductance**

The existence of Si substrate that surrounds the TSV is lossy in nature and has a substantial effect on the substrate loss of a via line due to the substrate conductivity ($\sigma_{\text{Si}}$) and that is the primary cause of substrate conductance ($G_{\text{Si}}$). Therefore, the $G_{\text{Si}}$ can be expressed as

$G_{\text{Si}} = g'_{\text{Si}} \times h_{\text{TSV}}$; where $g'_{\text{Si}}$ is the *p.u.l.* conductance of Si substrate as shown in Fig. 2.

The expression of $G_{\text{Si}}$ can be obtained as

$$G_{Si} = \frac{\pi \times \sigma_{Si} \times h}{\cosh^{-1}\left(\frac{p_{\text{TSV}}}{d_{\text{TSV}}}\right)} \tag{5}$$

where $h = h_{\text{TSV}} - h_{\text{IMD}}$

**Via Capacitance**

TSVs are associated with different kinds of capacitive parasitics that can be described as follows:

*Substrate Capacitance*

The semiconducting characteristic of the Si substrate that acts as an insulator, causes a capacitive coupling between the outer TSV shell and neighboring TSV. Hence, the substrate capacitance ($C_{\text{Si}}$) is a critical parasitic to consider. In general, $C_{\text{Si}}$ can be expressed as

$C_{\text{Si}} = c'_{\text{Si}} \times h_{\text{TSV}}$; where $c'_{\text{Si}}$ is the *p.u.l.* Si substrate capacitance as shown in Fig. 2.

However, the expression of $C_{\text{Si}}$ can be obtained as:

$$C_{\text{Si}} = \frac{\pi \times \varepsilon_0 \varepsilon_{r,Si} \times h}{\cosh^{-1}\left(\frac{p_{\text{TSV}}}{d_{\text{TSV}}}\right)} \tag{6}$$

*Equivalent TSV Capacitance*

An equivalent TSV capacitance ($C_{\text{TSV}}$) constitutes a parallel combination of liner capacitance ($C_{\text{liners}}$) due to the different liner layers placed in a tubular dielectric medium and a bump capacitance $\left(C_{\text{Bump(Btw)}}\right)$ that is due to the separation between inner and outer surfaces of bump because of the presence of a tubular dielectric medium. In general, $C_{\text{TSV}}$ can be defined as

$C_{\text{TSV}} = c'_{\text{TSV}} \times h_{\text{TSV}}$; where $c'_{\text{TSV}}$ is the *p.u.l.* TSV equivalent capacitance as shown in Fig. 2.

$$C_{\text{TSV}} = C_{\text{Bump(Btw)}} + C_{\text{liners}} \tag{7}$$

$$C_{\text{liners}} = 2\pi \times \varepsilon_0 \varepsilon_{r,\text{SiO}_2} \varepsilon_{r,\text{polymer liner}} \times \frac{h}{\ln\left(\frac{d_{\text{inner}}/2 + t_{\text{Total}}}{d_{\text{inner}}/2}\right)} \tag{8}$$

$$C_{\text{Bump(Btw)}} = 2\pi \times \varepsilon_0 \varepsilon_{r,\text{SiO}_2} \varepsilon_{r,\text{polymer liner}} \times \frac{h}{\ln\left(\frac{d_{\text{Bump,inner}}/2 + t_{\text{Total}}}{d_{\text{Bump, inner}}/2}\right)} \tag{9}$$

*Insulating Capacitance*

An insulating capacitance ($C_{\text{ins}}$) is the series combination of oxide and depletion capacitances $\left(C_{\text{ox}}\text{ and } C_{\text{dep}}\right)$ that can be obtained as

$C_{\text{ins}} = c'_{\text{ins}} \times h_{\text{TSV}}$; where $c'_{\text{ins}}$ is the *p.u.l.* insulating capacitance as shown in Fig. 2.

$$C_{\text{ins}} = \frac{C_{\text{ox}} \times C_{\text{dep}}}{C_{\text{ox}} + C_{\text{dep}}} \tag{10}$$

However, the expressions of $C_{\text{ox}}$ and $C_{\text{dep}}$ are provided in [21].

*Bump Capacitances*

Bump-to-silicon substrate capacitances $\left(C_{\text{Bump1}}\text{ and } C_{Bump2}\right)$ exist between the upper and lower bumps and the Si substrate due to the IMD and bottom oxide layers, respectively. These capacitances can be expressed as

$C_{\text{Bump1}} = c'_{\text{Bump1}} \times h_{\text{TSV}}$ and $C_{\text{Bump2}} = c'_{\text{Bump2}} \times h_{\text{TSV}}$; where $c'_{\text{Bump1}}$ and $c'_{\text{Bump2}}$ are the *p.u.l.* upper and lower bump-to-silicon substrate capacitances as shown in Fig. 2.

Although, the expressions of $C_{\text{Bump1}}$ and $C_{\text{Bump2}}$ can be obtained from [20].

*Underfill, IMD, and Bottom Capacitance*

The underfill and IMD capacitances ($C_{\text{Underfill}}$ and $C_{\text{IMD}}$) occur between the upper bump and the Si substrate due to the underfill and IMD layers, respectively. Similarly, a bottom capacitance ($C_{\text{Bottom}}$) forms the lower bump and Si substrate due to the bottom oxide layer. So, $C_{\text{Underfill}}$, $C_{\text{IMD}}$, and $C_{\text{Bottom}}$ can be expressed as: $C_{\text{Underfill}} = c'_{\text{Underfill}} \times h_{\text{TSV}}$, $C_{IMD} = c'_{\text{IMD}} \times h_{\text{TSV}}$, and $C_{\text{Bottom}} = c'_{\text{Bottom}} \times h_{\text{TSV}}$; where $c'_{\text{Underfill}}$, $c'_{\text{IMD}}$, and $c'_{\text{Bottom}}$ are the *p.u.l.* underfill, IMD, and bottom capacitances (as shown in Fig. 2). The equations of $C_{\text{Underfill}}$ and $C_{\text{IMD}}$ can be obtained from [22] and $C_{\text{Bottom}}$ is provided in [20].

Table 5 presents the via parasitics for a coaxial TSV at 32 nm technology based on technology-dependent physical parameters of the TSV.

## 3 Impact of a Tubular Dielectric Medium on a Coaxial TSV

This section demonstrates the impact of a tubular dielectric medium for the proposed *RLGC* model by using a DVL setup at 32 nm technology. Figure 3 represents a 2-line DVL setup employed for circuit-level simulation using the coaxial TSV employing a *T*-type network. The via line in the DVL of Fig. 3 primarily represents the *RLGC* parasitics of the TSV as depicted in Fig. 2. The via lines are driven by the CMOS drivers with the supply voltage of 0.9 V at 32 nm technology and each line is terminated with a load capacitance of $C_{\text{load}} = 0.1$ aF.

**Table 5** Parasitic values of a coaxial TSV for different TSV heights and inner shell radius

| Inner Radius (a) for $h_{TSV}$ = 20 μm | Polymer liner material | Parasitic values of a Coaxial shaped TSV using a combination of $SiO_2$ and different polymer liners in a tubular dielectric medium | | | | | | | | | | | | |
|---|---|---|---|---|---|---|---|---|---|---|---|---|---|---|
| | | $C_{TSV}$ (pF) | $C_{ins}$ (pF) | $G_{Si}$ (mΩ) | $C_{Si}$ (fF) | $C_{Bottom}$ (fF) | $C_{IMD}$ (fF) | $C_{underfill}$ (fF) | $C_{Bump2}$ (pF) | $C_{Bump1}$ (pF) | $L_{via,in}$ (pH) | $L_{via,out}$ (pH) | $R_{via,in}$ (Ω) | $R_{via,out}$ (Ω) |
| $a =$ 1.35 μm | BCB | 0.1161 | 0.0626 | 0.3576 | 3.7682 | 0.0024 | 0.4222 | 1.2234 | 0.1118 | 0.0006 | 6.5363 | 3.2630 | 0.2483 | 0.0568 |
| | PPC | 0.1271 | 0.0626 | 0.3576 | 3.7682 | 0.0024 | 0.4222 | 1.2234 | 0.1118 | 0.0006 | 6.5363 | 3.2630 | 0.2483 | 0.0568 |
| | PI | 0.1534 | 0.0626 | 0.3576 | 3.7682 | 0.0024 | 0.4222 | 1.2234 | 0.1118 | 0.0006 | 6.5363 | 3.2630 | 0.2483 | 0.0568 |
| $a =$ 1.8 μm | BCB | 0.0965 | 0.0604 | 0.3576 | 3.7682 | 0.0032 | 0.4222 | 1.2234 | 0.0836 | 0.0006 | 5.8171 | 3.2630 | 0.1752 | 0.0568 |
| | PPC | 0.1056 | 0.0604 | 0.3576 | 3.7682 | 0.0032 | 0.4222 | 1.2234 | 0.0836 | 0.0006 | 5.8171 | 3.2630 | 0.1752 | 0.0568 |
| | PI | 0.1274 | 0.0604 | 0.3576 | 3.7682 | 0.0032 | 0.4222 | 1.2234 | 0.0836 | 0.0006 | 5.8171 | 3.2630 | 0.1752 | 0.0568 |
| $a =$ 2.25 μm | BCB | 0.1161 | 0.0583 | 0.3576 | 3.7682 | 0.0041 | 0.4222 | 1.2234 | 0.0666 | 0.0006 | 5.2593 | 3.2630 | 0.1353 | 0.0568 |
| | PPC | 0.1271 | 0.0583 | 0.3576 | 3.7682 | 0.0041 | 0.4222 | 1.2234 | 0.0666 | 0.0006 | 5.2593 | 3.2630 | 0.1353 | 0.0568 |
| | PI | 0.1534 | 0.0583 | 0.3576 | 3.7682 | 0.0041 | 0.4222 | 1.2234 | 0.0666 | 0.0006 | 5.2593 | 3.2630 | 0.1353 | 0.0568 |
| $a =$ 2.7 μm | BCB | 0.1161 | 0.0564 | 0.3576 | 3.7682 | 0.0049 | 0.4222 | 1.2234 | 0.0553 | 0.0006 | 4.8035 | 3.2630 | 0.1103 | 0.0568 |
| | PPC | 0.1271 | 0.0564 | 0.3576 | 3.7682 | 0.0049 | 0.4222 | 1.2234 | 0.0553 | 0.0006 | 4.8035 | 3.2630 | 0.1103 | 0.0568 |
| | PI | 0.1534 | 0.0564 | 0.3576 | 3.7682 | 0.0049 | 0.4222 | 1.2234 | 0.0553 | 0.0006 | 4.8035 | 3.2630 | 0.1103 | 0.0568 |
| Inner Radius (a) for $h_{TSV} = 50$ μm | | | | | | | | | | | | | | |
| $a =$ 1.35 μm | BCB | 0.3484 | 0.1879 | 0.0011 | 0.0113 | 0.0024 | 0.4222 | 1.2234 | 0.1118 | 0.0006 | 14.623 | 7.4219 | 0.5713 | 0.13 |
| | PPC | 0.3813 | 0.1879 | 0.0011 | 0.0113 | 0.0024 | 0.4222 | 1.2234 | 0.1118 | 0.0006 | 14.623 | 7.4219 | 0.5713 | 0.13 |
| | PI | 0.4602 | 0.1879 | 0.0011 | 0.0113 | 0.0024 | 0.4222 | 1.2234 | 0.1118 | 0.0006 | 14.623 | 7.4219 | 0.5713 | 0.13 |

(continued)

**Table 5** (continued)

| Inner Radius (a) for $h_{TSV}$ = 20 μm | Polymer liner material | Parasitic values of a Coaxial shaped TSV using a combination of $SiO_2$ and different polymer liners in a tubular dielectric medium | | | | | | | | | | | | |
|---|---|---|---|---|---|---|---|---|---|---|---|---|---|---|
| | | $C_{TSV}$ (pF) | $C_{ins}$ (pF) | $G_{Si}$ (mΩ) | $C_{Si}$ (fF) | $C_{Bottom}$ (fF) | $C_{IMD}$ (fF) | $C_{underfill}$ (fF) | $C_{Bump2}$ (pF) | $C_{Bump1}$ (pF) | $L_{via,in}$ (pH) | $L_{via,out}$ (pH) | $R_{via,in}$ (Ω) | $R_{via,out}$ (Ω) |
| $a$ = 1.8 μm | BCB | 0.2895 | 0.1812 | 0.0011 | 0.0113 | 0.0032 | 0.4222 | 1.2234 | 0.0836 | 0.0006 | 13.041 | 7.4219 | 0.4023 | 0.13 |
| | PPC | 0.3168 | 0.1812 | 0.0011 | 0.0113 | 0.0032 | 0.4222 | 1.2234 | 0.0836 | 0.0006 | 13.041 | 7.4219 | 0.4023 | 0.13 |
| | PI | 0.3824 | 0.1812 | 0.0011 | 0.0113 | 0.0032 | 0.4222 | 1.2234 | 0.0836 | 0.0006 | 13.041 | 7.4219 | 0.4023 | 0.13 |
| $a$ = 2.25 μm | BCB | 0.3484 | 0.1751 | 0.0011 | 0.0113 | 0.0041 | 0.4222 | 1.2234 | 0.0666 | 0.0006 | 11.814 | 7.4219 | 0.3105 | 0.13 |
| | PPC | 0.3813 | 0.1751 | 0.0011 | 0.0113 | 0.0041 | 0.4222 | 1.2234 | 0.0666 | 0.0006 | 11.814 | 7.4219 | 0.3105 | 0.13 |
| | PI | 0.4602 | 0.1751 | 0.0011 | 0.0113 | 0.0041 | 0.4222 | 1.2234 | 0.0666 | 0.0006 | 11.814 | 7.4219 | 0.3105 | 0.13 |
| $a$ = 2.7 μm | BCB | 0.3484 | 0.1693 | 0.0011 | 0.0113 | 0.0049 | 0.4222 | 1.2234 | 0.0553 | 0.0006 | 10.811 | 7.4219 | 0.2529 | 0.13 |
| | PPC | 0.3813 | 0.1693 | 0.0011 | 0.0113 | 0.0049 | 0.4222 | 1.2234 | 0.0553 | 0.0006 | 10.811 | 7.4219 | 0.2529 | 0.13 |
| | PI | 0.4602 | 0.1693 | 0.0011 | 0.0113 | 0.0049 | 0.4222 | 1.2234 | 0.0553 | 0.0006 | 10.811 | 7.4219 | 0.2529 | 0.13 |

*$a$ = inner shell radius

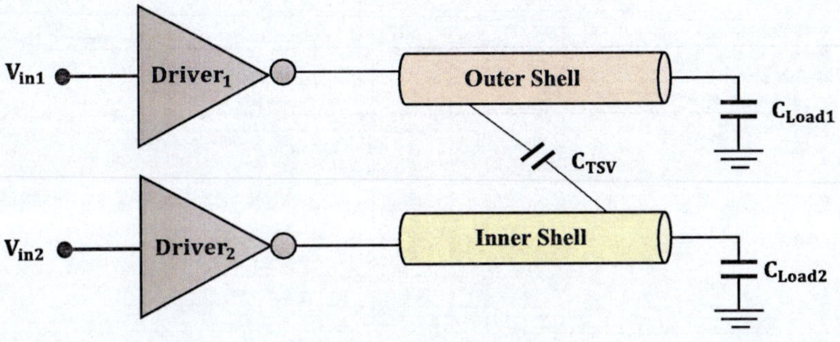


**Fig. 3** Schematic view of the Driver-Via-Load (DVL) configuration

The CMOS driver is employed due to its operation in both the linear and saturation regions whereas a resistive driver can only be operational in the linear region [21].

Using the aforementioned DVL configuration shown Fig. 3, the consequent sub-sections examined the influence of a tubular dielectric medium on the overall reliability of the coaxial TSV in terms of crosstalk induced delay and peak noise for different TSV heights and inner radius.

## *3.1 Impact of Tubular Dielectric Medium on Peak Noise*

Using the DVL configuration illustrated in Fig. 3, this sub-section examines the impact of a tubular dielectric medium on the via performance of coaxial TSV, considering the noise effect in terms of peak voltage for various via heights and inner shell radius of the coaxial TSV. The quantitative values of the TSV parasitics (indicated in Table 4) are used to investigate the impact of peak noise under the influence of tubular dielectric medium.

Noise coupling exists in the substrate region whenever fast signal transitions occur in neighboring TSVs. This noise coupling phenomenon is similar to that which occurs in the inner and outer conductors of coaxial TSVs due to the presence of a tubular dielectric medium with oxide and polymer insulating layers. The switching behaviors of coupled via lines can be used to assess the performance of the TSV in terms of peak noise and crosstalk interference. Peak noise is observed when the aggressor line is excited with a pulse input and the victim line is grounded. Under this scenario, the victim line is subjected to a voltage spike in the form of peak noise based on functional crosstalk. It exhibits unintended spikes on the victim line that appears to be the source of faults in digital circuit [22]. By varying the inner shell radius of the coaxial TSV, Figs. 4a and b present the peak noise for different via heights of $h_{TSV} = 20\ \mu\text{m}$ and $h_{TSV} = 50\ \mu\text{m}$, respectively, at 32 nm technology considering the impact of a tubular dielectric medium with different polymer liners.

It can be noticed that the coupling capacitance between inner and outer shells of coaxial TSV becomes significant for a reduced dielectric area. Hence, the dielectric

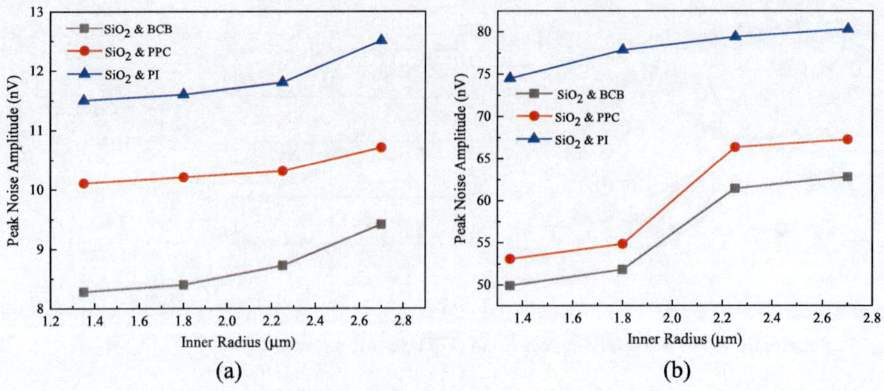


**Fig. 4** Peak noise of a coaxial TSV for via heights of **a** $h_{TSV} = 20$ μm and **b** $h_{TSV} = 50$ μm at 32 nm technology

medium related capacitive noise coupling is one of the primary reliability issue in case of a coaxial TSV. In this regard, from Fig. 4, it can be investigated that the unintentional spike in voltage in the victim line increases significantly for higher TSV and larger inner radius due to its *p.u.l.* increase in the parasitic values. Hence, the overall increase in peak noise is predominantly contributed to the cumulative effect of coupling capacitance and via resistance. It can be investigated that the percentage increase in peak voltage of the coaxial TSV using BCB polymer dielectric with $a = 2.7$ μm compared to $a = 1.35$ μm is 13.64% and 25.85% at via heights of 20 μm and 50 μm, respectively at 32 nm technology. Apart from this, using Fig. 4, it can also be inferred that compared to those other polymer dielectrics such as PPC and PI, the inclusion of BCB in the tubular dielectric medium of coaxial TSV yields the minimum voltage spike in the victim line. This is because of employing BCB as a dielectric medium to isolate the via lines primarily results in a lower coupling capacitance and hence lesser crosstalk interference. It is due to the lower dielectric constant of BCB ($\varepsilon_r = 2.6$) compared to PPC ($\varepsilon_r = 2.9$) and PI ($\varepsilon_r = 3.5$). Furthermore, BCB has a higher CTE (= 42 ppm/K) than other polymer dielectrics that can sustain severe thermal stress owing to high temperatures, results in reduced leakage between the inner and outer shells, leading to substantially lower peak noise. The significant reduction in peak noise of coaxial TSV using BCB polymer with $a = 1.35$ μm compared to PPC and PI are 21.98% and 38.88%, respectively, at via height of 20 μm. Similarly, these percentage reductions are observed as 6.12% and 49.18%, respectively, at via height of 50 μm.

## 3.2 Impact of Tubular Dielectric on Crosstalk Induced Delay

This sub-section demonstrates the performance of coupled shells of a coaxial TSV under the influence of a tubular dielectric medium associated with different polymer liners. The analysis is carried out for the presented model using DVL setup as depicted in Fig. 3 at 32 nm technology. In TSV, each signal line experiences significant delay as the signal passes through it.

When all signals are in the opposite switching transitions from one other, out-phase dynamic crosstalk occurs, causing worst-case crosstalk interference with a considerable Miller Coupling Factor (MCF). Under this situation, one of the signal lines acts as a victim that experiences more delay when compared with the other line, referred as an aggressor.

Using different via parasitics introduced in Tables 4, an efficient analysis is observed in terms of crosstalk induced delay by varying the inner shell radius as depicted in Fig. 5 a and b at via heights of 20 μm and 50 μm, respectively, at 32 nm technology. It is evident (shown in Fig. 5) that the coaxial TSV with BCB polymer dielectric exhibits lesser crosstalk induced delay than the TSV with PPC and PI dielectrics irrespective of the via height and inner shell radius under consideration. This is due to the fact that the BCB polymer material has a lesser ability to withstand electrostatic forces between the inner and outer shells of the coaxial TSV, resulting in reduced coupling capacitance ($C_{TSV}$) of BCB than PPC or PI. The percentage improvement in crosstalk induced delay of the coaxial TSV using BCB polymer compared to PI can be observed as 1.98% and 2.5% at $h_{TSV} = 20$ μm and $h_{TSV} = 50$ μm, respectively, considering $a = 1.35$ μm. Similarly, these improvements can be observed as 7.62% and 4.81% when $a = 2.7$ μm. It is due to the lower relative permittivity of BCB ($\varepsilon_r = 2.65$) compared to PPC ($\varepsilon_r = 2.9$) and PI ($\varepsilon_r = 3.5$).

Further, considering the crosstalk induced delay of a coaxial TSV with different scenario, Table 6 presents the percentage reduction in crosstalk induced delay

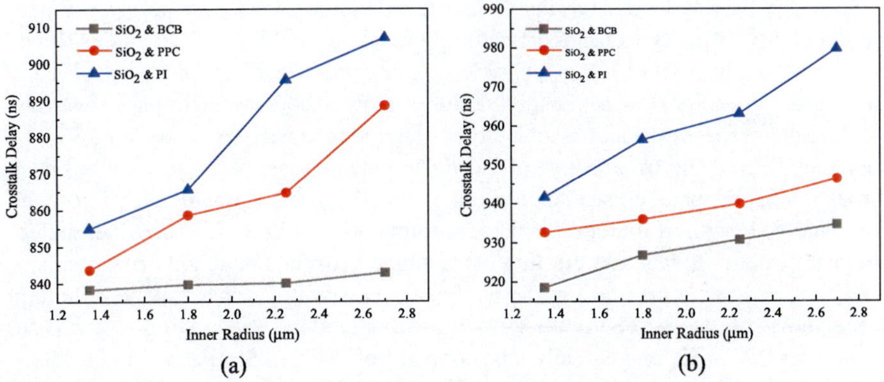


**Fig. 5** Crosstalk induced delay of a coaxial TSV at **a** $h_{TSV} = 20$ μm and **b** $h_{TSV} = 50$ μm, respectively, at 32 nm technology

**Table 6** % reduction in crosstalk induced delay employing various polymer dielectrics w.r.t. BCB polymer dielectric

| Inner radius (a) (μm) | $h_{TSV} = 20$ μm | | $h_{TSV} = 50$ μm | |
|---|---|---|---|---|
| | % reduction in delay employing various liners | | | |
| | PPC (%) | PI (%) | PPC (%) | PI (%) |
| 1.35 | 0.63 | 1.94 | 1.52 | 2.51 |
| 1.8 | 2.25 | 3.08 | 0.981 | 3.19 |
| 2.25 | 2.91 | 6.60 | 0.977 | 3.47 |
| 2.7 | 5.40 | 7.61 | 1.25 | 4.81 |

using different polymer dielectrics compared to the BCB. It is observed that the crosstalk delay significantly decreases for lower via height and inner shell radius (also presented in Fig. 5) irrespective of the polymer liner used in tubular dielectric medium. Considering $a = 1.35$ μm, it can be observed that the significant percentage improvement in crosstalk induced delay of the coaxial TSV using BCB polymer dielectric compared to the PPC and PI are 0.63% and 1.94%, respectively, (as presented in Table 5) at the via height of 20 μm, whereas these percentage improvements are 1.52% and 2.51%, respectively, at via height of 50 μm. The key reason behind this fact is that the quantitative values of via resistance and coupling capacitance parasitics are substantially lower for a reduced via dimension, as summarized in Table 4. This is because the lesser quantitative values of via resistance and coupling capacitance due to lower structural parameters of TSVs (ex. height and diameter of TSV) resulting in improved crosstalk induced delay.

## 4 Conclusion

In this paper, the impact of a tubular dielectric medium with different polymer dielectric liners are critically examined for the presented novel *T*-type equivalent *RLGC* model of a coaxial TSV at 32 nm technology. The presented *RLGC* model is validated for different via height and inner shell radius in order to address the impact of various polymer dielectrics on peak noise and crosstalk induced delay of a coaxial TSV at a frequency of 20 GHz. Irrespective of the via height and inner shell radius, it has been witnessed that the peak noise and crosstalk induced delay are considerably improved in coaxial TSV with a tubular dielectric medium comprising BCB as polymer dielectric that introduces the least coupling capacitance between the neighboring shells. It is investigated that, considering $a = 1.35$ μm, the significant percentage reduction in peak noise of coaxial TSV with BCB liner compared to PPC and PI dielectrics are 21.98% and 38.88%, respectively, at via height of 20 μm. Similarly, the considerable percentage improvement in crosstalk induced delay of coaxial TSV with BCB compared to the PPC and PI dielectrics are 0.63% and 1.94%, respectively, at via height of 20 μm. Furthermore, regardless of the polymer liner employed in the tubular dielectric medium, the peak noise and crosstalk induced delay is reduced considerably for lower via height and inner shell radius. Therefore, a coaxial shaped TSV with the BCB polymer in its tubular dielectric medium can be considered as the best suited TSV configuration for improved electrical performance in terms of peak noise and crosstalk interference for the future VLSI TSVs in 3D ICs.

## References


1. Zhao WS, Zheng J, Hu Y, Wang G, Dong L, Yu L, Yin WY (2016) High-frequency analysis of Cu-carbon nanotube composite through-silicon vias. IEEE Trans Nanotechnol 15(3):506–511
2. Liu B, Li C, Li C, Zhang S (2021) Effect of temperature and single event transient on crosstalk in coupled single-walled carbon nanotube (SWCNT) bundle interconnects. Int J Circuit Theory Appl 49(10):3408–3420
3. Jin J, Zhao WS, Wang DW, Chen HS, Li EP, Yin WY () Investigation of carbon nanotube-based through-silicon vias for PDN applications. IEEE Trans Electromagn Compat 60(3):738–746 92018
4. Nabil A, Bernardo JA, Ma Y, Abouelatta M, Shaker A, Bouchet LF, Ragai H, Gontrand C (2020) Electrical modeling of tapered TSV including MOS-field effect and substrate parasitics: analysis and application. Microelectron J 100:104797
5. Hu QH, Zhao WS, Fu K, Wang DW, Wang G (2021) Electrical modeling of carbon nanotube-based shielded through-silicon vias for three-dimensional integrated circuits. Int J Numer Model Electron Netw Devices Fields 34(3):e2842
6. Qian L, He X, Qian K, Xia Y (2018) Study of silicon core coaxial through silicon via for three dimensional integration. In: International proceedings on symposium on circuits and systems (ISCAS). IEEE, Florence, Italy, pp 1–4
7. Qian L, Xia Y, Qian K, Wang J (2018) Electrical modeling and characterization of silicon-core coaxial through-silicon vias in 3-D integration. IEEE Trans Comp Packag Manufact Technol 8(8):1336–1343
8. Xu Z, Lu JQ (2012) Three-dimensional coaxial through-silicon-via (TSV) design. IEEE Electron Device Lett 3(10):1441–1443
9. Chen H, Shi X, Wang J, Hu Y, Wang Q, Cai J (2021) Development of hybrid bonding process for embedded bump with Cu-Sn/BCB structure. In: 71st International proceedings on electronic components and technology conference (ECTC). IEEE, San Diego, CA, USA, pp 476–480
10. Xue C, Cheng Z, Chen Z, Yan Y, Cai Z, Ding Y (2018) Elimination of scallop-induced stress fluctuation on through-silicon-vias (TSVs) by employing polyimide liner. IEEE Trans Device Mater Reliab 8(2):266–272
11. Basha SJ, Kumar VR (2020) Design of MWCNT based through silicon vias with polymer liners to reduce the crosstalk effects. ECS J Solid State Sci Technol 9(4):041002
12. Zhang Z, Ding Y, Xiao L, Cai Z, Yang B, Wu Z, Su Y, Chen Z (2021) Development of Cu seed layers in ultra-high aspect ratio through-silicon-vias (TSVs) with small diameters. In: 71st International proceedings on electronic components and technology conference (ECTC). IEEE, San Diego, CA, USA, pp 1904–1909
13. Lee M, Jung DH, Kim H, Cho J, Kim J (2016) High-frequency temperature-dependent through-silicon-via (TSV) model and high-speed channel performance for 3-D ICs. IEEE DesignTest 33(2):17–29
14. Su J, Wang F, Zhang W (2015) Capacitance expressions and electrical characterization of tapered through-silicon vias for 3-D ICs. IEEE Trans Compon Packag Manuf Technol 5(10):1488–1496
15. Fu K, Zhao WS, Wang G, Swaminathan M (2018) Modeling and performance analysis of shielded differential annular through-silicon via (SD-ATSV) for 3-D ICs. IEEE Access 6:33238–33250

16. Su J, Zhang W, Yao C (2019) Partial coaxial through-silicon via for suppressing the substrate noise in 3-dimensional integrated circuit. IEEE Access 7:98803–98810
17. Chandrakar S, Gupta D, Majumder MK (2021) Role of through silicon via in 3D integration: impact on delay and power. J Circuits, Syst Comput 30(3):2150051
18. Khan NH, Alam SM, Hassoun S (2011) Mitigating TSV-induced substrate noise in 3-D ICs using GND plugs. In: 12th International proceedings of international symposium on quality electronic design. IEEE, Santa Clara, CA, USA, pp 1–6
19. Khan NH, Alam SM, Hassoun S (2013) GND plugs: a superior technology to mitigate TSV-induced substrate noise. IEEE Trans Compon Packag Manufact Technol 3(5):849–857
20. Kim J, Pak JS, Cho J, Song E, Cho J, Kim H, Song T, Lee J, Lee H, Yang S, Suh MS, Byun KY, Kim J (2011) High-frequency scalable electrical model and analysis of a through silicon via (TSV). IEEE Trans Compon Packag Manufact Technol 1(2):181–195
21. Sahu CC, Kumbhare VR, Majumder MK Compact AC modeling of eddy current for cylindrical through silicon via. IETE J Res 1–13 92021
22. Sahu CC, Anand S, Majumder MK (2021) An analysis of the eddy effect in through-silicon vias based on Cu and CNT bundles: the impact on crosstalk and power 20(6):2456–2470

# Novel Approach for the Reduction of Critical Paths in Static Timing Analysis Without Degradation in QOR

**K. Ranjit Kannan and G. Lakshminarayanan**

**Abstract** This paper focuses on grading the timing paths for static timing analysis (STA) without any compromise in Quality of Results (QOR). STA is an exhaustive methodology to verify timing aspect of the design. Typically, in synchronous digital designs, all logic timing paths covering various scenarios are included for STA run. But this is very pessimistic and often leads to overdesign, more power/area and increased time and effort. The proposed technique is used for systematically grading the timing paths followed by the implementation methodology to exclude invalid critical paths for STA and optimization, which could otherwise pose challenges or be a deterrent for design closure. Statistically estimated path delay results of undetected faults show that in a typical design, more than 50% of the critical paths are not valid for timing. An application-specific algorithm has also been proposed for efficient grading of timing paths.

**Keywords** Static timing analysis · Engineering change order · STA path grading

K. Ranjit Kannan (✉)
Analog Devices, Bangalore, India
e-mail: ranjit.kannan@analog.com

G. Lakshminarayanan
National Institute of Technology Tiruchirappalli, Tiruchirappalli, India
e-mail: laksh@nitt.edu



C. Giri et al. (eds.), *Emerging Electronic Devices, Circuits and Systems*,
Lecture Notes in Electrical Engineering 1004,
https://doi.org/10.1007/978-981-99-0055-8_17

# 1 Introduction

Static timing analysis (STA) is a proven and established methodology to ensure timing and timing related functional sanctity of standard cell digital design and complements exhaustive functional simulations needed to catch real issues in design. With advanced technologies, more and more logic gates are accommodated in the same chip. This means that in the design phase there are numerous paths for timing closure. Often few critical timing paths are difficult to close and would require multiple Engineering Change Order (ECO) iterations and, in many cases, additional manual effort. A significant part of physical design cycle time is spent in fixing timing critical paths with adverse impact on time to market.

A valid question would be: Are all those timing paths functionally valid? Depending on primary inputs combinations, various choices of local inputs (driver flops output) are available for the targeted combinational logic between launch registers and capture register. STA is an exhaustive methodology that covers all possible scenarios. Static Timing Analysis with all possible timing paths sounds robust but in actual case is pessimistic. The extent to which the timing paths can be trimmed are significant and the benefits in terms of reduction in design cycle time and efforts for design closure are substantial.

# 2 Proposed Technique

## *2.1 Path Delay Analysis*

STA evaluates timing paths (estimates total delay through combination logic) from every launch flop to each possible capture flop. The total delay from a launch path to a capture flop through combinational logic will include clock-q delay of launch flop (flop delay), cell delay of the gates in that path corresponding to the cell timing arc and net delay. For a path from source to destination flop to be active and dominate in terms of delay, the value at source should propagate to the target destination through the combinational gates in its path. This implies that the combinational gates in this path should have their other input(s) in a state which will allow the mainstream path input to pass through.

In practical cases, it is high likely that, no possible combination of values at primary flops exist to allow data propagation through the timing path under consideration. There are several reasons, the main being overlaps/conflicts/redundancy in combinational logic [1].

An indirect but an effective approach was chosen to statistically estimate and show that on an average 50% of timing critical paths are not really timing critical as they do not under any realistic circumstance can be exercised (cannot be sensitized). This was done through deterministic robust test pattern generation for path delay faults [1–3] targeting critical paths. Unlike stuck at fault pattern generation, the pattern is

**Table 1** Path delay coverage in typical digital designs

| S. No. | Project name | Path delay coverage (%) | Pattern count | Timing simulations |
|---|---|---|---|---|
| 1 | Hxxx-Rev2 | 29.17 | 9 | PASS |
| 2 | Mxxx | 20 | 40 | PASS |
| 3 | Hxxx-Rev6-T40 Package | 33.33 | 17 | PASS |
| 4 | Hxxx-Rev6-T48 Package | 88.2 | 12 | PASS |
| 5 | Hxxx-Rev6-T56 Package | 37.50 | 15 | PASS |
| 6 | Axxx-Rev-2 | 30.03 | 7 | PASS |
| 7 | Bxxx-Rev-1 | 34.26 | 25 | PASS |
| 8 | Axxx-Rev-1 | 39.50 | 364 | PASS |
| 9 | Axxx-Rev-1 | 45.90 | 421 | PASS |
| 10 | Axxx-Rev-1 | 14.50 | 23 | PASS |

not shifted directly through scan chain but through the fan-in combinational logic in the preceding cycle. This implies that, with full controllability (through scan shift) at the $n - 1$ stage, the test pattern is generated for $n + 1$ stage. Real functional scenario will involve multiple stages, the initial being primary inputs. Undetected faults due to no feasible path delay test vector implies that the path is not exercisable. Tool limitation is ignored for this evaluation. Table 1 the results of this investigation.

## 2.2 *Timing Analysis of a Path and Data Propagation*

Timing analysis report for critical paths provides in detail the data path through which the delay is calculated. The details include the logic gates in sequence, the timing arcs (causal relation from input to output), output rise/fall, the delay associated, etc. A sample timing path through a combination logic is shown in Fig. 1.

In order to propagate the value (as defined in STA report for the path) through each logic gate in the path, the other inputs should have a supporting value to pass the mainstream value. This can be extended to Stuck-At-Fault (SAF) pattern generation technique used in Design For Testability (DFT) [4]. D-Algorithm is very popular test pattern generation algorithm, and many industry standard tools use this for test generation for fault detection.

The concept of Propagation D Cube (PDC), part of D algorithm, can be used for identification of the supporting values in other input(s) of a gate. Propagation D cubes (PDC) of AND logic gate is shown in Fig. 2.

```
Point                                                        Incr        Path
------------------------------------------------------------------------------
clock SYS_2x_CLK (rise edge)                                 0.000       0.000
clock network delay (propagated)                             2.846 *     2.846
I_ORCA_TOP/I_RESET_BLOCK/sys_2x_rst_n_buf_reg/CP (sdcrq1)
                                                             0.000       2.846 r
I_ORCA_TOP/I_RESET_BLOCK/sys_2x_rst_n_buf_reg/Q (sdcrq1)
                                                             0.418 *     3.264 r
I_ORCA_TOP/I_RESET_BLOCK/U92/B1 (aor21d2)                    0.000 *     3.264 r
I_ORCA_TOP/I_RESET_BLOCK/U92/Z (aor21d2)                     0.637 *     3.901 r
I_ORCA_TOP/I_RESET_BLOCK/sys_2x_rst_n (RESET_BLOCK)          0.000 *     3.901 r
I_ORCA_TOP/I_RISC_CORE/reset_n (RISC_CORE)                   0.000 *     3.901 r
I_ORCA_TOP/I_RISC_CORE/I_ALU/reset_n (ALU)                   0.000 *     3.901 r
I_ORCA_TOP/I_RISC_CORE/I_ALU/Neg_Flag_reg/CDN (sdcrb1)
                                                             0.073 *     3.974 r
data arrival time                                                        3.974
```

**Fig. 1** Timing path reported in STA tool (PT—synopsys)

| a | b | out |
|---|---|---|
| **1** | D/D’ | D/D’ |
| **D/D’** | 1 | D/D’ |
| **D/D’** | D/D’ | D/D’ |

**Fig. 2** PDC of AND logic gate

*D* can be either be 1 or 0; and $D'$ is complement of *D*. To propagate a value ($D/D'$) from input to output, a value for the other input(s) corresponding to the gate as described from the PDC table has to be chosen. In short, this is a value which will allow the input (mainstream) to reflect a change in output as in STA report. The alternate input value for each gate in the path can be estimated as per mainstream value in the STA report. The alternate value in each gate is backtracked till appropriate values for primary flop inputs are established or results in conflict with other backtracked values.

## *2.3 Application-Specific Algorithm*

A more application-specific algorithm using STA, Propagation D cube and Boolean algebra is developed to grade timing paths for static timing analysis. The primary objective of this hybrid algorithm is to find out if any real-case test pattern(s) exist for primary critical paths.

To estimate the test pattern(s) of the primary critical path(s), Boolean expressions of the gate logic corresponding to the required value of the supporting inputs can be derived for every gate in the path [4]. This is expressed as Sum Of Products (SOP) of their primary inputs (Flip-Flops Outputs/primary inputs). Common term(s) in the SOP expressions of all supporting inputs correspond to feasible test vector(s). If no common term exists, no feasible real-case pattern for the path is possible.

In the event of few (limited to one/two) feasible test vectors, the validity of these test vectors can further be scrutinized by graph-based modeling techniques.

If no feasible real-case test pattern is found, then this path should be excluded for STA by adding appropriate timing exception constraints. It is more efficient to include only timing critical paths for this grading.

## 3 Implementation Methodology

With Multi-Mode, Multi-Corner scenarios (MMMC), the number of timing views for timing design closure has gone up and any reduction in invalid critical paths would speed up the design process and the flow.

The methodology applies to two flows.

1. Post-Synthesis physical design—Proactive Approach
2. Post-Layout ECO timing fix—Reactive Approach.

### 3.1 *Post-synthesis Physical Design Flow*

This is a proactive flow, which can reduce good amount of design (optimization) effort and time. Static timing analysis postsynthesis gives a good idea on timing critical paths which may pose a challenge for timing closure during physical design.

A good starting point would be to exclude those timing paths which are not likely to be exercised in practice. If logic optimization through logic restructuring is enabled during physical design, then the approach is not to be applied on this flow and should be confined to post-layout ECO timing fix flow only. The flowchart is shown in Fig. 3.

### 3.2 *Post-layout ECO Timing Fix*

This flow is meant for designs which have gone through physical implementation but require ECO(s) for fixing timings on some critical paths. Fixing such paths usually need more effort, as in many designs, fixing a critical path usually disturbs others leading to loops which are iterative. In Post-layout ECO timing fix flow, the ECO timing iterations and subsequent STA will exclude invalid critical paths thereby reducing the effective load. Figure 4 shows the flowchart for Post-Layout ECO fix.

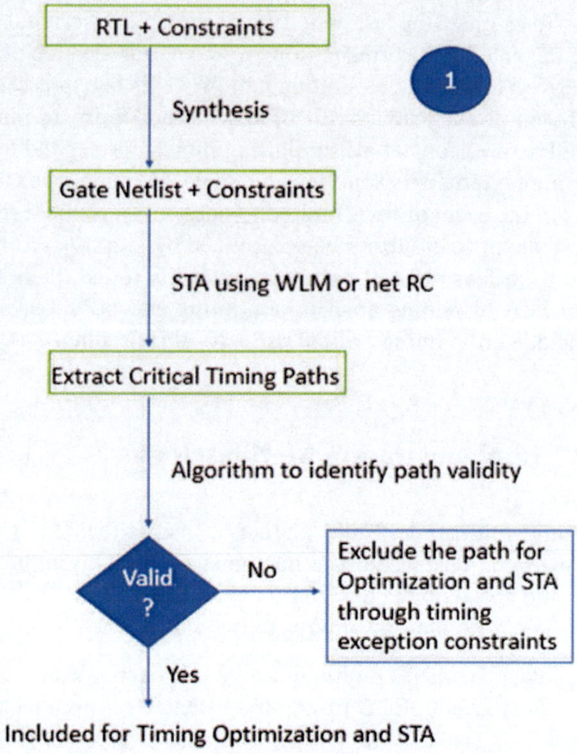


**Fig. 3** Flowchart of post-synthesis physical design flow

Both the flows provide ample flexibility and can be applied as per need. The application-specific algorithm aims to reduce the effective time needed to identify the invalid timing paths and also can be applied in parallel mode in multiple machines.

## 4 Conclusion

This paper presents a novel yet practical and deterministic solution to minimize overdesign without any compromise in QOR by excluding practically invalid timing paths for STA and optimization. The benefits include reduction in design effort and cycle time apart from area and leakage power savings. An application-specific algorithm is also briefly discussed to improve efficiency. Future work will focus on graph based path grading techniques for static timing analysis.

**Fig. 4** Post-layout ECO timing fix flow

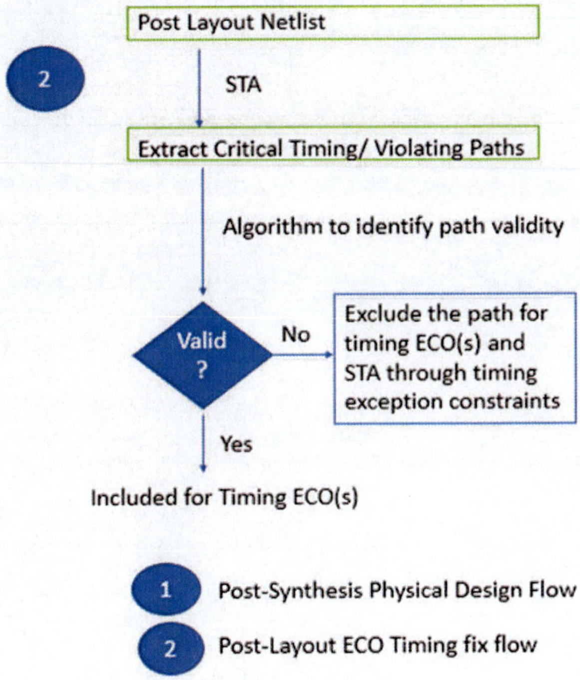


## References


1. Fuchs K, Pabst M, Rossel T (1994) RESIST: a recursive test pattern generation algorithm for path delay faults considering various test classes. IEEE Trans Comput Aided Des Integr Circuits Syst 13(12):1550–1562
2. Tragoudas S, Karayiannis D (1999) A fast nonenumerative automatic test pattern generator for path delay faults. IEEE Trans Comput Aided Des Integr Circuits Syst 18(7):1050–1057
3. Schulz MH, Fuchs K, Fink F (1989) Advanced automatic test pattern generation techniques for path delay faults. In: The nineteenth international symposium on fault-tolerant computing. Digest of papers, pp 44–51
4. Bhattacharya D, Agrawal P, Agrawal VD (1992) Delay fault test generation for scan/hold circuits using Boolean expressions. In: Proceedings 29th ACM/IEEE design automation conference, pp 159–164

# Coupling Transition Reduction on On-Chip Buses Using Adaptive Bus Encoding (ABE)

**Sumantra Sarkar, Ayan Biswas, Anindya Sundar Dhar, and Rahul M. Rao**

**Abstract** This paper presents an adaptive encoding framework for the reduction of coupling transition activity in on-chip data buses. The technique relies on the observation of data characteristics over fixed window sizes and formation of cluster with bit-lines. The proposed method utilizes redundancy in space and time to prevent loss of information while retrieving data. Unlike off-chip buses, the focus is to decrease coupling transitions which consume higher power than switching transitions in this scenario. We present analytical and experimental results which demonstrate the activity reduction of our encoding scheme for various data sets.

**Keywords** On-chip bus · Bus encoding schemes · Low power VLSI design

## 1 Introduction

The major challenges for digital systems today are power requirement, thermal management and reliability. With every new technology generation, density of the logic circuits increases, and performance improves. Initially, the industry focus primarily was to improve performance; but this approach did not sustain for long as thermal and reliability limits were reached for some ICs [1]. This led to increased focus on power efficient design and power optimization at different levels of design

S. Sarkar (✉)
AMD India Private Limited, Bengaluru, India
e-mail: sumantraster@gmail.com

A. Biswas
EECS, UC Berkeley, Berkeley, USA
e-mail: ayan-biswas@ece.iitkgp.ernet.in

A. S. Dhar
IIT Kharagpur, Kharagpur, India
e-mail: asd@ece.iitkgp.ernet.in

R. M. Rao
IBM, Bengaluru, India
e-mail: rahulmrao@in.ibm.com

 
C. Giri et al. (eds.), *Emerging Electronic Devices, Circuits and Systems*,
Lecture Notes in Electrical Engineering 1004,
https://doi.org/10.1007/978-981-99-0055-8_18

hierarchy for semiconductor industry. As minimum feature size reduces, the influence of interconnects on total power becomes increasingly important. In today's design, more than 50% of total power can come from interconnects. This is expected to increase as we enter into the era of deep or ultra-deep nanometer designs. One of the major problems of scaling process technologies is the increase of coupling capacitance to the neighboring wires, since aspect ratio (defined as ratio of line thickness to width) increases. In [2], Sylvester et al. has shown the rise in ($C_c/C_{\text{total}}$) with advancement of technology generation. It is observed that 70% of overall capacitance ($C_{\text{total}}$) comes from coupling capacitance ($C_c$). For 90 nm technologies, the ratio of the coupling capacitance of an interconnect to its ground capacitance is nearly 5.5 which is 85% of the total capacitance [3]. This illustrates the increased dominance of power dissipation by coupling capacitances. Therefore, reduction of the dynamic power due to coupling transitions becomes crucial in modern digital design.

Overall average power consumption of $N$-bit on-chip bus ($P_d$) per clock cycle can be modeled by $P_d = P_{dg} + P_{dc}$. Here $P_{dg} = C_L f\ V_{DD}^2 \sum_{i=1}^{N} \alpha_i^{cl}$ denotes dynamic power consumption due to self-capacitance of the bit-lines, and $P_{dc} = C_C f\ V_{DD}^2 \sum_{i=1}^{N} \sum_{j \epsilon \text{coupled line}} \alpha_i^{cl}$ denotes the power consumption due to coupling capacitance between adjacent bit-lines. Here $C_L$ and $C_c$ are the load and coupling capacitances, $V_{DD}$ is supply voltage and $f$ is the clock frequency. $\alpha_i^{cl}$ and $\alpha_{ij}^{cc}$ define the average self-switching activity and coupling transitional activity, respectively, for the $i$th bit-line. Since $C_c \gg C_L$, $P_d$ can be minimized by reducing $\alpha_{ij}^{cc}$, at the expense of increasing $\alpha_i^{cl}$ if necessary.

In this paper, we are focusing on reducing coupling transitional activity ($\alpha_{i,j}^{cc}$) on data streams whose statistics are not known a priori and switching characteristics change spatially (across the width of the bus) and temporally. We propose an adaptive clustering technique by observing data over time with the aid of spatio-temporal redundancy. The rest of the paper is organized as follows. Related works are briefly discussed in Section 2. In Section 3, we define the problem and present overview of application for the proposed method. An extensive theoretical study of the proposed algorithm is presented along with probabilistic analysis. Various experimental results and comparisons with existing encoding schemes are reported in Section 4. It is shown that the proposed adaptive technique provides reduction in coupling transition count in on-chip buses in all scenarios and outperforms prevalent encoding techniques when switching characteristics vary randomly among bit-lines. Finally, Section 5 summarizes our findings.

## 2 Related Work and Their Contributions

In this section, we discuss previous work on on-chip power minimization by the reduction of coupling transition. Coupling transition can be reduced without encoding by means of using shielding techniques [4, 5] which require ground or VDD between every two adjacent wires on the bus, or by increasing line-to-line spacing [6, 7] and repeater insertion [8]. All these techniques are efficient but significantly increase the chip area. There are different data encoding schemes which reduce the coupling transition either by using extra control lines [7, 9–13], or by incorporating more switching in the bit-lines. For example, the encoding scheme in [9] requires 71% extra bit-lines which increases the routing congestion along with area. Encoding techniques like gray code [14], T0 [15], and T0-XOR [16] are not suitable for on-chip buses since those techniques target reduction of the self-switching but ignore the coupling transitions which may lead to increase of coupling power. Tiehan et al. [17] proposed a dictionary-based adaptive encoding scheme for data buses which depends upon the patterns in the transmitted data. Bus invert coding (BIC) [18] reduces the coupling by decreasing the switching transitions but does not consider coupling switching while encoding. In [19], coupling transition is minimized by performing both odd and even inversion, and then transmitting the one which provides less switching activity on the bus. One hop-by-hop encoding approach is proposed in [10], which uses 5 bits for 4 bits to reduce the coupling switching activity. Even though the encoding ensures no Type II transition in 2 adjacent wires, it ends up with higher bus energy consumption due to higher data transfer time. H, HF, and odd, even, full invert (OEF) encoding highlighted in [20] reduces both self and coupling switching activities. In [21], an adaptive off-chip bus encoding scheme is presented which reduces self-switching on the bus by exploiting correlation among bit-lines in each observation window. The basic windowing principle used in the current paper is similar to that used in [21].

## 3 Adaptive On-Chip Bus Encoding

The objective function is to reduce the total number of coupling transitions on the bus. This is because a coupling transition incurs higher power consumption than a switching transition. One bit-line will be chosen as the basis line, and a set of bit-lines will be chosen as the corresponding cluster, such that xor-ing all the bit-lines in the cluster with the basis leads to maximal reduction of coupling transitions over the entire bus. The choice of the basis line and the corresponding cluster has to be done efficiently with minimal hardware area and power consumption, such that the encoding and decoding circuitry consume considerably less power than the power saved from the bus by application of the algorithm and subsequent reduction of coupling transitions. The cluster and basis information is sent to the decoder using temporal redundancy between 2 consecutive windows, and spatial redundancy in the form of an extra bit-line, respectively, as highlighted in [21]. The difficulty is that the occurrence of a coupling transition at any instant between 2 adjacent bit-lines depends on the joint switching characteristics of both the lines. In the off-chip encoding scheme in [21], the choice of a basis and presence of a bit-line in the cluster depended only on the switching characteristic of the basis and the bit-line. But now, these will depend on the switching characteristic of the preceding and succeeding bit-lines as well. As such, the computation of optimal basis and cluster becomes vastly more complex due to the inherent interdependency where the presence of every bit-line in the cluster depends on every other bit-line in the cluster.

## 3.1 Mathematical Background

An attempt is made to determine the relation between the switching probability of 2 bit-lines and the coupling probability between them, to determine whether coupling savings can be expected when the proposed adaptive encoding technique is adopted.

Let the switching probability of two adjacent bit-lines be $p_a$ and $p_b$. Table 1 lists the different types of coupling transitions between 2 adjacent bit-lines, and their probabilities of occurrence. The case where both lines switch can be further subdivided into two equi-probable sub-cases:

(1) If both lines switch in the same direction (low to high or high to low), then there is no coupling.
(2) If both lines switch in opposite directions, then the coupling power consumption is twice the power consumption when only one bit-line switches. This is modeled by considering the number of coupling transitions to be two when this situation occurs.

Therefore, the expected number of coupling transitions between the 2 bit-lines is:

$$C_{ab} = 1\,x\,(p_a(1 - p_b) + (1 - p_a)p_b) + 2\,x\,(p_a p_b/2)$$

or,

$$C_{ab} = p_a + p_b - p_a p_b \tag{1}$$

Now suppose the basis line has a switching probability $p_s$. Then the switching probabilities of the 2 lines after being xor-ed with the basis are:

$$p'_a = p_a(1 - p_s) + (1 - p_a)\,p_s \tag{2}$$

$$p'_b = p_b(1 - p_s) + (1 - p_b)\,p_s \tag{3}$$

**Table 1** Types of coupling between 2 adjacent BIT-lines

| Scenario | Situation | Probability | Coupling types |
|---|---|---|---|
| I | Both lines don't switch | $(1 - p_a)(1 - p_b)$ | No coupling |
| II | Only one-line switches | $(1 - p_a)p_b + p_a(1 - p_b)$ | Type I |
| III | Both lines switch | $p_i p_j$ | No coupling/type II |

The expected number of couplings between the 2 modified bit-lines is (using Eqs. 1, 2 and 3):

$$C_{a'b'} = C_{ab} + p_s(1 - p_a - p_b) + (1 - 2p_a)(1 - 2p_b)\, p_s(1 - p_s) \quad (4)$$

Instead, if only one bit-line is xor-ed with the basis, then the expected number of couplings is:

$$C_{ab'} = C_{ab} + (1 - p_a)(1 - 2p_b)\, p_s \quad (5)$$

$$C_{a'b} = C_{ab} + (1 - p_b)(1 - 2p_a)\, p_s \quad (6)$$

Some important inferences can be made from Eqs. (4), (5) and (6):

(1) If only one bit-line is xor-ed with the basis, then the switching probability of that line should be greater than 0:5 to have coupling savings. E.g., $(1 - 2p_b) < 0$ to have $C_{ab'} < C_{ab}$ (Eq. 5).
(2) If both lines are xor-ed with the basis (Eq. 4), then coupling savings are assured if $(1 - p_a - p_b) < 0$ and $(1 - 2p_a)(1 - 2p_b) < 0$. If both $p_a, p_b < 0.5$, then final coupling count is more. If only one of $(1 - p_a - p_b) < 0$ and $(1 - 2p_a)(1 - 2p_b) < 0$ is satisfied, then the value of $p_s$ determines whether savings are possible.

### 3.2 *Theoretical Description of the Method*

The optimal basis line and the corresponding cluster have to be found by an exhaustive search technique. If the bus width is *N*, then *N* iterations are necessary. At the *i*th iteration, the *i*th line is chosen as the basis. The lines from 1 to $(i - 1)$ form the up tree, termed here as'up Trellis', while the lines from $(i + 1)$ to *N* form the down tree, termed here as 'down Trellis'. The analysis of each Trellis tree is outlined below.

For simplicity of explanation, consider the case when bit-line $b_1$ is basis. Assume that bus width is 5. We compute the xor-ed bit-lines $b_i \oplus b_1$, which form the nodes of the Trellis tree as shown in Fig. 1. Now we compute all the relevant values of coupling transitions between the nodes, as shown by the edges in Fig. 1. The values are stored in four registers. Regs1 stores the values for the pair $b_{i-1} \oplus b_1$ and $b_i \oplus b_1$, reg s2 stores the values for the pair $b_{i-1}$ and $b_i \oplus b_1$, reg s3 for the pair $b_{i-1} \oplus b_1$ and $b_i$, and reg s4 for the pair $b_{i-1}$ and $b_i$. The topmost node in Fig. 1 is essentially a ground line, while the bottommost node is the basis line which has to be sent as an extra line over the bus, thereby incurring coupling transitions with $b_5 \oplus b_1$ or $b_5$, as the case may be.

Now the Trellis tree has to be traversed from the ground line to the other end to find the path that incurs the minimum coupling transitions. Two registers (Reg1a and Reg2a) are used to store the cumulative value of coupling transitions as we traverse the 'down' tree downwards (The 'up' tree would have been traversed upwards). The

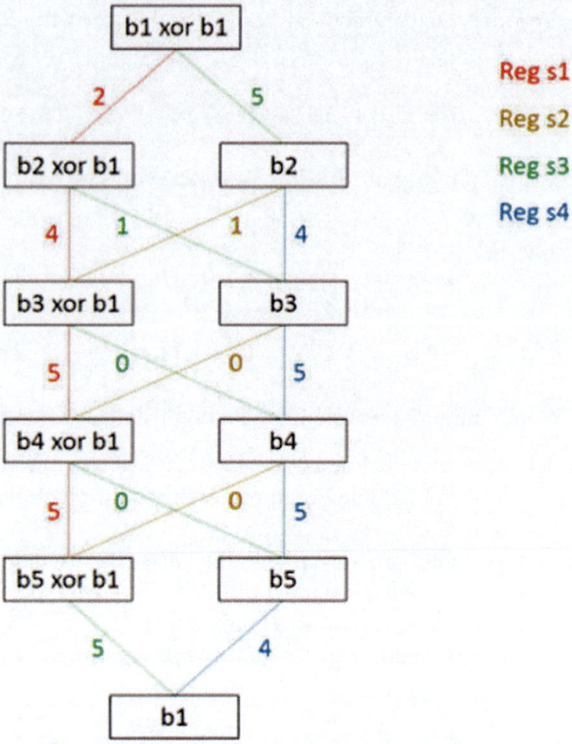


**Fig. 1** Trellis tree with 5 bit-lines, with $b_1$ as basis

number of coupling transitions incurred on reaching $b_i \oplus b_1$ by the optimal path is stored in Reg1a, while the corresponding value for $b_i$ is stored in Reg2a. Two more registers (Reg1b and Reg2b) are used to store the optimum path traversed to reach $b_i \oplus b_1$ and $b_i$, respectively. At the $i$th stage, if $b_i \oplus b_1$ is reached via $b_{i-1} \oplus b_1$, then 1 is concatenated to the path in Reg1b, and value in reg s1 is added to Reg1a. Otherwise if $b_i \oplus b_1$ is reached via $b_{i-1}$, then Reg1b gets the path stored in Reg2b concatenated with 1, and value in reg s2 is added to Reg1a. Reg2b is updated in a similar fashion with 0 concatenation, with Reg2a using s3 and s4.

Figure 2 shows the values of the 4 registers at every step while traversing the entire tree. Once the bottommost node is reached, tracing back the optimal path is trivial: since node $b_1$ is reached through $b_5$, the optimal path is in Reg2b, and the number of coupling transitions incurred in that path is in Reg2a (If $b_1$ was reached through $b_5 \oplus b_1$, then use corresponding values in Reg1b and Reg1a). These provide the number of coupling transitions and the corresponding cluster taking $i$th line as basis.

The entire process has to be iterated taking all $N$-bit-lines as basis. Each iteration has a complexity of $O(N)$, so overall complexity is $O(N^2)$. Finally, the $N$ values of coupling transitions are compared to find which bit-line incurs minimum number

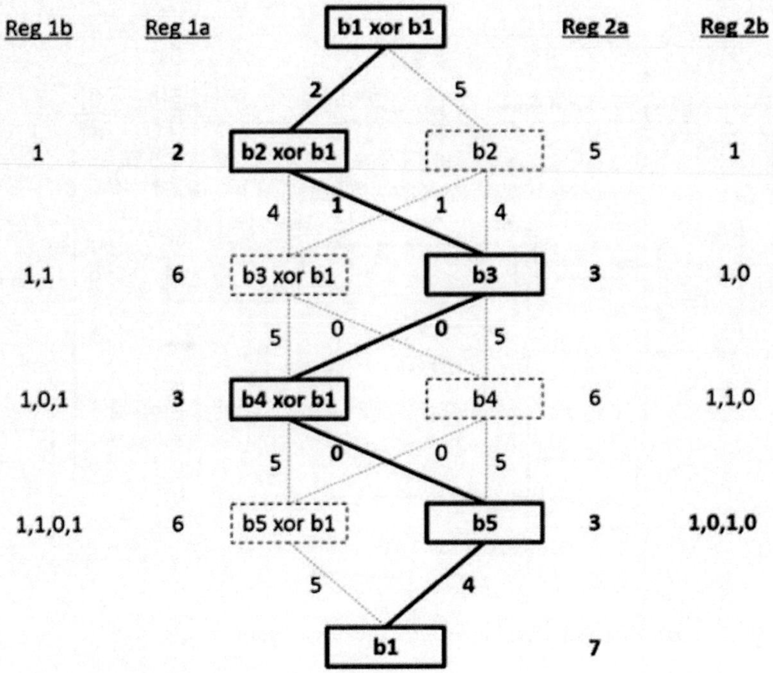


**Fig. 2** Optimal path to traverse the trellis tree

of coupling transitions when chosen as the basis. Thus, the entire algorithm can be summarized as follows:

(1) For $i = 1$ to $N$

  (a) Choose $i$th line as basis line
  (b) Determine the cluster to have minimum number of couplings ($\rho_i$) in that observation window using the Trellis tree technique described earlier

(2) End
(3) If $\rho_k = \text{argmin}_i(\rho_i)$, $1 < i < N$ the $k$th bit-line is the basis line for that observation window.

The entire algorithm can be summarized by the signal flow diagram in Fig. 3.

## 4 Results

The proposed algorithm is implemented in MATLAB and evaluated for various data sets. The comparison is done with respect to different prevalent encoding techniques. The performance of the algorithm depends on the inclusion of temporal redundancy after every observation window, and the selection of the window size.

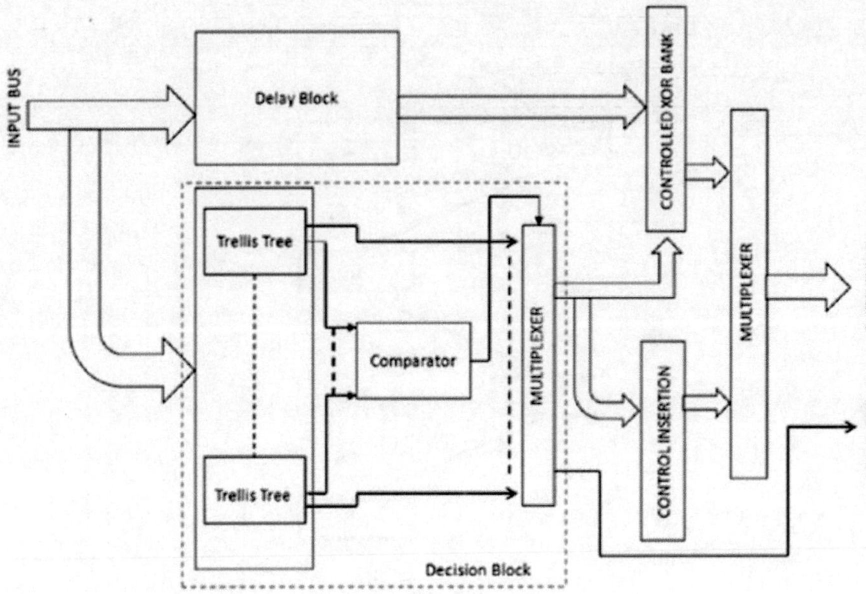


**Fig. 3** Block diagram showing signal flow of the proposed algorithm

## *4.1 Impact of Observation Window Size on Coupling Savings*

Various data sets, with fixed switching probabilities, as well as randomly varying switching probabilities, are sent through the different encoding techniques. Figure 4 shows the impact of variation in window size on coupling savings. Here, File_r contains bit-lines with randomly varying switching probability in both spatial and temporal direction. In File $p$ where ($p = 0{:}3; 0{:}5; 0{:}8$), the switching probability of each bit-line is uniformly $p$.

It is found that number of coupling transitions as well as number of switching transitions are minimized when observation window size is 16, for majority of the data sets. It is also found that the optimal value of 16 does not depend on the bus width. Systems that have stringent performance requirement, demand the increase in clock frequency to eliminate the impact of temporal redundancy while retrieving back the original data. The reason for choosing 16 clock cycles in one observation window for the adaptive encoding can be justified as follows:

1. With increase in window size (i.e., higher number of transitional clock cycles), the local correlation between the lines will decrease, reducing the efficiency of the adaptive clustering technique and the overall coupling savings obtained from the data.
2. With decrease in window size (i.e., fewer number of transitional clock cycles), more temporal redundancy is necessary to convey the clustering information, which may lead to reduction in coupling savings.

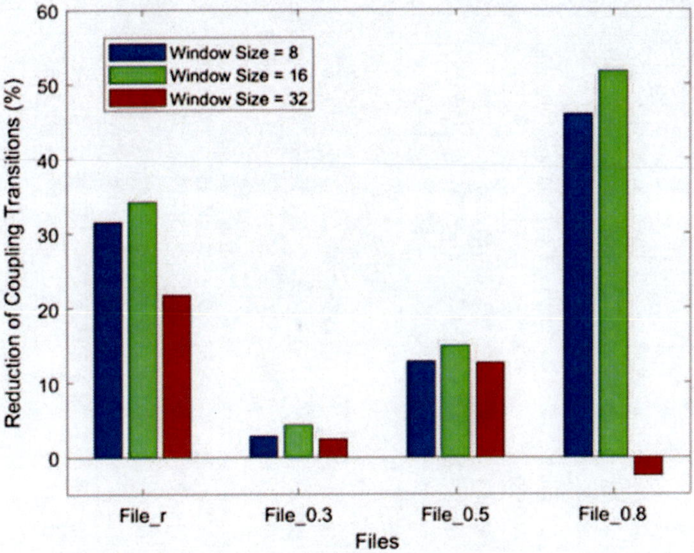


**Fig. 4** Variation of coupling savings with window size for different files

## *4.2 Comparison with Pre-existing Encoding Schemes*

The coupling savings for the different files using the proposed algorithm are compared with the coupling savings from other existing algorithms, namely BIC, Schemes 1, 2, 3, FOC, FTC, and FPC schemes, since these encoding techniques provide better coupling and self-switching savings as compared to other existing encoding techniques with minimal increase in the number of control lines. Table 2 shows the percentage of coupling savings.

It can be seen that the proposed algorithm consistently gives extremely high coupling savings (10–30%) as compared to other encoding techniques for a wide variety of data sets. Figure 5 shows the coupling savings for different data sets and provides the region of application for the proposed scheme.

It has to be noted that some of the encoding techniques decrease coupling transitions at the expense of increase in switching transitions. Table 3 shows the impact of the different encoding techniques on switching transitions. Proposed method ensures no increase of self-transition, rather reduces the self-transition.

The coupling savings and switching savings have to be taken together, with proper weighting factors, to determine the best encoding scheme. Thus, it is evident that, when there is a variation of switching probability (File_r), we get much higher coupling and switching transition savings using the proposed method as compared to other existing techniques. This provides the region of application for our proposed on-chip encoding scheme.

**Table 2** Comparison of coupling savings using different encoding techniques

| File name | Initial | Our method | BIC | ODD Invert (Scheme 1) | Odd + Full invert (Scheme 2) | Odd + Even + full invert (Scheme 3) | FOC encoding | FTC encoding | FPC encoding |
|---|---|---|---|---|---|---|---|---|---|
| | Coupling | Coupling savings | Coupling savings | Coupling savings | Coupling savings | Coupling savings | Coupling savings | Coupling savings | Coupling savings |
| Lena.bmp | **1,032,017** | **108,778 (10.54%)** | **– 10,450 (– 1.01%)** | **28,574 (2.77%)** | **40,282 (3.90%)** | **52,654 (5.10%)** | **– 57,153 (– 5.54%)** | **101,934 (– 9.88%)** | **9093 (0.91%)** |
| Configure.txt | 2,108,653 | 243,212 (11.53%) | – 80,075 (– 3.80%) | 115,259 (5.47%) | 99,835 (4.73%) | 91,091 (4.32%) | – 98,283 (– 4.66%) | – 363,918 (– 17.26%) | – 33,546 (– 1.59%) |
| Linux.exe | 427,395 | 68,339 (15.99%) | – 4529 (– 1.06%) | 19,017 (4.45%) | 42,517 (9.95%) | 27,323 (6.39%) | – 13,613 (– 3.19%) | – 91,273 (– 21.36%) | 43,432 (10.16%) |
| MD5SUMS.txt | 7,449,477 | 618,989 (8.31%) | – 321,459 (– 4.32%) | 770,043 (10.34%) | 703,987 (9.45%) | 549,885 (7.38%) | – 59,393 (– 0.80) | – 1,087,6 61 (– 14.60%) | – 45,052 (– 0.60%) |
| Music.mp3 | 16,612,576 | 2,381,819 (14.34%) | 1,222,029 (7.36%) | 2,036,946 (12.26%) | 2,771,474 (16.68%) | 2,994,303 (18.02%) | – 93,729 (– 0.56%) | – 904,486 (– 5.44%) | 1,235,116 (7.43%) |
| File_0.5 | 509,289 | 72,963 (14.33%) | 46,827 (9.19%) | 40,656 (7.98%) | 74,415 (14.16%) | 60,621 (11.90%) | – 16,698 (– 3.28%) | – 38,841 (–7.63%) | 38,478 (7.56%) |
| File_r | **473,200** | **133,848 (28.29%)** | **36,166 (7.64%)** | **40,560 (8.57%)** | **80,444 (17%)** | **81,458 (17.21%)** | **– 20,956 (– 4.43%)** | **– 40,898 (– 8.64%)** | **43,264 (9.14%)** |

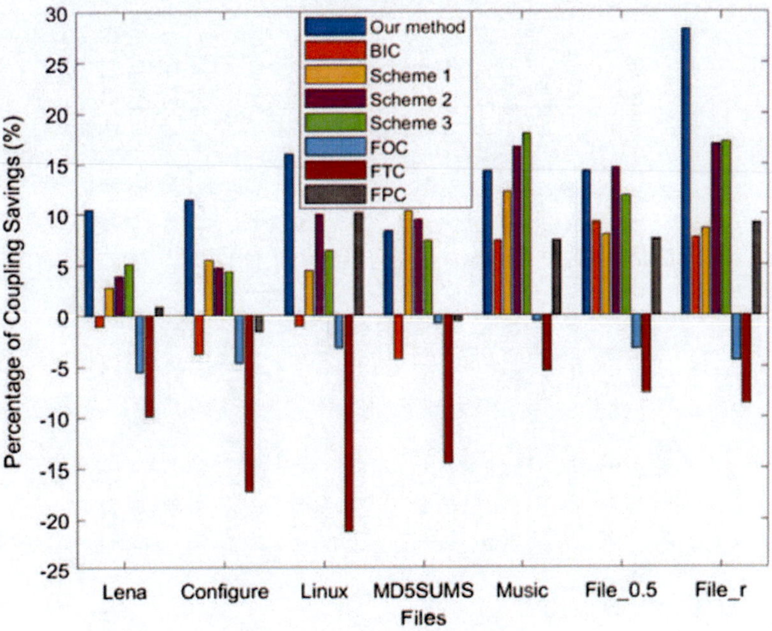


**Fig. 5** Coupling savings from different encoding schemes

## 5 Conclusion

In conclusion, an adaptive on-chip bus encoding technique is proposed which does not require prior knowledge about the signal characteristics. The proposed algorithm selectively encodes a cluster of lines within the fixed observation window, by performing XOR operation with a basis bit-line, selected based on the switching and coupling characteristics of the bit-lines in that window, with the objective of minimizing the number of coupling switching transitions. Hence the clustering is adaptive over every observation window. The formation of the cluster depends on the coupling transitions, which in turn depend on the switching activity of adjacent pairs of bit-lines, so the probability of one bit-line being present in the cluster inherently depends on the status of all the other bit-lines in that observation window. The coupling savings from the proposed algorithm and from existing bus encoding techniques are enumerated, and it is found that our algorithm consistently gives high coupling savings for different types of input patterns. The best performance is achieved when the data on the bus changes spatially and temporally across the bit-lines, which provides the region of application for the proposed on-chip encoding scheme.

**Table 3** Comparison of switching savings using different encoding techniques

| File name | Initial | Proposed method | BIC | ODD Invert (Scheme 1) | Odd + Full Invert (Scheme 2) | Odd + Even + full invert (Scheme 3) | FOC Encoding | FTC Encoding | FPC encoding |
|---|---|---|---|---|---|---|---|---|---|
| | Switching | Switching savings | Switching savings | Switching savings | Switching savings | Switching savings | Switching savings | Switching savings | Switching savings |
| Lena.bmp | **723,490** | **9818 (1.36%)** | **58,704 (8.11%)** | **12,658 (1.75%)** | **28,156 (3.89%)** | **24,175 (3.34%)** | **– 112,688 (– 15.5%)** | **– 94,229 (– 13.02%)** | **– 443,315 (– 61.27%)** |
| Configure.txt | 1,409,399 | 166 (0.01%) | 42,799 (3.04%) | 53,696 (3.81%) | 29,888 (2.12%) | 27,161 (1.93%) | – 218,447 (– 15.50%) | – 337,590 (– 23.95%) | – 932,966 (–66.20%) |
| Linux.exe | 347,584 | – 18,444 (– 5.31%) | 26,996 (7.77%) | – 3312 (0.95%) | –19,904 (–5.73%) | – 15,912 (– 4.58%) | – 37,102 (– 10.67%) | – 87,854 (– 25.28%) | – 207,498 (– 59.70) |
| MD5SUMS.txt | 4,666,588 | – 191,930 (4.11%) | 63,882 (1.37%) | 317,766 (6.81%) | 255,268 (5.47%) | 273,454 (5.86%) | – 1,013,482 (– 21.72%) | – 968,182 (– 20.75%) | – 3,263,439 (69.93%) |
| Music.mp3 | 11,760,625 | 413,681 (3.52%) | 1,703,467 (14.48%) | 269,511 (2.29%) | 976,739 (8.31%) | 1,016,772 (8.65%) | – 1,111,758 (– 9.45%) | – 1,417,256 (– 12.05%) | – 6,640,163 (56.46%) |
| File_0.5 | 366,992 | 10,164 (2.77%) | 59,532 (16.22%) | – 363 (– 0.10%) | 32,307 (8.80%) | 39,930 (10.88%) | – 21,417 (– 5.48%) | – 41,745 (– 11.37%) | – 199,650 (– 54.40%) |
| Flle_r | **363,688** | **99,710 (27.42%)** | **62,868 (17.29%)** | **– 7098 (– 1.95%)** | **51,714 (14.22%)** | **37,180 (10.22%)** | **– 20,280 (–5.58%)** | **– 21,970 (– 6.04%)** | **– 200,096 (– 55.02%)** |

## References

1. International technology roadmap for semiconductors, 2.0:2015
2. Sylvester D, Hu C (2001) Analytical modeling and characterization of deep submicrometer interconnect. Proc IEEE 89:634–664
3. Muroyama M, Ishihara T, Yasuura H (2008) Analysis of effects of input arrival time variations on on-chip bus power consumption. In: 18th international workshop, PATMOS, pp 62–71
4. Vittal A, Marek-Sadowska M (1997) Crosstalk reduction for VLSI. IEEE Trans Comput Aided Des Integr. Circuits Syst 16(3):290–298
5. Ghoneima M, Ismail YI, Khellah MM, Tschanz JW, De V (2006) Formal derivation of optimal active shielding for low power on-chip buses. IEEE Trans Comput Aided Des Integr Circuits Syst 25(5):821–836
6. Macchiarulo L, Macii E, Poncino M (2002) Wire placement for crosstalk energy minimization in address buses. In: Proceeding of 18th design, automation and test in Europe conference and exhibition, pp 158–162
7. Ayoub R, Orailoglu A (2005) A unified transformational approach for reductions in fault vulnerability, power and crosstalk noise and delay on processor buses. In: Proceedings of the Asia and South Pacific design automation conference, vol 2, pp 729–734
8. Banerjee K, Mehrotra A (2002) A power optimal repeater insertion methodology for global interconnects in nanometer designs. IEEE Trans Electron Devices 49(11):2001–2007
9. Lyuh CG, Kim T (2006) Low power bus encoding with crosstalk delay elimination. IEE Proc Comput Digit Tech 153(2):93–100
10. Pande PP, Ganguly A, Zhu H, Grecu C (2006) Energy reduction through crosstalk avoidance coding in networks on chip. In: Proceedings of the 9th Euromicro conference on digital system design architectures, methods and tools, pp 689–695
11. Ki KW, Hyun BK, Shanbhag N, Liu CL, Sung KM (2000) Couplingdriven signal encoding scheme for low power interface design. In: Proceedings of the IEEE/ACM international conference on computer-aided design, pp 318–321
12. Rung-Bin L (2008) Inter-wire coupling reduction analysis of bus invert coding. IEEE Trans Circuits Syst I Reg Papers 55(7):1911–1920
13. Khan Z, Arslan T, Erdogan AT (2006) Low power system on chip bus encoding scheme with crosstalk noise reduction capability. IEE Proc Comput Digit Tech 153(2):101–108
14. Su CL, Tsui CY, Despain AM (1994) Saving power in the control path of the embedded processors. IEEE Des Test Comput 11(4):24–31
15. Benini L, De Micheli G, Macii E, Sciuto D, Silvano C (1997) Asymptotic zero-transition activity encoding for address buses in low power microprocessor-based systems. In: Proceedings of the GLS-VLSI-97: IEEE great lakes symposium VLSI, Urbana, IL, pp 77–82
16. Fornaciari W, Polentarutti M, Sciuto D, Silvano C (2000) Power optimization of system-level address buses based on software profiling. In: Proceedings of the 8th international workshop on hardware/software codesign, pp 29–33
17. Lv T, Henkel J, Lekatsas H, Wolf W (2003) A dictionary based en/decoding scheme for low power data buses. IEEE Trans VLSI Syst 11(5):943–951
18. Stan MR, Burleson WP (1995) Bus-invert coding for low-power I/O. IEEE Trans VLSI Syst 3(1):49–58
19. Yan Z, Lach J, Skadron K, Stan MR (2002) Odd/even bus invert with twophase transfer for buses with coupling. In: Proceeding of the international symposium on low power electronics and design, pp 80–83
20. Jafarzadeh N, Palesi M, Khademzadeh A, Kusha AA (2014) Data encoding techniques for reducing energy consumption in network-on-chip. IEEE Trans VLSI Syst 22(3):675–685
21. Sarkar S, Biswas A, Dhar AS, Rao RM (2017) Adaptive bus encoding for transition reduction on off-chip buses with dynamically varying switching characteristics. IEEE Trans VLSI Syst 25(11)

# Modelling and Analysis of Confluence Attack by Hardware Trojan in NoC

**Sachin Bagga, Ruchika Gupta, and John Jose**

**Abstract** Over the years, system on chip (SoC) designs have evolved extensively sophisticated in order to fulfil the need of increasing complexity of running applications driven by the advancement of VLSI technology. The next generation multiprocessor system on chip (MPSoC) integrates hundreds of processing elements on a single chip, expected to achieve high performance, low latency, and low-power consumption. Tiled chip multicore processors (TCMP) with network on chip (NoC) have become a foundation for the computation critical embedded and real-time systems. Faster time-to-market restraint and business competition compelled manufacturers to look for the prospects of manufacturing SoCs integrated with various third party intellectual property (3PIP). Usage of 3PIP gave rise to exploit the underlying interconnect while adding some unwanted malicious circuit known as hardware trojan (HT), making NoC vulnerable to get attacked. A tiniest manipulation of any communication attribute by HT can degrade the overall behaviour of the system significantly while impacting NoC performance metrics. In this paper, we present one of such study considering HT malign behaviour of manipulating the output port of each incoming flit from specific ports once genuine route computation takes place and re-directs all to one port causing disruption in tile communication.The proposed HT is intermittent in nature and activates for few cycles. We study the behaviour of proposed confluence attack and analyze its impact over network level. The empirical evaluation exhibits the misbehaviour of the packet communication in terms of wrong usage of VCs and extra hop distance overhead. To validate the proposed work, various performance metrics like buffer utilization, virtual channel utilization, number of flits processed, link utilization, and the like are analyzed. We also show that such

S. Bagga (✉) · R. Gupta
Chandigarh University, Mohali, Punjab, India
e-mail: sachin8510@gmail.com

R. Gupta
e-mail: gupt009@rnd.iitg.ac.in

R. Gupta · J. Jose
Indian Institute of Technology Guwahati, Assam, India
e-mail: johnjose@iitg.ac.in



C. Giri et al. (eds.), *Emerging Electronic Devices, Circuits and Systems*,
Lecture Notes in Electrical Engineering 1004,
https://doi.org/10.1007/978-981-99-0055-8_19

HTs are difficult to detect due to marginal increase at malicious port traffic leaving no trace of malicious conduct.

**Keywords** Network on chips · Hardware trojans · Confluence attack

## 1 Introduction

In the last decade, momentous expansion in consumer electronics gadgets has resulted in the emergence of powerful tiled chip multicore processor (TCMP) with network on chip (NoC) as the underlying communication framework [1]. To reduce the overall system design cost, the use of third party intellectual property (IP) blocks is becoming a common practice. Such inexpensive and potentially insecure IPs put forth crucial data security challenges [2, 3]. Due to the exposure of complete data travel information and the IP connectivity, NoC becomes a natural choice for the adversaries to exploit the vulnerability and mount the attacks. One such exploit is the HT attack, the addition of unwanted circuitry into the IP design that can result in creating malicious activity including denial of service (DoS), information leakage, and data-stealing attacks [4, 5].

Figure 1 shows internal architecture of a 4 × 4 mesh NoC-based TCMP in which a tile comprises of a processing element (PE), private L1 caches, a share of L2 cache bank, and other relevant logic units required to carry out the communication. In TCMP, NoC packets are generated when a cache miss encounters, and the required data needs to be brought from the remote tile. NoC packets travel in the form of smaller flow control units known as flit across the network. Cache miss request packets consist of single head flit with mandatory attributes for routing like source

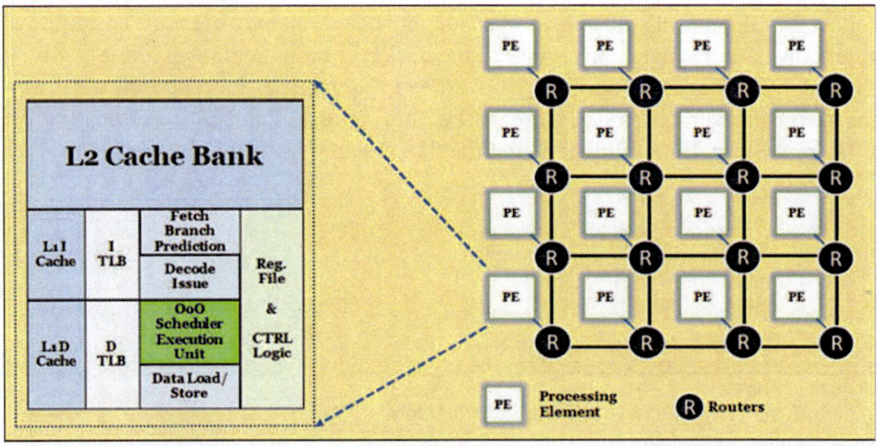


Fig. 1 4×4 mesh NoC-based SoC

address and destination address while cache miss reply packets are multi-flit packets containing a head flit followed by multiple body flit and a tail flit carrying the data/payload. Wormhole switching is followed by the packets such that all the body flits and tail flit of a head flit automatically follow the same path as that of the head flit. NoC router has input buffers, route computation unit, virtual channel allocator, switch allocator, and a crossbar. Route computation is done at every intermediate router for forwarding the packet to its destination based on the underlying routing technique. The mere presence of an unwanted circuit behaving maliciously is enough to bring down the system performance drastically. In this paper, we propose one of such malign HTs sitting at NoC router altering the computed output port every time to one output port irrespective of incoming direction calling a confluence attack. We make the following contributions in the paper:

(i) We identify a suitable location to place a HT such that there is no amendment in the packet header information.
(ii) We propose the feasibility of HT over one of the nodes assuming HT is intermittent and gets activated with a probability 'p'.
(iii) Modelling the proposed confluence attack HT.
(iv) Implementation and analysis of the confluence attack induced by HT over NoC level parameters.

## 2 Related Work

HT enabling packet misrouting to cause denial of service, delay of service, and injection suppression is modelled, and based upon the impact, shielding technique is proposed to decrease the HT impact [3]. Various performance metrics like effective average deflected packet latency, effective average packet latency, and throughput are discussed. Corresponding to the mitigation techniques proposed, analysis in terms of hardware and timing overhead is also discussed. Another HT residing inside the route computation unit resulting in misrouting of the flits creating the impacts like deadlock, decrease in packet injection, delay of service, and denial of service is modelled while further, a dynamic shielding technique is proposed to isolate HT infected IP [4]. Validation by performing analysis of the latency of packets is also performed. A potential threat model that alters the NoC packet and leads to creation of dead flit in a router buffers is also discussed in which impact analysis of dead flit with two variants: one modifies the head flit to body flit, and another modification from a body flit to head flit is studied. The impact analysis in terms of variation of the instruction per cycle, average buffer occupancy, and cache miss penalty is also gathered [6].

There can be a possibility of HT mounted on the input buffers of NoC routers while changing the destination address field of chosen NoC packets. Such possibility is proposed and modelled while the impact of HT at network, cache, and core level is captured [7]. In this work, HT significantly impacts the L1 cache and is capable

of bringing an application to a complete halt. The analysis in terms of assumption of re-transmission of the impacted packet is also done while the assumptions of re-transmission are later considered to be unacceptable because of high latency overhead. Importance of electromagnetic(EM) radiation analysis for the purpose of hardware security is highlighted [8]. A novel hardware security solution is proposed which is based on the various analysis related to EM. Another side channel-aware detection technique using test generation approach working on the principle of multiple excitation of rare switching is also proposed [9]. The proposed work significantly increases the sensitivity of HT, thus helps in easy detection of HT using side channel analysis techniques.
Adaptive routing technique having non-interference characteristics is proposed in the literature to secure NoC from the timing attacks [10]. The work prevents the information leakage with 2–20% improved routing performance with power penalty of 1.84%.

## 3 Architecture Details for Baseline TCMP

A tile broadly includes processor, L1 instruction cache, L1 data cache, L2 cache, and a network adapter. L2 cache is uniformly shared amongst all the tiles and is accessible in a sequential manner. Crossbar, switch allocator, virtual channel allocator, routing unit, and input port buffers are some of the major components of the modern router [7]. The components of a router are given as follows:

(i) Buffering of incoming flits (BW)
(ii) Route computation (RC)
(iii) Virtual channel allocation (VA)
(iv) Switching packets from input port to output port (SA)
(v) Switch traversal (ST)
(vi) Link traversal (LT)
(vii) Management of power including link scaling.

Proposed work uses wormhole routing in which the routing is performed with the following characteristics:

(i) Route computation is performed only once per packet.
(ii) Virtual channel allocation is done once per packet and is embedded in the head flit of the packet.
(iii) Head flit acts as a parent node while the body flit and tail flit inheriting the head flit information following the same path.

## 4 Threat Model

In this study, we propose a malicious implant called HT that resides in the NoC router. Out of the various pipeline stages in router architecture, viz., BW, RC, VA, SA, ST, and LT, the HT is activated during RC: route computation stage as shown in the Fig. 2.

If an HT flicker happens, malicious behaviour is triggered by changing the legitimate output port always to the south direction, and thus, all the flits confluence to one output port. This HT behaviour leads to undesired service delays and service denial scenario. Cache miss requests, replies, evicted cache blocks, and coherence messages are all carried in NoC packets. An exploited NoC router with the suggested HT can misroute these packets causing latency-critical applications to function poorly at the application level. Such HTs can be added to a NoC IP at any point during IC life cycle, including the specification, design, fabrication, and manufacturing phases [3].

In this paper, we assume that the suggested HT is deployed in NoC IP during the pre-silicon stage, either by an attacker with access to the system design or through an untrustworthy 3PIP. In the proposed work, HT is randomly activated with a probability of 0.1, i.e. 10% probability along with the following two conditions existing simultaneously:

(i) Router-ID 5 must come in routing path of the flit.
(ii) Incoming port of the flit must be either east or west.

Since proposed HT is internally triggered and behaves intermittently malicious, targeting only few flits, thus catching such HT with verification/code, electronic design automation (EDA) tools will be difficult.

Figure 3 shows an instance of HT residing at Router-ID 5 and behaving maliciously. Any source–destination pair passing through HT node 5 gets deflected from the genuine route and confluence to a single direction (south). Example shows source–destination pairs 4-7 and 7-4 in which confluence attack is captured due to the fact of a packet coming from east and west input port, respectively. However, the 13-1 and 1-13 source–destination pairs even if the confluence attack hits would not get impacted for the obvious reason of being in the south direction. By the virtue of the source–destination travelling combination with XY routing, there is no effect on flits coming from the north input port or south input port. As a result, all the flits trapped by Router-Id 5 during activation as a HT will have to cover the extra distance

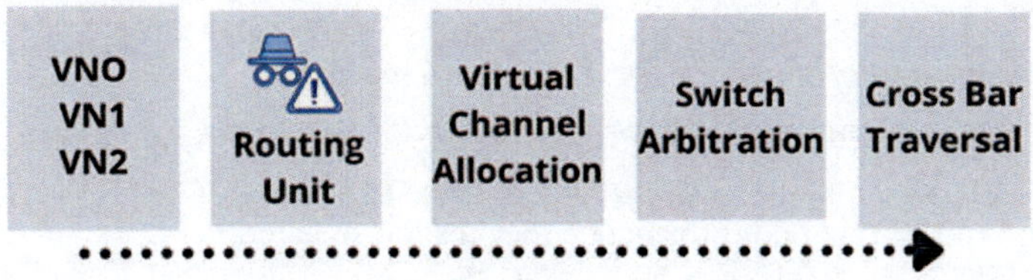


**Fig. 2** HT embedded inside the RC unit resulting in flit misrouting to wrong output port

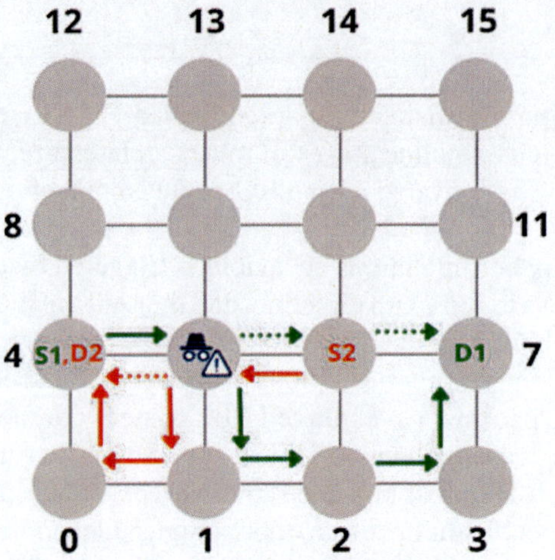


**Fig. 3** HT modelling for deflection of flits coming from east and west direction

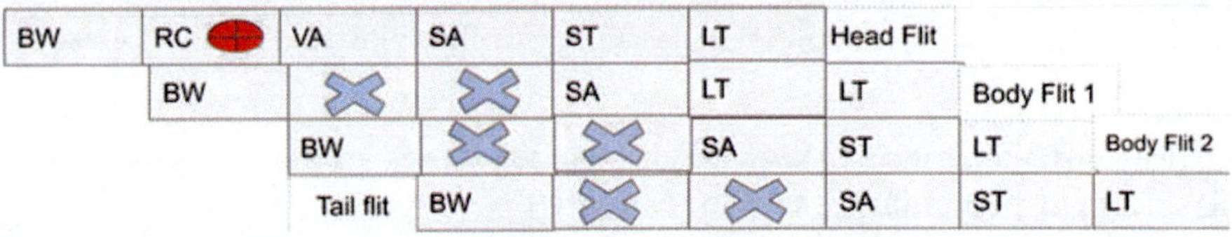


**Fig. 4** HT triggered for wormhole routing during route computation

before reaching the final destination. Flit can only reach the target if a particular node usually acts, and the real output port is obtained. The effect of the HT can be seen in terms of delay of service as the flits are misrouted hence delayed and can only be reached to the destination by traversing extra hops. This leads to delay of service attack and end up utilizing unavoidably more resources for packet transmission and reception. Figure 4 shows the instance when HT is triggered during the wormhole routing. Misrouting of the head flit during route computation by HT results in misrouting of all the subsequent flits of that packet. Depending upon the injection rate value with the same probability (0.1), the number of flits deflected may increase or decrease.

## 4.1 Performance Metrics for Impact Analysis

To conduct the empirical evaluation and to observe the HT impact, the following network and router-related performance metrics are selected:

(i) **Virtual Channel Availability**: To acquire the statistics for the traffic of a certain router, we calculate the number of VCs utilized dynamically within a stipulated cycle period.
(ii) **Injection Rate**: It is represented as a number of packets injected per node per cycle, and its value varies from 0 to 1.
(iii) **Router Load**: It indicates the router participation in the processing of total network traffic, more number of flits passing through the router indicates more router load.
(iv) **Flits processed by router**: It represents the total count of flits transmitted through a certain router.
(v) **Buffer Utilization**: When packets or flits cannot be forwarded to output links right away, they are stored in buffers. Buffer utilization measures how full a buffer is. Back-pressure and overload of existing connections can be detected using them.
(vi) **Internal Link Utilization(ILU)**: All outbound links to nearby routers and the network adapter at the local port are monitored for link utilization. To determine the available bandwidth, link utilization monitors are employed during resource allocation. Internal links are one-way and form mesh connections. It connects the routers to create a particular topology. It gives insights into the total amount of bandwidth used for transmission.

## 5 Experimental Setup and Results

We use an event driven simulator gem5 to model and implement the 16 tiles $4 \times 4$ mesh NoC [11]. Inside gem5, Garnet 2.0 [12] is an interconnection network model simulated using the parameters shown in Table 1. Gem5's ruby memory system model provides the topology and routing infrastructure for Garnet 2.0. The implementation of the micro-architectural system for an on-chip network router is validated with the Garnet 2.0 module [13]. The simulation runs for the injection rate 0.1 and 0.2. To make the HT difficult to detect by common metrics like power and energy consumption, the deployment of malicious activity is modelled only in a single router.

In traffic pattern synthetic uniform random, all CPUs have an equal chance of being randomly selected as a source or destination node, and flits begin travelling accordingly for the selected source–destination pair, resulting in an unbiased examination for the proposed work. In the proposed work, simulation is executed under the two conditions:

(i) **Baseline Case (B)**: With no HT activation (ideal case)
(ii) **Hardware Trojan Case (T)**: HT is activated.

**Table 1** Simulation parameter

| Parameter | Description |
|---|---|
| Network | Garnet 2.0 |
| CPU count | 16 |
| Topology | Mesh-XY |
| Mesh rows | 4 |
| Sim cycles | 5000 |
| HT probability | 0.1 |
| Injection rate | 0.1, 0.2 |
| VC per vnet | 2 |
| Traffic | Synthetic uniform random |
| Execution status | Baseline(B), hardware trojan(T) |

To study the impact of HT, percentage change is calculated for the HT case with respect to the baseline or idle condition as shown in Eq. 1 below

$$\text{Percentage\_Change} = \frac{\text{T\_Value} - \text{B\_Value}}{\text{B\_Value}} * 100 \tag{1}$$

## 5.1 *Effect of HT on Virtual Channel Utilization*

A virtual channel (VC) is a distinct queue in the router that allows numerous VCs to share the physical wires (physical link) between two routers. Head-of-line blocking can be minimized by associating numerous distinct queues with each input port. On a cycle-by-cycle basis, virtual channels arbitrate physical link bandwidth. Every VC has its control buffer, which contains the following values: packet length (PL), status (S), virtual channel identifier (VCID), and output port (OP). When a flit arrives at a router, the input port demultiplexer extracts the VCID from the incoming flit's common prefix and stores it in the appropriate VC. Figure 5 shows the analysis related to used virtual channels corresponding to each router of mesh for (B) and (T) at injection rate 0.1 and 0.2. The percentage change is calculated for (B) and (T) of each router at injection rates 0.1 and 0.2. During analysis, it is figured out that there is a positive percentage increase of up to 5, 13, and 4% in the usage of virtual channels for Router-ID 0, 1, and 2. The effect of HT is clearly visible on an increase in virtual channel utilization, it will result in earlier network saturation, and more number of flits dropping will be there.

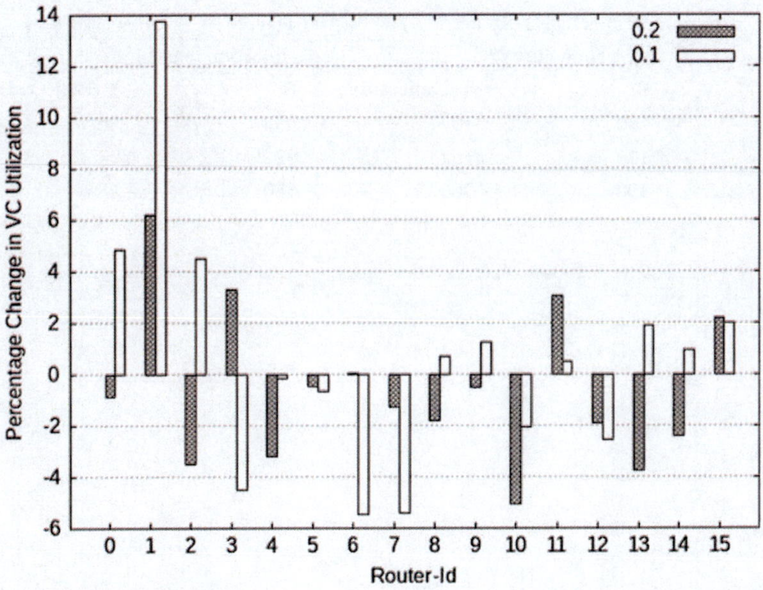


**Fig. 5** Percentage change in average virtual channel utilization for complete mesh at 0.2 and 0.1 injection rate

## 5.2 *Effect of HT on Flits Processing and Flits Deflection Count*

When a message is sent over the network, it is first divided into a data packet, which will then be divided into fixed-length flits, or flow control units. To compute the next outgoing port, the route computation unit extracts the destination ID from the head flit, and the output port is changed correspondingly. As a result, after a head flit's routing is complete, the output port stores the next outgoing port information for all following flits of that packet. Table 2 shows flit processed by Router-ID 5 during cycle times, when HT is triggered at injection rates 0.2 and 0.1. As HT is deflecting the flits coming from east and west port, for injection rate 0.2 in total 365 flits are misrouted by Router-ID 5, out of which 197 were entering from east port and 168 were entering from west port, for injection rate 0.1 in a total of 180 flits are misrouted by Router-ID 5, out of which 86 were entering from east port and 94 were entering from west port. The deflected flits result in increasing the network traffic in that particular link, there is no effect on the flits coming from north or south port. Figure 6 shows the analysis corresponding to the flits processed by each router of mesh for (B) and (T) at injection rates 0.1 and 0.2. The percentage change is calculated for (B) and (T) of flits processed at injection rates 0.1 and 0.2. During the analysis, it is figured out that there is a positive percentage increase in flits processed of up to 5, 30, 5, and 5% for routers 0, 1, 2, and 4. The increase in the number of flits

**Table 2** Misrouting of flits by HT(Router-Id 5) when got triggered

| Direction | Injection rate 0.2 | | Injection rate 0.1 | |
|---|---|---|---|---|
| | B | T (Deflected) | B | T (Deflected) |
| East | 230 | 197 | 88 | 86 |
| West | 170 | 168 | 92 | 94 |
| Total | 400 | 365 | 180 | 180 |

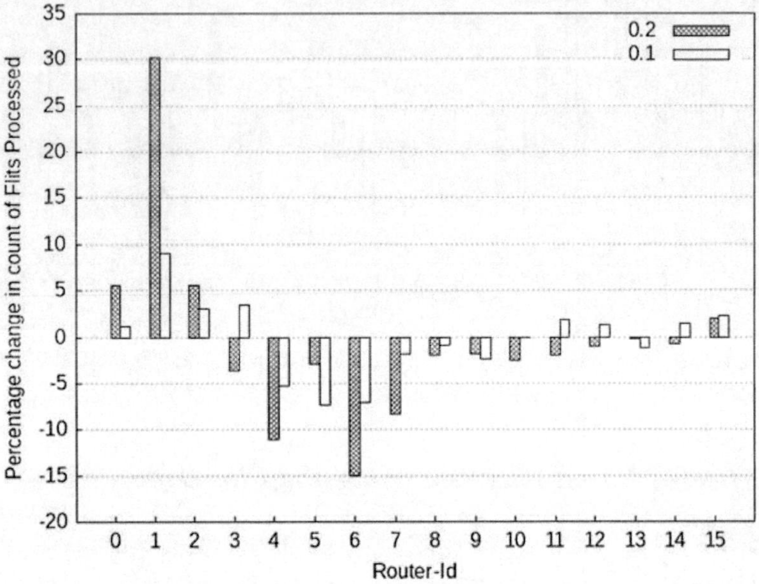


**Fig. 6** Change in number of flits processed by each router of mesh for (B) and (T) for injection rate 0.2 and 0.1

processed by a certain number of routers gives clear indications of the presence of HT, as the number of flits injected, VCs and injection rate were the same still certain routers processed more number of flits.

## *5.3 Effect of HT on Link Utilization*

If two candidate ports have the same number of available VCs, link utilization is considered for the selection process. Then, within the current monitoring period, the port with the lowest link utilization is chosen. In the uncommon event that both ports have the same link utilization, the first port is used.

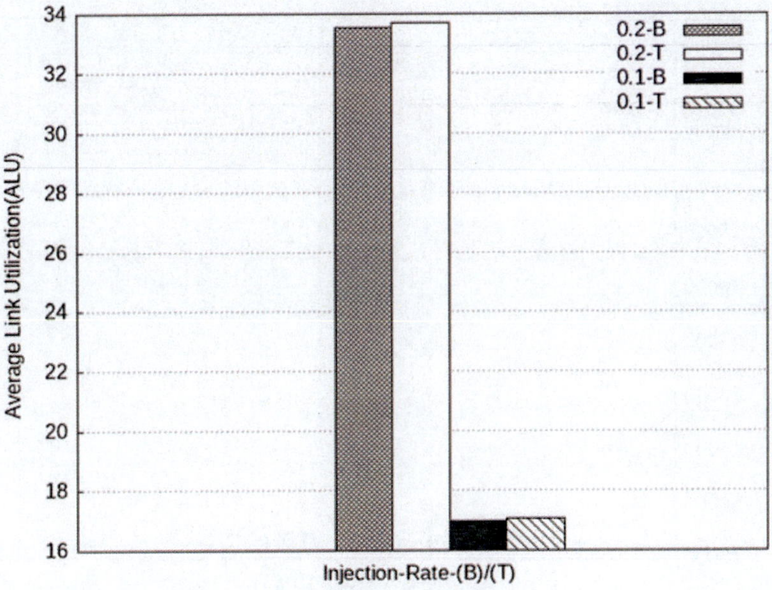


**Fig. 7** Average link utilization for total number of flits injected

(i) **Average Link Utilization**

This parameter takes into consideration all the links like external(IN & OUT) which are bidirectional and internal that are unidirectional. The link utilization is calculated as per Eqs. 2 and 3. Activity mentioned in Eq. 2 is a count of how many times a particular link is utilized. Time delta is calculated as the difference of curCycle() and start cycle(). curCycle() gives the current simulation cycle time, and start cycle() is the starting simulation time.

$$\text{AverageLinkUtilization}+=\text{Activity}/\text{TimeDelta} \tag{2}$$

$$\text{TimeDelta}=\text{curCycle()}-\text{StartCycle()} \tag{3}$$

Figure 7 briefs about the effect of HT on average link utilization by the flits for the same source and destination in the case of (B) and (T). Overall, there is a 47% increase in the average link utilization for the (T) case in comparison with the (B).

(ii) **Internal Link Utilization**

Internal link utilization is the count of activities in a particular link between two routers. The link utilization is calculated as per Eq. 4.

$$\text{IntLinkUtilization}+=\text{activity}; \tag{4}$$

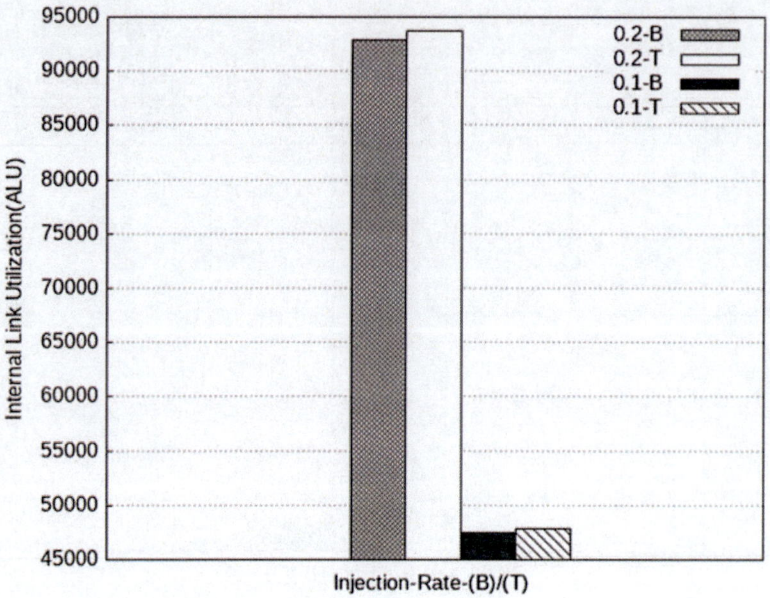


**Fig. 8** Internal link utilization for total number of flits injected

Figure 8 briefs about the effect of HT on average internal link utilization by the flits for the same source and destination in case of (B) and (T). There is a 0.84% increase in the internal link utilization for the (T) case in comparison with the (B) as HT is deflecting more number of flits in one particular direction.

## 5.4 *Effect of HT on Router Load in Terms of Count of Activation of Each Router of Mesh*

In the proposed work, a counter is deployed inside the input unit of each router that keeps on increasing every time a router is activated. Figure 9 shows the analysis corresponding to each router activation of mesh for (B) and (T) at injection rates 0.1 and 0.2. During the analysis, it is figured out that there is a positive percentage increase in router activation of up to 10, 29, 9, and 6% for Router-ID 0, 1, 2, and 5. Also from the given figure, it can be seen that for the baseline case, the flits processed are similar for Router-ID (0, 3, 12, and 15),(1, 2, 4, 7, 8, 11, 13, and 14) , and (5, 6, 9, and 10) as analysis is done for uniform random traffic, and equal chances are given to every router for flits processing. But when the HT is active, this symmetrical relation is disrupted, which gives signals of the presence of some malicious activity in the chip.

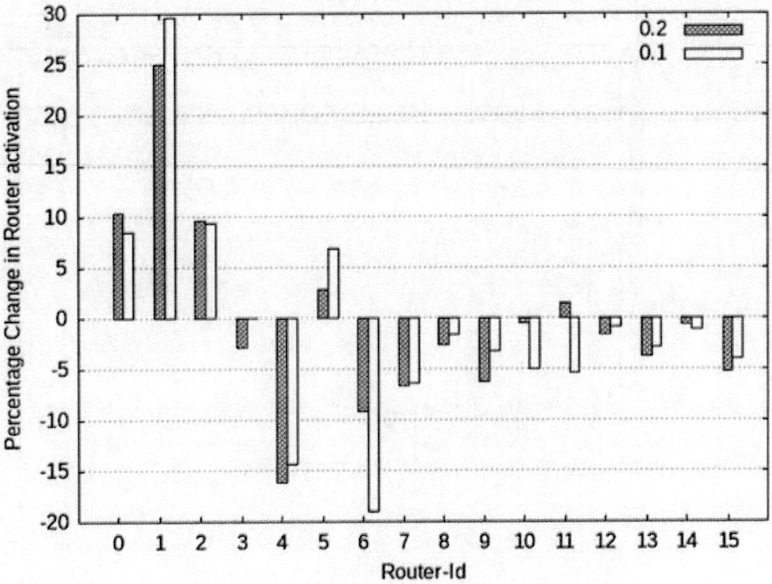


**Fig. 9** Change in router load of for (B) and (T) at injection rate 0.2 and 0.1

## 5.5 *Effect of HT on Variation in Buffer Reads for Each Router of Mesh*

When packets or flits cannot be forwarded to output links right away, they are stored in buffers. On both the input and output ports, flits can be buffered. When the switch's allocation rate is higher than the channels, output buffering occurs. Proposed work uses the wormhole routing having a provision that a packet must not be completely received for flit transmission, and the subsequent router does not need to have buffer space available for the entire packet, which results in minimal buffer needs and reduces the delay. But HT affects the buffer requirements up to a great extent, and its effect is exclusively studied in the proposed work.

Figure 10 shows the analysis related to buffer reads corresponding to each router of mesh for (B) and (T) at injection rates 0.1 and 0.2. There is a positive percentage increase in the buffer read for Router-ID 0, 1 , 2 , 3, and 5 with value 1.88, 6.39, 3.00, 1.7, and 0.52 at injection rate 0.1, respectively, and with respect to injection rate 0.2, change is 20.21, 6.32, 1.73, 1.32, and 1.46. The excessive usage of a buffer can result in earlier saturation and will result in the delay of service.

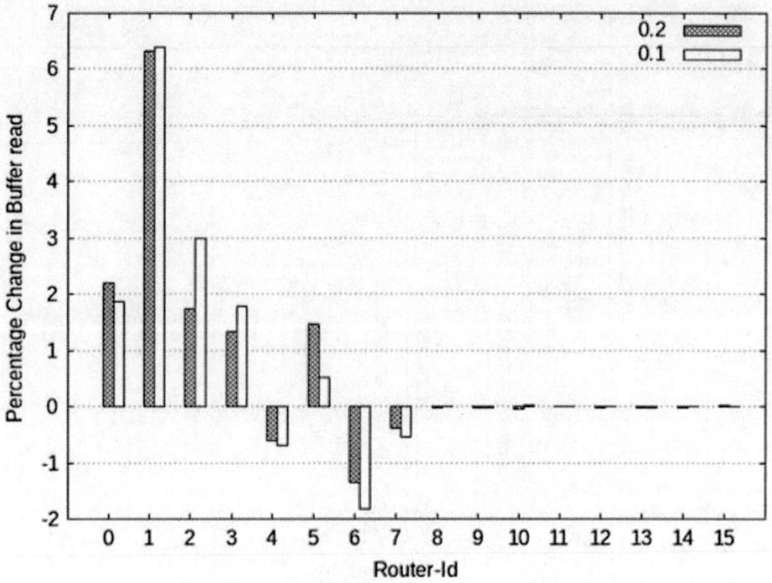


**Fig. 10** Change in buffer utilization for (B) and (T) at 0.2 and 0.1 injection rate

### 5.6 *Effect of HT on Hop Count, Network Latency, and Queuing Latency*

Proposed modelled HT will redirect the flits in the wrong output port that will further result in covering extra hops for the same source and destination, which further effect the network latency. Table 3 shows the analysis done for the flits passing through Router-ID 5 in terms of hop count, network latency, and queuing latency. There is approximately a 7% increase in the sum of hops, a 2% increase in network latency, and a 5.16% decrease for the flits passed through Router-ID 5 as a result of HT activation.

From the above analysis, it is clear that out of the complete chip, our region of interest (ROI) has been confined to the Router-ID 0, 1, 2, 3, and 5. These routers have shown anomalies in the performance metrics related to network statistics, flits processing, router load, and buffer reads. But based upon the above analysis, it will be very difficult to commit exactly which is the malicious Router-ID in the given region. The nature of the HT like activation for a very limited time and deflection of flits only in particular ports hides the malicious node very well. HT showing unusual variation in power consumption, and energy can easily lead to detection, but the proposed HT modelled is internally triggered and intermittently kept it hidden.

**Table 3** Analysis for the flits passing through Router-ID 5 in terms of hop count, network latency, and queuing latency for (B) and (T) at injection rates 0.1 and 0.2

| Parameter | Injection rate 0.2 | | Injection rate 0.1 | |
|---|---|---|---|---|
| | B | T | B | T |
| Count_Flits | 400 | 365 | 180 | 180 |
| Hops_Count_Sum | 1160 | 1246 | 572 | 75 |
| Hops_Count_Avg | 2.9 | 342 | 3.17 | 759 |
| Network_latency_Sum | 7434 | 7584 | 2445 | 2995 |
| Network_latency_Avg | 18.63 | 20.84 | 13.58 | 16.64 |
| Queuing_latency_Sum | 864 | 815 | 360 | 360 |
| Queuing_latency_Avg. | 2.16 | 2.23 | 2 | 2 |

## 6 Conclusion and Future Work

In the proposed work, the usual 3-stage pipelined input buffered router is implemented. HT is embedded inside the route computation and triggered with the probability 'p'. Suggested HT contrives a novel confluence attack by altering the genuine output port always into a fixed specific output port. The HT notably influenced the cores surrounding the HT router node in a 16 tiles $4 \times 4$ mesh NoC. The HT impact has clearly shown the potential of HT attack to impair the system performance by keeping the resources busy unnecessarily. In the proposed study, various types of analyzes are performed that are helpful to detect the presence of malicious activity in the network. Since HT is activated for a smaller duration and act maliciously for specific directions only, it becomes extremely difficult to trace the exact location of the HT in the network; however, the potential suspicious region can be filtered out. Localization of the HT infected Router-ID responsible for misrouting of the packets while creating confluence attack holds the promising scope as the future work.

## References


1. Charles S, Logan M, Mishra P (2020) Lightweight anonymous routing in NoC based SoCs. In: Design, automation & test in Europe conference & exhibition (DATE). IEEE, pp 334–337
2. Charles S, Lyu Y, Mishra P (2019) Real-time detection and localization of DoS attacks in NoC based SoCs. In: Design, automation & test in Europe conference & exhibition (DATE). IEEE, pp 1160–1165
3. Manju R, Das A, Jose J, Mishra P (2020) Sectar: Secure noc using trojan aware routing." In: 14th international symposium on networks-on-chip (NoCs). IEEE, pp 1–8
4. Rajan M, Das A, Jose J, Mishra P (2021) Trojan aware network-on-chip routing. In: Network-on-chip security and privacy, p 277
5. Das A, Babu S, Jose J, Jose S, Palesi M (2018) Critical packet prioritisation by slack-aware re-routing in on-chip networks. In: Twelfth international symposium on networks-on-chip (NoCS), IEEE, pp 1–8

6. Khan MH, Gupta R, Jose J, Nandi S (2021) Dead flit attack on NoC by hardware trojan and its impact analysis. In: Proceedings of the 14th international workshop on network on chip architectures, pp 10–15
7. Kulkarni VJ, Manju R, Gupta R, Jose J, Nandi S (2011) Packet header attack by hardware trojan in NoC based TCMP and its impact analysis. In: 15th international symposium on networks-on-chip (NoCS), IEEE, pp 21–28
8. He J, Guo X, Tehranipoor M, Vassilev A, Jin Y (2021) Em side channels in hardware security: attacks and defenses. IEEE Design Test, p 1
9. Huang Y, Bhunia S, Mishra P (2018) Scalable test generation for trojan detection using side channel analysis. IEEE Trans Inf Forensics Secur 13(11):2746–2760
10. Boraten TH, Kodi AK (2018) Securing NoCs against timing attacks with non-interference based adaptive routing. In: Twelfth IEEE/ACM international symposium on networks-on-chip (NOCS) IEEE, pp 1–8
11. https://www.gem5.org/documentation/learning_gem5/introduction/ Last accessed 26 January 2022
12. https://www.gem5.org/documentation/general_docs/ruby/garnet-2/ Last accessed 26 January 2022
13. Binkert N, Beckmann B, Black G, Reinhardt SK, Saidi A, Basu A, Hestness J, Hower DR, Krishna T, Sardashti S et al (2011) The gem5 simulator. ACM SIGARCH Comput Archit News 39(2):1–7

# Investigating the Impact of Ge-Quantum Well Width in $Si/SiO_2/Ge/SiO_2/Pt$ Resonant Tunneling Device with NEGF Formalism

**Nilayan Paul, Basudev Nag Chowdhury, and Sanatan Chattopadhyay**

**Abstract** In this work, a $Si/SiO_2/Ge/SiO_2/Pt$ resonant tunneling device (RTD) with an asymmetric double barrier has been modeled by adopting NEGF formalism. The impact of Ge-quantum well widths below, equal, and above its excitonic Bohr radius (EBR ~ 25 nm) on resonant tunneling current is investigated at room temperature. The tunneling current peaks are observed to appear for decreasing the well width to equal or less than the EBR of Ge. Such peak values increase with downscaling of the well width up to a certain value and then it decreases with further miniaturization. The maximum peak current is obtained to be ~ 13 $mA/cm^2$ for Ge-well width of 17 nm. The corresponding maximum peak-to-valley current ratio (PVCR) is estimated to be ~ 18 at room temperature, which is larger in order than the conventional RTDs. Therefore, the current work may provide the route for fabrication of Si/Ge-based high performance resonant tunneling devices operational at room temperature.

**Keywords** Resonant tunneling device (RTD) · Asymmetric double barrier · NEGF formalism · Quantum well width · Local density of states (LDOS) · Peak-to-valley current ratio (PVCR)

## 1 Introduction

Resonant tunneling phenomenon in semiconducting device structures has been exploited for multifaceted applications including quantum cascade lasers [1], Terahertz (THz) oscillators [2, 3], THz imaging [4], high speed memory [5], and digital circuits [6], as well as for qubit generation in quantum computing [7, 8] and random number generation [9] in recent times. Such fascinating applications of the phenomenon have been possible through their implementation in novel nanoscale

N. Paul · B. Nag Chowdhury · S. Chattopadhyay (✉)
Department of Electronic Science, University of Calcutta, 92 A.P.C. Road, Kolkata 700009, India
e-mail: scelc@caluniv.ac.in

S. Chattopadhyay
Centre for Research in Nanoscience and Nanotechnology (CRNN), JD Block, Sector III, Saltlake City, Kolkata 700098, India



C. Giri et al. (eds.), *Emerging Electronic Devices, Circuits and Systems*,
Lecture Notes in Electrical Engineering 1004,
https://doi.org/10.1007/978-981-99-0055-8_20

electronic devices including the resonant tunneling diode [10], resonant tunneling bipolar transistors and FETs [11, 12], TFETs [13, 14], and double quantum dots [7, 15].

Several approaches have been attempted to theoretically model such an interesting physical phenomenon, with varying degrees of validity level and accuracy. Such methods include Easki-Tsu formula [16], density matrix [17] and Wigner function [18] formulations, equivalent circuit simulation through SPICE [19], and Non-Equilibrium Green's function (NEGF) formalism [20].

Among such approaches, direct calculation of transmission coefficients is conceptually simple, however, it is a tedious process to calculate or compute, and also based on the assumption of only pure states to be present in the device. This contradicts the practical carrier transport in resonant tunneling which is a non-equilibrium process and therefore, may have mixed states. This is resolved by adopting the density matrix and Wigner-function approaches, which are based upon quantum statistical formulation of the transport processes. However, it is difficult to implement such methods for computational modeling, with the only possible simplification being a semi-classical Boltzmann limit. Equivalent circuit simulations on the other hand can proficiently depict the overall behavior of a resonant tunneling device, however, cannot provide exact results originated from the actual quantum phenomena. In this context, NEGF approach is capable of modeling the non-equilibrium transport of charges in active quantum device coupled to reservoirs, where all the statistical and phase breaking processes are incorporated through corresponding self-energy matrices. Therefore, the NEGF approach has emerged as the most promising method for modeling quantum transport in electronic and optoelectronic devices [21–24].

Predominantly, RTDs have been fabricated using III–V materials, particularly the InP, InGaAs, InAlAs-based systems [25]. Such devices show negative differential resistance (NDR) at low bias voltages, however, exhibit small peak-to-valley ratio (PVCR) of around 3–4 [25]. Moreover, the device fabrication techniques using such materials are complicated and expensive compared to the conventional silicon/germanium-based mature technologies. Further, for the implementation of real devices it is highly challenging to achieve two symmetric tunnel-barriers, and therefore modeling of RTD with asymmetric barriers is crucial since it can provide the realistic technological flexibility in terms of fabrication in practice.

Therefore, the current work deals with theoretical investigation on the impact of quantum well width in an asymmetric double barrier resonant tunneling device (DB-RTD), based on non-equilibrium Green's function (NEGF) formalism, with second quantization field operators for electrons. Self-consistency between the electrostatic and quantum conditions is achieved through the simultaneous solution of the equivalent Schrodinger–Poisson equations. The DB-RTD is considered to comprise of a germanium (Ge) thin film sandwiched in-between two asymmetric silicon dioxide ($SiO_2$) insulating barriers, placed on a silicon (Si) substrate. The transport characteristics of the device are investigated at room temperature by varying the thickness of germanium film and its PVCR is studied.

## 2 Scheme of the Device

The $Si/SiO_2/Ge/SiO_2/Pt$ double barrier RTD structure has been considered for modeling in the current work and its schematic has been depicted in Fig. 1a. The thickness of bottom and top $SiO_2$ barriers is assumed to be 5 nm and 15 nm, respectively. A 20 nm thick germanium (Ge) film is considered to be sandwiched in-between such insulating layers leading to the formation of a Ge-quantum well consisting of 2D-electron gas. The Ge and Si layers are assumed to be intrinsic, and the thickness of the Ge layer is chosen around its excitonic Bohr radius (EBR ~ 25 nm) so that the energy spacing of quasi-bound states in Ge-potential well is higher than the carrier thermal energy at room temperature (~25 meV), which further makes the impact of electron–phonon scattering negligible. Pt is assumed to be the metal contact in such device. The relevant band diagram of the device structure in unbiased condition is depicted in Fig. 1b. The carrier effective masses in the Si substrate, $SiO_2$ barriers, and Ge-quantum well are taken from Refs. [26–28].

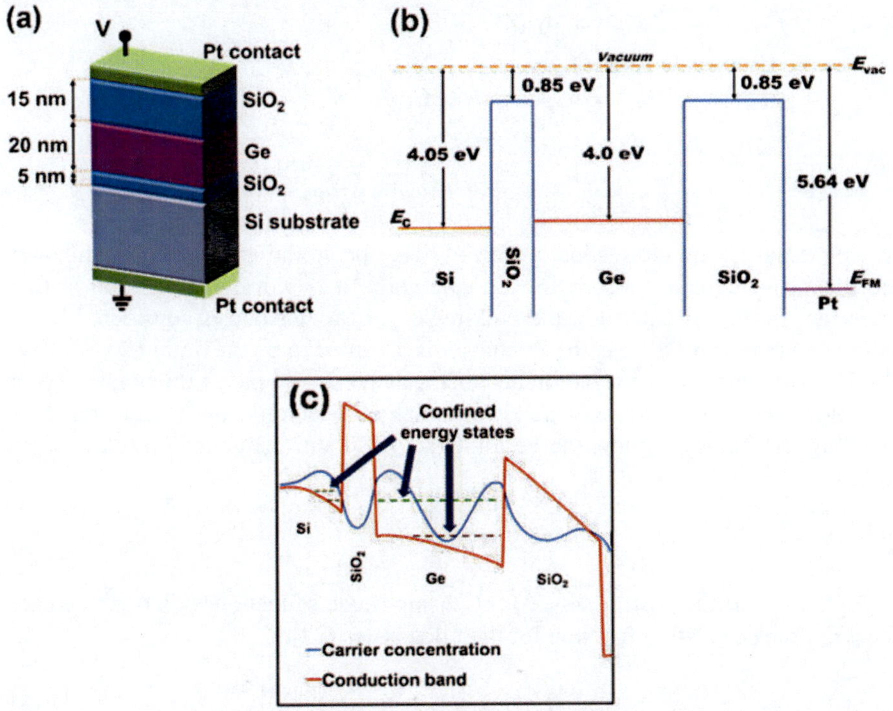


**Fig. 1** **a** Schematic diagram of the resonant tunneling device considered in this work. **b** Schematics of band structure of the $Si/SiO_2/Ge/SiO_2/Pt$ device in equilibrium condition. **c** Schematic of the confined electron energy states in the Ge-quantum well and at the $Si/SiO_2$ interface, and the carrier concentration in the device

It is imperative to note that, electrons in the Ge-quantum well considered in the present RTD are free to move in the two transverse directions, whereas their motion is restricted along longitudinal direction due to the double barrier structure. On application of a positive bias at the top metal contact, downward bending of the conduction band also forms a quantum well at the $Si/SiO_2$ interface due to surface quantization, as shown schematically in Fig. 1c. Resonant tunneling occurs when the energy level(s) of electrons confined in the potential well at the $Si/SiO_2$ interface get tuned with the energy level(s) of quasi-bound state(s) in the Ge-quantum well and can be observed as a peak in the current–voltage characteristics.

## 3 Theoretical Modeling

The analytical model of electron transport in the resonant tunneling structure is developed through the self-consistent solution of Schrodinger–Poisson equations in non-equilibrium low energy field theoretical regime. The relevant equations for describing the system are given by [21–24]:

$$i\hbar\frac{\mathrm{d}}{\mathrm{d}t}C_i = H_{\mathrm{ISO}}C_i + \tau_{ir}C_r \tag{1a}$$

$$i\hbar\frac{\mathrm{d}}{\mathrm{d}t}C_r = H_R C_r + \tau_{ri}^* C_i \tag{1b}$$

where $C_i$ and $C_r$ are the second quantized field operators for electron, $i$ indicating the active device and $r$ stands for the reservoirs. It is worthy to mention at this point that in the current work, the bulk region of Si substrate is considered to be source ($S$) reservoir whereas the Pt contact is assumed to be the drain (D). $H_{\mathrm{ISO}}$ in Eq. (1a) represents isolated Hamiltonian of the active device (i.e., without interaction with S/D) while $H_R$ represents the Hamiltonian of reservoirs, and $\tau$ indicates their coupling. All the $C$'s follow the Fermi–Dirac (FD) anti-commutation relations as follows:

$$\{C_i, C_k^+\} = \delta_{ik} \tag{2}$$

Equation (1a) and (1b) are solved for $C_i$ using Green's function ($G(E)$) to calculate the two-time correlation function for the filled states [29],

$$n_{ij}(t, t') = \left\langle C_j^+(t')C_j(t)\right\rangle \tag{3}$$

The Fourier transform of Eq. (3) into energy domain in steady-state leads to matrix,

$$[n(E)] = [G(E)][\Sigma_R(E)]\left[G^+(E)\right] \tag{4}$$

where the device Green's function for electrons is given by

$$G(E) = (E - H_{\mathrm{ISO}} - \Sigma_R)^{-1} \tag{5}$$

and the carrier inflow/outflow from/to the reservoirs can be estimated from the corresponding self-energy matrices as given by,

$$[\Sigma_R(E)] = [\tau][n_R(E)]\left[\tau^+\right] \tag{6}$$

Such self-energy physically implies a "disturbance" to the active device Hamiltonian from the reservoirs during non-equilibrium transport of carrier under biased condition. Subsequently, the outflow matrix is given by,

$$[\Gamma_R(E)] = i\left[\Sigma_R(E) - \Sigma_R^+(E)\right] \tag{7}$$

that leads to the transmission coefficient from source ($R = 1$) to drain ($R = 2$) to be,

$$T(E) = \Gamma_1(E)G(E)\Gamma_2(E)G^+(E) \tag{8}$$

It may be noted that the Green's function obtained from Eq. (5) is put into Eq. (4), which on integration over energy gives rise to the carrier density ($\rho$). Such carrier distribution, on putting into Poisson's equation,

$$\frac{\mathrm{d}}{\mathrm{d}z}\left(\varepsilon(z)\frac{\mathrm{d}}{\mathrm{d}z}\right)\varphi(z) = e\rho(z) \tag{9}$$

leads to obtain the potential $\varphi(z)$, which is then put into the device Hamiltonian,

$$H_{\mathrm{ISO}} = -\frac{\hbar^2}{2}\frac{\mathrm{d}}{\mathrm{d}z}\left(\frac{1}{m^*(z)}\frac{\mathrm{d}}{\mathrm{d}z}\right) + E_C(z) + (-e\varphi(z)) \tag{10}$$

where $E_C$ denotes the conduction band minimum and $m^*$ represents the effective mass of electrons. At this point, consistency needs to be checked between Eqs. (5) and (9) to obtain the quantum-electrostatic simultaneous solution of the potential and charge distribution. Once self-consistency is achieved, the tunneling current is calculated from Landauer formula given by,

$$I = \frac{e}{h}\int \mathrm{d}E\,T(E)(f_1(E) - f_2(E)) \tag{11}$$

It is imperative to mention that the integral over energy must be performed by considering all possible momentum states along the two transverse directions.

## 4 Results and Discussion

To investigate the impact of Ge-quantum well width on resonant tunneling through the $Si/SiO_2/Ge/SiO_2/Pt$ device, the variations of transmission coefficient with energy and applied bias are plotted in Fig. 2a–c, for different well widths of 15 nm, 20 nm and 25 nm, respectively. It is apparent from such figures that the transmission coefficient exhibits several peaks indicating resonant tunneling in the proposed device. Such peaks are observed to shift toward lower energy values with increasing bias for all the well widths, indicating a corresponding alignment of quantized states between $Si/SiO_2$ interfacial well and Ge-quantum well. However, with the increase of well width, the number of resonance peaks increases in the same energy and voltage range and reaches toward saturation for widths of the order of EBR. Further, a close inspection of such transmission peaks shows the broadening of resonant energy levels to decrease with decreasing well width indicating a corresponding sharper resonance. Such countering effects of downscaling the Ge-quantum well width, such that the decreasing number of eigenstates contributing to resonance and their increasing sharpness results to a maximum in the tunneling current peak for a well width of ~ 20 nm.

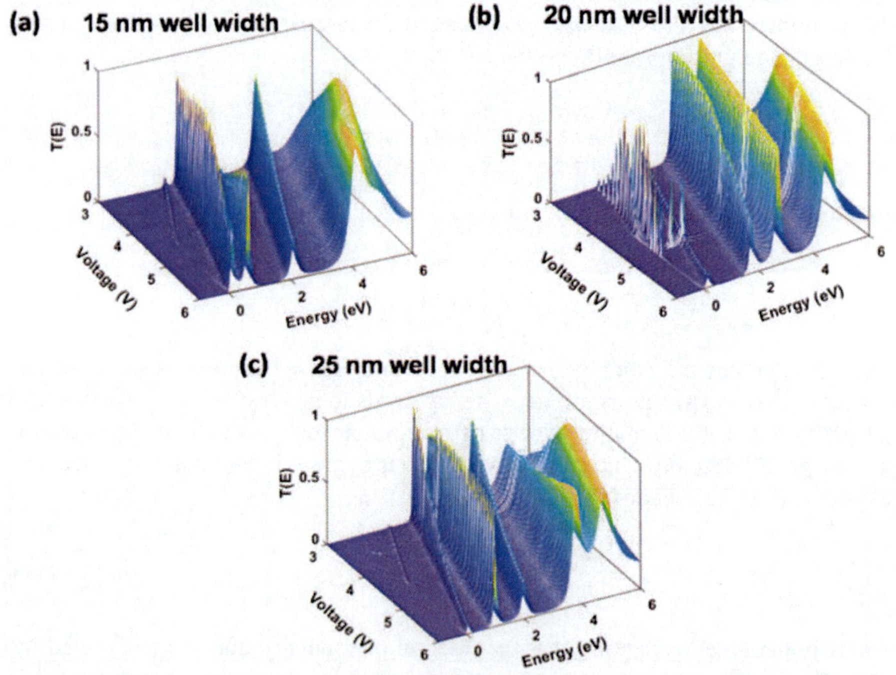


**Fig. 2** Variation of transmission coefficient with variation of applied bias voltage. **a** For device with 15 nm thick Ge-quantum well layer. **b** For device with 20 nm thick Ge-quantum well layer. **c** For device with 25 nm thick Ge-quantum well layer

Further, the comparative analyses of resonant tunneling phenomena in the present device for similar quantum well widths for applied bias and no bias conditions are carried out by plotting the respective local density of states (LDOS) in Fig. 3a–f. It is apparent from the plots of Fig. 3a, c, and e for zero bias that with increasing well width, the Ge-quantum well allows more bound states (i.e., 1st and 2nd excited states along with ground state) that can contribute to resonance with the states of Si. The similar LDOS plots are represented accordingly in Fig. 3b, d, and f for an applied bias of 4.5 V. It is evident from such figures that the tuning between different eigenstates of Si/$SiO_2$ surface quantized well and the Ge-quantum well changes for different Ge-widths, where the maximum level tuning is observed from 20 nm well width.

The resulting current–voltage characteristics of the present RTD are plotted in Fig. 4 for varying the Ge-quantum well width from 30 nm (above EBR) to 10 nm (far below EBR). It is apparent from Fig. 4 that the resonant tunneling current peaks are observed to shift toward higher voltages for reducing the Ge-well width. This is attributed to the energy increase of ground state and other excited states with downscaling of the quantum well width. Further, as mentioned earlier, the countering phenomena of decreasing number of eigenstates and their increasing resonance sharpness with miniaturization of the quantum well width lead to a maximum value of tunneling current peak (13.36 mA/$cm^2$) and the PVCR (~ 18) for a well width of ~ 17 nm. The PVCR obtained for such device structure and dimensions is significantly higher (Table 1) than the conventional RTDs [25].

## 5 Conclusion

A Si/$SiO_2$/Ge/$SiO_2$/Pt asymmetric double barrier resonant tunneling device has been modeled by obtaining the quantum-electrostatic self-consistent solution through NEGF formalism. The variation of resonant tunneling current with applied bias and Ge-quantum well width at room temperature is investigated in detail. The results indicate that significant resonant tunneling requires higher voltages for lower Ge-well width. For reducing the Ge-well width from above EBR to a sub-EBR value, the tunneling current peak and PVCR exhibits a maximum value at a well width of 17 nm. Such tunneling current peak value and the PVCR reaches up to 13.36 mA/$cm^2$ and ~ 18, respectively, in the present device at room temperature, which are larger in order than the conventional RTDs. Therefore, the current work may provide a route for cost-effective fabrication of Si/Ge-based high performance resonant tunneling device operational at room temperature.

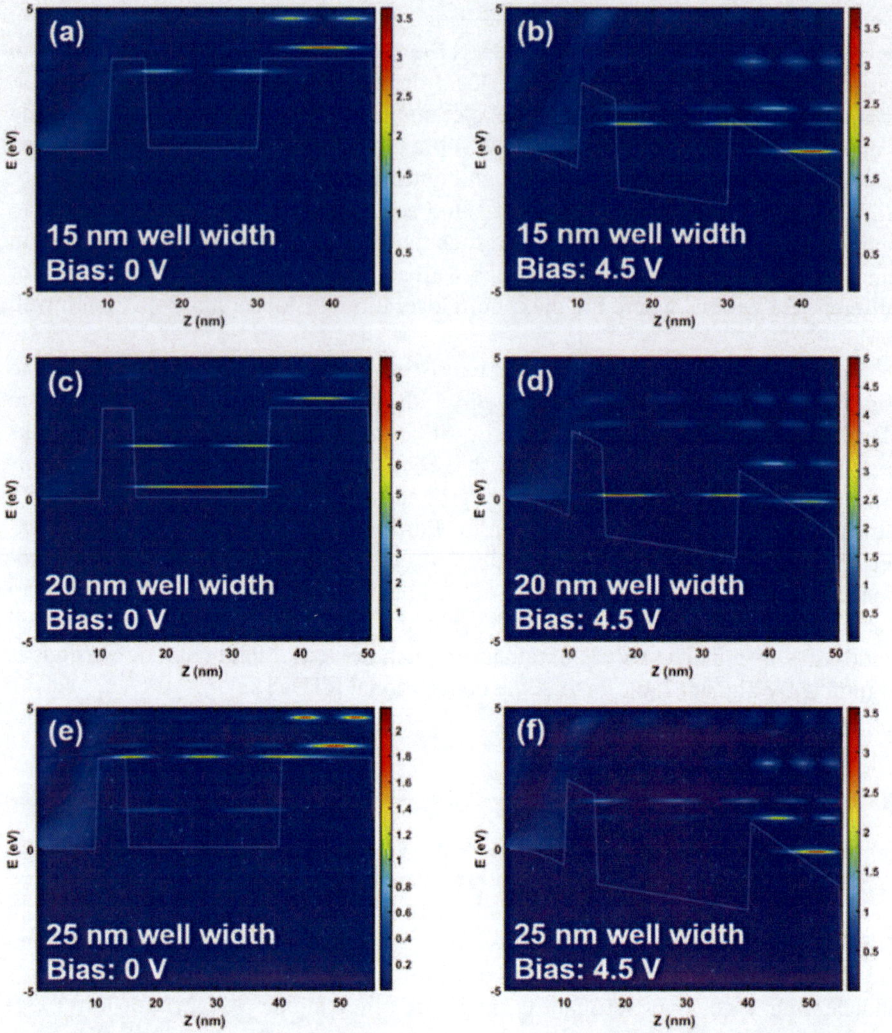


**Fig. 3** Plot of the density of states (DOS) obtained from self-consistent solution of Schrodinger–Poisson equations in the RTD structure; **a** 15 nm thick Ge layer in equilibrium; **b** 15 nm thick Ge layer in resonant condition (4.5 V); **c** 20 nm thick Ge layer in equilibrium; **d** 20 nm thick Ge layer near resonant condition (4.5 V); **e** 25 nm thick Ge layer in equilibrium; **f** 25 nm thick Ge layer near resonant condition (4.5 V)

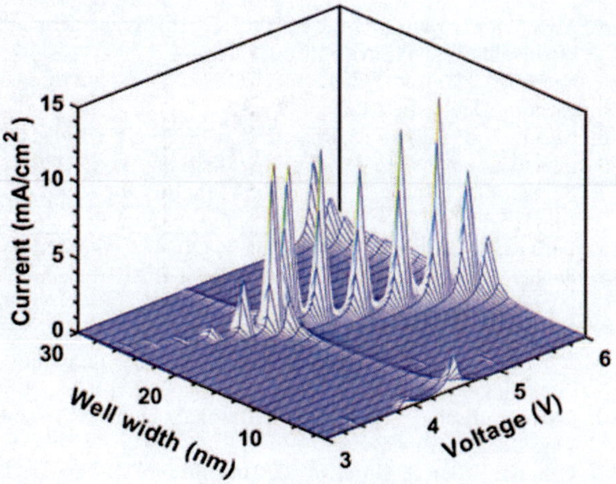


**Fig. 4** Current valley characteristics of the device for varying Ge-quantum well width

**Table 1** Peak current (excited state) of RTDs for different Ge-quantum well width

| Well width (nm) | Vg (V) | Peak current (mA/cm$^2$) | PVCR |
|---|---|---|---|
| 25 | 3.30 | 0.12 | 3.75 |
| 24 | 3.53 | 0.13 | 8.50 |
| 23 | 3.75 | 0.76 | 8.96 |
| 22 | 4.00 | 3.58 | 10.44 |
| 21 | 4.23 | 11.01 | 10.09 |
| 20 | 4.65 | 11.30 | 17.86 |
| 19 | 4.98 | 9.55 | 17.38 |
| 18 | 5.33 | 11.68 | 16.76 |
| 17 | 5.65 | 13.36 | 12.47 |
| 16 | 5.88 | 8.26 | 14.03 |
| 15 | 6.00 | 3.86 | 5.65 |

**Acknowledgements** The authors would like to acknowledge the Department of Electronic Science, University of Calcutta for providing infrastructural support to conduct this work.

## References

1. Sirtori C, Capasso F, Faist J, Hutchinson AL, Sivco DL, Cho AY (1998) Resonant tunneling in quantum cascade lasers. IEEE J Quantum Electron 34(9):1722–1729
2. Maekawa T, Kanaya H, Suzuki S, Asada M (2016) Oscillation up to 1.92 THz in resonant tunneling diode by reduced conduction loss. Appl Phys Exp 9(2):024101
3. Asada M, Suzuki S (2021) Terahertz emitter using resonant-tunneling diode and applications. Sensors 21(4):1384

4. Miyamoto T, Yamaguchi A, Mukai T (2016) Terahertz imaging system with resonant tunneling diodes. Jpn J Appl Phys 55(3):032201
5. Seabaugh AC, Kao YC, Yuan HT (1992) Nine-state resonant tunneling diode memory. IEEE Electron Device Lett 13(9):479–481
6. Mazumder P, Kulkarni S, Bhattacharya M, Sun JP, Haddad GI (1998) Digital circuit applications of resonant tunneling devices. Proc IEEE 86(4):664–686
7. Shinkai G, Hayashi T, Hirayama Y, Fujisawa T (2007) Controlled resonant tunneling in a coupled double-quantum-dot system. Appl Phys Lett 90(10):103116
8. Fujisawa T, Hayashi T, Hirayama Y (2004) Controlled decoherence of a charge qubit in a double quantum dot. J Vac Sci Technol B Microelectron Nanometer Struct Process Meas Phenom 22(4):2035–2038
9. Bernardo-Gavito R, Bagci IE, Roberts J, Sexton J, Astbury B, Shokeir H, McGrath T, Noori YJ, Woodhead CS, Missous M, Roedig U (2017) Extracting random numbers from quantum tunnelling through a single diode. Sci Rep 7(1):1–6
10. Chang L, Esaki L, Tsu R (1974) Resonant tunneling in semiconductor double barriers. Appl Phys Lett 24(12):593–595
11. Capasso F, Sen S, Gossard AC, Hutchinson AL, English JH (1986) Quantum well resonant tunneling bipolar transistor operating at room temperature. In: International electron devices meeting. IEEE, pp 282–285
12. Sen S, Capasso F, Beltram F, Cho AY (1987) The resonant-tunneling field-effect transistor: a new negative transconductance device. IEEE Trans Electron Devices 34(8):1768–1773
13. Leburton JP, Kolodzey J, Biggs S (1988) Bipolar tunneling field-effect transistor: a three-terminal negative differential resistance device for high-speed applications. Appl Phys Lett 52(9):1608–1620
14. Krishnamohan T, Kim D, Raghunathan S, Saraswat K (2008) Double-gate strained-ge heterostructure tunneling FET (TFET) with record high drive currents and $\ll$60 mV/dec subthreshold slope. In: 2008 IEEE international electron devices meeting. IEEE, pp 1–3
15. Chen G, Klimeck G, Datta S, Chen G, Goddard WA III (1994) Resonant tunneling through quantum-dot arrays. Phys Rev B 50(11):8035
16. Tsu R, Esaki L (1973) Tunneling in a finite superlattice. Appl Phys Lett 22(11):562–564
17. Grubin HL (1995) Density matrix simulations of semiconductor devices. In: Quantum transport in ultrasmall devices. Springer, Boston, MA
18. Frensley WR (1987) Wigner-function model of a resonant-tunneling semiconductor device. Phys Rev B 36(3):1570
19. Neculoiu D, Tebeanu T (1996) SPICE implementation of double barrier resonant tunnel diode model. In: International semiconductor conference, 1996, CAS'96 proceedings, vol 1, 19th edn. IEEE, pp 181–184
20. Datta S (2000) Nanoscale device modeling: the Green's function method. Superlattices Microstruct 28(4):253–278
21. Sikdar S, Chowdhury BN, Chattopadhyay S (2021) Design and modeling of high-efficiency ga as-nanowire metal-oxide-semiconductor solar cells beyond the Shockley-Queisser limit: an negf approach. Phys Rev Appl 15(2):024055
22. Nag Chowdhury B, Chattopadhyay S (2014) Investigating the impact of source/drain doping dependent effective masses on the transport characteristics of ballistic Si-nanowire field-effect-transistors. J Appl Phys 115(12):124502
23. Sikdar S, Chowdhury BN, Ghosh A, Chattopadhyay S (2017) Analytical modeling to design the vertically aligned Si-nanowire metal-oxide-semiconductor photosensors for direct color sensing with high spectral resolution. Phys E 87:44–50
24. Chowdhury BN, Chattopadhyay S (2016) Unusual impact of electron-phonon scattering in Si nanowire field-effect-transistors: a possible route for energy harvesting. Superlattices Microstruct 97:548–555
25. Paul DJ (2003) Nanoelectronics. In: Meyers RA (ed) Encyclopedia of physical science and technology, 3rd edn. Academic Press

26. Neophytou N, Paul A, Lundstrom MS, Klimeck G (2008) Bandstructure effects in silicon nanowire electron transport. IEEE Trans Electron Devices 55(6):1286–1297
27. Sikdar S, Chowdhury BN, Chattopadhyay S (2019) Understanding the electrostatics of top-electrode vertical quantized Si nanowire metal–insulator–semiconductor (MIS) structures for future nanoelectronic applications. J Comput Electron 18(2):465–472
28. Niquet YM, Allan G, Delerue C, Lannoo M (2000) Quantum confinement in germanium nanocrystals. Appl Phys Lett 77(8):1182–1184
29. Keldysh LV (1965) Diagram technique for nonequilibrium processes. Sov Phys JETP 20(4):1018–1026

# Comparative Analysis of Normal and Anemic RBC by Employing Impedimetric and Voltammetric Studies

**Debopam Bhattacharya, Aindrila Roy, Chirantan Das, Basudev Nag Chowdhury, Anupam Karmakar, and Sanatan Chattopadhyay**

**Abstract** The present work reports the Electrochemical Impedance Spectroscopy (EIS) and cyclic voltammetry (CV) technique-based qualitative analysis of normal and anemic erythrocytes (RBCs). Theoretical understanding of the effect of an external applied electric field on such RBCs has been developed by considering their morphological differences. CV measurements exhibit relatively less oxidation current for anemic RBC suspensions when compared to an equivalent hematocrit (Hct) suspension of normal RBC. EIS analysis suggests an increasing trend of impedance with the increase of Hct count for anemic RBC suspensions; however, opposite trend has been observed for normal RBCs. Such reverse trends in electrochemical parameters can further be exploited for the identification and quantitative detection of anemic blood samples.

**Keywords** Electrochemical impedance spectroscopy · Cyclic voltammetry · Anemia

## 1 Introduction

Anemia, also known as erythrocytopenia, is an illness in which blood shows a quantitative reduction of hemoglobin and consequently a reduced ability to transport oxygen. The onset of the above ailment can be due to various reasons, specifically a lower than a normal number of RBC counts or reduction in its oxygen-carrying capacity. According to the World Health Organization (WHO), anemia is a serious global public health crisis that must be considered with utmost priority [1]. The evaluations of the WHO show that a significant proportion of children below 5 years of age and equally a noteworthy proportion of pregnant women all over the world

D. Bhattacharya · A. Roy · C. Das · B. Nag Chowdhury · A. Karmakar · S. Chattopadhyay (✉)
Department of Electronic Science, University of Calcutta, 92 A.P.C. Road, Kolkata 700009, India
e-mail: scelc@caluniv.ac.in

S. Chattopadhyay
Centre for Research in Nanoscience and Nanotechnology (CRNN), JD Block, Sector III, Salt Lake City, Kolkata 700098, India

 
C. Giri et al. (eds.), *Emerging Electronic Devices, Circuits and Systems*,
Lecture Notes in Electrical Engineering 1004,
https://doi.org/10.1007/978-981-99-0055-8_21

are plagued with anemia. India has a severe prevalence of anemia in children of age group 6–59 months and women of reproductive age (15–49 years) [1]. A wide range of symptoms like exhaustion, weakness, shortness of breath, unrhythmic heartbeats, chest pain, headaches [2] is associated with anemia, including an increased risk of maternal and child mortality [1].

Normal human RBCs exhibit a biconcave-discoid shape known as discocytes; however, in certain physiological conditions, such shape can be deformed that significantly affects the quantity and oxygen-carrying capacity of hemoglobin. Among such shape deformations, particularly in the most common instance of spherocytosis, the RBCs are distorted almost to a spherical shape from their normal biconcave discoid shape. Hereditary spherocytosis (HS), a type of congenital inherited condition, is reported to be the most prevalent cause of spherocytosis, though in some cases, the disorder may be generated due to autoimmune hemolytic anemia (AIHA), an acquired disorder [3, 4]. Thus, fast and accurate detection of anemia has profound consequences to its early treatment.

Medical diagnosis techniques in recent times have achieved robust, simple, and cost-effective strategies due to various point-of-care electronic devices dedicated to measuring bio-system parameters [5]. In this context, digital microfluidics has emerged as a superior technique for bio-analytical measurements due to the reduced requirement of test samples, rapid processing, enhanced sensitivity, and assimilation of multiple processes in a single device (Lab-on-Chip) [6–8].

Several methods have been proposed for the quantitative and qualitative detection of such analytical measurements and detection of anemia. Complete blood count (CBC) [9], mean corpuscular volume (MCV) [10], bone marrow examinations [11], high-performance liquid chromatography (HPLC) [12], cryohemolysis [13], osmotic fragility test [14] are routinely used in laboratory diagnostic processes for the detection of anomalies related to blood. However, most diagnostic procedures are complex, expensive, involve highly skilled labor, require substantial sample volume (200–500 μl), time-consuming (24–36 h for the results to be analyzed), and demonstrate high laborious processes. In this context, the EIS technique has gained interest recently due to the growing need for rapid, feasible, point-of-care diagnosis. EIS allows monitoring and sensitive detection of hematocrit (Hct) (the volume percentage of RBCs in a blood volume and is evaluated based on cell number and size) and erythrocyte sedimentation rate (ESR) [15]. Our previous research works show an on-wafer two-electrode setup-based quantitative [16] and qualitative [17] detection of anemia using impedimetric techniques. In vivo voltammetric analysis has also been implemented for electrochemical detection of hemoglobin in red blood cells [18].

The current work demonstrates a scheme of detecting anemia by employing electrochemical techniques to comparatively study the normal and anemic RBCs. A comprehensive understanding of the effect of an external electric field on normal and anemic RBC suspension is developed based on their morphological differences. Such conceptual framework estimating the dependence of electrochemical parameters on RBC morphology and counts is utilized for the detection of anemia by

**Table 1** Required packed and normal saline volume to obtain the desired hematocrit percentage

| Hct (%) | Packed cell volume (μl) | Normal saline (μl) |
|---|---|---|
| 10 | 100 | 900 |
| 20 | 200 | 800 |
| 30 | 300 | 700 |

analyzing the voltammograms and impedance spectroscopy results of normal and anemic RBCs. The proposed method can be further used to detect and analyze other similar biological samples.

## 2 Materials and Methods

### *2.1 Sample Preparation*

A heparinized vial (15 U/ml) is used to extract blood from an antecubital vein of healthy and anemic female volunteers at ambient room temperature (25 ± 0.5 °C). The subjects had required diet and rest and the instructions to extract blood are followed according to the hemorheological laboratory techniques [19]. Packed RBC is prepared by washing whole blood three times with normal saline (0.90% w/v of NaCl) and is centrifuged at 9000 rpm at 4 °C. Solutions of different Hcts (10, 20, and 30%) are prepared from the packed cells by diluting with normal saline. Such diluted samples are mixed thoroughly to obtain homogeneous mixtures. The required packed cell and normal saline volume for Hct dilution are illustrated in Table 1. This research program is performed with required consent from the Biosafety and Ethics Committee of the University of Calcutta.

### *2.2 Theoretical Representation*

A comprehensive theoretical representation of normal and anemic RBCs is developed on the basis of their morphological dissimilarities. A suspension of both normal and anemic RBCs with and without an applied electric field is illustrated in Fig. 1a, b, respectively. In general, normal RBCs are biconcave discoid in nature and have a 'trinity' of potentials (transmembrane potential, dipole potential, and surface potential) [20] around the cell. The phospholipid bilayer of the cell membrane consists of lipid and protein dipoles [21], leading to a net dipole moment of each RBC [17]. However, a suspension of normal RBCs does not exhibit any particular polarizability due to the random orientation of such RBC diploes. On application of an external electric field, such randomly distributed dipoles tend to align themselves in the direction of the applied field, thereby increasing the conductivity of the normal RBC suspension.

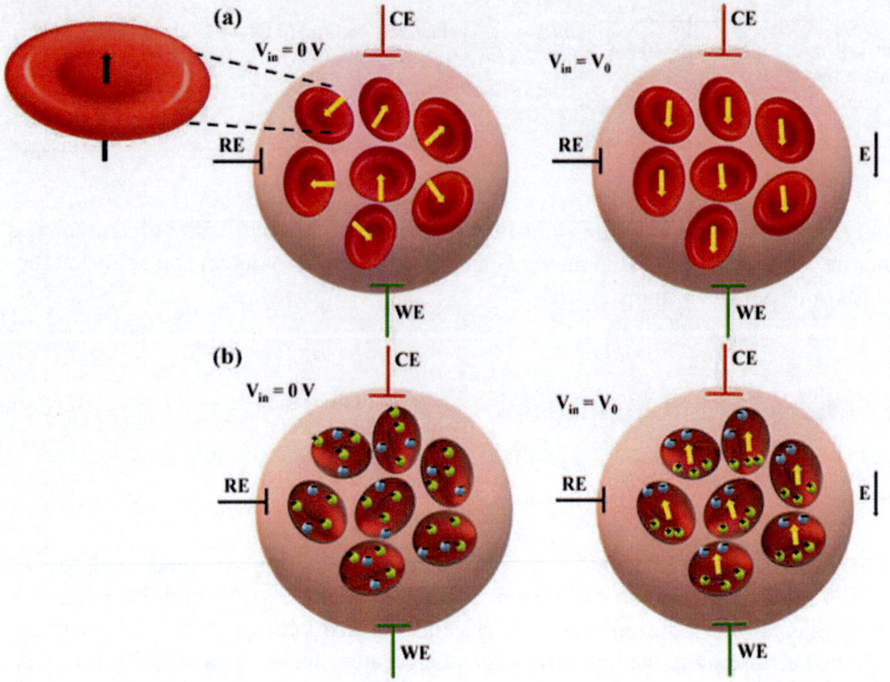


**Fig. 1** Theoretical representation of RBCs of both **a** normal and **b** anemic in a suspension with and without an applied electric bias

However, spherocytes have abnormal cell morphology due to defects and deficiencies of proteins in the cell membrane [22] and have inadequate adherence amid the cytoskeleton and the phospholipid bilayer [3]. This leads to a spherically symmetric distribution of ions inside the anemic RBCs resulting in no predefined dipoles. Thus, with an applied external bias, in the cytoplasm of such cells, positive and negative ions align themselves according to their polarity in the electric field, creating an electric dipole to each RBC in opposite direction to the external field. This is eventually expected to reduce the net conductivity of the anemic RBC suspension.

# 3 Experimental Details

## 3.1 *Electrochemical Measurements*

Impedimetric and voltammetric analyses of the RBC suspensions with different hematocrit counts of normal and anemic blood samples are performed employing Electrochemical Workstation (CH Instruments, Model CHI660E) and Three-Screen Printed Electrode (Zensor, Part Number: TE100) as the sensing platform. A single cycle (two segments) voltammetry is performed in the voltage range of − 1 V–1 V, with a positive scan, where the scan rate is kept at 0.1 V/s with a sensitivity of 0.1 mA/V. Further, impedance spectroscopy is performed using a 5 mV AC voltage (peak to peak) in the frequency range of 50 Hz–1 MHz. The schematic of the experimental setup and the sample's real-time image under test are depicted in Fig. 2a, b, respectively.

Figure 2c represents the circuit diagram of the current and impedance measurement. A potentiostat set up with three electrodes, i.e., counter electrode (CE), reference electrode (RE), and the working electrode (WE), is used for the electrochemical measurements of the RBC suspension. The potential is controlled and the current is measured at the WE. An input signal Vi applied to the '+' pin of the op-amp results in an output current flowing from CE to WE through the cell suspension and is measured across it. The potential difference between the WE and RE is measured using an electrometer and is feedback into the op-amp.

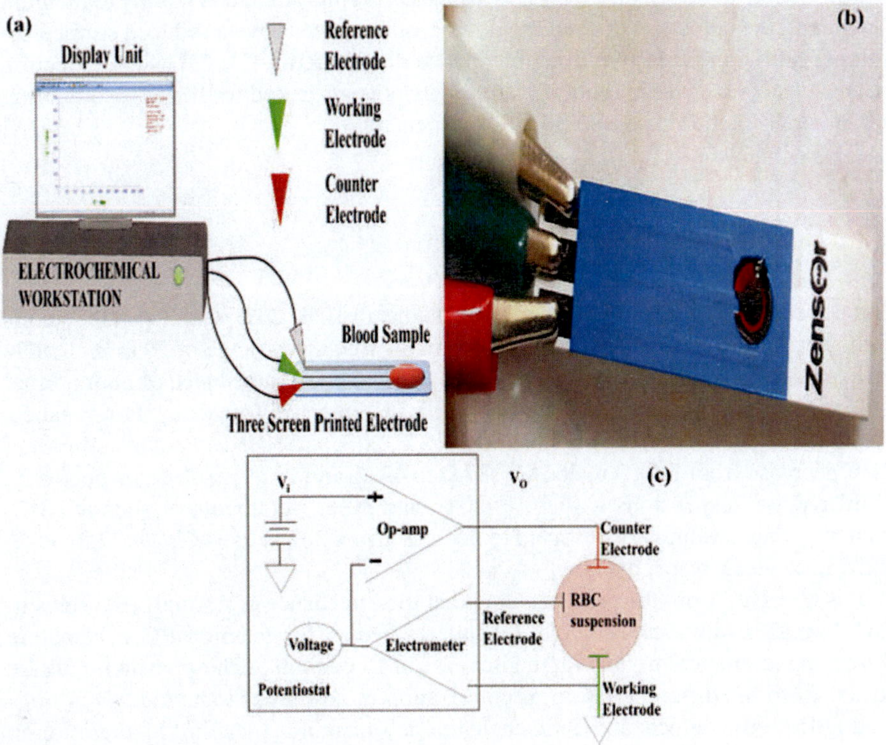


**Fig. 2** **a** Schematic representation of the experimental setup, **b** real-time image of the sample under test, **c** circuit of the experimental setup

# 4 Results and Discussions

## 4.1 Voltammetric Analysis

The cyclic voltammograms (from – 0.5 V to 0.5 V) of normal and anemic RBC suspensions for 10%, 20%, and 30% Hct counts are illustrated in Fig. 3a–c, respectively. The oxidation peak of normal RBC suspension is observed to appear at – 0.012 V, – 0.07 V, and – 0.036 V for 10%, 20%, and 30% Hct counts, respectively. It is evident from the figures that the suspensions with anemic RBCs have less oxidation current than the suspensions with normal RBCs. The peak oxidation currents for the suspension with 10%, 20%, and 30% Hct of normal RBC are 2.588 μA, 2.712 μA, and 2.685 μA, respectively. However, such oxidation currents for the suspensions with 10%, 20%, and 30% Hct of anemic RBC are 2.105 μA, 2.154 μA, and 2.156 μA, respectively, indicating the suspensions with anemic RBCs to exhibit less oxidation compared to that of normal RBCs. At this point, it is worthy to mention that due to the defects in the cell membrane, spherocytes have a reduced surface-to-volume ratio compared to normal biconcave discoid RBCs [3, 23]. Such reduction of the surface-to-volume ratio of anemic RBCs is attributed to the lesser oxidation rate of such solution than normal RBC suspension.

## 4.2 EIS Analysis

The impedimetric response of RBC suspensions in 10%, 20%, and 30% Hct counts obtained from normal and anemic persons in the frequency range of 50 kHz–1 MHz is illustrated in Fig. 4a, b, respectively. It is observed that impedance of both normal and anemic RBC suspensions decreases with increment in frequency. For instance, the impedance values of suspensions with 10%, 20% and 30% Hct counts of normal RBC decrease from 1.4 kΩ to 265 Ω, 187 Ω to 49 Ω, and 182 Ω to 35 Ω, respectively. Similarly, for suspensions with 10%, 20%, and 30%, Hct counts of anemic RBC, such impedance values are observed to decrease from 262 Ω to 34 Ω, 294 Ω to 60 Ω, and 532 to 99 Ω, respectively.

It is observed from the plots of Fig. 4 that the impedance of normal RBC suspension decreases with increasing the Hct count, whereas, for anemic RBC, such nature shows an incremental trend with the increase of Hct count. Such opposite trends are attributed to the different nature of polarization in normal and anemic RBCs originating from the defects and deficiencies of cell membrane leading to alteration in shapes and surface morphology of anemic RBC (spherocytes). As mentioned earlier, the normal RBCs are of biconcave discoid in nature and thus exhibiting randomly oriented finite dipole moments, which on application of an external bias align themselves along the electric field, thereby behaving as a paraelectric material. Such dipole alignment further enhances the net field in the solution, and therefore, its conductivity increases with Hct count. However, spherocytes are devoid of any pre-existing

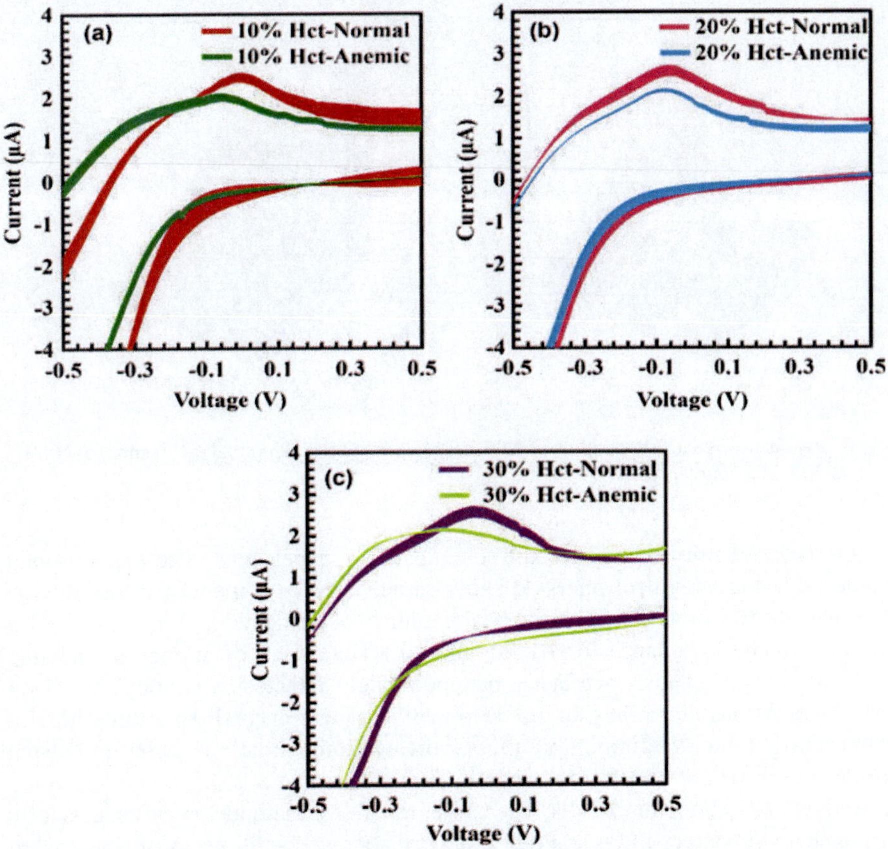


**Fig. 3** Comparative plots showing cyclic voltammograms of normal and anemic RBC suspensions with **a** 10%, **b** 20%, and **c** 30% Hct count in the voltage range − 0.5–0.5 V

dipoles due to their spherical symmetry of ion distribution. Therefore, with an applied electric bias, such ions inside the cytoplasm of such cells align themselves according to their opposite polarity with reference to the electric field, forming an equivalent dielectric material. With increasing Hct counts in anemic RBC suspension, such dielectric property enhances which eventually reduces its resultant conductivity.

## 5 Conclusion

The present work sought to develop an EIS-based scheme for detection of anemia by a comparative study of normal and anemic RBC suspensions. A conceptual framework of the electrochemical behavior of such suspensions has been developed on the basis of their morphological differences, which predicts an opposite nature of variation

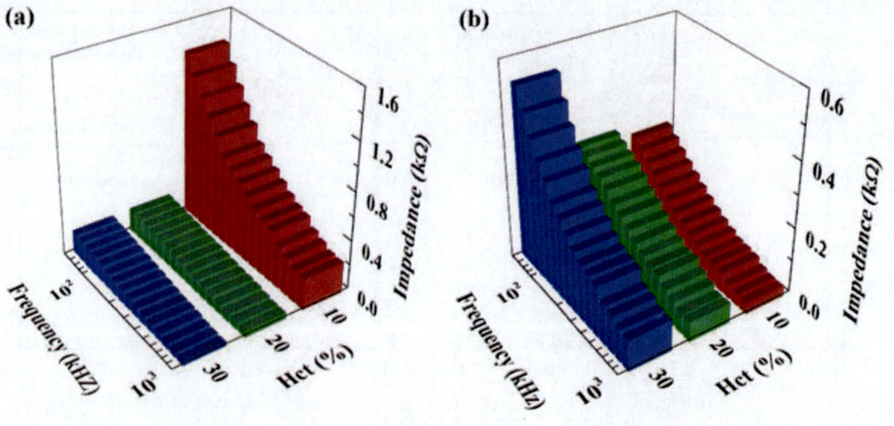


**Fig. 4** Impedance spectroscopy plots of RBC suspensions of both **a** normal and **b** anemic with 10, 20, and 30% Hct counts

for the polarizability of normal and anemic RBC suspensions. The experimental results of electrochemical analyses show satisfactory agreement with the theory. The relevant EIS plots exhibit an opposite nature (decremental and incremental) of variation of the impedance for 10, 20, and 30% Hct counts of normal and anemic RBC suspensions. Thus, anemia can be qualitatively detected by employing EIS on RBC suspensions depending on its decremental or incremental variation with Hct count. Further, the voltammogram plots indicate more oxidation current in normal RBCs than those obtained for the anemic ones which suggests the oxygen-carrying capacity to fall in anemic RBCs. Therefore, the above method can be successfully implemented to detect and predict the extent of anemic conditions by analyzing and comparing the voltammograms and impedance values of anemic RBC suspensions for rapid, precise, and point-of-care detection of anemia.

**Acknowledgements** Mr. Debopam Bhattacharya likes to acknowledge SERB, Department of Science and Technology (DST, GoI), for funding the project titled 'Point-of-care Electronic Diagnosis of Anemic Diseases by Employing Impedimetric Techniques', CRG/2018/001922. The authors would also like to acknowledge the funding support from DST-PURSE (Ref. no. DST-PURSE/FUND ALLOCATION/PH I/005/7502; dated 04.08.2012), Government of India, for developing the Electrical Characterization Laboratory.

The authors would like to acknowledge Dr. Roshnara Mishra and Anusua Singh from Department of Physiology, University of Calcutta, India, for providing normal and anemic RBC suspensions.

## References


1. World Health Organization, https://www.who.int/health-topics/anaemia
2. Mayo Clinic, https://www.mayoclinic.org/diseases-conditions/anemia
3. Zamora EA, Schaefer CA (2022) Hereditary spherocytosis. In: StatPearls [Internet]. StatPearls Publishing, Treasure Island, FL. Updated 20 July 2021. Available from: https://www.ncbi.nlm.nih.gov/books/NBK539797/. PMID: 30969619
4. Da Costa L, Mohandas N, Sorette M, Grange MJ, Tchernia G, Cynober T (2001) Temporal differences in membrane loss lead to distinct reticulocyte features in hereditary spherocytosis and in immune hemolytic anemia. Blood 98(10):2894–2899. https://doi.org/10.1182/blood.v98.10.2894. PMID: 11698268
5. Chattopadhyay S, Chakraborty S, Das C, Saha R (2015) Recent progresses on micro- and nano-scale electronic biosensors: a review. In: Nanospectrum: a current scenario. Allied Publishers Pvt. Ltd., pp 19–40 (Chapter 5). ISBN: 978-93-85926-06-8
6. Dixon C, Lamanna J, Wheeler A (2020) Direct loading of blood for plasma separation and diagnostic assays on a digital microfluidic device. Lab Chip 20. https://doi.org/10.1039/D0LC00302F
7. Sista RS, Ng R, Nuffer M, Basmajian M, Coyne J, Elderbroom J, Hull D, Kay K, Krishnamurthy M, Roberts C, Wu D, Kennedy AD, Singh R, Srinivasan V, Pamula VK (2020) Digital microfluidic platform to maximize diagnostic tests with low sample volumes from newborns and pediatric patients. Diagnostics (Basel) 10(1):21. https://doi.org/10.3390/diagnostics10010021. PMID: 31906315; PMCID: PMC7169462 (2020)
8. Yap BK, Soair SNM, Talik NA, Lim WF, Mei IL (2018) Potential point-of-care microfluidic devices to diagnose iron deficiency anemia. Sensors (Basel) 18(8):2625. Published 10 Aug 2018. https://doi.org/10.3390/s18082625
9. Gulati GL, Hyun BH (1994) The automated CBC. A current perspective. Hematol Oncol Clin North Am 8(4):593–603. PMID: 7961282
10. Coyer SM (2005) Anemia: diagnosis and management. J Pediatr Health Care 19(6):380–385
11. Bridges KP, Howard A (2008) Principles of anemia evaluation. In: Anemias and other cell disorders, 1st edn. The McGraw-Hill, USA, pp 4–18
12. Khera R, Singh T, Khuana N, Gupta N, Dubey AP (2015) HPLC in characterization of hemoglobin profile in thalassemia syndromes and hemoglobinopathies: a clinicohematological correlation. Indian J Hematol Blood Transfus 31(1):110–115. https://doi.org/10.1007/s12288-014-0409-x. Epub 5 Jun 2014. PMID: 25548455; PMCID: PMC4275515
13. Iglauer A, Reinhardt D, Schröter W, Pekrun A (1999) Cryohemolysis test as a diagnostic tool for hereditary spherocytosis. Ann Hematol 78(12):555–557. https://doi.org/10.1007/s002770050557. PMID: 10647879
14. Roper D, Layton M (2006) Investigation of the hereditary haemolyticanaemias: membrane and enzyme abnormalities. https://doi.org/10.1016/B0-44-306660-4/50014-3
15. Zhbanov A, Yang S (2017) Electrochemical impedance spectroscopy of blood for sensitive detection of blood hematocrit, sedimentation and dielectric properties. Anal Methods 9(22):3302–3313. https://doi.org/10.1039/C7AY00714K
16. Chakraborty S, Das S, Das C, Chandra S, Sharma KD, Karmakar A, Chattoapadhyay S (2020) On-chip estimation of hematocrit level for diagnosing anemic conditions by impedimetric techniques. Biomed Microdevices 22(2):38. https://doi.org/10.1007/s10544-020-00493-5. PMID: 32430696
17. Chakraborty S, Das C, Saha R, Karmakar A, Chattoapadhyay S, Das S, Mishra R, Mishra R (2018) Bio-dielectric variation as a signature of shape alteration and lysis of human RBCs: an on-chip analysis. In: 2018 international symposium on devices, circuits and systems (ISDCS), pp 1–4. https://doi.org/10.1109/ISDCS.2018.8379645
18. Toh RJ, Peng WK, Han J, Pumera M (2014) Direct in vivo electrochemical detection of haemoglobin in red blood cells. Sci Rep 4:6209. https://doi.org/10.1038/srep06209. PMID: 25163492; PMCID: PMC4147368

19. Baskurt OK, Boynard M, Cokelet GC, Connes P, Cooke BM, Forconi S, Liao F, Hardeman MR, Jung F, Meiselman HJ, Nash G, Nemeth N, Neu B, Sandhagen B, Shin S, Thurston G, Wautier JL (2009) International expert panel for standardization of hemorheological methods. New guidelines for hemorheological laboratory techniques. Clin Hemorheol Microcirc 42(2):75–97. https://doi.org/10.3233/CH-2009-1202. PMID: 19433882
20. O'Shea P (2003) Intermolecular interactions with/within cell membranes and the trinity of membrane potentials: kinetics and imaging. Biochem Soc Trans 31(Pt 5):990–996. https://doi.org/10.1042/bst0310990. PMID: 14505466.
21. Brockman H (1994) Dipole potential of lipid membranes. Chem Phys Lipids 73(1–2):57–79. https://doi.org/10.1016/0009-3084(94)90174-0. PMID: 8001185
22. Eber S, Lux SE (2004) Hereditary spherocytosis–defects in proteins that connect the membrane skeleton to the lipid bilayer. Semin Hematol 41(2):118-141. https://doi.org/10.1053/j.seminhematol.2004.01.002. PMID: 15071790
23. Bissonnette B, Luginbuehl I, Marciniak B, Dalens BJ (eds) (2006) Hereditary spherocytosis. In: Syndromes: rapid recognition and perioperative implications. McGrawHill. https://accessanesthesiology.mhmedical.com/content.aspx?bookid=852&sectionid=49517683

# Differential Fault Analysis of Trivium Using Artificial Neural Network on SoC Platform

**Arijit Tewary, Swagata Mandal, Amlan Chakrabarti, Debasri Saha, and Avishek Adhikari**

**Abstract** Stream ciphers are highly frequent in the realm of secret key cryptography because of their low-computational complexity and latency. Though variety of attack models for testing the robustness of various stream ciphers are available, the differential fault analysis (DFA) model is the most popular. DFA is used not only to disclose the identity of any unknown cipher but also to recover the secret KEY and the cipher's internal state. Deep learning approaches extract patterns from large amounts of cryptographic data, which aids in crypt-analysis. Here, single bit and multi-bit faults are injected into the internal register of trivium cipher, and artificial neural network (ANN) is used to identify the injected fault location without any handcrafted features of keystream unlike other existing state-of-the-art fault analysis techniques. In order to enhance the speed of fault analysis, we have used system on chip (SoC) platform where ANN-based classifier has been placed in the programmable logic area as an accelerator. This enhances the speed of fault analysis by 9% for single bit fault, 12% for two adjacent bit faults, and 13% for three adjacent bit faults compared to their software models. To the best of the author's knowledge, this is the first of kind research where ANN-based hardware accelerator has been used for fault location identification on SoC platform.

**Keywords** Fault analysis · ANN · XOR-DKS · SoC · Hardware · FPGA

A. Tewary (✉) · A. Chakrabarti · D. Saha
University of Calcutta, Kolkata, India
e-mail: ari.bitmca@gmail.com

S. Mandal
Jalpaiguri Government Engineering College, Jalpaiguri, India

A. Adhikari
Presidency University, Kolkata, India

 
C. Giri et al. (eds.), *Emerging Electronic Devices, Circuits and Systems*,
Lecture Notes in Electrical Engineering 1004,
https://doi.org/10.1007/978-981-99-0055-8_22

# 1 Introduction

Fast development of the online business and exchanges increases the security risk in the computerized world. Through various cryptography techniques such as encryption, hashing [1], and data protection from the digital intruders can be enhanced. In 1917, Gilbert Vernam introduced the concept of stream cipher [2] in cryptography which encrypts plain text into cipher text using variable length messages. The stream cipher concept has grown in prominence, and it now plays a significant part in symmetric key encryption. Stream ciphers such as trivium [3] are very prevalent in modern cybersecurity.

In general, linear feedback shift register (LFSR) and nonlinear feedback shift register (NFSR) are used in stream ciphers (NFSR). Various methods of crypt-analysis have been used on stream ciphers in recent literature. Crypt-analysis entails a number of statistical and algebraic analysis that not only assess the mathematical power of ciphers but also reveal their internal architecture. Based on the availability of the cipher, attacks can be classified into two broad categories [4]: invasive and noninvasive attack. When the ciphers are physically available to crypt-analysts, they can inject fault into the cipher, measure timing, etc., which is known as invasive attack. On the other hand, availability of only cipher text instead of cipher leads to noninvasive attack.

Traditional attack models on stream ciphers include known plain text attack [5], cipher text only attack, chosen plain text attack, etc. Outside of the typical paradigm, however, additional techniques exist, such as the side channel attack [6], memory attack, cache attack, and so on. Side channel attack involves fault analysis, power analysis, and timing analysis. In side channel attack, crypt-analysts try to analyze the ciphers from the side channel information.

A popular and effective crypt-analysis technique is the fault attack or differential fault analysis (DFA) [7]. In DFA, crypt-analysts inject fault into the internal registers of cipher which generate the faulty output. Fault can be either transient or permanent [8]. In transient fault, faults are injected within the cipher for one clock cycle only. Whereas injected fault will permanently damage a register of the cipher in permanent fault. Crypt-analysts can inject either single bit or multi-bit fault into the cipher. In this work, transient or temporary multi-bit faults are considered. Faults can be injected in the internal register of the cipher in various ways [9] like clock glitch, power spike, electromagnetic induction, etc. Here, clock glitch is used to inject the fault in the internal registers of the cipher as this method is less costly and quite convenient. Analyzing the difference between non-faulty output and faulty output, the internal state of the ciphers along with the initial key can be recovered. In [10], authors proposed DFA on trivium using single bit fault. Multi-bit fault injection technique has been proposed by the authors in [11] for DFA on trivium in any random key generation round which makes the model more generic compared to other attack models.

DFA faulty and non-faulty keystream are XOR-ed together to produce XOR differential keystream. The attacker will try to find out the patterns in the XOR differential output of the cipher which gives the signature of the cipher. The majority of current literature attempts to manually extract numerous signatures [12] from the XOR differential keystream, which is a complicated and time-consuming procedure. In this work, deep learning techniques will be employed to find out the patterns from a set of XOR differential keystreams generated from the trivium cipher without calculating the signature manually.

Various machine learning techniques can be used in the crypt-analysis to find the hidden patterns from huge data. In [13], proposed a classification method for encrypted traffic using machine learning. Hospodar et al. proposed a method of side channel analysis of advanced encryption standard (AES) cipher using least squares support vector machine (LS-SVM) in [14], where power consumption is taken as the parameter for the analysis. Besides machine learning, other deep learning methods based on neural networks [15] are also used for crypt-analysis. In machine learning-based crypt-analysis, computation complexity is one of the main challenges due to the usage of huge amount of data. For the analysis of large amounts of data, a combination of CPU and graphic processing unit (GPU) is commonly used. In general, the presence of several cores in a GPU allows for parallel processing, which improves the speed of the data analysis. The fundamental issue with GPUs is that they cannot run in a standalone manner, necessitating the use of a host CPU.

Another computing platform is the field programmable gate array (FPGA), which can be used as standalone platform and provides data parallelism, on-field reconfigurability. Due to fault injection flexibility and simultaneous data processing, crypt-analysis employing fault injection on an FPGA platform [16] is gaining popularity in recent literature. DFA on a variety of stream ciphers, such as grain [12] and trivium [17], has already been implemented on FPGA platform. In order to enhance the flexibility we have used system on chip (SoC) platform for crypt-analysis which integrate processing system or CPU with FPGA. In this work, we have proposed a neural network-based deep learning technique for DFA of trivium on SoC platform. The key contributions can be summarized as follows:

- Proposal of method of deep learning-based differential fault analysis of trivium using single and multi-bit fault injection.
- Implementation of the proposed method on SoC platform.

The organization of this paper is as follows: Section 2 describes the background of trivium and ANN for crypt-analysis. Section 3 proposes the model for differential fault analysis on trivium. The hardware implementation is described in details in Sect. 4. Section 5 studies the results and performance analysis. Finally, Sect. 6 concludes the presented work.

## 2 Background Studies

### 2.1 *Basics of Trivium Cipher*

Christophe De Canniere and Bart Preneel had proposed trivium [3] in the year 2005 which had been chosen as winner in the eSTREAM project. It is a hardware-oriented synchronous stream cipher which also provides a speedy execution scenario in the software platform.

Trivium uses 80 bit KEY and 80 bit initialization vector (IV). Using the KEY and IV, trivium generates the keystream which will be used to encrypt the message. There are three phases of keystream generation.

- Initialization phase: Internal state of the cipher is loaded with the KEY and IV.
- Update phase: Updated the internal state of the registers without generating any keystream.
- Keystream generation: In this phase, keystream is generated serially.

Trivium has 288-bit internal state and denoted as $(a_1 \ldots a_{288})$. Initialization phase loads the key $(k_1 \ldots k_{80})$ and IV $(\mathrm{IV}_1 \ldots \mathrm{IV}_{80})$ to the inner state of registers as follows:

$$
\begin{aligned}
&(a_1 \ldots a_{93}) \leftarrow (k_1 \ldots k_{80}, 0, \ldots, 0),\\
&(a_{94} \ldots a_{177}) \leftarrow (\mathrm{IV}_1 \ldots \mathrm{IV}_{80}, 0, \ldots, 0),\\
&(a_{178} \ldots a_{288}) \leftarrow (0, \ldots 0, 1, 1, 1).
\end{aligned}
$$

In the second phase, the internal state of the registers will rotate by four cycles, i.e., $4 \times 288 = 1152$ rounds without generating any keystream as follows: Here, XOR is denoted by "+", and AND is denoted by "·"

for i=1 to 1152 do:

$$
\begin{aligned}
&t_1 \leftarrow a_{66} + a_{91} \cdot a_{92} + a_{93} + a_{171}\\
&t_2 \leftarrow a_{162} + a_{175} \cdot a_{176} + a_{177} + a_{264}\\
&t_3 \leftarrow a_{243} + a_{286} \cdot a_{287} + a_{288} + a_{69}\\
&(a_1 \ldots a_{93}) \leftarrow (t_3, a_1, \ldots, a_{92})\\
&(a_{94} \ldots a_{177}) \leftarrow (t_1, a_{94}, \ldots, a_{176})\\
&(a_{178} \ldots a_{288}) \leftarrow (t_2, a_{178}, \ldots, a_{287})
\end{aligned}
$$

In third phase, keystream bits $z_1, \ldots, z_N$ where $N \leq 2^{64}$ are generated as follows:

for i=1 to N do:

$$
\begin{aligned}
&t_1 \leftarrow a_{66} + a_{93}\\
&t_2 \leftarrow a_{162} + a_{177}\\
&t_3 \leftarrow a_{243} + a_{288}\\
&z_i \leftarrow t_1 + t_2 + t_3\\
&t_1 \leftarrow t_1 + a_{91} \cdot a_{92} + a_{171}\\
&t_2 \leftarrow t_2 + a_{175} \cdot a_{176} + a_{264}\\
&t_3 \leftarrow t_3 + a_{286} \cdot a_{287} + a_{69}\\
&(a_1 \ldots a_{93}) \leftarrow (t_3, a_1, \ldots, a_{92})
\end{aligned}
$$

$(a_{94} \dots a_{177}) \leftarrow (t_1, a_{94}, \dots, a_{176})$
$(a_{178} \dots a_{288}) \leftarrow (t_2, a_{178}, \dots, a_{287})$

## 2.2 ANN for Crypt-Analysis

Artificial neural network (ANN) is a productive tool to explore implicit features of the data set by tuning its various parameters. After completion of the training with user data, the testing happens for the desired output. In classical machine learning approaches, users have to find out the features (handcrafted features) using various mathematical techniques which are used to train the machine learning model. When the finding of features from the data set is difficult, the deep learning techniques like ANN may become very useful tools. ANN aids in the formulation of an appropriate model directly from a variety of input data without any handcrafted features. This helps to formulate the model without any manual feature extraction. In crypt-analysis, a huge sum of data must be analyzed in order to determine the pattern in the input data set. Since ANN uses a large quantity of data to develop the model, it is an excellent choice for crypt-analysts to analyze the data sets.

# 3 Proposed Method

Proposed method consists of two phases: offline and online phase. In the offline phase, data set will be generated, and ANN model will be developed using this data set. In the online phase, ANN model will be used to identify the fault location from the XOR differential keystream.

## 3.1 Data Set Generation

Length of the internal register of the trivium cipher is 288 bit. The cipher produces a single keystream for each KEY and IV pair. On the other hand, injecting a fault into the internal registers just before the keystream generation produces a faulty keystream. Operation of trivium cipher involves multiple rounds [3], and here, we will consider single round for generation of both faulty and non-faulty keystream. Let $O = \{O_1, O_2, \dots, O_i \dots\}$ be the non-faulty keystream. After giving fault, the bit value will change, and faulty keystream will be generated. Let $O' = \{O'_1, O'_2, \dots, O'_i \dots\}$ be the faulty keystream. XOR operation between faulty and non-faulty keystream generates differential keystream(XOR-DKS). Let $O''_i$ is the XOR-DKS generated due to the fault injection in the $i$th location which can be represented as $O''_i = O_i \bigoplus O'_i$. $k$ bit long XOR-DKS can be represented as $O''^{,k}$

where $O^{\prime\prime,k} = \{O_1^{\prime\prime}, O_2^{\prime\prime} \dots O_k^{\prime\prime}\}$. In order to create the data set, the XOR-DKS is used along with the fault position numbers. Generated data set will consist of $288 \cdot 2^{12}$ XOR-DKS each having the length of 288 bit.

## 3.2 Parameters Selection

Performance of the proposed method depends on various parameters of the fault model like number of injected fault, nature of the injected fault, and dimension of the XOR-DKS which are listed below

- Nature of injected fault: Here, both single and multi-bit faults are considered for fault analysis of trivium cipher. For multi-bit fault, we have considered only two and three adjacent bit faults. Random multi-bit faults may generalize the fault model but it increases the computational complexity.
- Length of the XOR-DKS: We have taken 288 bits as the length of the XOR-DKS. The DKS length can be increased or decreased but length of the internal register of trivium cipher fixed the length of XOR-DKS to 288 bits.
- Number of XOR-DKS for each fault: Number of XOR-DKS for each fault location is $2^{12}$.
- Less than $2^{12}$ XOR-DKS decreased accuracy rate, while greater than $2^{12}$ XOR-DKS value increase computational complexity.

## 3.3 Model Training

Data set generated using fault injection technique is used to train the ANN model. For single bit fault, two adjacent, and three adjacent bit faults, we have developed three different ANN models using three distinct data sets. The parameter of these ANN models like number of hidden layers, epoch size, and batch size is illustrated in Table 1. We have chosen different learning parameters since the combination of these parameters provided the best accuracy. The adaptive moment estimization (ADAM) [18] optimization function is used for ANN training. ADAM-optimizer provides faster optimization compared to other existing optimization techniques like AdaGrad, AdaDelta [18], etc. Each data set is partitioned into 70:30 proportion for training and testing. Figure 1 illustrates the workflow of data set generation using XOR-DKS and formulation of ANN model. Trained ANN model can figure out random fault location from any randomly generated XOR-DKS. The internal state of the encryption will be extracted using the SAT solver [19, 20] after the faulty location has been identified.

**Table 1** Parameters of ANN performance analysis

| Fault in no. of bits | Hidden layers | Epoch size | Batch size |
|---|---|---|---|
| Single bit | 2 | 20 | 1800 |
| Consecutive two bits | 2 | 20 | 1600 |
| Consecutive three bits | 2 | 20 | 1500 |

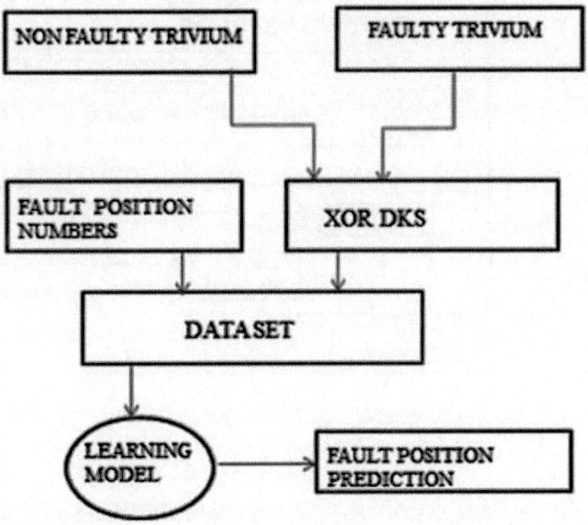


**Fig. 1** Differential fault analysis on trivium using ANN

## 4 Hardware Implementation

Proposed deep learning-based accelerator for crypt-analysis has been implemented on ZCU104 [21] SoC using Xilinx Vivado 2017.4 platform, and results obtained from the implemented model are verified using behavioral simulations. Figure 2 shows the detailed internal architecture of the proposed method on the SoC platform. The proposed crypt-analysis has two stages: offline and online. In offline, for each key and IV pair, a XOR differential keystream is generated on the reconfigurable platform. XOR-DKS for single bit, two, and three adjacent bits is generated in parallel. The generated XOR-DKS for each fault location will be used to train the ANN model on the Tesla 30C GPU platform. After the completion of the training operation, an abstraction of the trained model will be transferred to the processing system of SoC through UART communication which will be stored in the on chip memory (OCM). During the online phase, weight vectors from the on chip memory will be transferred to the line buffer in the processing logic through advanced eXtensible interface (AXI) bus as illustrated in Fig. 2.

Weight vector stored in the line buffer will be transferred to the weight vector storage memory and will be read by the multiplier module as required. In the next step, fault injector module will inject fault in the internal register of the trivium which

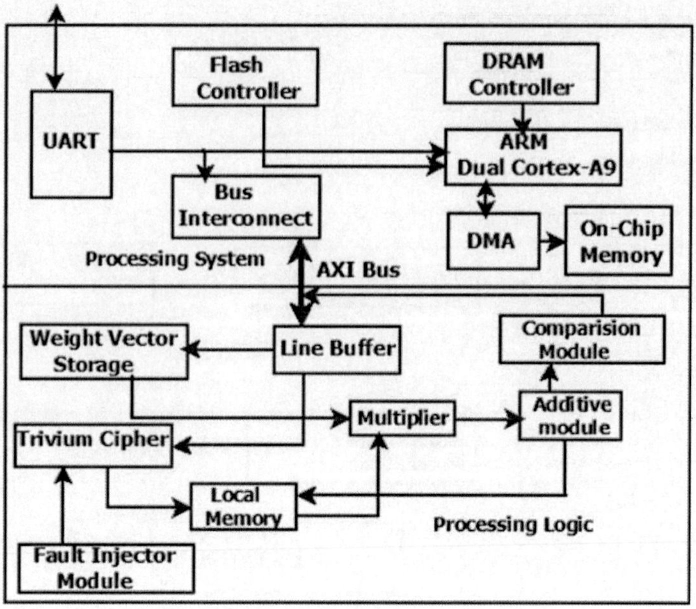


**Fig. 2** Internal architecture of ANN accelerator for crypt-analysis on trivium cipher

will generate XOR-DKS. The generated differential keystream will be stored in to the local memory and will be input of the trained ANN. The multiplier, additive module, and comparison module will all be activated successively throughout the fault detection operation. Following the fault location identification, the detected fault location will be delivered to the processing system via the AXI bus, and then to the computer via universal asynchronous receiver transmitter (UART).

Fault injector module shown in Fig. 2 uses clock glitched technique. Originally, clock is generated from a 200 MHz SiT9102AI oscillator which phase shifted by a phase lock loop (PLL). Then, the shifted clock is AND-ed with the original clock to generate the glitched clock. The IP core developed for ANN using Vivado HLS and integrated with Zynq processor system as shown in Fig. 3. Processor system reset module shown in Fig. 3 used to reset both ANN module and zynq processor.

## 5 Result and Performance Analysis

Data set has been generated for single bit and multi-bit faults using XOR-DKS on programmable logic unit of SoC. A multi-bit fault analysis is carried out using successive two bit faults and successive three bit faults. An example of a adjacent two bit fault would be $(1, 2)$, $(2, 3)$, $(3, 4) \ldots (287, 288)$, and adjacent three bit fault would be $(1, 2, 3)$, $(2, 3, 4) \ldots (286, 287, 288)$. Generated data set will be transferred

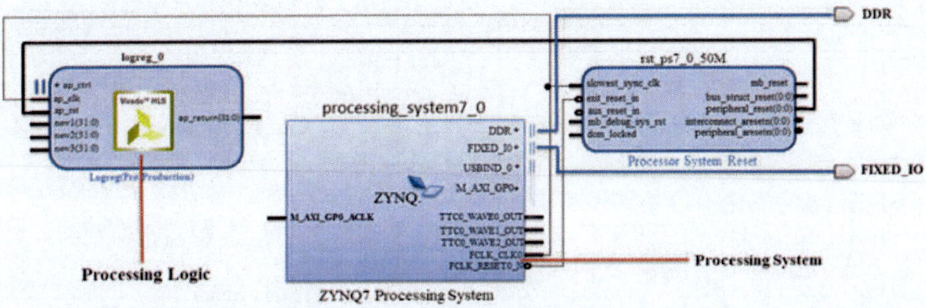


**Fig. 3** Interfacing of zynq processor with ANN accelerator

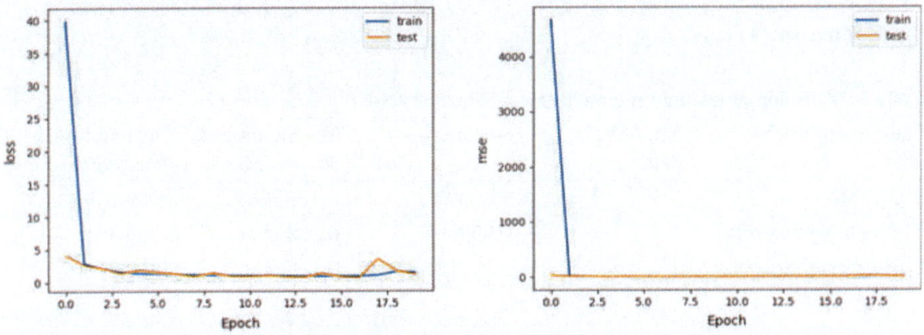


(a) Single Bit Loss vs Epoch Graph (b) Single Bit MSE vs Epoch Graph

**Fig. 4** Single bit fault graph

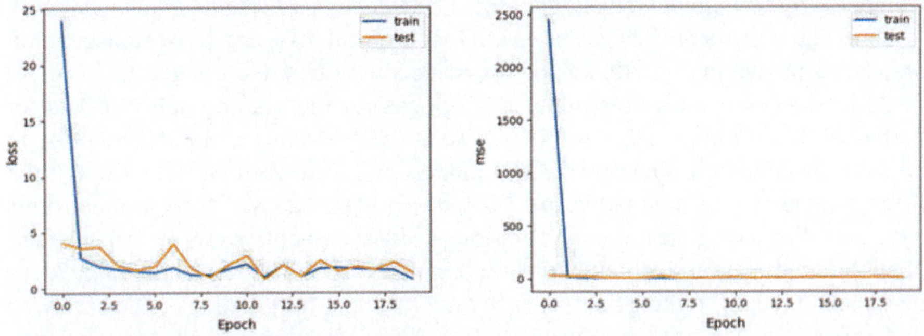


(a) Double Bit Loss vs Epoch Graph (b) Double Bit MSE vs Epoch Graph

**Fig. 5** Double bit faults graph

to the local computer through UART and will be stored into a CSV file. The CSV file will be used to develop the ANN model on GPU platform.

Figures 4a, 5a and 6a show loss versus epoch graph for the single bit, two bits, and three bits fault, respectively. The loss versus epoch graph illustrates how the data

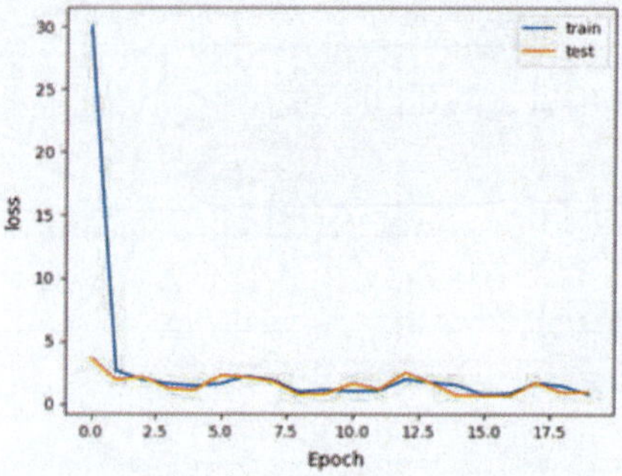


(a) Three Bit Loss vs Epoch Graph

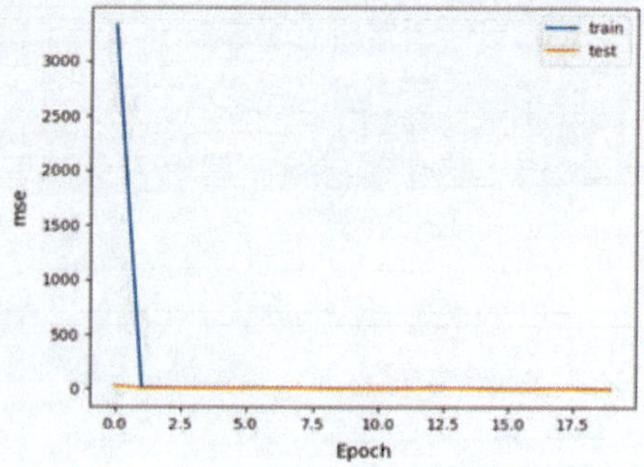


(b) Three Bit MSE vs Epoch Graph

**Fig. 6** Three bit faults graph

**Table 2** Variation of testing time on hardware and software platform

| Fault in no. of bits | No. of training data | Accuracy | Testing time in software (s) | Testing time in hardware (s) |
|---|---|---|---|---|
| Single bit | $2^{12}$ | 1.00000 | 0.10526 | 0.0967 |
| Consecutive two bits | $2^{12}$ | 0.99963 | 0.10894 | 0.09768 |
| Consecutive three bits | $2^{12}$ | 0.99918 | 0.11584 | 0.1023 |

model learns and behaves during the training operation. The value of loss function decreased rapidly with epoch value as shown in the loss versus epoch graph. The small difference between training and testing value indicates that ANN models have been properly formulated for our data set.

Mean square error (MSE) versus epoch for single bit, two, and three adjacent fault models are shown in Figs. 4b, 5b and 6b, respectively. These three graphs illustrate the MSE deviation versus the number of epochs for the training and test data for the model. Decreasing nature of MSE curve at the beginning and finally settling to a fixed low value indicate trained ANN models are performed well. In Table 2, the differences in testing time between FPGA-based hardware and software platforms are shown. Following the training operation, we have two options for testing the fault location identification method: hardware or software. As we can see in Table 2, for the same number of training data for each fault location, testing time on the hardware platform is less than on the software platform for all single bit faults, two, and thee adjacent faults. This is because operations on each link among layers inside the hardware platform can be executed in parallel. Accuracy rate for single bit faults are 1, two adjacent faults are 0.99963, and three adjacent faults are 0.99918. $R$ squared ($R^2$) values of the model are greater than 0.9, and this implies that our neural model prediction is very good. LUT requirement for implantation of trained ANN model on hardware platform is 3225 for single bit fault, 3145 for two adjacent bit faults, and 3102 for three adjacent bit faults.

## 6 Conclusion and Future Work

In this work, an artificial neural network approach has been presented for differential fault analysis on trivium on FPGA-based SoC platform. The proposed method will be used to extract the trivium cipher's key and internal states. The attack model injects single bit and multi-bit faults into the cipher's internal register, making fault injection more realistic. When compared to their software counterparts, FPGA-based ANN accelerators improve the speed of defect site detection. We intend to expand the work to include other stream ciphers in the future.

## References

1. Peng T, Leckie C, Ramamohanarao K (2007) An attack-resistant hashing scheme. In: 2007 Australasian telecommunication networks and applications conference, pp 307–310. https://doi.org/10.1109/ATNAC2007.4665307
2. Gorbenko I, Kuznetsov A, Lutsenko M, Ivanenko D (2017) The research of modern stream ciphers. In: 2017 4th international scientific-practical conference problems of info-communications. Science and technology (PIC S&T), pp 207–210. https://doi.org/10.1109/INFOCOMMST2017.8246381
3. De Cannière C (2006) Trivium: a stream cipher construction inspired by block cipher design principles. In: Katsikas SK, López J, Backes M, Gritzalis S, Preneel B (eds) Information security. Springer, Berlin, pp 171–186
4. Osuka S, Fujimoto D, Hayashi YI, Homma N, Beckers A, Balasch J, Gierlichs B, Verbauwhede I (2018) Fundamental study on non-invasive frequency injection attack against RO-based TRNG. In: 2018 IEEE international symposium on electromagnetic compatibility and 2018 IEEE Asia-Pacific symposium on electromagnetic compatibility (EMC/APEMC), pp 8 (2018). https://doi.org/10.1109/ISEMC.2018.8394008
5. Wang F, Sang J, Liu Q, Huang C, Tan J (2021) A deep learning based known plaintext attack method for chaotic cryptosystem. CoRR abs/2103.05242. https://arxiv.org/abs/2103.05242
6. Sravani MM, Ananiah Durai S (2019) Side-channel attacks on cryptographic devices and their countermeasures–a review. In: Tiwari S, Trivedi MC, Mishra KK, Misra AK, Kumar KK (eds) Smart innovations in communication and computational sciences. Springer, Singapore, pp 209–226
7. Vasyltsov I, Saldamli G (2012) Fault detection and a differential fault analysis countermeasure for the montgomery power ladder in elliptic curve cryptography. In: Advanced theory and practice for cryptography and future security. Math Comput Modell 55(1):256–267. https://doi.org/10.1016/j.mcm.2011.06.017
8. D'Angelo S, Metra C, Sechi G (1999) Transient and permanent fault diagnosis for FPGA-based TMR systems. In: Proceedings 1999 IEEE international symposium on defect and fault tolerance in VLSI systems (EFT'99), pp 330–338. https://doi.org/10.1109/DFTVS.1999.802900
9. Korczyc J, Krasniewski A (2012) Evaluation of susceptibility of FPGA-based circuits to fault injection attacks based on clock glitching. In: 2012 IEEE 15th international symposium on design and diagnostics of electronic circuits systems (DDECS), pp 171–174. https://doi.org/10.1109/DDECS.2012.6219047
10. Hojs'ık M, Rudolf B (2008) Differential fault analysis of trivium. In: Nyberg K (ed) Fast software encryption. Springer, Berlin, pp 158–172
11. Dey P, Adhikari A (2014) Improved multi-bit differential fault analysis of trivium. In: Meier W, Mukhopadhyay D (eds) Progress in cryptology–INDOCRYPT 2014. Springer, Cham, pp 37–52

12. Dey P, Chakraborty A, Adhikari A, Mukhopadhyay D (2015) Improved practical differential fault analysis of grain-128. In: 2015 design, automation test in Europe conference exhibition (DATE), pp 459–464. https://doi.org/10.7873/DATE.2015.0921
13. Alshammari R, Zincir-Heywood AN (2009) Machine learning based encrypted traffic classification: identifying SSH and skype. In: 2009 IEEE symposium on computational intelligence for security and defense applications, pp 1–8 (2009). https://doi.org/10.1109/CISDA.2009.5356534
14. Hospodar G, Gierlichs B, Mulder ED, Verbauwhede IMR, Vandewalle J (2011) Machine learning in side-channel analysis: a first study. J Cryptogr Eng 1:293–302
15. Alani MM (2012) Neuro-cryptanalysis of DES and triple-DES. In: Huang T, Zeng Z, Li C, Leung CS (eds) Neural information processing. Springer, Berlin, Heidelberg, pp 637–646
16. Al-Shara'a S, Ibraheem RK, Bayat O (2019) Implementation of cryptanalysis based on FPGA hardware using AES with SHA-1. In: 2019 international conference on smart applications, communications and networking (SmartNets), pp 1–7. https://doi.org/10.1109/SmartNets482252019.9069786
17. Potestad-Ordónez FE, Jiménez-Fernández C, Valencia-Barrero M (2016) Fault injection on FPGA implementations of trivium stream cipher using clock attacks
18. Kingma DP, Ba J (2015) Adam: a method for stochastic optimization. In: The International conference on learning representation, San Diego
19. Mironov I, Zhang L (2006) Applications of sat solvers to cryptanalysis of hash functions. In: Biere A, Gomes CP (eds) Theory and applications of satisfiability testing—SAT 2006. Springer, Berlin, pp 102–115
20. SAGE, Open source mathematical software system. http://www.sagemath.org/
21. XILINX (2019) Zcu104 evaluation board. In: UG1267 (v1.1), 9 Oct 2018

# Investigation of Adders for Retinal Neuromorphic Circuits

**Payal Shah and Surendra Singh Rathod**

**Abstract** Neuromorphic circuits are extensively being researched worldwide to emulate biological responses in electronic circuits. Integrated circuits to mimic the behavior of retina are also being modeled and developed. One of the important functions of retina is to identify the moving objects. Starburst amacrine cell is one of the cells in retina which is responsible for detection of differential motion. Researchers have modeled starburst amacrine cell that requires adders as an implicit element. Adders are primarily used for adding responses received from many bipolar cells. This paper is to investigate various types of adders and to identify the adder which would be more suitable for modeling starburst amacrine cell in neuromorphic applications. In this paper, investigation of classical CMOS analog adder, differential analog adder, low power adder, nonlinear adder and op-amp-based adders is carried out. The primary objective is to identify from these adders which would be more suitable in case of differential motion detection and should give near biological response. All the simulations are done using TSMC 180 nm technology.

**Keywords** Starburst amacrine cell · Retina · Differential motion · Adder · Neuromorphic

## 1 Introduction

The retina word comes from the Latin word rete which refers to a network of nerves. The function of retina is to receive the light from the focused lens and convert it into neural signals and finally to send these signals to brain through optical nerve for visual recognition. The retina contains light sensitive cells and image processing cells. All the cells in retina are divided in three layers. According to Kolb [1]; the

P. Shah (✉) · S. S. Rathod
Sardar Patel Institute of Technology, Mumbai, India
e-mail: payal_shah@spit.ac.in

S. S. Rathod
e-mail: surendra_rathod@spit.ac.in

 
C. Giri et al. (eds.), *Emerging Electronic Devices, Circuits and Systems*,
Lecture Notes in Electrical Engineering 1004,
https://doi.org/10.1007/978-981-99-0055-8_23

first layer of retina contains the photoreceptors which are of two types; rod and cone; to detect dim and bright light. So electrically, they are equivalent to transducer. The signals from photoreceptors integrated and regulated through the first cell of second layer called horizontal cells. The horizontal cells can be of four types, and electrically, they perform spatial filtering. The next cell in the second layer is bipolar cell. There are 11 types of bipolar cells. The bipolar cells provide direct pathway between input and output signals. From the bipolar cell, the signal passes to the third layer of cells which includes approximately 30 types of amacrine cells and 20 types of ganglion cells. The ganglion cell basically generates action potential and send it to human brain through visual cortex. The amacrine cells combine, regulate and insert time domain aspects to the signals. The biological function of wide field and medium field amacrine cells is still not known, and hence, it is still under research. The narrow field amacrine cells create functional subunits in receptive field of ganglion cells. Due to which ganglion cells are able to detect movement of the object. The starburst amacrine cell is one of the types of narrow field amacrine cell. The responses of amacrine cell are transient and sustained in nature. The transient amacrine cells give on and off depolarizing outputs to light available at receptive field. Hence, they act as an amplifier. The sustained amacrine cell behaves in the same manner as the horizontal cell [2]. Thus, input to amacrine cell is from five to six on and off bipolar cells [3]. Amacrine cells are mainly responsible for converting excitatory signals from bipolar cells to inhibitory signals and relaying these signals to the bipolar cells of the central receptive field. The sub-linear addition is done at amacrine cell to get the output in amplified form. Hence, starburst amacrine cell which is responsible for detecting differential motion and can be considered as sub-linear adder. Thus, it is necessary to identify which configuration of the adder gives biologically plausible response.

## 2 Literature Survey

There are many adders published in the literature. However, adder designs are primarily being focused on signal processing concepts, majorly neglecting the fundamental aspects of neuromorphic circuits like adaptable nonlinear output based on the variation in the parameters of the input signals. Popular adder designs mostly used in neuromorphic circuits are reported by Chaoui [4], Alejandro Diaz-Sanchez and Ramirez-Angulo [5], Al-Nsour and Abdel-Aty-Zohdy [6] and Zhou et al. [7]. Woo and Sarpeshkar [8]; proposed a circuit of analog adder with spiking neurons which has features of scalability and precision. The authors claimed the designed analog adder can be used for digital computation also. The dendritic computation in neurons is nonlinear in nature, and it also plays vital role on neural behavior. Mamdouh and Parker [9, 10]; proposed adder circuit to mimic this characteristic of dendrite.

Barzegarjalali and Parker [11]; built neuromorphic circuit for short term memory using hybrid approach. The neurons are mimicked using CMOS and synapses using CNT technology. They used concept of remembrance of location and appearance of an object for a longer time based on strengthening the synapses or increasing the new neural connections. However, this approach is based on hybrid technology that might be difficult for fabrication. A modified synapse model based on leaky integrated fire model and two state kinetic models was developed by Kazemi et al. [12]. The designed circuit is suitable for both analog and digital implementation. Also, it increases the scalability for large biological neural connections. But this model is verified for cortical neurons and not the retinal cells.

## 3 Classical CMOS Analog Adder

The classical adder which is basic building block of many neuromorphic circuits was proposed by Chaoui [4]. It is self-biased circuit that computes sum of two voltages.

### *3.1 Circuit Implementation of Classical CMOS Analog Adder*

The schematic of classical CMOS analog adder is shown in Fig. 1. The principle of this adder is based on the current mode technique in which linear voltage to current converters and then current to voltage converters are used. Readers interested to know more about principle of working and mathematical modeling of this adder can refer Chaoui [4].

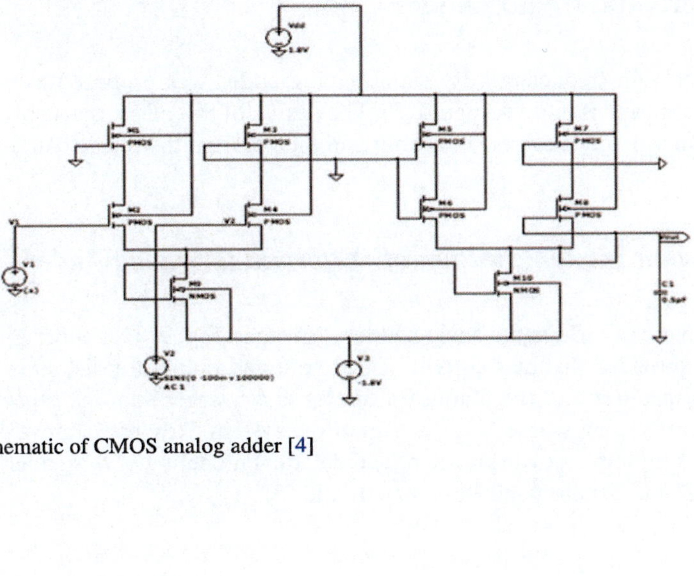

**Fig. 1** Schematic of CMOS analog adder [4]

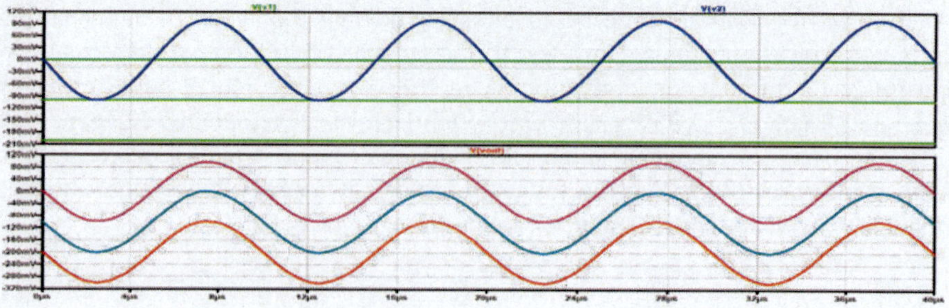

**Fig. 2** Response of classical CMOS analog adder

### 3.2 *Simulation Results of Classical CMOS Analog Adder*

This adder is simulated with TSMC 180 nm technology, and the response of adder is shown in Fig. 2. When simulations are carried out with TSMC 180 nm, with ± 1.8 V supply voltage, it has been observed that adder is properly doing the linear addition with less than − 52 dB of total harmonic distortion and power consumption of less than 40 μW. From ac analysis, it is observed that the frequency response is well behaved with bandwidth of 35 MHz. Figure 2 shows the results when input V1 is stepped as 0 V, − 90 mV and − 200 mV while keeping another sinusoidal voltage to added, i.e., V2 same.

## 4 Differential Analog Adder

A compact high frequency differential analog adder was proposed by Alejandro Diaz-Sanchez and Ramirez-Angulo [5]. The design of this adder is also based on the principle of voltage to current converter and then current to voltage converter.

### 4.1 *Circuit Implementation of Differential Analog Adder*

The schematic of differential analog adder is given in Fig. 3. This adder gives linear operation provided that bias current of the input and output transistors is the same and also aspect ratio of the output transistors is $N^2$ larger than the aspect ratio of the input transistors, where $N$ is the gain of the system. Readers interested to know more about principle of working and mathematical modeling of this adder can refer Alejandro Diaz-Sanchez and Ramirez-Angulo [5].

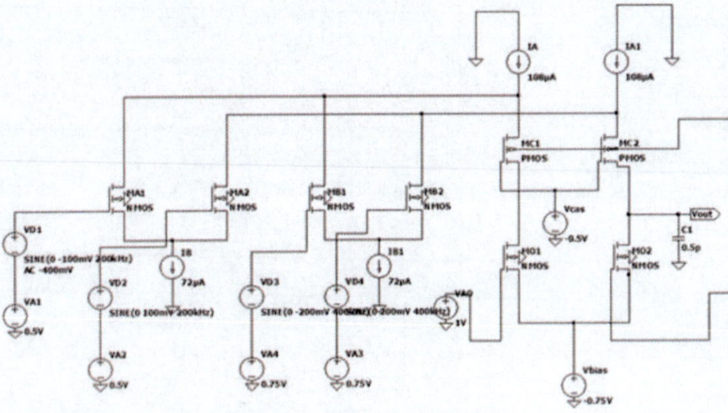

Fig. 3 Schematic of differential analog adder [5]

## 4.2 *Simulation Results of Differential Analog Adder*

The simulations are carried out using TSMC 180 nm technology, and simulation results of addition of two sinusoidal signals are shown in Fig. 4. It is observed that total harmonic distortion is slightly more as compared to the classical CMOS adder and the bandwidth obtained from ac analysis is 20 MHz. In this case, more number of current sources are required, still average power dissipation for this adder is also below 40 $\mu$W.

# 5 Low Power Analog Adder

Only NMOS-based analog adder was reported by Al-Nsour and Abdel-Aty-Zohdy [6]. This adder was proposed to increase accuracy by eliminating the need for voltage to current and current to voltage converter circuits.

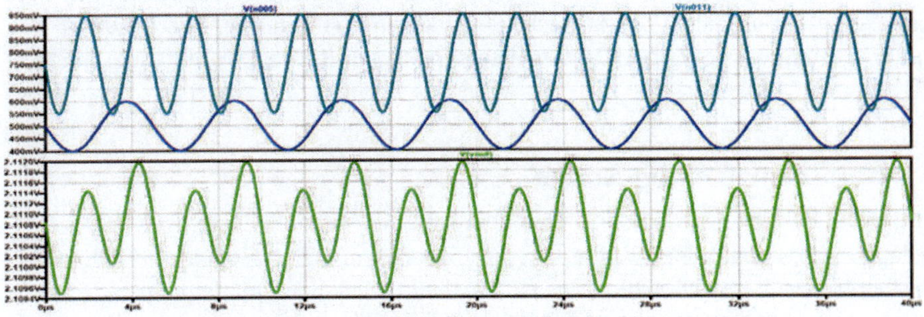

Fig. 4 Two sinusoidal inputs given to differential analog adder and their sum

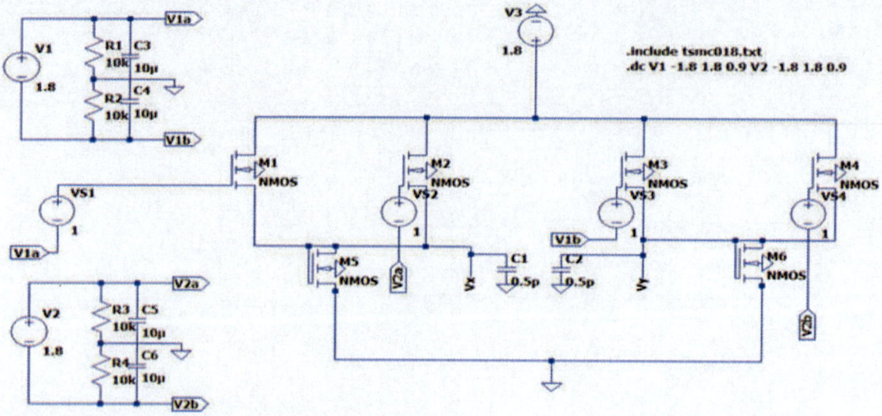


**Fig. 5** Schematic of low power analog adder [6]

## 5.1 *Circuit Implementation of Low Power Analog Adder*

The schematic of low power analog adder is shown in Fig. 5. From mathematical analysis, output of this adder is $(V_x - V_y) = (V_1 + V_2)/2$. Readers interested to know more about principle of working and mathematical modeling of this adder can refer Al-Nsour and Abdel-Aty-Zohdy [6].

## 5.2 *Simulation Results of Low Power Analog Adder*

Extensive circuit simulations are carried out with TSMC 180 nm technology with $V_{\mathrm{dd}} = 1.8$ V. As shown in Fig. 6, it is found that differential output voltage remains below $\pm$ 1 V for the input voltage ranges of $\pm$ 1.8 V. Also, ac analysis revealed that it has well behaved frequency response giving large bandwidth of 200 MHz with less than 40 $\mu$W of power dissipation and total harmonic distortion less than $-$ 17 dB. Figure 7 shows that expected sum of input sinusoidal voltages (same input) (red) is same as actual sum $(V_x - V_y)$ (green) observed after simulation. The benefit of this adder is that it uses only NMOS transistors, however, it also gives linear addition.

# 6 Nonlinear Analog Adder

Adaptive nonlinear adder was reported by Zhou et al. [7]. This adder is also based on the classical CMOS analog adder; however, it has been made adaptive to fully emulate computations within dendritic branches. This adder automatically adjusts neuron responses based on the magnitudes of EPSP potentials.

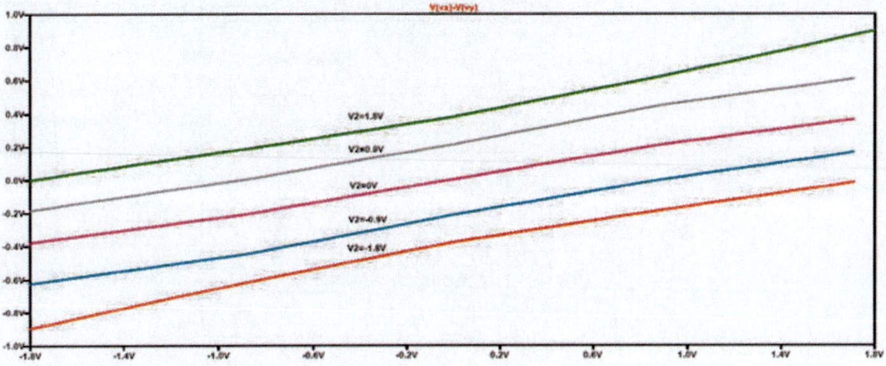


**Fig. 6** Differential output voltage for input ranges from ± 1.8 V

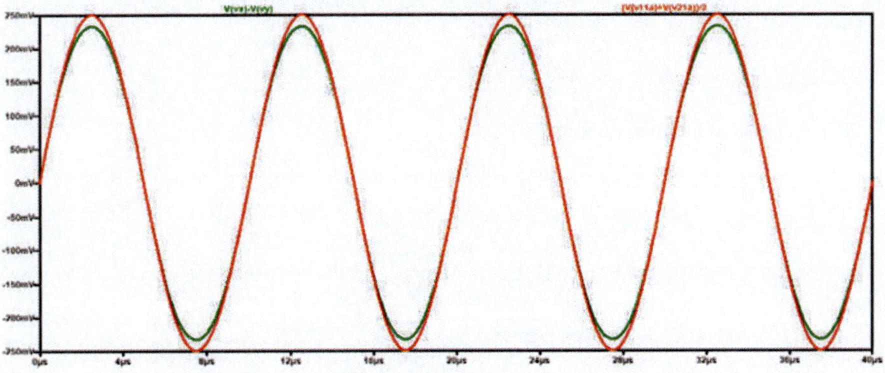


**Fig. 7** Simulation results for input sinusoidal voltages

## 6.1 Circuit Implementation of Nonlinear Analog Adder

The nonlinear adder shown in Fig. 8 consists of classical CMOS analog adder along with control comparators and pseudo-inverters. Figure 9 shows the implementation of sub-linear and super-linear control circuits. It has been observed that without delay circuit, the response of the circuit is not degraded and hence, delay circuits proposed by Zhou et al. [7] are not implemented in this work. Readers interested to know more about principle of working of this adder can refer Zhou et al. [7].

## 6.2 Simulation Results of Nonlinear Analog Adder

Simulations are carried out with TSMC 180 nm at ± 1.8 V supply voltage. This adder works as a linear adder for weak EPSP, i.e., input voltages smaller that 36 mV (2% of

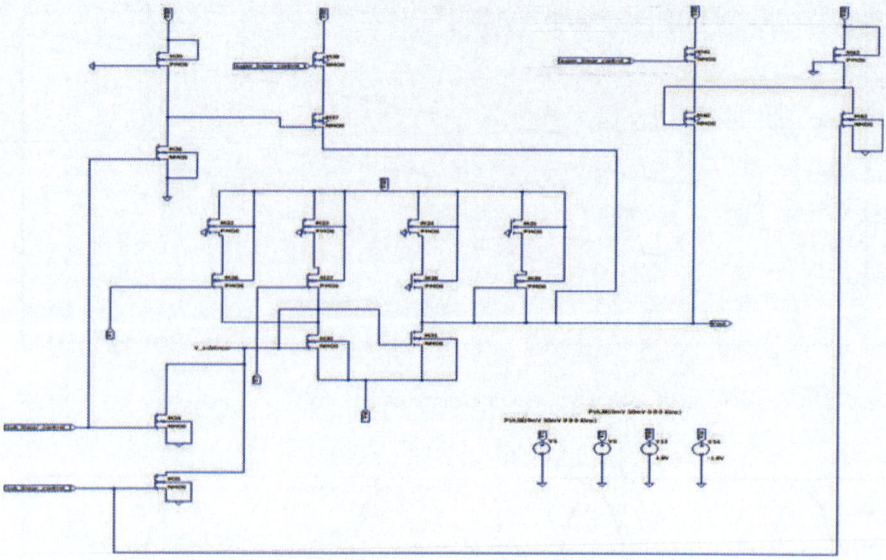

Fig. 8 Schematic of nonlinear adder [7]

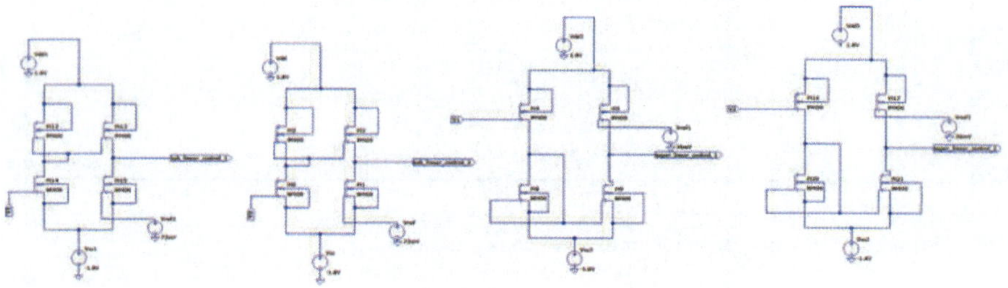

Fig. 9 Sub-linear and super-linear control circuits [7]

$V_{dd}$), as a super-linear adder for medium EPSP, i.e., input voltages between 36 and 72 mV (2–4% of $V_{dd}$) and as a sub-linear adder for strong EPSP, i.e., input voltages above 72 mV (larger than 4% of $V_{dd}$) [7]. This combination of linear, super-linear and sub-linear is observed in the simulations of input voltages 10 mV, 50 mV and 150 mV, respectively, as shown in Figs. 10, 11 and 12. Figure 10 shows that arithmetic sum $V_1 + V_2$ (20 mV) and output obtained $V_{out}$ (20 mV) is same justifying the linear response. Figure 11 shows that obtained output voltage $V_{out}$ (>100 mV) is more than arithmetic sum $V_1 + V_2$ of input voltages (100 mV) justifying super-linear response of the adder. Finally, Fig. 12 shows that obtained output voltage $V_{out}$ (< 300 mV) is less than arithmetic sum $V_1 + V_2$ of input voltages (300 mV) justifying the sub-linear response.

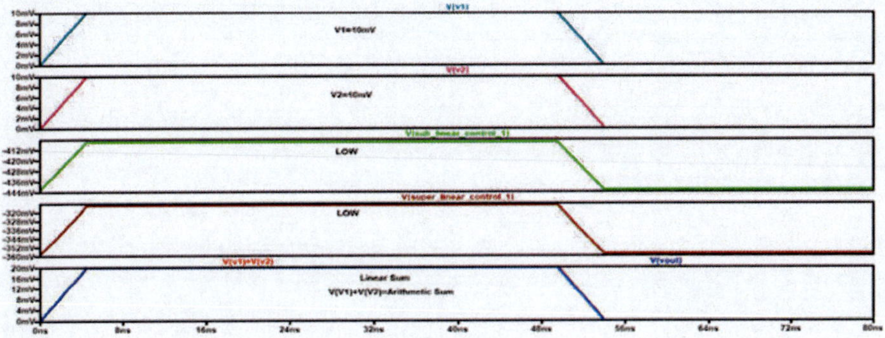


**Fig. 10** Linear addition with weak input signals of 10 mV

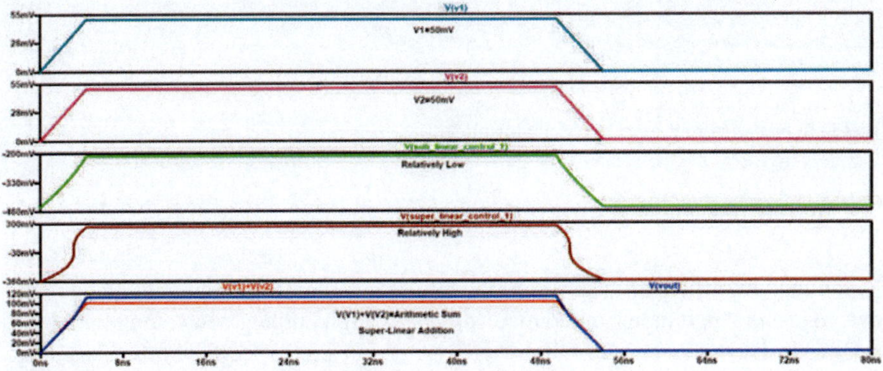


**Fig. 11** Super-linear addition with medium input signals of 50 mV

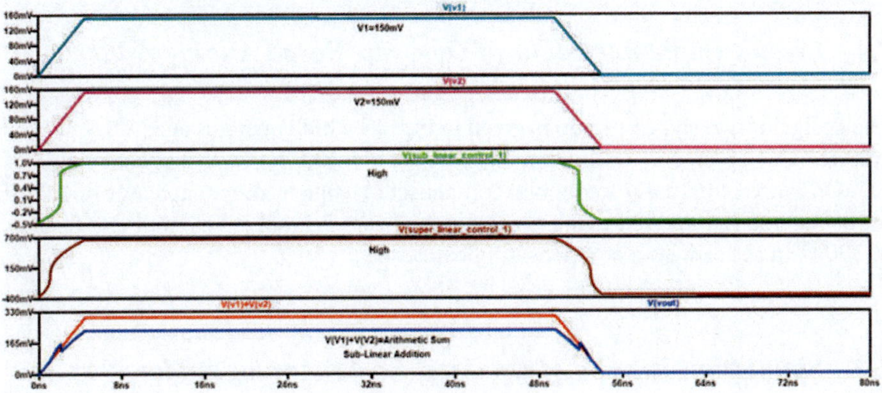


**Fig. 12** Sub-linear addition with very strong signal inputs of 150 mV

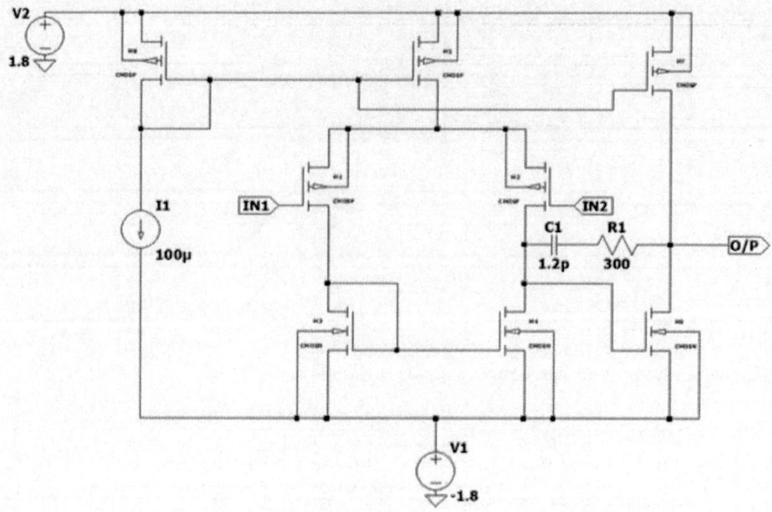


**Fig. 13** Schematic of op-amp-based analog adder

# 7 Op-Amp-Based Analog Adder

Operational amplifiers are basic components of many analog circuits offering many advantages like high input impedance, high gain, high output swing along with large bandwidth. Op-amps can be tailor-made for the desired specifications. Two stage op-amp can be used for summation of analog signals.

## 7.1 *Circuit Implementation of Op-Amp-Based Analog Adder*

The design of two stage op-amp is given in Fig. 13 which operates at ± 1.8 V supply voltage. This op-amp is designed to have a gain of 1000 V/V with a phase margin of 75°. Op-amp is internally frequency compensated using a Millar capacitor of 1.2 pF for better stability and phase lag created by the zero is resolved by adding a resistor of 300 Ω in series with a compensating capacitor.

## 7.2 *Simulation Results of Op-Amp-Based Analog Adder*

Two stage op-amp is simulated with TSMC 180 nm technology for addition of two input voltage signals In1 and In2. Op-amp is connected in non-inverting configuration with feedback resistance $R_f = 4$ KΩ and $R_1 = 1$ K resistance connected between

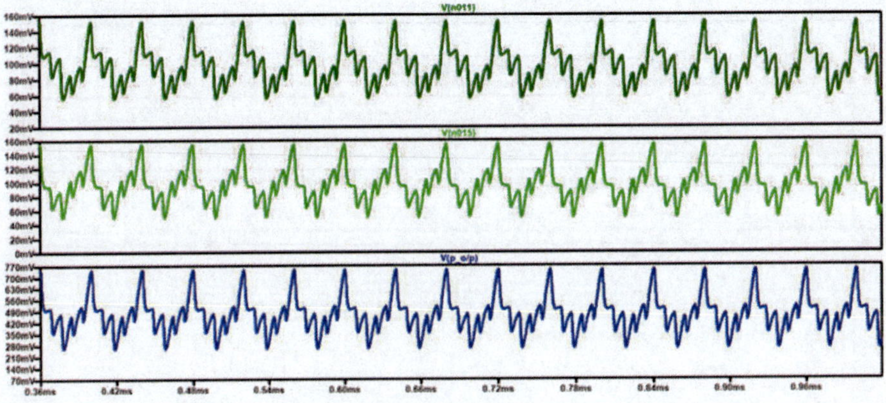


**Fig. 14** Summation of two input signals

inverting terminal and ground. Input signals connected to op-amp are specially generated from bipolar cells. It has been found that output of the op-amp is as expected, i.e., $V_{out} = (1 + R_f/R_1)(\mathrm{In}_1 + \mathrm{In}_2)/2$, e.g., here if $\mathrm{In}_1 = \mathrm{In}_2 = 150$ mV, then output $V_{out} = 750$ mV as seen from Fig. 14.

## 8 Application of Adders for Differential Motion Detection

### *8.1 Application of Classical CMOS Analog Adder in Starburst Amacrine Cell*

Classical CMOS adder is used to sum outputs of bipolar cells for the differential motion detection as described in Tseng et al. [13]. Differential motion detection is improved by using starburst amacrine cell circuitry which has wave-shaping circuit, delay circuit and an adder circuitry. Actually, direction selectivity is done using reciprocal synapse circuitry. When stimulus is passing from intermediate to distal compartment, circuit will evoke response of intermediate compartment. This signal is added to the distal compartment after adding some delay. Meanwhile input stimulus reaches distal compartment. This results in a larger response at distal compartment. Outputs of the bipolar cell (topmost signals in Fig. 15) are applied to two adders adder-1 and adder-2 as shown. The recorded response shows that amacrine cell properly differentiates between signals moving from left to right or right to left.

Generally, more than two bipolar cells could be connected to an amacrine cell for addition of responses, for example 5–6 ON-bipolar cells to one amacrine cell or 5–6 OFF-bipolar cells to one amacrine cell. It means if we are creating circuits for emulating the behavior of amacrine cell, then we need to connect at least five responses and verify the functionality of an adder. From the first-hand look, it can

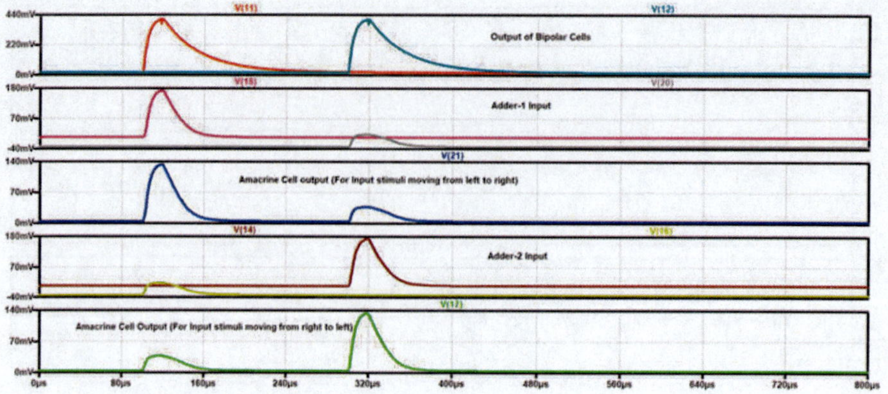


**Fig. 15** Summation of two input signals

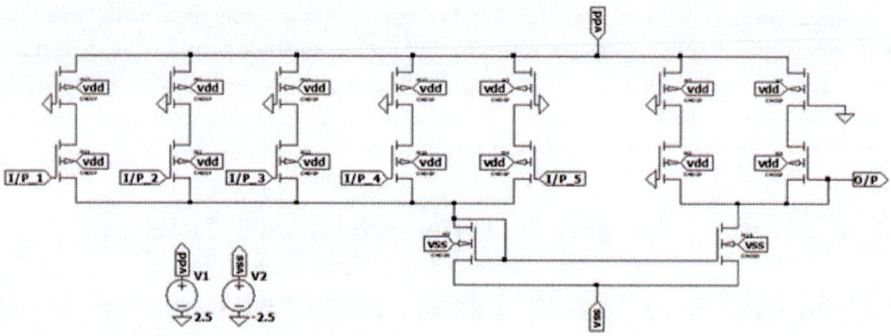


**Fig. 16** Voltage addition of five inputs

be thought that by extending classical adder to five inputs as shown in Fig. 16 will suffice, however, if we carefully observe the amacrine cell construction, then we need to create a cascaded network of classical adders as shown in Fig. 17 for correctly emulating the behavior. This adds to the complexity in the circuit and actually a delayed response if we use classical adder for modeling amacrine cell.

Also, classical adder, low power adder and differential adders give linear response, while in reality the response of biological amacrine cell is adaptive in nature, i.e., nonlinear. Nonlinear added would be the better solution to classical adder.

## 8.2 *Application of Op-Amp Adder in Starburst Amacrine Cell*

As discussed in the previous session, starburst amacrine cell will have more than two inputs coming from the bipolar cell. We have designed five input op-amp as shown in Fig. 18 and applied five inputs from the bipolar cells to emulate the response of

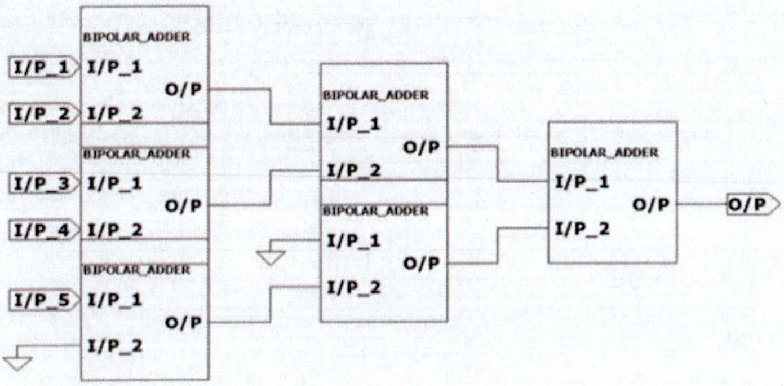


**Fig. 17** Cascaded network required for voltage addition of five inputs with classical CMOS analog adder

**Fig. 18** Application of op-amp for summation of five signals from bipolar cells

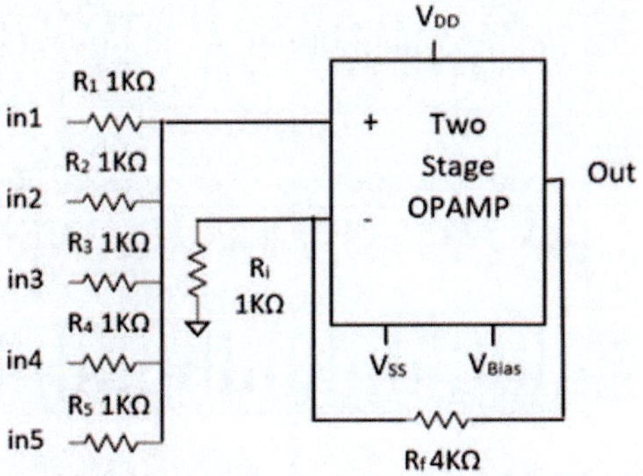


central receptive field and peripheral receptive field. We got the expected output as shown in Fig. 19, from both the op-amp-based adders used for central receptive field and peripheral receptive field. Better response can be noted from Fig. 20 showing enhancement and suppression of signal from adder. Thus, op-amp can also be used for sub-linear addition to mimic the adaptive biological signals.

### *8.3 Development of Differential Motion Detector with Op-Amp Adder*

For further verification of applicability and scalability of op-amp-based adder in development of differential motion detector circuit, an entire system consisting of 7 by 70 photoreceptor matrix is implemented. There are 490 photoreceptors and 10 bipolar cells in the full circuit as shown in Fig. 21. The 490 photoreceptors are divided into 10 blocks of 49 photoreceptors. The seven photoreceptors in a column are parallelly connected to each other, and the linearly summed output is applied to

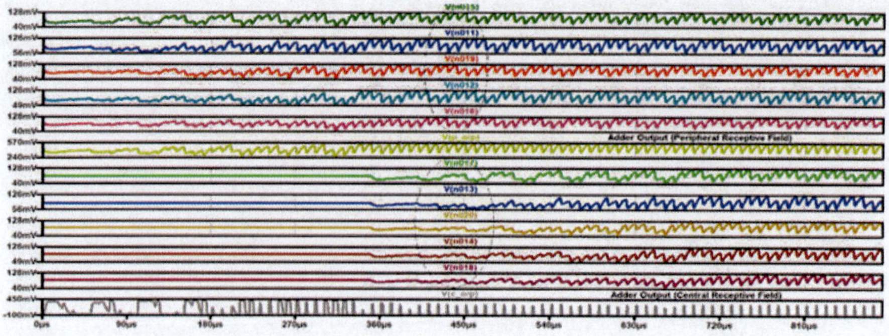


**Fig. 19** Response of op-amp-based adder (central as well as peripheral) to the five inputs from bipolar cells

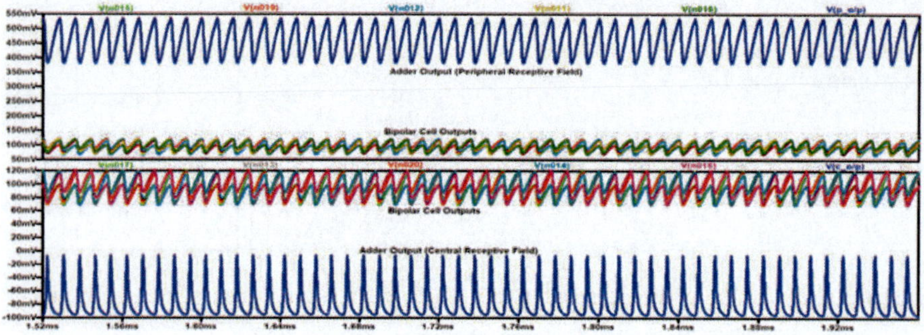


**Fig. 20** Adder output for same frequency of signals (enhanced output in case of peripheral receptive field and suppressed output in case of central receptive field) from five bipolar cells

bipolar cells. Five bipolar cell outputs are added to create central receptive field and five to create peripheral receptive field. The simulation results are shown here for two cases. The results shown in Fig. 22 indicate that final output of the differential motion detector is suppressed due to suppression of central receptive field adder output indicating that there is no differential motion as inputs have same frequency. While Fig. 23 indicate that differential motion is detected in the final output because there is addition of signals due to variation in frequency. Thus, op-based adder can be scaled and it dynamically adjusts the output voltage.

## 9 Conclusion

In this paper, classical CMOS analog adder, differential analog adder, low power adder, nonlinear adder and op-amp-based adders are investigated. It has been observed that all the adders can be used for linear addition; however, neuromorphic

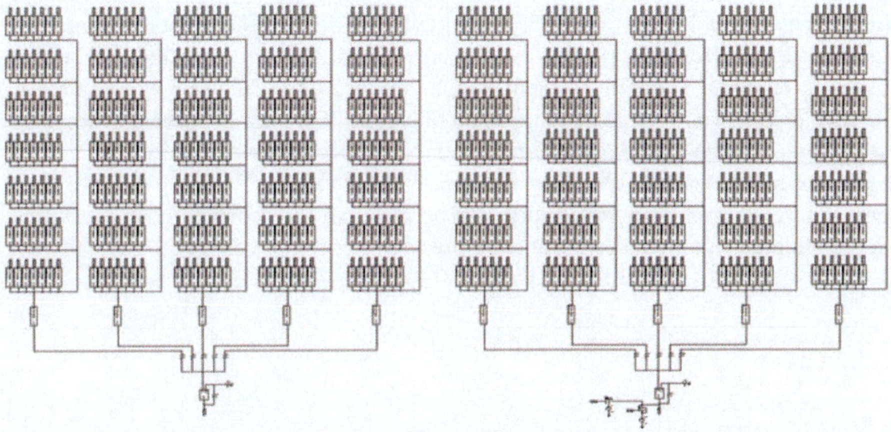

**Fig. 21** Photoreceptor matrix along with bipolar cells, op-amp-based adder, inhibiting transistor, rectifying transistor to generate final output of amacrine cell

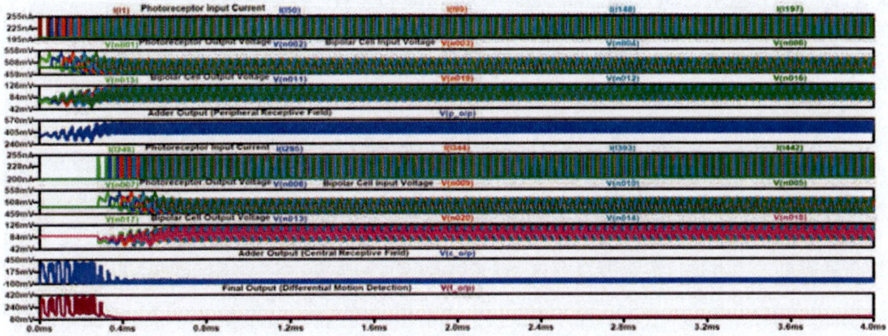


**Fig. 22** Suppressed response of the entire system (shows suppression of adder output of central receptive field due to input signals moving with the same frequency)

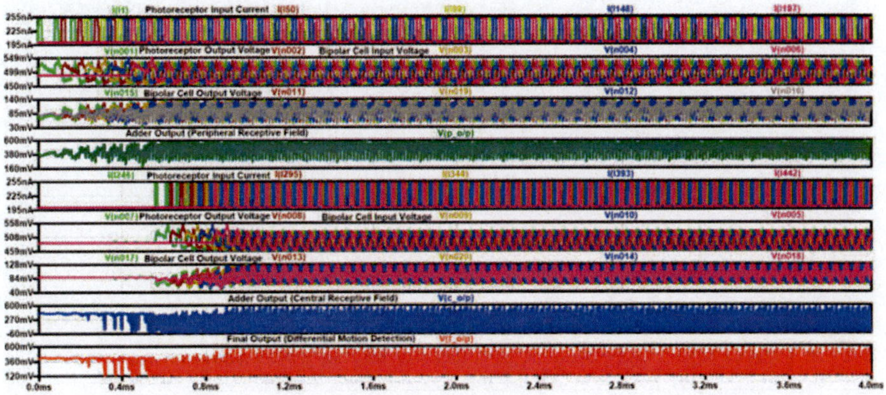


**Fig. 23** Differential motion detected in the final output

circuit demands nonlinear adaptable summation of signals. Starburst amacrine cell that is used for differential motion detection needs scalable adders with good performance parameters. There is a tradeoff among all the adders in terms of complexity, linearity, stability and range of summation of signals. On most of these performance parameters, op-amp-based adders are found to be more suitable for addition of more number of inputs without changing the design of the op-amp. In this paper, we have demonstrated that op-amp-based adder can be successfully incorporated in emulating a large scalable retinal neuromorphic circuit.

## References

1. Kolb H (2003) How the retina works. Am Sci 91:28–35
2. Dowling JE (2015) Retina: an overview. In: Encyclopedia of the human brain
3. Tsukamoto Y, Omi N (2017) Classification of mouse retinal bipolar cells: type-specific connectivity with special reference to rod-driven all amacrine pathways. Front Neuroanat 11
4. Chaoui H (1995) CMOS analogue adder. Electron Lett 31(3):180–181
5. Diaz-Sanchez A, Ramirez-Angulo J (1997) A compact high frequency VLSI differential analog adder. In: Proceedings of the 39th Midwest symposium on circuits and systems, pp 21–24
6. Al-Nsour M, Abdel-Aty-Zohdy HS (1999) Simple low power analogue MOS voltage adder. Electron Lett 35(7)
7. Zhou X, Guo Y, Parker AC, Hsu C, Choma J (2013) Biomimetic non-linear CMOS adder for neuromorphic circuits. In: 6th annual international IEEE EMBS conference on neural engineering, pp 876–879
8. Woo SS, Sarpeshkar R (2013) A spiking-neuron collective analog adder with scalable precision. In: IEEE international symposium on circuits and systems, pp 1620–1623
9. Mamdouh P, Parker AC (2017) A power-efficient biomimetic intra-branch dendritic adder. In: International joint conference on neural networks, pp 3946–3952
10. Mamdouh P, Parker AC (2017) A switched-capacitor dendritic arbor for low-power neuromorphic applications. In: IEEE international symposium on circuits and systems
11. Barzegarjalali S, Parker AC (2016) A neuromorphic circuit mimicking biological short-term memory. In: 38th annual international conference of the IEEE engineering in medicine and biology society, pp 1401–1404
12. Kazemi A, Ahmadi A, Alirezaee S, Ahmadi M (2016) A modified synapse model for neuromorphic circuits. In: IEEE 7th Latin American symposium on circuits and systems, pp 67–70
13. Tseng KC, Parker AC, Joshi J (2011) A directionally-selective neuromorphic circuit based on reciprocal synapses in Starburst Amacrine cells. In: 2011 annual international conference of the IEEE engineering in medicine and biology society, EMBC, pp 5674–5677

# Sputtered $HfO_2/ZrO_2$ Induced Interfacial Ferroelectric HZO Layer for Negative Capacitance Applications

**Ankita Sengupta, Basudev Nag Chowdhury, Bodhishatwa Roy, Subhrajit Sikdar, and Sanatan Chattopadhyay**

**Abstract** The current work reports fabrication of a ferroelectric HZO layer within the gate stack of Pt/$ZrO_2$/HZO/$HfO_2$/p-Si MOS device for achieving 'negative capacitance'. Such HZO layer is formed by RF-magnetron sputtering with in-situ substrate heating followed by post-deposition annealing. The thicknesses of $HfO_2$, HZO and $ZrO_2$ layers are obtained to be ~ 10 nm, ~ 6 nm and ~ 9 nm, respectively. The negative slope of polarization–electric field (P–E) curve extracted from capacitance–voltage (C–V) measurements indicates negative capacitance appearing in the device. The slope of such P–E curve is obtained to be ~ 0.3 pF/cm which suggests a significant 'negative capacitance' in comparison with the other reported values. Therefore, the current work provides a route for fabricating high speed planar MOSFETs with sub-threshold swing (SS) beyond the classical limit.

**Keywords** HZO layer · Ferroelectrics · Negative capacitance · $HfO_2/ZrO_2$ double layer

## 1 Introduction

The design and implementation of negative capacitance by using ferroelectric materials are gaining tremendous attention for sustaining gradual performance improvement of metal–oxide–semiconductor field-effect transistors (MOSFETs) in terms of speed and power [1]. The incorporation of such 'negative capacitance' (NC) in MOS devices has emerged to be an alternative approach to obtain sub-threshold swing (SS) beyond the classical limit (i.e., 60 mV/dec at room temperature) [2–4]. Several ferroelectric materials such as, $BaTiO_3$, $PbTiO_3$, $PbZrO_3$ and PVDF have shown promising negative capacitance effect; however, their lack of compatibility with the

A. Sengupta · B. Nag Chowdhury · B. Roy · S. Sikdar · S. Chattopadhyay (✉)
Department of Electronic Science, University of Calcutta, Kolkata, India
e-mail: scelc@caluniv.ac.in

S. Chattopadhyay
Centre for Research in Nanoscience and Nanotechnology (CRNN), University of Calcutta, Kolkata, India

 
C. Giri et al. (eds.), *Emerging Electronic Devices, Circuits and Systems*,
Lecture Notes in Electrical Engineering 1004,
https://doi.org/10.1007/978-981-99-0055-8_24

existing CMOS technology is a major concern [5, 6]. In this regard, $HfO_2$ has emerged as the potential alternative CMOS-compatible material which shows significant ferroelectricity [7–9]. The use of such material was first reported in 2011 with opening up a novel route to incorporate negative capacitor gate stack into the existing planar-CMOS technology since $HfO_2$ has already been developed as a matured high-*k* gate dielectric technology [10]. Reports are available on stabilizing the ferroelectricity in $HfO_2$ by performing $HfO_2$-thickness variation and doping with different materials such as, Al, Zr and Si [11–13]. It has been observed that incorporation of 50% Zr in $HfO_2$ exhibited promising ferroelectric effect due to the formation of hafnium zirconate (HZO) [13, 14]. Such approach provided the possibility of integrating HZO-based negative capacitance effect into the gate stack of CMOS devices [15]. Several technologies have been adopted to grow HZO including atomic layer deposition (ALD) and pulsed laser deposition (PLD) techniques on Si-substrates [16–19]. However, such technologies involve complex fabrication procedures to grow HZO thin films, and thus, it requires an industry friendly process for large scale production. It is also to be mentioned that only 'negative capacitance' does not improve the speed and power of the MOS devices since presence of the hysteresis loop in polarization–electric field (P–E) curve induces an intrinsic power loss in the devices [20]. Therefore, it is essential to fabricate 'loop-free' 'negative capacitance' gate stacks with relatively higher 'negative slope' to achieve high switching speed in NC-MOSFETs [21].

In this context, the current work deals with fabrication of the dielectric/ferroelectric/dielectric ($ZrO_2$/HZO/$HfO_2$) gate stacks by employing RF-sputtering technique, where the HZO layer is formed at the interface of $HfO_2$/$ZrO_2$ during in-situ substrate heating followed by high temperature annealing. The thicknesses of different gate stack layers are obtained from cross-sectional FESEM images and verified from spectroscopic ellipsometry measurements. The formation of interfacial HZO layer is confirmed from the XRD-study. Finally, C–V measurements have been performed by employing Keysight 4980AL. The P–E profile curve is then extracted from the measured C–V characteristics to study the slope of P–E curve in the NC-region, which indicates a significant 'negative capacitance' in the present device that is capable of lowering the SS-values in MOSFETs beyond its classical limit.

## 2 Experimental Details

In the current work, a (100) p-silicon (p-Si) substrate (resistivity 0.1–0.5 Ω-cm) is used for fabrication of the Pt/$ZrO_2$/HZO/$HfO_2$/p-Si MOS device structure. Prior to the growth of such oxide layers, the p-Si wafer is cleaned by standard RCA-1 and RCA-2 protocols followed by a bathe in 1% HF solution for 2 min to remove the residual native oxide. The cleaned p-Si substrate is then placed into the sputtering chamber for sequential deposition of $HfO_2$ and $ZrO_2$ layers. A 2 in. 99.99% pure $HfO_2$ and $ZrO_2$ targets are used for this purpose. During sputtering, initially the base pressure is kept at ~ $5 \times 10^{-6}$ mbar and then, a constant Ar gas flow is maintained at 50 sccm throughout the deposition process. $HfO_2$ is deposited at 110 W RF-power for 10 min with sputter pressure maintained at ~ $3.2 \times 10^{-2}$ mbar. Subsequently, the $ZrO_2$ layer is deposited at 90 W for 30 min (~ $3.2 \times 10^{-2}$ mbar sputter pressure) with in-situ substrate heating at 600 °C inside the chamber. It is to be mentioned that the oxide targets are pre-sputtered for 10 min before each of the deposition processes to remove any surface impurities. The fabricated samples are then subjected to post-deposition annealing at 800 °C in Ar environment for 30 min. The top and back contacts for the present MOS device are made by DC-sputtering platinum where a shadow mask of area $1.26 \times 10^{-3}$ $cm^2$ is used for top-contact. The thicknesses of different oxide layers are estimated from the cross-sectional FESEM (ZEISS AURIGA) images and then verified by performing spectroscopic ellipsometry (SE-850) measurements. The crystallographic orientations of different oxide layers are studied from XRD profile (PANalyticalX'Pert Powder). Finally, capacitance–voltage (C–V) characteristics of the fabricated devices are obtained by employing LCR meter (Keysight E4980AL).

## 3 Results and Discussion

Figure 1a shows the schematic of $Pt/ZrO_2/HZO/HfO_2$/p-Si MOS device structure considered in the current work, and the relevant cross-sectional FESEM image (placed on 90° tilted stub) of the fabricated device is shown in Fig. 1b. It is to be mentioned that the FESEM image is taken at relatively lower EHT of 3 kV and relatively higher magnification of 116 kX. Three distinct layers of different oxides on Si substrate with distinguished interfaces in-between are observed from the FESEM image. The thicknesses of different oxide layers are estimated from such FESEM image for $HfO_2$ and HZO + $ZrO_2$ to be ~ 10 nm and ~ 16 nm, respectively.

Figure 2a shows plot of the XRD spectra for $ZrO_2/HZO/HfO_2$/p-Si multi-layer device which indicates the crystallographic planes of [200]-$HfO_2$ (JCPDS-83–0808),

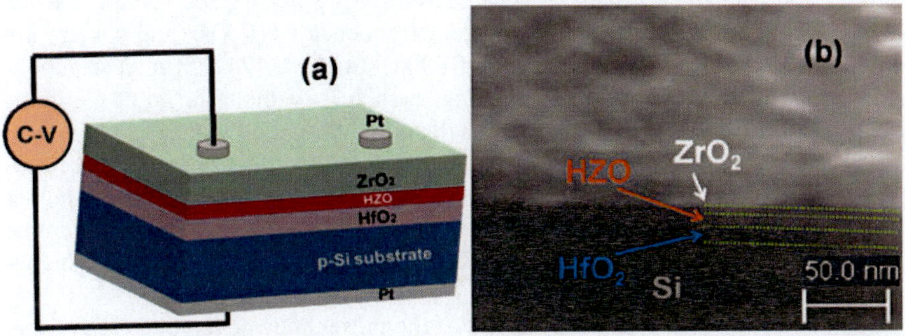


**Fig. 1** **a** Schematics of the fabricated dielectric/ferroelectric/dielectric device structure. **b** Cross-sectional FESEM image of the fabricated device placed on a 90° tilted stub

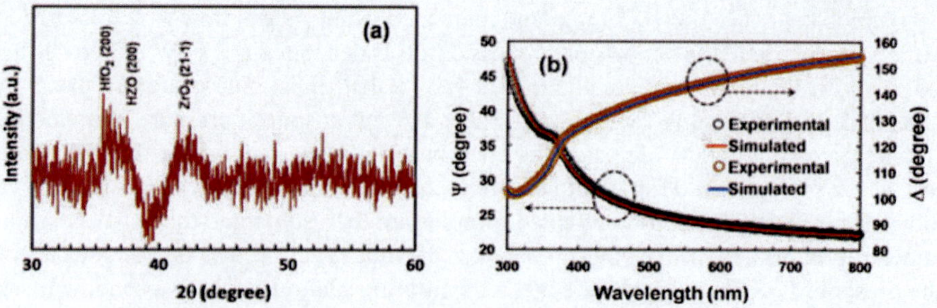


**Fig. 2** **a** Plot of XRD profile for the $ZrO_2$/HZO/$HfO_2$ multi-layer device. **b** Plot of experimental (symbol) and simulated (lines) values of amplitude ratio (Ψ) and phase difference (Δ) in the wavelength region of 300–800 nm

[200]-HZO (JCPDS-83-0809) and [21-1]-$ZrO_2$ (JCPDS-18-0599) at the diffraction peaks appearing at 35.5°, 36.6° and 42.65°, respectively. Further, thicknesses of different sputter-deposited oxide layers have been calculated from the spectroscopic ellipsometry data. The corresponding variation of relative amplitude ratio (Ψ) and phase difference (Δ) (at 60° incidence angle) in the wavelength region of 300–800 nm for the multi-layer device (without metal contacts) has been plotted in Fig. 2b along with the relevant simulated data. An air/$ZrO_2$/HZO/$HfO_2$/p-Si model is considered for such simulation purpose and Fig. 2b shows a satisfactorily good agreement between the simulated data with the experimentally obtained values. The thicknesses of different oxide layers are calculated from such simulated data and are obtained to be 9.23 nm, 6.63 nm and 10 nm for $ZrO_2$, HZO and $HfO_2$, respectively, which also corroborate the estimated thicknesses from the FESEM image. Thus, XRD and ellipsometry studies confirm the presence of an interfacial HZO ferroelectric layer in-between $ZrO_2$ and $HfO_2$. The formation of such interfacial HZO layer may be attributed to the elemental inter-diffusions of $ZrO_2$ and $HfO_2$ during *in-situ* substrate heating assisted sputtering followed by post-deposition annealing at relatively higher temperatures [22, 23].

The C–V characteristics of the fabricated stack gated MOS device (Pt/$ZrO_2$/HZO/$HfO_2$/p-Si) for two different frequencies (10 kHz and 5 kHz) are shown in Fig. 3. The single wing 'butterfly like structure' [24, 25] indicates loop-free negative capacitance region appearing in such C–V plot (shown at the inset) which meets the criteria for developing high speed MOSFETs with relatively lower intrinsic losses. It is also observed from Fig. 3 that the effect of interfacial ferroelectric HZO layer is more prominent (shown at the inset) at relatively higher frequency (i.e., at 10 kHz). It is worthy to mention that such C–V hump cannot be attributed to the interface traps since the effect of interface traps gets reduced at higher frequencies [26].

Figure 4a shows the equivalent circuit of the present device to obtain P–E curve from the measured C–V characteristics. Such circuit contains a ferroelectric capacitor (i.e., HZO in the present case) encapsulated between two dielectric capacitors (i.e., $ZrO_2$ and $HfO_2$ in the current device). The potential drop across the entire oxide

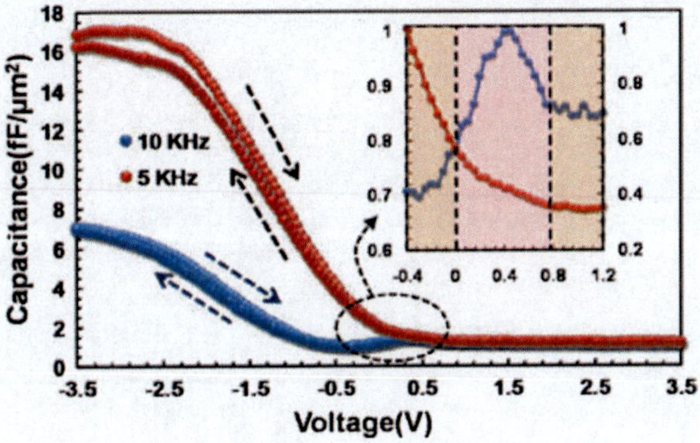


**Fig. 3** Plot of capacitance–voltage (C–V) characteristics of the $Pt/ZrO_2/HZO/HfO_2$/p-Si MOS structure; Inset shows the enlarged portion of negative capacitance region

stack ($ZrO_2/HZO/HfO_2$) is considered to be ($V_{app} - \Psi_S$) where the surface potential $\Psi_S$ is calculated from high and low frequency C–V measurements [26]. It is worthy to mention that ferroelectric materials possess a spontaneous electric polarization even at zero bias, which leads to the generation of a depolarizing electric field due to incomplete screening of the bound charges in the adjacent layers [27–29]. Such depolarizing field determines the strength of 'negative capacitance' since it controls the slope of P–E curve in the NC-region. It has already been reported that such phenomenon is very unstable and adding a dielectric layer in series with a ferroelectric layer stabilizes this effect since the total capacitance becomes positive [30].

It is apparent from the equivalent circuit diagram of Fig. 4a that the voltage drop across the three capacitors in series can be given by

$$\begin{cases} \dfrac{Q_1}{C_1} = V - V_1 \\ \dfrac{Q_f - PA}{C_f} = V_1 - V_2 \\ \dfrac{Q_2}{C_2} = V_2 \end{cases} \quad (1)$$

Using the electrical neutrality of all central nodes in series at zero external bias, the total charge on ferroelectric capacitor is obtained to be

$$Q_f = \frac{PA}{C_f\left(\frac{1}{C_1} + \frac{1}{C_f} + \frac{1}{C_2}\right)} \quad (2)$$

Thus, the total electric field across the ferroelectric layer is calculated to be

$$E_f = \frac{V_{ox}}{d} - \frac{\frac{PA}{C_{eq}}}{C_f d_f\left(\frac{1}{C_{eq}} + \frac{1}{C_f}\right)} \quad (3)$$

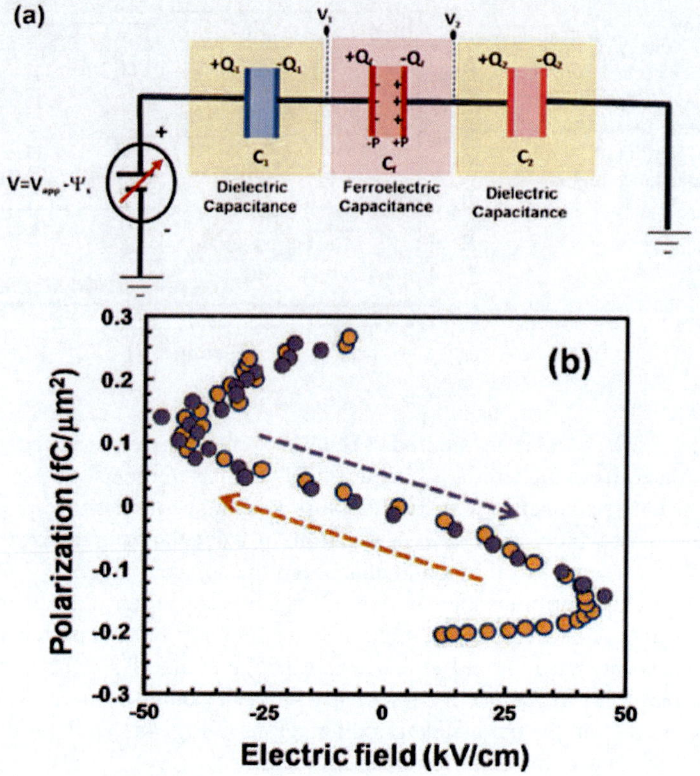


**Fig. 4** **a** Schematic diagram of equivalent circuit of Pt/$ZrO_2$/HZO/$HfO_2$/p-Si MOS device considered for calculating the P–E curve. **b** Plot of variation of polarization with electric field for the fabricated device ($HfO_2$/HZO/$ZrO_2$)

where the thickness of ferroelectric layer is $d_f$ and the total oxide thickness is $d$; $C_{eq}$ is the equivalent capacitance developed in the dielectric stack layer at zero bias. The total polarization $(P)$ in the ferroelectric capacitor is given by

$$P = \frac{1}{A}\int\left(C - \frac{\varepsilon_0 A}{d}\right)\mathrm{d}V \tag{4}$$

The P–E curve is obtained using such equations from C–V plots of Fig. 3 for the $HfO_2$/HZO/$ZrO_2$/p-Si stack gate MOS device and is plotted in Fig. 4b. It is apparent from the negative slope of P–E curve that the present device exhibits significant ‘negative capacitance’ along with its ‘loop-free’ nature. The modulus of such negative slope obtained in the current work is ~ 0.3 pF/cm which is much lower than the other reported values as summarized in Table 1. It is to be mentioned that, less the modulus value of P–E slope, more is the strength of negative capacitance. Further, the absence of any loop in P–E curve indicates that the energy dissipation due to hysteresis effect is negligible. Thus, the significant loop-free negative capacitance obtained in the

**Table 1** Summary of the values of 'negative slope of P–E' curves for fabricated negative capacitors

| S. No. | Device structure | Fabrication method | Slope of P–E curve (pF/cm) | References |
|---|---|---|---|---|
| 1 | TiN/HZO/$Al_2O_3$/TiN | ALD | ~ 5 | [21] |
| 2 | TiN/HZO/$Ta_2O_5$/TiN | ALD | ~ 15 | [31] |
| 3 | TiN/HZO/$Al_2O_3$/TiN | ALD | ~ 3.33 | [31] |
| 4 | HZO FET | ALD | ~ 10 | [32] |
| 5 | Pt/$ZrO_2$/HZO/$HfO_2$/p-Si | Sputtering | ~ 0.3 | This work |

current work suggests that MOSFETs with such $HfO_2$/HZO/$ZrO_2$ gate stack has the potential to attain much lower SS beyond its classical limit (< 60 mV/dec) at room temperature.

## 4 Conclusion

In the current work, a $ZrO_2$/HZO/$HfO_2$ stack layer has been fabricated on p-Si substrate by employing RF-magnetron sputtering technique. The intermediate HZO layer is formed by simultaneous impact of *in-situ* substrate heating and post-deposition annealing at relatively higher temperatures of 600 °C and 800 °C, respectively. The thickness of such layers is obtained from FESEM image and spectroscopic ellipsometry measurements to be ~ 10 nm, ~ 6 nm and ~ 9 nm for $HfO_2$, HZO and $ZrO_2$, respectively. XRD profile confirms the formation of [200]-plane of HZO in-between crystalline $HfO_2$ and $ZrO_2$ layers. The C–V characteristics of fabricated Pt/$ZrO_2$/HZO/$HfO_2$/p-Si MOS device show loop-free ferroelectric capacitance, which is further analyzed by extracting P–E curve with a negative slope. Such slope indicating 'negative capacitance' region is obtained to be ~ 0.3 pF/cm, which is significantly lower in comparison with the relevant reports available in the literature. Thus, the results suggest the fabricated $ZrO_2$/HZO/$HfO_2$ gate stack which is capable of developing low power, high speed planar MOSFETs with SS beyond the classical limit at room temperature.

**Acknowledgements** The authors would like to thank Department of Electronic Science, Centre of Excellence (COE) and Centre for Research in Nanoscience and Nanotechnology (CRNN), University of Calcutta for providing the infrastructure to pursue the current research work.

## References


1. Salahuddin S, Datta S (2008) Use of negative capacitance to provide voltage amplification for low power nanoscale devices. Nano Lett 8(2):405–410
2. Jain A, Alam MA (2014) Stability constraints define the minimum subthreshold swing of a negative capacitance field-effect transistor. IEEE Trans Electron Devices 61(7):2235–2242
3. Liu X, Liang R, Gao G, Pan C, Jiang C, Xu Q, Luo J, Zou X, Yang Z, Liao L, Wang ZL (2018) $MoS_2$ negative-capacitance field-effect transistors with subthreshold swing below the physics limit. Adv Mater 30(28):1800932
4. Alam MN, Roussel P, Heyns M, Van Houdt J (2019) Positive non-linear capacitance: the origin of the steep subthreshold-slope in ferroelectric FETs. Sci Rep 9(1):1–9
5. Izyumskaya N, Alivov Y, Morkoc H (2009) Oxides, oxides, and more oxides: high-κ oxides, ferroelectrics, ferromagnetics, and multiferroics. Crit Rev Solid State Mater Sci 34(3–4):89–179
6. Das D, Khan AI (2021) Ferroelectricity in CMOS-compatible hafnium oxides: reviving the ferroelectric field-effect transistor technology. IEEE Nanatechnol Mag 15(5):20–32
7. Park MH, Lee YH, Kim HJ, Kim YJ, Moon T, Kim KD, Mueller J, Kersch A, Schroeder U, Mikolajick T, Hwang CS (2015) Ferroelectricity and antiferroelectricity of doped thin $HfO_2$-based films. Adv Mater 27(11):1811–1831
8. Böscke TS, Müller J, Bräuhaus D, Schröder U, Böttger U (2011) Ferroelectricity in hafnium oxide thin films. Appl Phys Lett 99(10):102903
9. Hoffmann M, Slesazeck S, Mikolajick T (2021) Progress and future prospects of negative capacitance electronics: a materials perspective. APL Mater 9(2):020902
10. Müller J, Polakowski P, Mueller S, Mikolajick T (2015) Ferroelectric hafnium oxide based materials and devices: assessment of current status and future prospects. ECS J Solid State Sci Technol 4(5):N30
11. Liu LB, Liu X, Cheng Y, Mao J (2017) Ferroelectricity in Al-doped $HfO_2$ on highly doped Si substrate. In: 2017 IEEE conference on electrical insulation and dielectric phenomenon (CEIDP). IEEE 2017, pp 70–73
12. Karbasian G, Tan A, Yadav A, Sorensen EM, Serrao CR, Khan AI, Chatterjee K, Kim S, Hu C, Salahuddin S (2017) Ferroelectricity in $HfO_2$ thin films as a function of Zr doping. In: 2017 international symposium on VLSI technology, systems and application (VLSI-TSA). IEEE 2017, pp 1–2
13. Lomenzo PD, Zhao P, Takmeel Q, Moghaddam S, Nishida T, Nelson M, Fancher CM, Grimley ED, Sang X, LeBeau JM, Jones JL (2014) Ferroelectric phenomena in Si-doped $HfO_2$ thin films with TiN and Ir electrodes. J Vac Sci Technol B Nanotechnol Microelectron Mater Process Meas Phenom 32(3):03D123
14. Müller J, Böscke, TS, Schröder U, Mueller S, Bräuhaus D, Böttger U, Frey L, Mikolajick T (2012) Nano Lett 12:4318
15. Luo Q, Cheng Y, Yang J, Cao R, Ma H, Yang Y, Huang R, Wei W, Zheng Y, Gong T, Yu J (2020) A highly CMOS compatible hafnia-based ferroelectric diode. Nat Commun 11(1):1–8
16. Kim SJ, Mohan J, Lee J, Lee JS, Lucero AT, Young CD, Colombo L, Summerfelt SR, San T, Kim J (2018) Effect of film thickness on the ferroelectric and dielectric properties of low-temperature (400 °C) $Hf_{0.5}Zr_{0.5}O_2$ films. Appl Phys Lett 112(17):172902
17. Nukala P, Antoja-Lleonart J, Wei Y, Yedra L, Dkhil B, Noheda B (2019) Direct epitaxial growth of polar $(1-x)\ HfO_2-(x)\ ZrO_2$ ultrathin films on silicon. ACS Appl Electron Mater 1(12):2585–2593
18. Coll M, Napari M (2019) Atomic layer deposition of functional multicomponent oxides. APL Mater 7(11):110901
19. Bégon-Lours L, Mulder M, Nukala P, De Graaf S, Birkhölzer YA, Kooi B, Noheda B, Koster G, Rijnders G (2020) Stabilization of phase-pure rhombohedral $HfZrO_4$ in pulsed laser deposited thin films. Phys Rev Mater 4(4):043401

20. Ryu TH, Min DH, Yoon SM (2020) Comparative studies on ferroelectric switching kinetics of sputtered $Hf_{0.5}Zr_{0.5}O_2$ thin films with variations in film thickness and crystallinity. J Appl Phys 128(7):074102
21. Hoffmann M, Max B, Mittmann T, Schroeder U, Slesazeck S, Mikolajick T (2018) Demonstration of high-speed hysteresis-free negative capacitance in ferroelectric $Hf_{0.5}Zr_{0.5}O_2$. In: 2018 IEEE international electron devices meeting (IEDM). IEEE 2018, pp 31–36
22. Ferrari S, Scarel G (2004) Oxygen diffusion in atomic layer deposited $ZrO_2$ and $HfO_2$ thin films on Si (100). J Appl Phys 96(1):144–149
23. Mueller MP, Pingen K, Hardtdegen A, Aussen S, Kindsmueller A, Hoffmann-Eifert S, De Souza RA (2020) Cation diffusion in polycrystalline thin films of monoclinic $HfO_2$ deposited by atomic layer deposition. APL Mater 8(8):081104
24. Hachemi MB, Salem B, Consonni V, Roussel H, Garraud A, Lefevre G, Labau S, Basrour S, Bsiesy A (2021) Study of structural and electrical properties of ferroelectric HZO films obtained by single-target sputtering. AIP Adv 11(8):085004
25. Liu H, Lu T, Li Y, Ju Z, Zhao R, Li J, Shao M, Zhang H, Liang R, Wang XR, Guo R (2020) Flexible quasi-van der Waals ferroelectric hafnium-based oxide for integrated high-performance nonvolatile memory. Adv Sci 7(19):2001266
26. Dalapati GK, Chattopadhyay S, Kwa KS, Olsen SH, Tsang YL, Agaiby R, O'Neill AG, Dobrosz P, Bull SJ (2006) Impact of strained-Si thickness and Ge out-diffusion on gate oxide quality for strained-Si surface channel *n*-MOSFETs. IEEE Trans Electron Dev 53(5):1142–1152
27. Bratkovsky AM, Levanyuk AP (2006) Depolarizing field and "real" hysteresis loops in nanometer-scale ferroelectric films. Appl Phys Lett 89(25):253108
28. Mai M, Zhu C, Liu G, Ma X (2018) Effect of dielectric layer on ferroelectric responses of P (VDF-TrFE) thin films. Phys Lett A 382(34):2372–2375
29. Mehta RR, Silverman BD, Jacobs JT (1973) Depolarization fields in thin ferroelectric films. J Appl Phys 44(8):3379–3385
30. Luk'yanchuk I, Tikhonov Y, Sené A, Razumnaya A, Vinokur VM (2019) Harnessing ferroelectric domains for negative capacitance. Commun Phys 2(1):1–6
31. Hoffmann M, Gui M, Slesazeck S, Fontanini R, Segatto M, Esseni D, Mikolajick T (2021) Intrinsic nature of negative capacitance in multidomain $Hf_{0.5}Zr_{0.5}O_2$-based ferroelectric/dielectric heterostructures. Adv Funct Mater 2108494
32. Tang YT, Su CJ, Wang YS, Kao KH, Wu TL, Sung PJ, Hou FJ, Wang CJ, Yeh MS, Lee YJ, Wu WF (2018) A comprehensive study of polymorphic phase distribution of ferroelectric-dielectrics and interfacial layer effects on negative capacitance FETs for sub-5 nm node. In: 2018 IEEE symposium on VLSI technology. IEEE 2018, pp 45–46

# Gain Flattening of Erbium-Doped Fiber Amplifier Using an In-Line M-S-M Fiber Structure

**Protik Roy and Partha Roy Chaudhuri**

**Abstract** We propose here an all-fiber device to flatten the gain profile of an erbium-doped fiber amplifier (EDFA). The design consists of a pair of identical multimode fibers (MMF) in which a short segment of a conventional single-mode fiber is spliced. Because of the core diameter mismatch, the fundamental core mode of the SMF interferes with the major cladding mode of the central SMF. We show through our analytical design that by adjusting the interference spectrum of the coupled modes, wavelength dependent loss in the device can be made to compensate the uneven variations in gain spectrum of EDFA. Using the in-line MMF-SMF-MMF concatenation structure, we demonstrate a flattened gain spectrum of EDFA within 0.87 dB over 32 nm long wavelength bandwidth around the center wavelength of 1546 nm.

**Keywords** MMF-SMF-MMF concatenation · Core diameter mismatch · Modal interference · Optical components · EDFA

## 1 Introduction

Fiber optic communication has evolved as the technology of choice in today's long distance communication having high capacity of internet data and unlimited channels of voice. As the demand for transmission of light signal information over the whole communication networks continues to expand, fiber optics have gained significant recognition and continued R&D thrust. In order to deliver a good quality signal at the receiver end in terms of shape/size while overcoming the attenuation over long distance, the optical amplifier came into being as a most needed device. Till date, many optical amplifiers have been demonstrated, however, erbium-doped fiber amplifiers (EDFAs) are widely used because they amplify wavelengths in the range of the low-attenuation around 1550 nm [1, 2]. EDFAs are also suitable for sensing [3] and amplification of multiple signals in WDM applications [4]. However, there is a non-uniformity in the gain profile over the whole C-band range (1530–1565 nm)

P. Roy · P. Roy Chaudhuri (✉)
Department of Physics, Indian Institute of Technology, Kharagpur, West Bengal, India
e-mail: roycp@phy.iitkgp.ac.in



C. Giri et al. (eds.), *Emerging Electronic Devices, Circuits and Systems*,
Lecture Notes in Electrical Engineering 1004,
https://doi.org/10.1007/978-981-99-0055-8_25

[1, 5]. Hence, to flatten their gain profile is of great importance in optical communications. In this article, we demonstrate a very flexible dynamic gain equalization of an EDFA by using an in-line multimode–single-mode–multimode (MSM) fiber structure. By appropriately tuning the parameters of the fibers in the MSM structure, the transmission loss in the interference spectrum can be adjusted to flatten the variations in the EDFA gain profile for a wide range of wavelength. Using this gain flattening filter (GFF) in concatenation with EDFA, a gain flatness of 0.87 dB over a broad wavelength range of 32 nm is achieved for a wide range of input signal and pump powers. In the following, we describe the analysis method of the proposed structure followed by the key results pertaining to our study.

## 2 Theory

In the WDM system, both the input signal and the 980 nm pump signal are multiplexed together in erbium-doped fiber by using WDM coupler. Then, the amplified output signal of the EDF is passed through MSM fiber circuit to flatten the non-uniform gain of EDFA. EDFA gain flattening circuit comprises by a laser source, WDM coupler, 980 nm pump laser, erbium-doped fiber, MSM fiber structure and output signal, as shown in Fig. 1.

Gain characteristics of a EDFA depend on material and structural parameters of EDF, i.e., absorption and emission cross-sections, Er3+ distribution radius, effective cross-sectional area, fiber length, signal wavelength, pump wavelength, signal power and pump power. By changing those parameters, the performance of EDFA gain at different operational conditions can be simulated. The output of EDF is passed through MSM fibers circuit structure which flattens non-uniform gain of EDFA.

The schematic of the proposed GFF is shown in Fig. 2. The structure consists of a small section of SMF of length $L$ spliced between a pair of identical conventional step-index MMFs. In this concatenated multimode–single-mode–multimode (MM–SM–MM) configuration, due to core diameter mismatch at the two splicing junction,

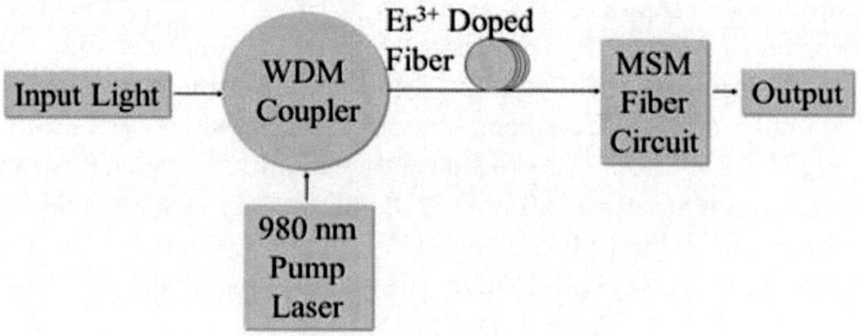


**Fig. 1** System model of gain flattened EDFA with MSM

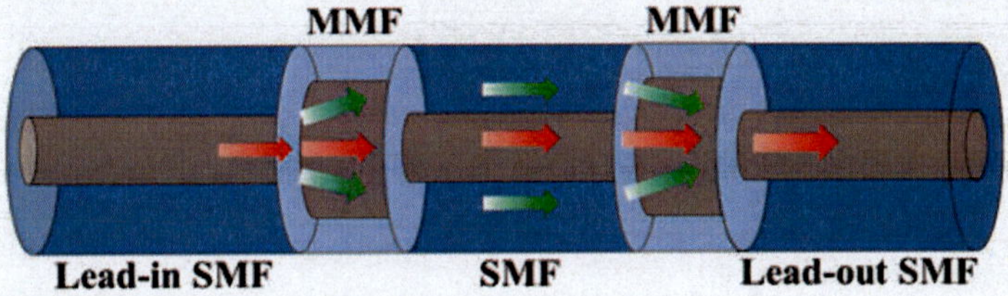


**Fig. 2** Schematic diagram of the MSM fiber

the light is guided by the cladding in presence of a surrounding medium of certain refractive index, a change of which modulates the transmitted light through the fiber circuit.

When the light traverses through the lead-in SMF falls on the first segment of the MMF, it excites the fundamental mode as well as several other modes, which we define as $E(r, 0)$ at the end of first segment of the MMF

$$E(r,0) = \sum_{q=1}^{m} b_q M_q \tag{1}$$

where $M_q$ are the modes through first segment of the MMF, $m$ is the total number of modes that can be accommodated within MMF and $b_q$ are the excitation coefficients of each modes in first segment of the MMF which can be calculated as

$$b_q = \frac{\int_0^\infty E(r,0) M_q r \, \mathrm{d}r}{\int_0^\infty M_q M_q r \, \mathrm{d}r} \tag{2}$$

Then, the light traverses through the first MMF at a distance of $L_1$ and the field distribution becomes

$$E(r, L_1) = \sum_{q=1}^{m} b_q M_q e^{i\beta_q L_1} \tag{3}$$

where the propagation constant of the $q$th mode is $\beta_q$. As the light propagates through the first MMF to the central SMF, modal power couple to both the fundamental mode and the cladding modes of the SMF. Assuming $S_p\ (r, 0)$ ($p = 1$ for core mode and $p > 1$ for cladding modes) as the $p$th mode of the central SMF, then the excitation coefficient $c_p$ of each mode can be calculated as

$$c_p = \frac{\int_0^\infty E(r, L_1) S_p r \, \mathrm{d}r}{\int_0^\infty S_p S_p r \, \mathrm{d}r} \tag{4}$$

After traveling through the central SMF, the amplitude of light mode can be written as

$$E_p(r,0) = c_p S_p e^{i\beta_p L} \tag{5}$$

where the propagation constant of the $p$th core/cladding mode is $\beta_p$. Therefore, after traversing the central SMF, the light intensity of the $p$th mode can be expressed as follows:

$$I_p = \left| E_p^*(r,0) \cdot E_p(r,0) \right| \tag{6}$$

Because of the phase difference between the cladding modes and the core mode, they interfere with each other in the last MMF, and the intensity of the interference pattern can be expressed as

$$I = I_1 + I_p + 2\sqrt{I_1 I_p}\cos\left(\frac{2\pi\,\Delta n_{eff}^p L}{\lambda}\right) \tag{7}$$

$$\Delta n_{\text{eff}}^p = n_{\text{core}} - n_{\text{clad}}^p \tag{8}$$

where the intensities of the core mode and the cladding mode are represented by $I_1$ and $I_p$, respectively. $n_{\text{core}}$ and $n_{\text{clad}}^p$ are the effective refractive indices of the core mode and the $p$th cladding mode, respectively. Although, the light traverses from first MMF to SMF, many cladding modes are excited in the central SMF. Out of which, only the major cladding modes interfere with the core mode of the central SMF in the last MMF, whereas the contribution from the other cladding modes is negligible on the interference loss. Here, in (7), only the main interference mode is considered in the calculation. The phase difference between the core and the cladding modes can be written as

$$\frac{2\pi\,\Delta n_{\text{eff}}^p L}{\lambda_{\text{dip}}} = (2m+1)\pi \tag{9}$$

Thus, $\lambda_{\text{dip}}$ can be calculated as

$$\lambda_{\text{dip}} = \frac{2\Delta n_{\text{eff}}^p L}{(2m+1)} \tag{10}$$

Total transmission ($T$) of the proposed MSM structure can be calculated as

$$T = \frac{\int_0^\infty I r\,\text{d}r}{\int_0^\infty |E(r,0)|^2 r\,\text{d}r} \tag{11}$$

The transmission loss of the interference spectrum of MSM can be adjusted to flatten the gain profile of the EDFA. Advantages of using MSM to flatten the gain in EDFA are that the entire data transmission system can be designed only by using optical fibers if parameters are optimized and then the MSM device can provide the best flattening of the gain spectrum of the EDFA. Interference between fundamental mode and cladding modes takes place in MSM which results loss in the transmission spectra of MSM device. This loss has wavelength band width of the order of 10 nm, therefore if the parameters of the MSM device are adjusted effectively in accordance with the gain peak of EDFA at resonance peak wavelength, we can flatten the gain response in the desired wavelength range. Here, the suggested work includes concatenating MSM with EDFA to flatten its gain by tuning MSM parameters at desired wavelength. The proposed gain flattening model is shown in Fig. 1.

## 3 Results and Discussion

We compute the transmission characteristics of MMF-SMF-MMF configurations analytically, shown in Fig. 3. For practical considerations, the identical MMFs have core/cladding diameters ($a_{m1}/a_{m2}$) as 19.9/125 μm and the piece of SMF has core/cladding diameters ($a_{s1}/a_{s2}$) as 8/125 μm. Here, RI's of SMF are, respectively, ns1 and ns2 and for MMF, nm1 and nm2, respectively. The RI of external medium is $n_{ex}$.

This nearly 10 dB interference loss can be used in series with EDFA, as shown in Fig. 1, to flatten a 980-nm-pumped EDFA with a peak at 1532-nm peak in the gain profile. The flattened gain spectrum with the usual gain of the EDFA at a pump power of 50 mW is shown in Fig. 4.

In conclusion, here we demonstrate a simple and adjustable method to flatten the dynamic gain spectrum of an EDFA. The working principle of our suggested device is based on modal interference loss between core mode and cladding, which can be adjusted to flatten the gain profile of an EDFA at desired wavelength.

It is easy to observe that the gain profile is flattened to 0.87 dB at the center wavelength 1546 nm over a bandwidth of 32 nm.

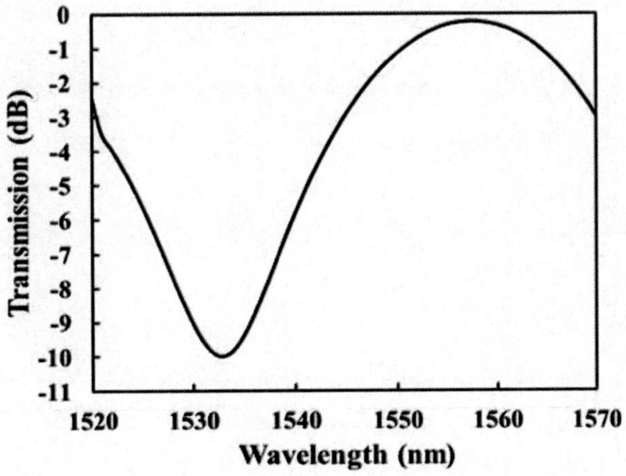


**Fig. 3** Transmission loss of the MSM structure

**Fig. 4** Gain profile of EDFA

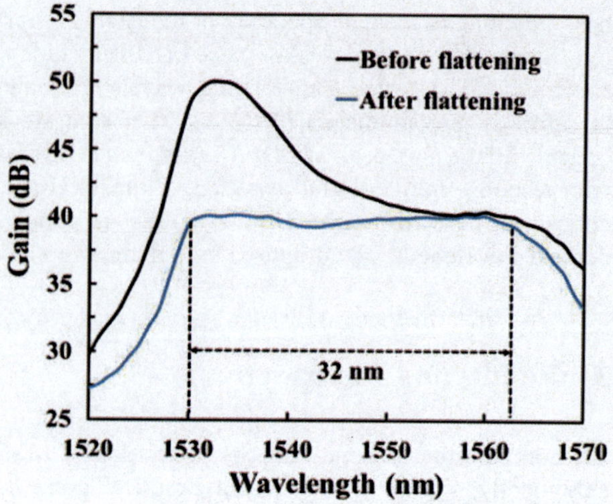


## References

1. Erdogan T, Sipe JE (1996) Tilted fiber phase gratings. J Opt Soc Am A 13(2):296–313
2. Desurvire E (1994) Erbium-doped fiber amplifier: principles and applications. Wiley, New York
3. Gloge D (1971) Weakly guiding fibers. Appl Opt 10(10):2252–2258
4. Ikhlef A, Hedara R, Chikh-Bled M (2012) Uniform fiber Bragg grating modeling and simulation used matrix transfer method. Int J Comput Sci 9(2):368–374
5. Erdogan T (1997) Fiber grating spectra. J Lightw Technol 15(8):1277–1294

# Experimental Demonstration of Electric Field Sensing Using Sagnac Loop Based Fiber Cantilever Configuration

**Isha Sharma and Partha Roy Chaudhuri**

**Abstract** We report here our experimental realization of all-fiber electric field sensing devices by devising a Sagnac mirror loop cantilever arrangement based on all-fiber cantilever beam deflection configuration. The reflected fiber exit beam is modulated with an electric field. Using cantilever–deflection–transmission analysis, the experimentally obtained results are theoretically modeled.

**Keywords** Fiber-optic sensor · Electric field sensing · Bismuth ferrite nanocomposite

## 1 Introduction

In the past few decades, fiber-optic sensors have undergone huge development. A variety of physical quantities, namely acoustic pressure, temperature, hydrostatic pressure, electric and magnetic field strengths, can be detected and measured by these sensors [1–4]. Optical fiber sensors are more preferred than other conventional technology due to its flexibility, safety, simple design, multiplexed operation, weight, low cost, etc. In terms of performance, the theoretical bandwidth technique is much faster than the corresponding conducting devices. These electric and magnetic field fiber-optic sensors have many applications in the defense sectors and nuclear reactors.

A magnetic field sensor which utilizes a magnetostrictive jacket over an optical fiber is proposed by Yariv and Winsor [5]. Using the cantilever configuration, magnetic field of nearly 2 mT has been detected by using a magnetostrictive material-coated optical fiber [6].

With this understanding, we attempt devising an all-optical fiber cantilever-based configuration which is able to detect an electric field in the surrounding with minimal system complexity. Our cascaded cantilever configuration having an all-fiber Sagnac loop (using 3 dB coupler forming a fiber loop mirror) acts as a perfect reflector in the form of fiber itself where all the injected powers go back to the same port into

I. Sharma · P. Roy Chaudhuri (✉)
Department of Physics, Indian Institute of Physics, Kharagpur, West Bengal, India
e-mail: roycp@phy.iitkgp.ac.in

 
C. Giri et al. (eds.), *Emerging Electronic Devices, Circuits and Systems*,
Lecture Notes in Electrical Engineering 1004,
https://doi.org/10.1007/978-981-99-0055-8_26

which light was originally injected. In our setup, this Sagnac mirror facilitates the forward and backward propagating light traversed the same identical path leading to greater sensitivity with more accuracy and repeatability.

In these kind of optical sensor, choosing a perfect probe material that can response to the external perturbation is a challenging task. We choose cobalt-doped bismuth ferrite nanoparticles as our electric probe material, since bismuth ferrite is the only multiferroic material that shows the coupling between magnetic and ferroelectric orders at room temperature. The advantage of choosing bismuth ferrite is that using the same fiber cantilever beam deflection platform, dual characterizations in terms of magnetic and electric properties of bismuth ferrite are possible as cobalt doping increases its magnetic properties.

## 2 Sagnac Mirror Loop Configuration

### *2.1 Experimental Setup*

Light from the He–Ne laser source (at 632 nm) is injected into one of the input ports of a 3 dB coupler using a 45× microscopic objective, and the output is taken from another port as shown in Fig. 1. One of the output ports acts as an input source for another 3 dB directional coupler-based Sagnac loop mirror formed. An index matching gel is applied on another output port of the first coupler to have minimal back reflection from the fiber end face [7]. The input port of the Sagnac loop is spliced with a coated single-mode fiber using an optimized composition of electric material ($BiFe_{0.9}Co_{0.1}O_3$) and kept head like a cantilever with coated length of 1 (± 0.1) cm.

Now, the reciprocity of the Sagnac loop is sustained in absence of any external factors in stationary and stable condition. Hence, the two counterpropagating light waves follow the same optical path and constructive interference occurs between two opposite propagating light waves after come back into the coupler. Here, the fiber loop mirror acts as a perfect reflector as all the input powers are now reflected back at the first port of the coupler. Also, a misalignment zone is created with the application of an external electric field due to bending of the fiber (toward the positive plate)

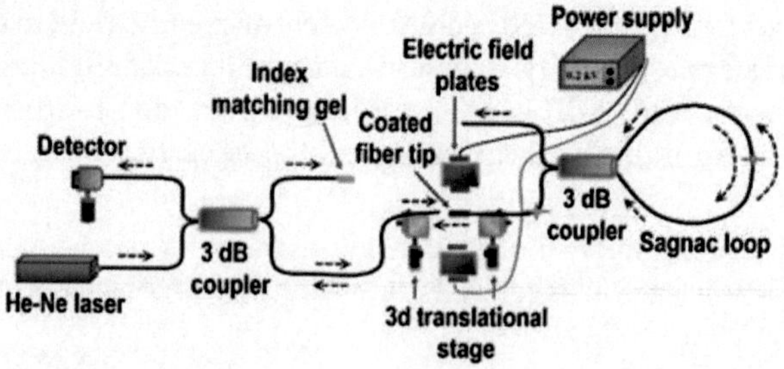


**Fig. 1** Schematic of experimental arrangement

as a response to induced polarization of probe material. Because of the mirror-like property of Sagnac loop, light transverses the identical path twice while traveling through the misalignment zone, resulting in a more sensitive configuration.

The light wave propagating in the backward recouples into the first 3 dB coupler after traversing the offset zone for a second time and is recorded by the detector.

## 3 Results Obtained from Our Model

The power variation with applied electric field is recorded for different cantilever lengths 3.1, 2.8, and 2.5 cm, respectively, and is shown in Fig. 2, keeping electric field plate distance at 1 cm. During the experiment, we used DC voltage for sensing DC static electric field as in AC voltage, the electric field will oscillate and the mechanical oscillation in no way can respond to the electrical frequency. Also, it is clear that for bigger cantilever length (3.1 cm), the sensitivity is higher. Ideally, more flexibility and deflection are exhibited by bigger cantilever.

### 3.1 *Theoretical Model*

Figure 3 illustrates the schematic of transverse misalignment between two optical fibers in Sagnac mirror loop configuration.

The variation of transmitted power coupled at the receiving output fiber with the transverse misalignment ($d_1$ and $d_2$) is given by [8]:

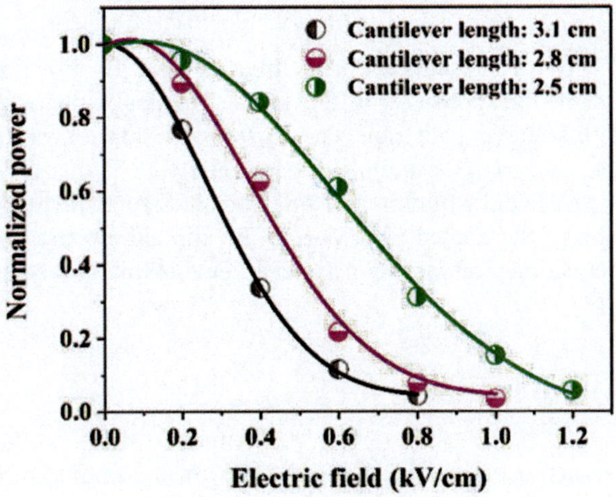


**Fig. 2** Response of Sagnac loop configuration at different cantilever lengths

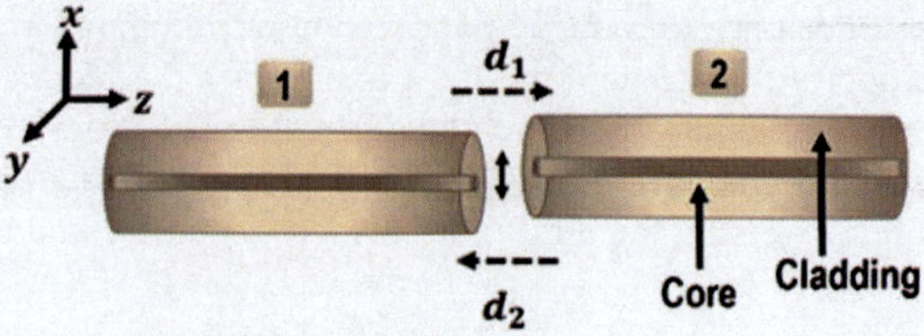


**Fig. 3** Schematic of transverse misalignment between two optical fibers

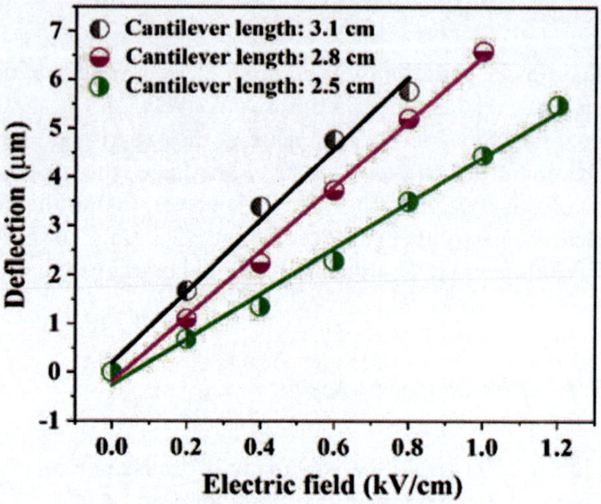


**Fig. 4** Deflection values estimated at different electric fields for varying cantilever lengths

$$T = \eta \left( \frac{4\omega_1^2 \omega_2^2}{\left(\omega_1^2 + \omega_2^2\right)^2} \right)^2 e^{-2\left[ \frac{d_1^2}{\omega_1^2 + \omega_2^2} + \frac{d_2^2}{\omega_2^2 + \omega_1^2} \right]} \quad (1)$$

where $\eta$ represents the loss effect of all 3 dB directional couplers, and $\omega_1$ and $\omega_2$ denote the spot sizes of fundamental modes of fiber 1 and fiber 2. Also, in our case, the cantilever path transverse by forward and backward propagating light is identical, i.e., $d_1 = d_2 = d$ and $\eta$ is taken as unity.

We fit our experimental data into our theoretical model (1) and estimate the deflection of the coated fiber due to the applied electric field. The calculated values of deflections for varying cantilever lengths (plates distance 1 cm) are shown in Fig. 4.

## 4 Conclusion

In this report, we present our experimental finding of designing an all-optical fiber electric field detection system based on fiber cantilever beam deflection principle. We demonstrated a Sagnac mirror loop configuration with coated fiber exit port.

The existence of an electric field is detected from the reflected power variation caused by the misalignment zone created between the optical fibers. For a precise understanding, our experimental results are theoretically analyzed.

## References


1. Lenz JE (1990) A review of magnetic sensors. Proc IEEE 6:973–989 (references)
2. Yu M, Balachandran B (2003) Acoustic measurements using a fiber optic sensor system. J Intell Mater Syst Struct 14:409–414
3. Lenz J, Edelstein AS (2006) Magnetic sensors and their applications. IEEE Sens 6:631–649
4. Alwis L, Sun T, Grattan KTV (2013) Optical fibre-based sensor technology for humidity and moisture measurement: Review of recent progress. Measurement 46(10):4052–4074
5. Yariv A, Winsor H (1980) Proposal for detection of magnetic fields through magnetostrictive perturbation of optical fibers. Opt Lett 5: 87–89
6. Pradhan S, Roy Chaudhuri P (2015) Experimental demonstration of all-optical weak magnetic field detection using beam-deflection of single-mode fiber coated with cobalt-doped nickel ferrite nanoparticles. Appl Opt 54(20):6269–6276
7. Lawrence CM, Nelson DV, Udd E, Bennett T (1999) A fiber optic sensor for transverse strain measurement. Exp Mech 39(3):202–209
8. Ghatak A, Thyagarajan K (2011) Introduction to fiber optics. Cambridge University Press

# An SMT-Based Reverse Engineering of Register Allocation in High-Level Synthesis

**Mohammed Abderehman and Chandan Karfa**

**Abstract** The variables of a high-level behaviour are mapped to hardware registers during register allocation (RA) in high-level synthesis (HLS). Reverse engineering such mapping is a challenging task due to possible many-to-many relations. In this paper, we propose an satisfiability modulo theory (SMT)-based reverse engineering framework to extract the register to variable mapping in HLS for efficient correlation between input C/C++ and the corresponding RTL generated by the HLS tool. In our method, we first extract a C-like behaviour, called RTL-C from RTL. Both the RTL-C and the input C are next converted to static single assignment (SSA) form. A satisfiability problem is then formulated to obtain the register to variable mapping from C and RTL-C. Finally, the registers in RTL-C are replaced by the corresponding variables in C using the mapping information to obtain the equivalent C code from RTL. The generated C code would be helpful for the algorithm developers to understand the RTL code and to reduce the RTL verification time as well. The proposed method is implemented and tested for several programmes. The experimental results demonstrate the usefulness of the proposed method.

**Keywords** High-level synthesis · Reverse engineering · RTL · Register allocation · FSMD · SMT

## 1 Introduction

High-level synthesis (HLS) is the process of translating high-level behaviours written in C/C++ into a register transfer level (RTL) structure that realizes the given behaviours [1]. Starting from the high-level description of an application, an HLS tool performs the following tasks: (i) prepossessing which transforms the input

M. Abderehman (✉) · C. Karfa
Indian Institute of Technology Guwahati, Guwahati, Assam 781039, India
e-mail: ma.adem@iitg.ac.in

C. Karfa
e-mail: ckarfa@iitg.ac.in

 
C. Giri et al. (eds.), *Emerging Electronic Devices, Circuits and Systems*,
Lecture Notes in Electrical Engineering 1004,
https://doi.org/10.1007/978-981-99-0055-8_27

description into an internal representation (IR), (ii) scheduling consists in assigning the operations to so-called control steps without violating data dependencies among operations, (iii) allocation and binding determines the selection of the types of hardware components and the number for each type to be included in the final implementation to satisfy the design constraints, and (iv) datapath and controller generation which the interconnections among the FUs and registers are determined. The RTL consists of a datapath and a controller FSM. The FSM decides the operations to be executed in the datapath in each time step of the controller FSM. Shorter design cycles, 10× fewer code sizes in a higher level of abstraction, and easy design space exploration make HLS an attractive design choice over the conventional RTL design process.

Although HLS comes with such huge advantages, the real use of HLS in commercial design houses is still limited. Who are the target users of the HLS tools—the RTL developers or the algorithm developers? The RTL developers are reluctant to use HLS because the RTL generated by an HLS tool is not as optimal as compared to the RTL they develop manually. On the other hand, the algorithm developers do not understand the RTL well. The quality of the RTL generated by the HLS tools greatly depends on the way one has written the C code. Moreover, they do not understand the real effects of all the hardware-related optimization parameters of the HLS tools. As a result, it is difficult for them to write HLS friendly C/C++ code and also choose the right set of optimization parameters during HLS to generate an RTL that meets the target design constraints. *A C equivalent of the RTL hardware would be helpful for the algorithm developers to understand/analyse the output RTL of a HLS tool and hence use HLS tool meaningfully.* The RTL co-simulation is the primary way to verify the correctness of the generated RTL of an HLS tool. Specifically, the test cases developed for the input C code are reused by the commercial HLS tools [2] for RTL simulation. Since the C-simulation is much faster than the RTL simulation [3], generating an equivalent C code from the corresponding RTL and using it for RTL verification would greatly reduce the verification time. One of the important steps in obtaining a high-level C behaviour from the RTL is to recover the mapping between registers in the RTL generated by HLS and variables in input C. It may be noted that the variables are mapped registers during the allocation and binding phase of HLS. *The number of registers in RTL usually is much less than the number of variables in input C. One register may store more than one variable provided their lifetimes do not overlap. Similarly, one variable may be split into more than one register for better mapping. Therefore, the relation between registers in the RTL and the variables in C is complex, and it is a challenging task to recover such mapping without taking any information from the HLS tool.*

In this work, we propose an automatic framework to extract the register to variable mapping information without taking any intermediate information from the HLS tool. In our approach, we first extract a high-level behaviour (RTL-C) from the RTL [3]. We then modelled both the input C code and RTL-C as a finite state machine with datapaths (FSMDs), called *C-FSMD* and *RTL-FSMD*, respectively. Both the *C-FSMD* and *RTL-FSMD* are then converted into to static single assignment (SSA) form [4]. A Satisfiability Modulo Theory (SMT)-based satisfiability problem is then formulated to obtain the register to variable mapping. With the information obtained from the mapping, we can rewrite the C-like behaviour, called *RTL-C* in terms of variables of input C, and finally generate an equivalent C code from the HLS generated RTL. The working of the proposed method is demonstrated on the RTLs generated by Vivado HLS tool [2] for a set of HLS benchmarks. This is the first to work that automatically reverse engineer the register to variable mapping in high-level synthesis *using a formal approach.*

The rest of the paper is organized as follows. Section 2 presents related works. Our reverse engineering approach is presented in Sect. 3. The SMT formulation of the register to variable reverse engineering is presented in Sect. 4. Experimental results are presented in Sect. 5. Section 6 concludes the paper.

## 2 Related Work

In this section, we overview some important RTL to C conversion works in the domain of HLS.

The method presented in [5] converts the RTL Verilog into the equivalent C code for equivalence checking between C and the Verilog RTL. In [6], a method is proposed to abstract RTL intellectual property (IP) blocks into C++ code. Most of the architectural characteristics of the behaviour are ignored during conversion to maintain its functionality. A method VeriIntel2C is presented in [7] to convert RTL behaviours written in Verilog into C code. To extract the functionality of the RTL designs and to generate a CDFG that captures the different structural forms present in RTL designs, the proposed method uses extended Hardware Petri Nets. The main goal of [6, 7] is to recover the of RTL functionality at higher abstraction level which enables HLS driven design space exploration (DSE). In [8], a methodology called VTOC is proposed to convert synthesizable Verilog into C for faster simulation. Parts of the Verilog language such as uncertain values, temporary bus fights, Verilog meta-language commands, and other simulation-level operations which are not normally compatible to hardware are not compiled to C by the VTOC compiler. The Verilator simulator [9] parses a Verilog code into equivalent behavioural description in C++ for fast simulation. Although over the years, Verilator has been incorporated with advanced optimizations to speedup simulations, it ignores the inherent FSMD framework of the HLS-based RTLs. Consequently, the generated C++ code is complex in terms of comprehension of code behaviour and incurs performance hampering redundancies. This impacts both simulation performance as well as debugging.

A recent work, FastSim [3], overcomes the limitations of Verilator. The paper proposes a framework that converts an HLS generated RTL to an equivalent C code similar to Verilator but takes advantage of the structure of the HLS generated RTL. The simulator automatically generates the behavioural FSMD in C code semantics from the HLS generated Verilog RTL while maintaining the state machine sequence of the synthesized RTL and also guarantees fast simulation, functional correctness of the RTL, cycle accuracy, accurate performance estimation, and generate a highly readable and debug friendly simulation code. However, extraction of the register to variable mapping is not targeted in FastSim. Recently, the authors have extended their FastSim work in [10] to extract the mapping between the variables and the registers using a Daikon-based invariant generation framework. This work utilized the C (RTL-C) code extracted from RTL by FastSim. In addition, it obtained the scheduled C code by decoding the scheduling information generated by the HLS tool after the scheduling phase in the HLS. It then combined both the behaviours and used Daikon to generate the invariant state-wise manner. From these invariants, it identifies the register to variable mapping. However, there is no formal guaranty of the correctness of the extracted mapping by this method.

Like [10], our work also uses RTL-C extracted by FastSim from the RTL generated by HLS. Instead of an invariant generator tool, we formulate the problem of extraction of register to variable mapping as a satisfiability problem and use Satisfiability Modulo Theory (SMT) solver to obtain the mapping information. Thus, register to variable mapping obtained in our method is always correct. Moreover, our work uses the input C code instead of the scheduled C code as in [10]. Therefore, our method can represent the RTL in terms of the variables of the original C code (instead of the variables in scheduled C code).

## 3 Register to Variable Reverse Engineering Approach

The primary motivation of this work is to extract the register to variable mapping in HLS for efficient correlation between input C/C++ and the corresponding HLS generated RTL. We rely on the technique presented in [3] to extract RTL-C code from the RTL generated by HLS tool. The overall flow of our approach is given in Fig. 1.

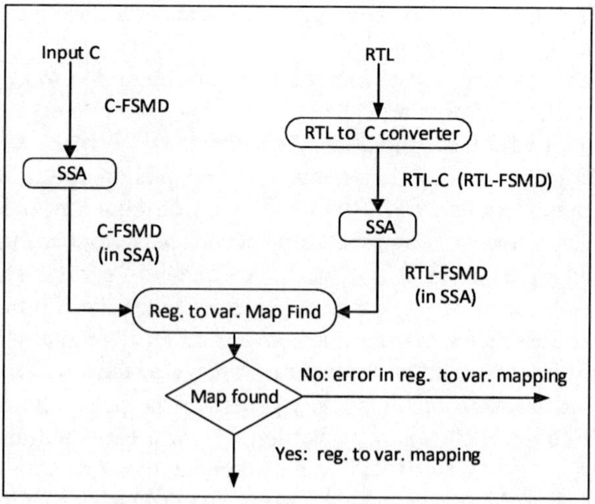


**Fig. 1** Register to variable reverse engineering flow

First, we formally modelled both the input C code and RTL-C as a finite state machine with datapaths (FSMD) [11], called C-FSMD and RTL-FSMD, respectively. Since the variables of the C specification are mapped to registers in the RTL, these two FSMDs, i.e. C-FSMD and RTL-FSMD are not comparable yet. We then convert both the FSMDs into a static single assignment (SSA) form in SSA step [4]. In SSA, we first convert all the operations in 3-address form, i.e. at most one operator in the right-hand side (RHS) expression of an operation. A variable in the C specification may be defined multiple times. As a result, such variables may be mapped to more than one register in the RTL. Also, multiple variables with a non-overlapping lifetime may be mapped to a single register which effectively indicates that a register is defined multiple times in RTL-FSMD. The SSA step on C-FSMD ensures that each variable is defined exactly once in the C specification. The SSA on RTL-FSMD ensures that each register is defined exactly once in the RTL. Since both these sets of variables capture the lifetimes of variables of the input specification, the number of variables in the C-FSMD and the number of registers in the RTL-FSMD will be the same after SSA step. In the next step, register to variable mapping is modelled as satisfiability problem and the mapping is obtained using SMT solver. In the following, we briefly discussed the FSMD and related theory. The SMT formulation is given in the next section. Due to limited space, we could not explain the RTL to C conversion method here. Interested readers can refer FastSim [3] for RTL to C conversion in detail.

## 3.1 *FSMDs and Related Theory*

The programmes are modelled as FSMD in this work. We briefly explain the model and background theory here. The detail can be found in [11]. An FSMD is an inherently deterministic model that can represent any hardware circuits [11]. An FSMD $M$ is defined as a 7-tuple $\langle Q, q_0, I, O, V, f, h \rangle$, where $Q$ is the finite set of states, $q_0 \in Q$ is the reset (initial) state, $I$ is the finite set of input variables, $O$ is the finite set of output variables, $V$ is the finite set of storage variables, $f : Q \times 2^S \rightarrow Q$ is the state transition function, and $h : Q \times 2^S \rightarrow U$ is the update function. Here, $S$ represents the set of relations over arithmetic expressions and Boolean literals and $U$ represents a set of storage and output assignments.

A *trace* of an FSMD is a finite walk from the reset state $q_0$ to itself, and $q_0$ should not occur in between. The *condition of execution* $c_\tau$ of a trace $\tau$ is a logical expression over $I$, which must be satisfied by the initial data state in order to traverse the path $\tau$. The *data transformation* $s_\tau$ of a trace $\tau$ over $O$ is an ordered tuple $\langle e_j \rangle$ of algebraic expressions over $I$ such that the expression $e_j$ represents the value of the output $o_j$ after execution of the trace in terms input variables. Computation of the condition of execution and data transformation can be obtained by by forward substitution [12]. The forward substitution method of finding data transformations is based on symbolic execution [13].

**Algorithm 1:** Register_to_variable_mapping (C-FSMD, RTL-FSMD)

**Result**: register to variable mapping

$V_c = \{\phi\}$, $V_r = \phi$, $I_c = \phi$, $I_r = \phi$, CTP = $\{\phi\}$.

**repeat**

  Generate next random input.

  Simulate the both C-FSMD and RTL-FSMD on the input to obtain the traces, $\tau_c$ and $\tau_r$, respectively.

  /* Collect variables and the registers that involved in $\tau_c$ and $\tau_r$ */

  $V_c = V_c \cup$ {set of all variables updated in $\tau_c$}.

  $V_r = V_r \cup$ {set of all variables updated in $\tau_r$}.

  $I_c = I_c \cup$ {set of all inputs involved in $\tau_c$}.

  $I_r = I_r \cup$ {set of all inputs involved in $\tau_r$}.

  CTP = CTP $\cup \langle \tau_c, \tau_r \rangle$.

**until** $(V_c \equiv V \wedge V_r \equiv R \wedge I_c \equiv I_r \equiv I)$;

$F$ = Formulate_SAT_problem(C-FSMD, RTL-FSMD, CTP).

Call Z3 for with all the formulas in $F$.

**if** *Z3 returns SAT* **then**

  **return** the register to variable mapping.

**else**

  Report error in register mapping.

## 4 SMT-Based Formulation of Register to Variable Reverse Engineering

An FSMD may contain many traces. Our first objective is to find the correspondence between the traces in C-FSMD and RTL-FSMD. The RTL-FSMD and the C-FSMD are simulated together to find the correspondence between the traces. The inputs and the outputs are the same for both the FSMDs. It may be noted that the mapping between the variables in C-FSMD and the registers in RTL-FSMD after SSA transformation will be one-to-one. Let us denote the set of variables in C-FSMD and the set of registers in RTL-FSMD as V and R, respectively. We model this mapping problem as the satisfiability problem between corresponding traces and use SMT solver Z3 [14] to solve this. A satisfiable instance will reveal the mapping between the registers and variables. The initial simulation runs with random inputs. The traces obtain in both the FSMDs for a given random input called as *corresponding trace pair (CTP)*. Let us assume that the CTP obtained by initial simulation is $\langle \tau_c, \tau_r \rangle$, where $\tau_c$ is a trace in C-FSMD and $\tau_r$ is the corresponding trace in RTL-FSMD. We shall collect the variables that get defined in $\tau_c$ in a set $V_c$. Similarly, all the registers get defined in $\tau_r$ are collected in a set $V_r$. Also, all the inputs that are involved in $s_{\tau_c}$ and in $s_{\tau_r}$ are collected in $I_c$ and $I_r$, respectively. We next generate the next trace using the concolic testing approach [15] and continue to do the above. This process will stop when $V_c = V$, $V_r = R$ and $I_c = I_r = I$. The idea here is to collect enough traces so that all variables, all registers, and all the inputs are involved.

**SAT formulation**: Let the set of variables and the set of registers be denoted as $V$ and $R$, respectively. As argued before, both these set contain the same number of elements. Let us assume each of them contains $N$ number of elements. Let us consider the Boolean variables $m_{vr}$, $\forall v \in V$, $\forall r \in R$. Let us assume that the Boolean variable $m_{vr} = 1$ if the variable $v \in V$ is mapped to register $r \in R$; otherwise, $m_{vr} = 0$. The mapping between the variables and the registers is one-to-one. Following two constraints are captures that fact.

(i) Each variable must be mapped to exactly one register.

$$\sum_{r=1}^{N} m_{vr} = 1, \ \forall v \in V \tag{1}$$

(ii) For each register, exactly one variable is mapped to that register.

$$\sum_{v=1}^{N} m_{vr} = 1, \ \forall r \in R \tag{2}$$

In addition, expression corresponding to each output in the traces C-FSMD will be rewritten in terms of $m_{rb}$ and $r$. Each variable $v_i$ in an output expression will be replaced by $\sum_{r=1}^{N} m_{v_i r} \times r$. We loosely denote this replacement as $v_i \circ m_{vr}$. For example, the expression corresponding to output $o_1$ is $3 \times v_1 + 4 \times v_2$ in a trace $\tau_c$ in C-FSMD. Assume that we have two registers $r_1$ and $r_2$ in RTL-FSMD. The $v_1$ will be rewritten as $m_{v_1 r_1} \times r_1 + m_{v_1 r_2} \times r_2$. This expression indicates that $v_1$ is replaced by the corresponding register. The corresponding register of $v_1$ is $r_1$ if $m_{v_1 r_1} = 1$; otherwise, $r_2$. Similarly, $v_2$ will be rewritten as $m_{v_2 r_1} \times r_1 + m_{v_2 r_2} \times r_2$. Let the output expression corresponding to $o_1$ in RTL-FSMD is $3 \times r_2 + 4 \times r_1$. The constrains will be passed to SMT solver for $o_1$ is $3 \times (m_{v_1 r_1} \times r_1 + m_{v_1 r_2} \times r_2)$ $+ 4 \times (m_{v_2 r_1} \times r_1 + m_{v_2 r_2} \times r_2) \equiv r_2 + r_1$. We will identify such constraints for each output for each trace pair identified in Algorithm 1. Let assume that the output set is denoted as $O$ in both the FSMDs. The third constrains is as follows.

(iii) For each corresponding trace pair selected in Algorithm 1, the outputs are equivalent.

$$\forall \langle \tau_c, \tau_r \rangle, \forall o \in O, s_{\tau_c}|_o \circ m_{vr} \equiv s_{\tau_r}|_o \tag{3}$$

Here, $s_{\tau_c}|_o / s_{\tau_r}|_o$ represents the expression of output $o$ in the data transformation $s_{\tau_c}/s_{\tau_r}$ of the trace $\tau_c/\tau_r$.

We call Z3 with all these constraints to find the values of $m_{vr}$ that satisfies all the constraints. If there is any solution, the SMT solvers will return the values of $m_{vr}$. The overall process is given as Algorithm 1. Using the value of $m_{vr}$, the exact mapping between variables and registers is obtained. All the registers in the RTL-FSMD can be replaced by the corresponding variable with $m_{vr}$.

***Example 1*** An example of identifying register mapping is given in Listing 27.1. Let us assume that there is only trace in each FSMD here. In this example, C code is out $= a - b$ which is mapped to $out = r2 - r1$ is RTL. The *Z3* returns $m11 = 0, m12 = 1, m21 = 1, m22 = 0$. Here, $m11 = 1$ indicates variable $a$ is mapped to register $r1$ and $m22 = 1$ indicates variable $b$ is mapped to register $r2$ and so on. Basically, the formulation recovers that the variable $a$ maps to register $r2$ and the variable $b$ maps to register $r1$.

**Listing 27.1** Example SMT code to obtain mapping

```
(declare-const m11 Int)
(declare-const m12 Int)
(declare-const m21 Int)
(declare-const m22 Int)
(declare-const out1 Int)
(declare-const out2 Int)
(declare-const r1 Int)
(declare-const r2 Int)
;values are either 0 or 1
(assert (or (= m11 1) (= m11 0)))
(assert (or (= m12 1) (= m12 0)))
(assert (or (= m21 1) (= m21 0)))
(assert (or (= m22 1) (= m22 0)))
;each var is mapped to one reg
(assert (= (+ m11 m12) 1))
(assert (= (+ m21 m22) 1))
;each reg is associated to exactly one var
(assert (= (+ m11 m21) 1))
(assert (= (+ m12 m22) 1))
(assert (= out2 (- r2 r1)))
(assert (= out1 (- (+ (* m11 r1) (* m12 r2))
              (+ (* m21 r1) (* m22 r2)))))
(assert (= out1 out2))
(check-sat)
(get-model)
```

In some scenarios, there may be more than one solution exists in our formulation. Specifically, when the output expression is commutative. Consider the output expression for the output $o_1$ is $v_1 + v_2$ in a trace $\tau_c$ in C-FSMD. Let us assume that the corresponding output expression in RTL-FSMD is $r2 + r1$. Since these expressions are commutative, an SMT solver can provide either $m11 = 0, m12 = 1, m21 = 1, m22 = 0$ (i.e. $v_1$ maps to $r_2$ and $v_2$ maps to $r_1$) or $m11 = 1, m12 = 0, m21 = 0, m22 = 1$ (i.e. $v_1$ maps to $r_1$ and $v_2$ maps to $r_2$) based on our formulation. Since, the output expressions $r2 + r1$ and $(m_{v_1r_1} \times r_1 + m_{v_1r_2} \times r_2) + (m_{v_2r_1} \times r_1 + m_{v_2r_2} \times r_2)$ are equivalent for both the mappings, both them are actually correct.

## 5 Experimental Results

The SMT-based register to variable reverse engineering framework is implemented in Python and is tested on a set of HLS benchmarks. We have used the Vivado HLS tool [2] to generate Verilog RTL for the benchmarks written in C. We then obtain the RTL-C (RTL-FSMD) from the Verilog RTL using FastSim [3] which uses the pyverilog parser [16] in background. Our tool invokes the SMT tool Z3 [14] as shown in Algorithm 1. A satisfiable instance will reveal the mapping between the variables and registers. With this mapping function, we can rewrite the RTL-FSMD in terms of variables and finally generate an equivalent C code from the RTL. One primary challenge was handling arrays. In Vivado HLS, arrays are mapped to block RAMs. For RAM, a separate module is created with address port, data port, write enable to access it. In each control state, the controller assigns appropriate values to these ports for RAM access. Based on these values, the rewriting method in FastSim identifies the actual memory read/write operation in each state.

The experiments have been performed on a machine with a CPU: Intel Core i7, 2.5GHz, and 8GB RAM. We evaluate our method on a variety of HLS benchmarks (Waka, Motion, DIFFEQ, maximum of three numbers (Max3), and auto-regressive lattice filter with branch (ARFWB) and without branch (ARFNB)), each of them written in C code. Table 1 presents the experimental results. The 1st column has the name of benchmarks used. The 2nd and 3rd columns are the lines and variables in the C code, respectively. The 4th, 5th, 6th, and 7th columns are the number of lines, number of variables, latency, and number of states of the RTL-FSMD in RTL code, respectively. It may be noted that the number of variables in C and the number of registers are not the same since Vivado HLS optimizes the register usage during register allocation. However, the number of variables in the C code after SSA and the number of registers in the RTL code after SSA are the same in all test cases. The

**Table 1** Experimental results for HLS benchmarks

| Bench | C code | | RTL code | | | | Our tool | | |
|---|---|---|---|---|---|---|---|---|---|
| | #lines | #var. | #lines | #reg. | latency | #States | #smtcode | SMT time (ms) | Time (ms) |
| Waka | 34 | 21 | 585 | 8 | 8 | 3 | 658 | 380 | 432.06 |
| Motion | 52 | 42 | 909 | 21 | 8 | 5 | 1920 | 1470 | 1540.54 |
| DIFFEQ | 14 | 15 | 389 | 7 | 6 | 5 | 578 | 34 | 92.55 |
| Max3 | 15 | 5 | 150 | 3 | 2 | 3 | 377 | 25 | 49.85 |
| ARFNB | 48 | 56 | 1185 | 21 | 10 | 7 | 1803 | 1380 | 1488.21 |
| ARFWB | 54 | 28 | 743 | 12 | 6 | 5 | 1547 | 1300 | 1397.06 |
| Matrixop | 33 | 90 | 3771 | 32 | 10 | 17 | 4803 | 2486 | 2798.57 |
| Parker | 51 | 23 | 330 | 12 | 4 | 2 | 547 | 27 | 39.06 |

8th and 9th columns show the number of lines in SMT format code and SMT time, respectively. The SMT time is the time required by Z3 for showing the satisfiability of the formulation. Finally, the 10th column reports the total (HDL parsing time + SMT code generation + Z3 run time) time spent by our system to produce the corresponding C code from the Verilog code generated by the Vivado HLS tool. The run times are not high and less than 3 seconds for all cases. The majority of time is taken by the SMT tool Z3. To check the correctness of our method, we have done simulation-based verification for reverse engineered C code and the input C code. The outputs match for all benchmarks used which confirm the correctness of our reverse engineering flow.

## 6 Conclusion

We have presented an SMT-based reverse engineering framework to obtain the register to variable mapping information between input C code and corresponding Verilog RTL generated by an HLS tool. Our method reconstructs high-level behaviour RTL-C from the RTL generated by the HLS tool. It then invokes SMT solver Z3 to find the register to variable mapping. We rewrite the RTL-C by replacing the register name (in RTL) with the corresponding variable name (of C) using the mapping obtained. We have tested with several HLS benchmarks for the Verilog generated by the Vivado HLS tool and found that the overall conversion time is reasonably small. The extracted mapping can be utilized by the algorithm developers to understand the RTL generated by HLS, can be used to debug any mismatch in RTL to C reverse engineering-based faster simulation framework [3] and can be useful in security analysis of register allocation [17].

## References

1. Gajski DD, Dutt ND, Wu AC, Lin SY (1992) High-level synthesis: introduction to chip and system design
2. Vivado high-level synthesis. http://xilinx.com/support/download.html
3. Abderehman M, Patidar J, Oza J, Nigam Y, Abdul Khader TM, Karfa C (2021) Fastsim: a fast simulation framework for high-level synthesis. IEEE TCAD
4. Cytron R, Ferrante J, Rosen BK, Wegman MN, Zadeck FK (1991) Efficiently computing static single assignment form and the control dependence graph. ACM Trans Progr Lang Syst 13(4):451–490
5. Mukherjee R, Tautschnig M, Kroening D (2016) V2c—a Verilog to C translator. In: 22nd TACAS, pp 580–586
6. Bombieri N, Liu H, Fummi F, Carloni L (2013) A method to abstract RTL IP blocks into C++ code and enable high-level synthesis. In: DAC, pp 1–9
7. Mahapatra A, Schäfer B (2018) Veriintel2c: abstracting RTL to C to maximize high-level synthesis design space exploration. Integration 64:08
8. Greaves DJ (2000) A Verilog to C compiler. RSP 2000, Paris, June 2000
9. Snyder W (2017) Verilator. http://www.veripool.org/wiki/verilator

10. Abderehman M, Gupta R, Karfa C (2021) Reverse engineering register to variable mapping in high-level synthesis. In: ISVLSI, pp 37–42
11. Gajski D, Ramachandran L (1994) Introduction to high-level synthesis. IEEE Trans Des Test Comput 44–54
12. Karfa C, Sarkar D, Mandal C, Kumar P (2008) An equivalence-checking method for scheduling verification in high-level synthesis. IEEE TCAD 27(3):556–569
13. Manna Z (1974) Mathematical theory of computation. McGraw-Hill Kogakusha, Tokyo
14. de Moura L, Bjørner N (2008) Z3: an efficient SMT solver. In: Ramakrishnan CR, Rehof J (eds) TACAS. LNCS, vol 4963, pp 337–340
15. Ahmed A, Mishra P (2017) Quebs: qualifying event based search in concolic testing for validation of RTL models. In: ICCD
16. pyVerilog. https://pypi.org/project/pyver/
17. Panigrahi P, Sahithya V, Karfa C, Mishra P (2022) Secure register allocation for trusted code generation. IEEE Embedded Syst Lett

# Hardware Primitives-Based Accelerator Architecture for NTRU-HRSS Scheme

**J. Mervin, Shabbir Darbar, and David Selvakumar**

**Abstract** This paper presents primitives-based accelerator for NTRU-HRSS701 encryption and decryption scheme. The basic primitives to construct the NTRU scheme are polynomial multiplication, polynomial lift operation, modulo converters, and random bit stream generator for random polynomials. We propose pipelined polynomial multipliers implemented using conventional method and generalized Karatsuba algorithm. The polynomial lift operation suitable for NTRU-HRSS701 is also implemented in hardware. The polynomial multiplier, poly-lift, modulo converters, and the NTRU accelerator have been implemented in RTL, synthesized for the Xilinx Virtex UltraScale FPGA, and the results are compared with existing works. The NTRU accelerator with conventional polynomial multiplier is found to be more efficient than other existing architectures which are either based on shift registers or number theoretic transform. Moreover, these realizations are for lower polynomial degree '$N$'. Further, it has been observed that the conventional approach is better than the generalized Karatsuba-based polynomial multiplier due to the large value of '$N$' ($N = 701$) for NTRU-HRSS and such a large value of '$N$' introduces more additions. Karatsuba algorithm reduces the number of multiplications and increases additions. This primitive-based architecture will further enable NTRU key generation, encapsulation, and de-capsulation in hardware.

**Keywords** NTRU-HRSS · Modular polynomial multiplier · Modular arithmetic · Polynomial lift · Karatsuba multiplier · Post-quantum crypto accelerators

---

J. Mervin (✉) · S. Darbar · D. Selvakumar
Secure Hardware and VLSI Design Group, Centre for Development of Advanced Computing, Bangalore, India
e-mail: mervinj@cdac.in

S. Darbar
e-mail: sdarbar@cdac.in

D. Selvakumar
e-mail: david@cdac.in

 
C. Giri et al. (eds.), *Emerging Electronic Devices, Circuits and Systems*,
Lecture Notes in Electrical Engineering 1004,
https://doi.org/10.1007/978-981-99-0055-8_28

# 1 Introduction

Post-quantum cryptography (PQC) is an emerging area of study on the aspects of designing crypto algorithms and schemes which are resistant to quantum attacks. The classical public-key algorithms such as RSA and ECC are based on the problems of factorization of large integers and discrete logarithm, which are hard problems for classical computers. With the advent of quantum machines which have, in principle, the superposition of states makes it feasible to factorize a large integer [1] and break conventional crypto schemes with quantum algorithms (Shor's and Grover's) approaches. Lattice-based post-quantum crypto algorithms are based on hard problems in lattices, viz., shortest vector problem (SVP) and closest vector problem (CVP), etc. [2], which cannot be solved with quantum machines.

NIST has initiated a process for standardization of PQC algorithms and released third-round candidates in July 2020. The list includes four public-key encryption and key-establishment algorithms and three digital signature algorithms [3]. NTRU (Nth degree truncated polynomial ring units) has been selected as a round 3 candidate which consists of different variants, namely NTRU-HRSS and NTRU-HPS. NTRU-HPS has been derived from the round 1 submission of NTRU-HRSS-KEM (Key encapsulation method) and the NTRUEncrypt scheme. Further, the NTRU Prime is an alternate candidate in round 3 submissions. NTRU-HPS and the NTRU-HRSS have minor algorithmic differences; however, most of the basic crypto computational primitives are similar. The NTRU-HRSS specifies a parameter set denoted as ntruhrss701. The NTRU-HPS has 3 parameter sets, namely ntruhps2048509, ntruhps2048677, and ntruhps4096821. The ntruhps2048509 and ntruhps4096821 variants are introduced to replace the ntru-pke-443 and ntru-pke-743 variants of NTRUEncrypt. The ntruhps2048677 was introduced as an alternative to the ntruhrss701. The terminologies adopted in this paper are as per the specification of NTRU provided as part of the NIST submission [4]. NTRU is a lattice-based scheme, and its security depends on SVP [5]. This work presents primitive built hardware accelerator architectures for NTRU-HRSS701-based ($N = 701$, $p = 3$, $q = 8192$) public-key encryption and decryption and demonstrated on FPGA. Key contributions of this work are summarized as follows:

1. A fully pipelined conventional polynomial multiplier with '$N$' stages linear array of '$N$' processing elements (Modular arithmetic unit/multiply accumulate unit).
2. Generalized Karatsuba algorithm-based polynomial multiplier as an alternate for conventional polynomial multiplier.
3. Hardware unit for polynomial lift operation required for ntruhrss701-based encryption and decryption.
4. NTRU-HRSS encryption and decryption schemes are fully in hardware with primitive units for polynomial multipliers (either conventional or Karatsuba) and polynomial lift instead of hardware/software co-design approach.

## 2 Related Works

This section discusses the published literatures on hardware architectures and design approaches for various primitives/building blocks, viz., modular polynomial multiplication, polynomial inversion, software/hardware co-design for NTRU key generation, encryption and decryption accelerators, high-level synthesis-based (HLS) design, etc., typically required/adopted for realizing NTRU schemes. The most popularly implemented variant of NTRU is the NTRUEncrypt [6, 7] which had been a candidate in Round 1 of the NIST standardization efforts for PQC. Most of the existing works have discussed mainly about polynomial multiplication in hardware since it is a time and resource-intensive operation in any lattice-based cryptographic scheme. Further, the prevailing works typically have adopted software/hardware co-design paradigm wherein the polynomial multiplication is handled in hardware, and the other NTRU computations are handled as software/firmware with an embedded microcontroller.

Wilhelm [6] discusses design considerations to realize the entire NTRU scheme illustrated in IEEE P1363.1 standard in hardware in VHDL and presents a testing and verification model. Various primitives considered in [6] for realizing IEEE 1363.1 NTRU are polynomial multiplication (textbook method with $N^2$ clock cycles), index generation function (IGF), blinding polynomial generation method (BPGM), and mask generation function (MGF) and polynomial inversion.

Liu et al. [7] presents a linear-feedback shift register-based (LFSR) architecture for encryption with '$N$' MAC units (basically modular arithmetic units with a scheme to exploit the ternary nature of one of the polynomial) and $N$ registers for NTRUEncrypt with different set of parameters for $N$, $p$, and $q$ where $N$ is a prime number defining the degree of the truncated polynomial, $q$ and $p$ need to be relatively prime, and $q$ (powers of two) should be considerably larger than $p$ (typically 3). Further, [7] presents the simulation and synthesis results of the design for a Cyclone IV FPGA and argues that design could achieve better delay (latency)-area-product as compared to other prevalent designs of that era for NTRUEncrypt. Basu et al. in [8] have adopted an HLS-based design approach and synthesized 11 PQC algorithm's (encapsulation/signature and de-capsulation/verification) software codes using HLS tools for FPGA (Xilinx Virtex-7) and ASIC with different synthesis constraints/optimization such as baseline, loop unrolling, and loop pipelining and analyzed the resulted hardware on the aspects of security level (1–5) versus latency-area-product (LAP). On NTRU-HRSS-KEM (which has relevance to this proposed paper), the article concludes that, with loop unrolling and for the security level of 1, the scheme exhibits low latency and low LAP.

Kamal et al. in [9] have implemented NTRUEncrypt ($N = 251, p = 3, q = 128$) with several design options for a conventional multiplier, utilizing the statistical properties of the distance between the non-zero elements in the polynomials, Mersenne prime number algorithm and look-up-table for integer coefficients reduction ($\bmod p$) and achieved different area-speed trade-offs. The polynomial multiplier has been implemented with 4 or 8-bit shift registers (one clock cycle to shift) rather than with large bit size barrel shifters thereby reducing the hardware complexity. Synthesized the realizations on Virtex-E FPGA, analyzed and concluded that Mersenne and look-up-table gives better throughput. Agrawal et al. [10] have adopted number theoretic transform (NTT) for polynomial multiplication for realizing oblivious transfer (OT) and zero-knowledge proof (ZKP) PQC primitives. Farahmand et al. in [11] implemented the NTRUEncrypt Short Vector Encryption Scheme (SVES), an IEEE P1363.1 standard compliant PQC scheme, with shift registers-based parallel high-speed polynomial (lengths 1499 and 1087) multiplier, BPGM and MGF functions, etc., in Virtex UltraScale FPGA. The results have been compared with Classic McEliece in terms of latency, frequency, and area. Liu in [12] proposes LFSR and an extended LFSR-based polynomial multiplication schemes for NTRUEncrypt basically as [7]. Further, a systolic array-based architecture with $N^2$ processing elements has been architected for polynomial multiplication in NTRUEncrypt and used in two efficient systolic array-based architectures for NTRU-based fully homomorphic encryption.

Article [13] presents a generalized classical Karatsuba multiplier scheme for polynomial multiplications with an arbitrary degree and recursive use. The same scheme has been adopted in this proposed article for realizing polynomial multiplication for NTRU-HRSS. Farahmand et al. in [14, 15] have adopted a software/hardware co-design approach with a shift register-based polynomial multiplier in programmable hardware (either as RTL or by adopting HLS method) of Zynq and rest of the units for NTRU on the processing system (Microcontroller) of Zynq on Xilinx Zynq UltraScale + MPSoC platform. Article [14] focuses on NTRUEncrypt (encryption and decryption) and [15] illustrates key encapsulation mechanisms of NTRUEncrypt, Streamline NTRU, NTRULPrime, NTRUPrime, and NTRU-HRSS. Throughput and area have been compared between co-design approaches versus software. Articles [16, 17] illustrate polynomial multiplier architectures based on NTT, viz., conventional, systolic array-based NTT and compare with convolution-based multipliers, viz., conventional and systolic array based. The NTT multipliers have been applied for the other PQC schemes, viz., NewHope Simple Key Exchange, CRYSTALS Dilithium digital signatures. Articles [7, 9, 11, 12] narrate the architectures for realizing specifically the NTRUEncrypt cryptosystem. Few other articles have adopted the hardware/software co-design approach typically with the time-consuming polynomial multiplication in hardware and rest of the NTRU computations in a processing system as software/firmware. Co-design approach affects throughput and the overall performance of encryption and decryption compared to 'all in' hardware approach adopted in this article.

This paper discusses our work on an efficient implementation of the NTRU-HRSS encryption and decryption based on pipelined polynomial multiplier, polynomial lift, and random bit generator. The design focuses on ntruhrss701 variant. The implemented design can be enhanced further to support other variants of NTRU. The primitives used in the design can also be used to implement other PQC algorithms.

# 3 NTRU-HRSS—Brief Description

## *3.1 NTRU-HRSS Encryption [4]*

The pseudocode for NTRU-HRSS encryption is given as Algorithm 1. Encryption in NTRU is performed by first multiplying the public-key polynomial 'h' with a randomly generated polynomial 'r' in field Rq (Step 2). Then, the input message 'm', which is defined in S3, is lifted using $\phi_1$.S3 ($\frac{m}{\phi_1}$) to obtain a lifted message polynomial in Rq (Step 3). The ciphertext 'c' is obtained by modular addition of both in Rq (Line4).

**Algorithm 1** NTRU-HRSS Encryption

**Input:** public key polynomial 'h' in Rq, message polynomial 'm' in S3, N, q
**Output:** ciphertext polynomial 'c' in Rq
1: Generate random polynomial 'r' in S3
2: h_mul_r ← h * r    mod Rq
3: mlift ← Lift(m)
4: c ← h_mul_r + mlift    mod Rq
5: return c

## *3.2 NTRU-HRSS Decryption [4]*

The pseudocode for NTRU-HRSS decryption is given as Algorithm 2. The decrypted message is obtained by first polynomial multiplication of the ciphertext polynomial 'c' and the private-key polynomial 'f' in field Rq (Step 1). The output of this multiplication is reduced using the polynomial $\phi_N$, and the base of the coefficients is converted from modulo-q to modulo-3 to obtain a polynomial in field S3. This output is multiplied in S3 with the private-key inverse polynomial 'finv' to obtain the decrypted message 'm' (Step 2).

**Algorithm 2** NTRU-HRSS Decryption

**Input**: Private key polynomial f in S3, Private key inverse polynomial finv in S3, ciphertext polynomial 'c' in Rq, N, q
**Output**: message polynomial 'm' in S3
1: c_mul_f ← c * f    mod Rq
2: m ← c_mul_f * finv    mod S3
3: return m

# 4 Proposed NTRU-HRSS Accelerator Architecture

The NTRU-HRSS encryption/decryption accelerators are constructed with the basic primitive hardware units, viz., polynomial multiplier (either conventional or Karatsuba based), polynomial lift, an integer modulo converter from modulo-q representation to modulo-3, and a deterministic random binary generator (DRBG) to generate random bit stream for random polynomial ‘r’. This section elaborates the proposed architectures for various basic primitives and the overall NTRU-HRSS accelerator for encryption and decryption.

## *4.1 Modular Polynomial Multiplier*

The polynomial multiplication is the most time-consuming operation in encryption and decryption schemes of all NTRU variants. In this work, the polynomial multiplier has been implemented based on two architectural approaches, and the performances of both have been compared with other published works. The proposed multiplier architectures support modular polynomial multiplication in Rq, R3, Sq, and S3.

### 4.1.1 Conventional Polynomial Multiplier

For conventional approach, ‘$N$’ processing elements (PEs) are used to compute $N$ coefficients of the output polynomial. Coefficients of the input polynomials are fed serially into the multiplier (part of PEs). The processing element computes and accumulates the partial products. Each coefficient of the output polynomial is computed in ‘$N$’ clock cycles. The algorithm requires $N^2$ multiplications and $N\ (N-1)$ additions. One input of the multiplier can be a large integer polynomial coefficients in modulo-q or a ternary polynomial coefficient. The second input to the multiplier must be a ternary polynomial coefficient. The polynomial multiplications in NTRU do not require multiplier of two large integer coefficients of the polynomials, but can be implemented by add/sub and shift operation. The block diagram for a sequential pipelined conventional polynomial multiplier with ‘$N$’ stages linear array of ‘$N$’ processing elements (Modular arithmetic unit/multiply accumulate unit) is shown in Fig. 1.

### 4.1.2 Generalized Karatsuba Polynomial Multiplier

The second method used for multiplication of polynomials is the Karatsuba algorithm as proposed in [13]. This approach is used because the degree of polynomial in NTRU-HRSS is a prime number. In general, a combination of the Karatsuba and

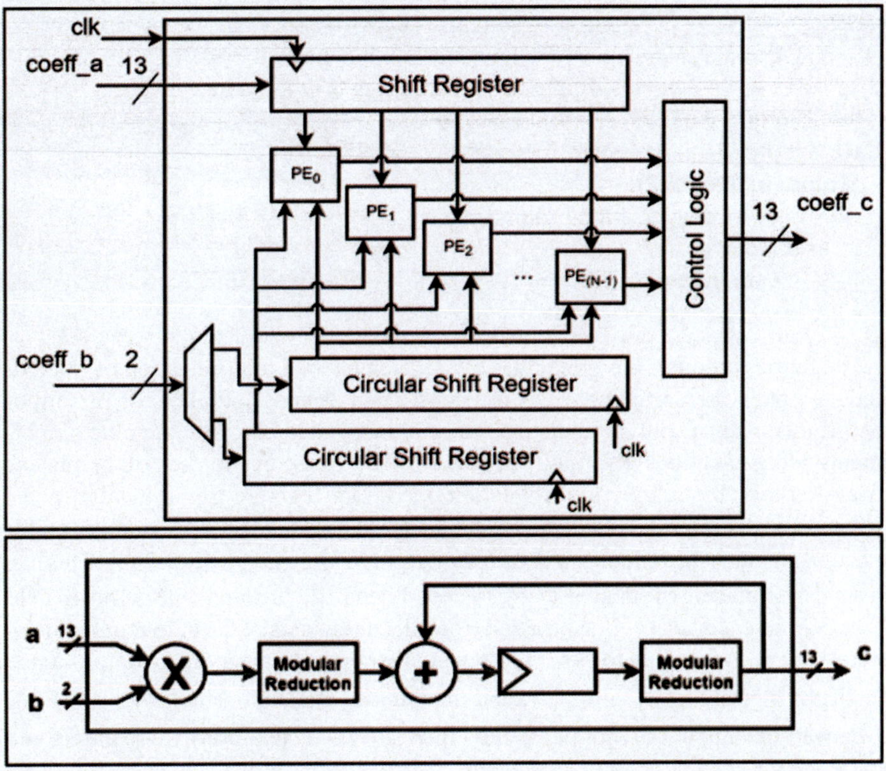


**Fig. 1** Sequential pipelined architecture for a Naïve modular polynomial multiplier and the processing element for modular arithmetic/multiply accumulate unit

the Toom–Cook algorithms must be applied along with '0' padding to the polynomials. Therefore, we implement the generalized Karatsuba algorithm for any arbitrary degree of polynomials proposed in [13] for multiplication and perform modular reduction to get the polynomial output in Rq, Sq, R3, or S3. The pseudocode for the polynomial multiplication in Rq using generalized Karatsuba algorithm is given in Algorithm 3.

**Algorithm 3** Modular polynomial multiplication using Generalized Karatsuba Algorithm

**Input**: Polynomial 'a' in Rq or R3 or Sq or S3, polynomial 'b' in S3 or R3, field (R or S)
**Output**: Polynomial 'a_mul_b' in Rq or R3 or Sq or S3
1: for i in 0 to (N-1)
3: for i in 1 to (2N-3)
4: $d_{s,t} \leftarrow (a_s + a_t)(b_s + b_t)$ where $s + t = i, 0 \leq s < t$
5: $c_0 \leftarrow d_0$

```
6: c_{2N-2} ← d_{N-1}
7: for i in 0 to (2N-3)
8:     c_i ← Σ d_{s,t} – (Σ d_s + Σ d_t) where s + t = i, 0 ≤ s < t
9:     if (i % 2 == 0)
10:        c_i ← c_i + d_{i/2}
11: for i in 0 to (N-2):
12:        c_i ← c_i + c_{i+N} mod (q or 3)
13: if (field = = S)
14:     for i in 0 to (N-1):
15:         c_i ← c_i – c_{N-1} mod (q or 3)
```

The partial products are generated by coefficient-wise multiplication of polynomials to obtain the partial products $d_i$ (steps 1 and 2) and multiplication of sum of coefficients of input polynomials $d_{s,t}$ (steps 3 and 4) and compute the output coefficients using equations given in [13] (steps 5–10). Since one of the polynomials is always ternary, the integer multiplication operations here are trivial involving just shift and add operations. However, this method performs the linear polynomial multiplication and must be reduced to obtain the output modulo $(x^N - 1)$ or $\frac{x^N-1}{x-1}$. This can be done by replacing the degree $x^N$ by 1 (steps 11 and 12). To obtain the output in field S, the coefficient of $x^{(N-1)}$ is subtracted from all coefficients of the output polynomial (steps 13–15). According to Ref. [13], the number of multiplication operations in the Karatsuba algorithm is $\left(\frac{N^2}{2} + \frac{N}{2}\right)$, and the number of additions is $\left(\frac{5N^2}{2} - \frac{7N}{2} + 1\right)$. To convert the output polynomial in field R or S, $(N - 1)$ additions are required, and thus, the total number of additions is $\left(\frac{5N^2}{2} - \frac{5N}{2}\right)$. The block diagram of the proposed architecture for Karatsuba multiplier is given in Fig. 2.

The cost ratio gives a ratio between the cost of a multiplication and an addition for which the Karatsuba algorithm is efficient. Since Karatsuba algorithm focuses on reducing the number of multiplications at the cost of more additions, the cost ratio would give a basis to decide whether the Karatsuba algorithm would be more efficient than the conventional method for polynomial multiplication [13].

The polynomial multiplication operations in NTRU are trivial since one of the polynomials is ternary. Since the generalized Karatsuba algorithm involves addition of coefficients of the polynomial before multiplication, it is more complex than the conventional approach for polynomial multiplication. However, we would ignore the difference for calculation of cost ratio. Let 't_m' and 't_a' be the cost of one addition and one multiplication, respectively.

For the Karatsuba algorithm, the total cost of computation is given by:

$$c_k = \left(\frac{1}{2} * N^2 + \frac{1}{2} * N\right) * t\_m + \left(\frac{5}{2} * N^2 - \frac{5}{2} * N\right) * t\_a \tag{1}$$

And for Naïve method, the cost of computation is given by:

$$c_s = N^2 * t\_m + N(N-1) * t\_a \tag{2}$$

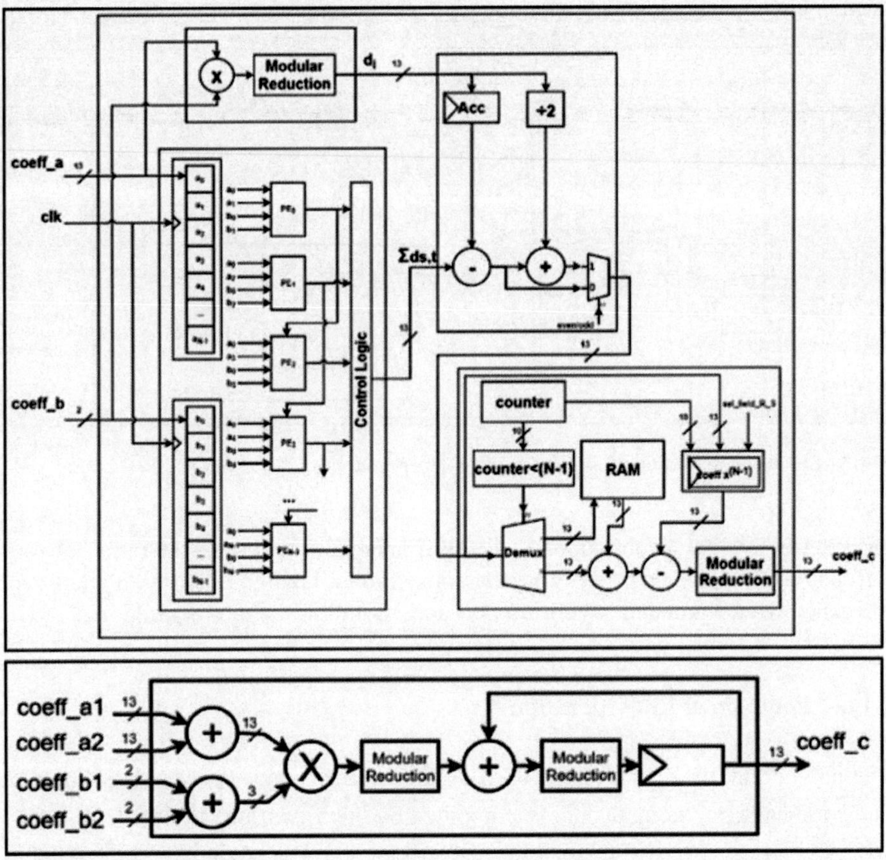


**Fig. 2** Generalized Karatsuba-based modular polynomial multiplier architecture and its processing elements (PEs) primitive

For efficient use of Karatsuba algorithm, $c_k < c_s$

$$\left(\frac{N^2}{2}+\frac{N}{2}\right)*t\_m+\left(\frac{5N^2}{2}-\frac{5N}{2}\right)*ta < N^2*t\_m+N(N-1)*t\_a \tag{3}$$

i.e., $\frac{t\_m}{t\_a} > \frac{(\frac{3}{2}*N^2-\frac{3}{2}*N)}{(\frac{1}{2}*N^2-\frac{1}{2}*N)}$.

By substituting value of $N$ as 701, we get,

$$\frac{t\_m}{t\_a} > \frac{(\frac{3}{2}*701^2-\frac{3}{2}*701)}{(\frac{1}{2}*701^2+\frac{1}{2}*701)} \rightarrow \frac{t\_m}{t\_a} > \frac{736{,}050}{245{,}350} \rightarrow \frac{t\_m}{t\_a} > 3 \tag{4}$$

Equation 4 implies that Karatsuba can be efficient if the cost of one multiplication should exceed the cost of 3 additions. However, the multiplication operation in the

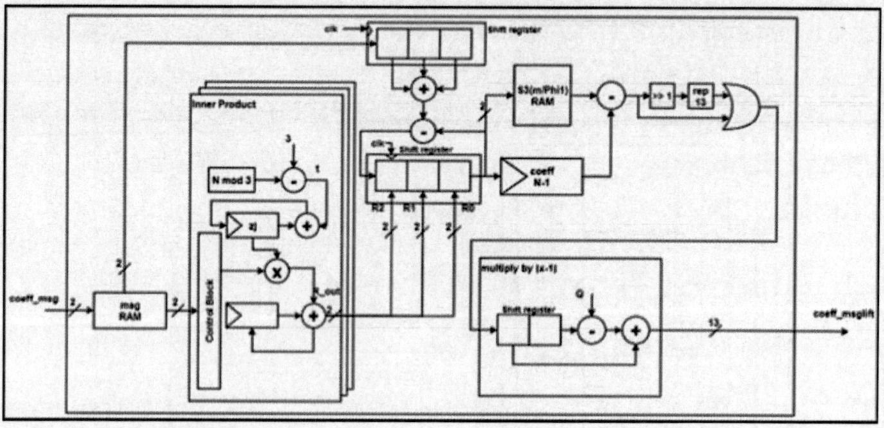


**Fig. 3** Hardware architecture for polynomial lift operation

worst case involves a subtraction and a shift operation, and thus, the ratio $\frac{t\_m}{t\_a} > 3$ will not hold true. This proves that the conventional method is more efficient than the generalized Karatsuba algorithm polynomial multiplication required in NTRU.

#### 4.1.3 Polynomial Lift Operation

The polynomial lift operation in NTRU-HRSS is defined as $\phi_1 * S3\left(\frac{m}{\phi_1}\right)$ where m is the message polynomial. This is computed by the algorithm defined in [18]. The inner product of the message polynomial is computed with the polynomials $z$, $z * x$, and $z * x^2$ where the polynomial $z$ is defined in [18]. The values of the inner products are the first 3 coefficients of the polynomial $\left(\frac{m}{\phi_1}\right)$, and the remaining coefficients of the polynomial are computed using Eq. 5,

$$c_i = c_{i-3} - (m_i + m_{i-1} + m_{i-2}) \tag{5}$$

The output polynomial is multiplied by polynomial $(x - 1)$ to obtain the lifted message output. The block diagram for the polynomial lift operation is given in Fig. 3.

#### 4.1.4 Modulo-*q* to Modulo-3 Converter

This module shifts the base of an integer from modulo-$q$ to modulo-3. An integer with base q cannot be directly converted into an integer with base 3 by performing a modulo-3 of the integer. So, the integer with base $q$ has to be normalized first before computing its modulo-3 value. The integer is first center lifted such that the value of the integer lies between $[3q - \frac{q}{2}, 3q + \frac{q}{2}]$. The modulo-3 of the center lifted value is computed to get the modulo-3 output of the converter.

### 4.1.5 Deterministic Random Binary Generator [19,20]

The deterministic random bit generator (DRBG) is used to generate the random polynomial used in NTRU encryption. The DRBG is implemented as per [20] which is based on NIST SP 800-90A [19], and the DRBG mechanism used is CTR-DRBG. This random number generation method fulfills the recommended scheme for random polynomial generation for round 3 NIST specification of NTRU.

### 4.1.6 NTRU Accelerator for Encryption and Decryption

The NTRU accelerator consists of the above basic primitives that are used to encrypt a message and decrypt a ciphertext in NTRU as shown in Fig. 4. The essential primitives used to construct the NTRU encryption and decryption algorithms are the polynomial multiplication, polynomial lift, modulo-$q$ to modulo-3 converter, a random number generator (DRBG), and storage elements (memory) that are used to store the intermediate outputs during computation. The control block consists of a finite state machine, which controls the flow of data in the accelerator.

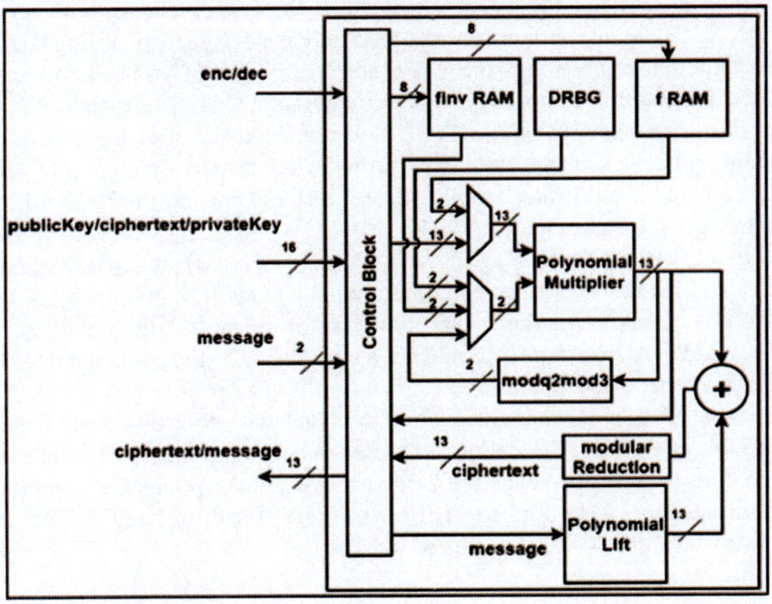


**Fig. 4** NTRU-HRSS accelerator architecture

To perform an encryption operation, the public key and the message polynomials are provided as inputs to the accelerator. The accelerator obtains the random polynomial coefficients from the DRBG and performs the polynomial multiplication of public key (h) and the random polynomial (r). The message (m) is lifted using message lift (polynomial lift) unit in the accelerator. The output of the polynomial multiplier and the message lift is added to obtain the ciphertext.

For a decryption operation, the private key is initialized into the private-key RAMs. This can be done by asserting the private-key configuration control signal and writing the private key and the private-key inverse polynomials via the 16-bit input data port. In each clock cycle, 4 coefficients of private key and 4 coefficients of the private-key inverse polynomials are provided as input via input data port. The private key must be initialized once before performing decryption operations.

To perform a decryption, the ciphertext is written to the accelerator via the input data port along with requisite control signals. The ciphertext is multiplied with the private-key polynomial using the polynomial multiplier. The output of this operation is a polynomial in Sq and the base of the coefficients the result is changed from modulo-$q$ to modulo-3. The resultant output is multiplied with the private-key inverse polynomial to obtain the decrypted message.

## 5 Verification of NTRU-HRSS Accelerator

The primitives required for the construction of the NTRU encryption/decryption accelerators, and the accelerator control block has been designed in Verilog HDL. The test bench for the verification of the NTRU accelerator (DUT) has been implemented in Python programming language using cocotb framework (A Python-based co-simulation environment) [21]. The NTRU algorithm has been implemented in Python programming language and is used to perform the key generation operation which is not implemented in hardware for this article. The block diagram for the verification environment is shown in Fig. 5.

The private key and the message polynomial are generated using a random number generator and fed to the DUT using configuration signals. The public key is derived from the private key using the Python function for generating the public key. The public key and the message polynomial are fed to the DUT along with control signals to indicate an encryption operation.

The result of encryption is again fed to the DUT with control signals indicating a decryption operation. The decrypted message is then compared to the original message. The scoreboard keeps track of successful encryptions and decryptions. This environment is also used to verify the design based on NIST NTRU-HRSS Known Answer Test (KAT) as specified in [4].

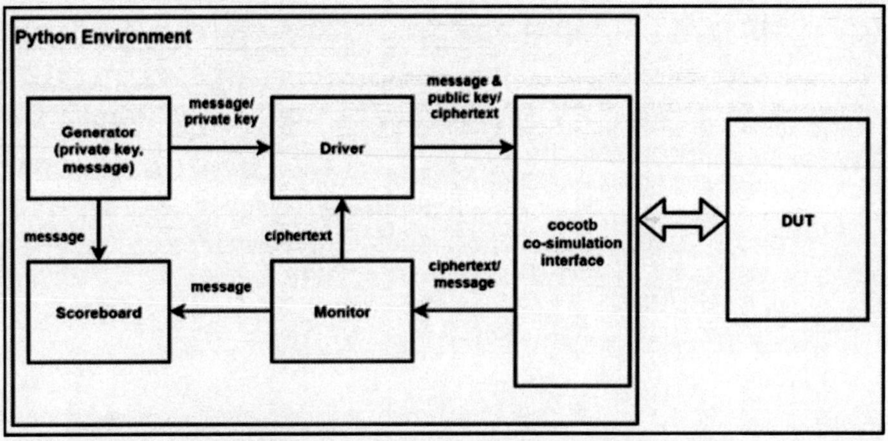


**Fig. 5** NTRU-HRSS verification environment based on cocotb [21]

# 6 Hardware Implementation and Resources Utilization of NTRU Primitives and NTRU-HRSS Accelerator

The NTRU primitives and the NTRU accelerator are implemented on Xilinx Ultra-Scale FPGA (xcvu095-ffva2104-2-e). The designs were synthesized, and the results are discussed in this section.

## *6.1 Polynomial Multipliers for NTRU-HRSS*

The polynomial multipliers were designed for the conventional method and the generalized Karatsuba algorithm based on [13]. The synthesis results are compared with existing implementations of only polynomial multipliers or NTRU hardware based on different kinds of polynomial multipliers. This is shown in Table 1.

From Table 1, it has been observed that the convolution-based polynomial multiplier implemented in [17] operates at very low frequency and requires a huge number of clock cycles. The bit shifter-based approach (4 or 8 bits) used in [9] is done for a small value of '$N$' as compared to NTRU-HRSS and operates at a very low frequency. The shift register-based approach in [11] is the most efficient architecture for polynomial multiplier for NTRUEncrypt cryptosystem. However, the concept of the multiplier in NTRUEncrypt is based on the property of ternary polynomials that contain a very high number of 0 s. This multiplier cannot be used for NTRU-HRSS. Also, the area utilized by the multiplier is very high compared to all other architectures since it is implemented as a parallel architecture.

The NTT-based architectures are defined for polynomials with degree of coefficients ($2^N - 1$). To apply this method for NTRU, zero padding must be done to

**Table 1** Comparison of polynomial multipliers with the proposed designs (conventional method and generalized Karatsuba)

| Design techniques | Target FPGA | $N$ | Cycles | Latency (μs) | BRAM | DSP | FF | LUT | Freq. (MHz) |
|---|---|---|---|---|---|---|---|---|---|
| CONV[17] | Zynq UltraScale + + | 1024 | 3720 | 17.4 | 2 | 4282 | 310,247 | 322,866 | 195.2 |
| HLS-NTRU encapsulation with loop unrolling [8] | Artix-7 | 701 | 22,594 | NA | 0 | 0 | 9035 | 65,356 | 66.7 |
| HLS-NTRU encapsulation with loop pipelining [8] | Artix-7 | 701 | 100,208 | NA | 0 | 0 | 12,225 | 75,141 | 66.7 |
| Poly Mul using bit shifter method with 8 shifts [9] | Virtex-7 | 251 | NA | NA | 251 | 0 | 5160 | 27,292 | 62.3 |
| Shift register-based poly Mul [11] | Virtex UltraScale | 1499 | 162 | NA | 0 | 0 | NA | 167,469 | 297 |
| Sequential NTT [17] | Zynq UltraScale + + | 1024 | 52,152 | 169.28 | 2 | 10 | 17,091 | 10,888 | 308.7 |
| Systolic array NTT [17] | Zynq UltraScale + + | 1024 | 18,636 | 101.84 | 29 | 58 | 3007 | 3140 | 183.9 |
| NTT with pipelined FFT [16] | Artix-7 | 1024 | 2671 | 17.89 | 57 | 55 | 8566 | 7760 | 149.3 |
| PISO-based multiplier for NTRUEncrypt [14] | Zynq UltraScale + MPSoC | 743 | NA | 29.6 | 1 (36 kb) | 0 | 49,674 | 76,972 | 300 |
| PISO-based poly Mul NTRU-HRSS encapsulation [15] | Zynq UltraScale + MPSoC | 701 | 6304 | 31.521 | 6 | 0 | 32,327 | 33,230 | 300 |
| Conventional-based multiplication [proposed work] | Virtex UltraScale | 701 | 1403 | 3.77 | 0 | 0 | 19,666 | 59,706 | 371.7 |
| Generalized Karatsuba-based multiplier [proposed work] | Virtex UltraScale | 701 | 1403 | 6.06 | 0 | 0 | 29,804 | 99,763 | 231.3 |

**Table 2** Hardware utilization of polynomial lift in Xilinx UltraScale for NTRU-HRSS

| Operation | $N$ | Cycles | Latency (us) | FF | LUT | Freq. (MHz) |
|---|---|---|---|---|---|---|
| Lift [this work] | 701 | 2115 | 4.29 | 3113 | 2429 | 492.4 |

the polynomials to make it compatible for NTT-based multiplication. The sequential NTT approach in [17] has the maximum frequency among all NTT implementations. However, the latency of the operation is very high as compared to others.

The proposed architecture for polynomial multiplication with conventional method has the maximum frequency among all implementations. The area utilized is also comparable to other architectures for value of '$N$' more than 700. The latency of the operation for conventional method is the least among all implementations. The generalized Karatsuba method of polynomial multiplication has a lower frequency as compared to the conventional method but outperforms most other implementations. The latency of the operation is also better compared to other implementations with the same value of '$N$'. In our NTRU encryption/decryption accelerator design, the conventional method of polynomial multiplication is used due to its low latency with lower clock cycle and high frequency.

## 6.2 *Polynomial Lift for NTRU-HRSS*

The polynomial lift operation for NTRU-HRSS is defined as $\phi_1 * S3\left(\frac{m}{\phi_1}\right)$ as discussed in Sect. 4. The polynomial lift operation is developed in Verilog HDL and implemented in Xilinx UltraScale FPGA. The hardware utilization detail for the polynomial lift operation is given in Table 2.

## 6.3 *NTRU-HRSS Accelerator–Encryption and Decryption*

The hardware implementation of NTRU-HRSS accelerator is compared with the existing literatures implementing NTRU encryption/decryption/encapsulation algorithms, and the details are shown in Table 3.

It is observed that the latency for encryption and decryption is higher for the proposed implementation than the LFSR-based implementations in [7, 12]. However, the LFSR and extended LFSR implementations are based on the NTRUEncrypt which has polynomials of degree less than that of NTRU-HRSS and exploit the property of fixed weight ternary polynomials. The NTRU-HRSS defined in Round 3 of NIST has arbitrary weight ternary polynomials which do not define fixed number of 1 s and −1 s for the coefficients of a ternary polynomial. The NTT-based PKC implementation in [10] provides only the number of clock cycles and the area utilization. It has been observed that the proposed implementation outperforms the given implementation

**Table 3** Comparison of various other NTRUs with the proposed architecture for NTRU-HRSS

| Design techniques | Target FPGA | *N* | Cycles | Latency (us) | BRAM | DSP | FF | LUT | Freq. (MHz) |
|---|---|---|---|---|---|---|---|---|---|
| LFSR-based encryptor [7] | Cyclone IV | 503 | 504 | 5.04 | 0 | 0 | NA | 16,603 LE | 99.8 |
| PKC with NTT-based Mul [10] | Zynq-7000 | 1024 | 48,128 (Enc)<br>24,576 (Dec) | NA | 3.5 | 26 | 1255 | 15,717 | NA |
| LFSR for NTRUEncrypt [12] | Cyclone IV | 401 | 402 (Enc) | 4.06 | 0 | 0 | 0 | 18,049 | 98.9 |
| Extended LFSR NTRUEncrypt [12] | Cyclone IV | 401 | 350 (Enc) | 3.58 | 0 | 0 | 8826 | 22,460 | 97.4 |
| NTRU-HRSS accelerator [proposed work] | Xilinx UltraScale | 701 | 178 (config)<br>2118 (Enc)<br>2816 (Dec) | 1.15 (config)<br>13.68 (Enc)<br>18.19 (Dec) | 2<br>2 (DRAM) | 0 | 23,686 | 62,762 | 154.7 |

in [12] in terms of area and number of clock cycles. Also, the proposed accelerator has a higher frequency among all implementations discussed in literature.

## 7 Conclusions and Future Works

In this paper, an efficient implementation of the NTRU-HRSS variant of the Round 3 submission of NTRU algorithm has been presented. The basic hardware primitives required to construct NTRU-HRSS encryptor/decryptor have been designed and implemented. These primitives are used to design an NTRU accelerator that can perform both NTRU-HRSS encryption and decryption. The NTRU accelerator with conventional polynomial multiplier is found to be efficient than other existing architectures, which are either based on shift registers or number theoretic transform, and moreover, these realizations are for lower polynomial degree '$N$', and the NTT has the disadvantage of '0' paddings to form the final '$N$' as the power of two. Further, it has been observed that the conventional approach is better than the generalized Karatsuba-based polynomial multiplier due to the large value of '$N$' ($N = 701$) for NTRU-HRSS and such large value of '$N$' introduces more additions. Karatsuba algorithm reduces the number of multiplications and increases additions.

These primitives proposed in this article can further be used to design the accelerators for NTRU-HPS variants. The key generation function of the NTRU can also be implemented in hardware as future work by implementing hardware for polynomial inversion and using the polynomial multiplier primitive proposed in this article to compute the public key. The DRBG implemented in the accelerator can also be used to generate the private key for NTRU. Further, other PQC algorithms can be implemented using this approach by architecting a more generic polynomial multiplier. Currently, the polynomial multiplier is targeted to a specific parameter set.

## References

1. Mavroeidis V et al (2018) The impact of quantum computing on present cryptography. Int J Adv Comput Sci Appl
2. Chi DP, Choi JW, Kim JS, Kim T (2015) Lattice-based cryptography for beginners. IACR Cryptol. ePrint Arch
3. NIST Round 3 Submissions PQC Standardization. https://csrc.nist.gov/projects/post-quantum-cryptography/round-3-submissions. Accessed 18 Mar 2022
4. Chen C, Danba O et al. NTRU algorithm specifications and supporting documentation. https://ntru.org/f/ntru-20190330.pdf. Accessed 18 Mar 2022
5. What is lattice-based cryptography why should you care. https://medium.com/cryptoblog/what-is-lattice-based-cryptography-why-should-you-care-dbf9957ab717. Accessed 18 Mar 2022
6. Wilhelm K (2008) Aspects of hardware methodologies for the NTRU public-key cryptosystem. Thesis
7. Liu B et al (2015) Efficient architecture and implementation for NTRUEncrypt system. In: 2015 IEEE 58th international Midwest symposium on circuits and systems (MWSCAS)

8. Basu K et al (2019) NIST post-quantum cryptography—a hardware evaluation study. IACR Cryptol. ePrint Arch
9. Kamal AA, Youssef AM (2009) An FPGA implementation of the NTRUEncrypt cryptosystem. In: 2009 international conference on microelectronics—ICM
10. Agrawal R, Bu L, Ehret A, Kinsy M (2019) Open-source FPGA implementation of post-quantum cryptographic hardware primitives. In: 29th international conference on field programmable logic and applications (FPL)
11. Farahmand F, Sharif MU, Briggs K, Gaj K (2018) A high-speed constant-time hardware implementation of NTRUEncrypt SVES. In: 2018 international conference on field-programmable technology (FPT)
12. Liu B (2015) Efficient architecture and implementation for NTRU based systems
13. Weimerskirch et al (2006) Generalizations of the Karatsuba algorithm for efficient implementations. IACR Cryptol. ePrint Arch
14. Farahmand F, Nguyen DT et al (2019) Software/hardware codesign of the post-quantum cryptography algorithm NTRUEncrypt using high-level synthesis and register-transfer level design methodologies. In: 2019 29th international conference on field programmable logic and applications (FPL)
15. Farahmand F et al (2019) Evaluating the potential for hardware acceleration of four NTRU-based key encapsulation mechanisms using software/hardware codesign
16. Thanawala N. Hardware acceleration of polynomial multiplication using pipelined FFT. UC Irvine. ProQuest ID: Thanawala_uci_0030M_17005
17. Nejatollahi H et al (2020) Exploring energy efficient quantum-resistant signal processing using array processors. In: ICASSP 2020—2020 IEEE international conference on acoustics, speech and signal processing (ICASSP)
18. Hülsing A et al (2017) High-speed key encapsulation from NTRU. International Association for Cryptologic Research, CHES
19. Random number generation using deterministic random bit generators. https://csrc.nist.gov/publications/detail/sp/800-90a/rev-1/final. Accessed 18 Mar 2022
20. Selvakumar D, Mervin J et al (2021) Formal verification and analysis of a pseudo random number generator. In: 25th international symposium on VLSI design and test
21. Cocotb co-simulation library. https://github.com/cocotb/cocotb. Accessed 18 Mar 2022

# Distributed Agent-Based Voltage Control Approach for Active Distribution Systems

**Ahmed Bedawy, Naoto Yorino, Yutaka Sasaki, Yoshifumi Zoka, Kihembo Samuel Mumbere, and Ryuta Kubo**

**Abstract** Nowadays, distribution systems are changing rapidly. Energy demand growing, reliability issues, and environmental rules which require a reduction in $CO_2$ emissions lead to increased penetration of distributed generation sources (DGs), particularly variable renewable energy sources (VRESs). Thus, the conventional voltage control strategies face several difficulties to solve the voltage violations problems. Therefore, proper voltage control techniques need to be developed to allow power delivery to consumers with high quality and reliability as well as increased DGs hosting capacity. This paper summarizes our recent voltage control work to minimize the voltage violations of the distribution systems with high DGs penetration. Firstly, a novel distributed control technique for multiple voltage regulators using a multi-agent control structure is presented where different types and configurations of voltage regulators are modeled. Then, a hierarchical strategy for voltage control is discussed, in which different voltage control techniques are cooperated to minimize the voltage violations of the whole system. In addition, the validation of the developed strategies is carried out using long-term analysis. The multi-agent control structure's autonomy and flexibility are utilized to develop the proposed voltage control technique. Comprehensive simulation studies are performed using the IEEE 123-bus test feeder. The simulation results demonstrate the effectiveness of the proposed technique on the mitigation of voltage violations for different DGs generation profiles, even in the case of long-term simulation.

**Keywords** Distributed generation · Multi-agent system · Variable renewable energy · Voltage control · Voltage violations

A. Bedawy (✉) · N. Yorino · Y. Sasaki · Y. Zoka · K. S. Mumbere · R. Kubo
Graduate School of Advanced Science and Engineering, Hiroshima University, Hiroshima 739-8527, Japan
e-mail: a.bedawy@ieee.org

A. Bedawy
Faculty of Engineering, South Valley University, Qena 83523, Egypt



C. Giri et al. (eds.), *Emerging Electronic Devices, Circuits and Systems*, Lecture Notes in Electrical Engineering 1004,
https://doi.org/10.1007/978-981-99-0055-8_29

## 1 Introduction

Recently, with increasing the importance of reducing carbon dioxide ($CO_2$) emissions, the penetration of renewable-based distributed generation resources (DGs) has been grown quickly in low-voltage distribution systems (LVDSs) [1]. As a result, voltage regulation problems have become more critical for the LVDSs because the renewable-based DGs with their generation intermittency lead to new operational challenges, such as large voltage fluctuations, excessive SVR tap changes, and voltage rise problems [2, 3]. Therefore, voltage control techniques can play a significant role in developing, operating, and maintaining the power quality of future LVDSs. Traditionally, regulating the voltage in the LVDSs is performed by the classic regulation devices such as on-load tap changers (OLTCs), shunt capacitors (SCs), and step voltage regulators (SVRs). Nevertheless, the characteristics of renewable-based DGs interrupt the operation of these voltage control devices [4, 5].

Implementing centralized voltage control schemes in the active LVDSs can perform the best performance for voltage regulation [6–8]. However, the centralized structure requires a complicated communication system, which outweighs the potential performance benefits. Furthermore, the reliability and speed of the communication system used in the centralized control degrade the control system's efficiency and performance. As a result, the high performance of the centralized control schemes can be achieved with high investment [9–11]. On the other hand, distributed control schemes are more reliable as they do not entirely depend upon communication for their operation. Therefore, they can be used to reduce the communication requirements of the voltage control systems [12]. The implementation of multi-agent as a flexible and robust control approach is described in [13, 14]. In addition, the authors have several works for voltage control based on multi-agent control structures, i.e., optimal voltage regulator management [15–17], reactive power control of the DGs [18], and cooperative voltage control approach [19, 20].

Several studies have been developed to mitigate the voltage violation problems in LVDSs with high DG installations. A distributed control scheme is proposed to mitigate the voltage rise by controlling the DG power injection [21]. In [22], a coordinated control scheme is formulated to manage the charge/discharge of battery energy storage systems (BESSs) by combining the local and the distributed control to regulate the system voltages. A coordinated voltage control strategy between the OLTC and the BESSs at medium and LVDSs is proposed in [23].

Considering the previous literature and in line with the present attempts to guarantee reliable voltage regulation in active LVDSs, this paper summarizes our proposed distributed voltage control techniques for the active unbalanced LVDSs based on the multi-agents control structure. The objective of the proposed techniques is to minimize the overall voltage violations of the LVDSs while decreasing the number of control operations of the voltage regulation equipment. The following are the main parts of the proposed voltage control techniques that will be discussed in this article:

1. An optimal decentralized control technique for managing the step voltage regulators is developed based on a multi-agent control structure.
2. A hierarchical control system is structured to coordinate the regulator control strategy in (1) and the local voltage control techniques, i.e., the reactive power control of DG inverters.

The positive feature of the proposed technique is ensuring autonomous control actions for minimizing the voltage deviations with a simple control system without using central management. The validation of the proposed strategy is conducted using different simulations, including several loads and DGs generation profiles.

## 2 Proposed Voltage Control Approach

The proposed voltage control approach considers LVDSs with radial configurations, which is frequent in most systems of many countries worldwide. The control structure is developed using a multi-agent system configuration, as presented in Fig. 1.

The LVDSs are characterized by an unbalanced configuration of loads, lines, and DG sources, while various structures of voltage regulators are installed. Therefore, the suggested voltage control approach can be represented as a multi-agent system structure consisting of different agents, as illustrated in Fig. 1. The main component of the proposed control system can be described below.

- **Blackboard**: Global memory for information exchange among different agents.
- **Local Agents**: Voltage regulators, DGs, and any voltage control equipment, i.e., SC, DSTATCOM, etc. The local agent can act according to the blackboard data.

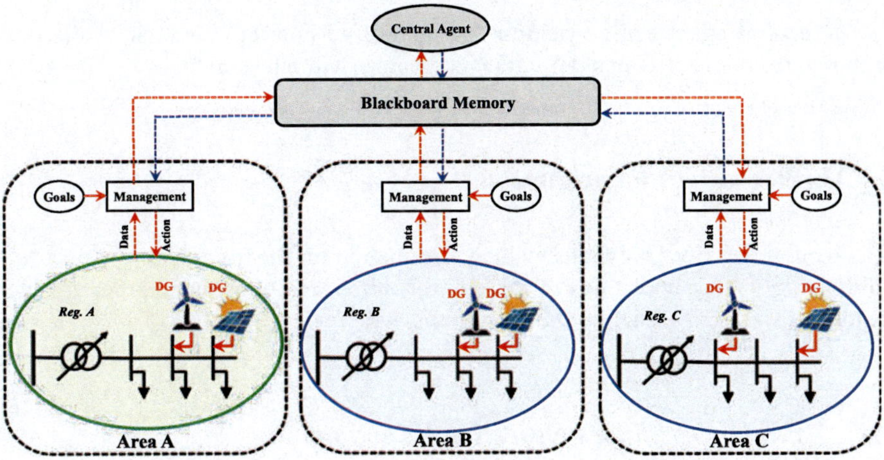


**Fig. 1** Proposed voltage control approach using multi-agent system structure

- **Central Agent**: A management agent which can be used in the case of centralized control for monitoring and real-time analysis.
- **Control Area**: Consists of multiple local agents. In this study, each area has one voltage regulator as the main controller, which acts based on the exchanged information from other agents and the blackboard.

In this study, we assume only two types of voltage control equipment in each control area: the voltage regulator and the reactive power of the DGs. The proposed hierarchical control structure for each control area consists of two stages as follows:

***First stage**: Local voltage control using DG agents*

- The DGs control their node voltages based on a local control algorithm using the reactive power capability of the DG inverter, which will be described in detail in the next section.
- After regulating the local node voltages, the DG agent shares its local measurements with the voltage regulator control system.

***Second stage**: Area voltage control using voltage regulators*

- The voltage regulator works independently to minimize voltage violations according to the observation nodes' data (including the DGs node information) and the blackboard data.
- Based on the obtained information, each regulator calculates a unique control parameter, hereinafter referred to as voltage/tap sensitivity index (*VTSI*).
- The *VTSI*s and all areas information are exchanged through the blackboard.
- Each area controller can act based on the status of the other areas, which are shared through the blackboard.
- For emergencies and lack of data, the control area can act autonomously to realize its objective using the available information.

The central agent can be helpful for centralized control operation, while our proposed technique acts in a decentralized manner without using it.

# 3 Mathematical Formulation

This section describes the mathematical formulation for the main control objective: minimizing voltage violations of the entire control area by managing the voltage regulators and controlling the DG reactive power.

## *3.1 Voltage Regulators Objective Function*

The voltage regulator objective is minimizing the voltage deviations' function $D$ for all areas included in the multi-agents' voltage control system.

$$\min D(\boldsymbol{V}(t)) = \sum_{A=1}^{\mathrm{NA}} W_A (V_{\mathrm{CA}}(t) - V_{\mathrm{RA}}(t))^2 \tag{1}$$

where

*D*: The sum of the voltage deviations of all controlled areas, calculated from the reference voltage values.

$D(V(t)) = [DV_1(t), DV_2(t), \ldots, DV_{NA}(t)]^T$: The voltage deviations vector for all controlled areas at time *t*.

NA: Number of controlled areas.

$V_{\mathrm{CA}}$: Center voltage of area *A*, calculated using the minimum and the maximum voltages of the observation points (*M*: no of observation nodes) in each area, as illustrated in (2) for area *A*.

$$V_{\mathrm{CA}} = \frac{V_{A,\min} + V_{A,\max}}{2} \tag{2}$$

$V_{\mathrm{RA}}$: Target center voltages of area *A*, representing the average of upper and lower voltage for each controlled area.

$W_A$: The weight coefficient of area *A*, which can be used to adjust the significance of observation points in each area.

The voltage *V* is the function of the regulator tap position Tp, the load power $P_L$, and the DG power $P_{\mathrm{DG}}$, as shown in (2).

$$\boldsymbol{V} = f(\boldsymbol{P_L}, \boldsymbol{P}_{\mathrm{DG}}, \boldsymbol{Tp}) \tag{3}$$

When regulator *A* tap position changes by ΔTp as in (4), the voltage deviations will change by $\Delta D(V(t))$ (5).

$$Tp_A(t+1) = Tp_A(t) + \Delta\, Tp_A(t),\ Tp_{A,\min} \le Tp_A \le Tp_{A,\max}$$
$$\Delta\, Tp_A(t) = T_{StepSize,A} \cdot T_{Status,A}(t),\ T_{\mathrm{Status},A}(t) = \begin{Bmatrix} +1 & \text{(increase)} \\ 0 & \text{(no change)} \\ -1 & \text{(decrease)} \end{Bmatrix} \tag{4}$$

$$\begin{aligned} \Delta D(\boldsymbol{V}(t)) &= D(\boldsymbol{V}(t+1)) - D(\boldsymbol{V}(t)) \\ &= \left[\frac{\partial \boldsymbol{D}}{\partial \boldsymbol{V}}\right] \cdot \left[\frac{d\boldsymbol{V}}{d\boldsymbol{Tp}}\right] \cdot \Delta\, \boldsymbol{Tp}(t) \\ &= \underbrace{\left[\frac{\partial \boldsymbol{D}}{\partial \boldsymbol{V}}\right] \cdot \left[\frac{d\boldsymbol{V}}{d\boldsymbol{Tp}}\right] \cdot \boldsymbol{T}_{\text{Step Size},}}_{VTSI} \cdot \boldsymbol{T}_{\text{Status}}(t) \\ &= \boldsymbol{VTSI}(t) \cdot \boldsymbol{T}_{\text{Status}}(t) = \sum_{A=1}^{NA} VTSI_A(t) \cdot T_{\text{Status},A}(t) \end{aligned} \tag{5}$$

$$\min D(V(t+1)) = D(V(t+1)) + \min\{\, \boldsymbol{VTSI}(t)\,.\,\boldsymbol{T}_{\text{Status}}(t)\} \tag{6}$$

The voltage/tap sensitivity index *VTSI* represents the sensitivity of area voltage deviations for the unit step change in the voltage regulator tap position. As noticed from (6), the *VTSI* is a vital control index, which can be used to realize an effective control performance to minimize the objective. Furthermore, the VTSI can be utilized to find the most effective regulator among the controlled areas to achieve the objective at time $t$. Based on the previous description, two control roles are formulated to achieve our main objective (1).

**Optimal Control Performance**

The optimal control condition of the proposed technique has been proved in [15]. The optimal performance of the proposed regulators' technique is achieved when one regulator act at the time instant $t$. The procedure for optimal control can briefly be described as follows:

1. Each control area calculates its *VTSI* value and shares it among the areas through the blackboard.
2. A comparison process is performed among the *VTSI*s (7).

$$VTSI_A(t) = \max\left|[\, VTSI_1(t), \ldots,\; VTSI_{NA}(t)]\right| > \alpha \tag{7}$$

3. The area with the highest *VTSI* should adjust its regulator tap position based on

$$\begin{array}{ll} \text{if } VTSI_A(t) < -\alpha & \text{Step up tap} \\ \text{if } VTSI_A(t) > \alpha & \text{Step down tap} \end{array} \tag{8}$$

where $\alpha$ is a predefined threshold value used to adjust the regulator's response according to the voltage deviation.

The previous formulation guarantees minimizing the overall system voltage deviations with an optimal tap setting for all the controlled areas.

**Suboptimal Control**

Suboptimal control is proposed to avoid the problems of comparison process used in the optimal control strategy. In the proposed suboptimal role, the regulator of each area can act independently based on the *VTSI* value only, as described in (9).

$$\begin{array}{ll} \text{if } VTSI_A(t) < -\alpha_0 & \text{Step up tap} \\ \text{if } VTSI_A(t) > \alpha_0 & \text{Step down tap} \\ \text{if } |\, VTSI_A(t)| < \alpha_0 & \text{No tap change} \end{array} \tag{9}$$

where $\alpha_0$ is a predefined common threshold value among all the regulators for suboptimal control operation. The suboptimal control role expects only one regulator to react simultaneously, which almost achieves the optimal control performance, as described in [15].

### 3.2 Reactive Power Control of the DG Inverter

The renewable-based DGs are usually connected to the LVDSs through electronic power inverters. In this work, we utilize the reactive power capability of inverter-based DGs to regulate local node voltage using sensitivity analysis. Furthermore, to avoid curtailing the DG active power $P_{DG}$, the reactive power of the DG inverter $Q_{DG}$ is constrained by the inverter's capacity $S_{DG}$, as illustrated in (10).

$$Q_{DG} = \sqrt{S_{DG}^2 - P_{DG}^2} \tag{10}$$

The reactive power required for regulating local node voltage is calculated using voltage/reactive power sensitivity analysis as described below.

$$\Delta Q_{DG,i} = \underbrace{\left[\frac{\partial v_i}{\partial Q_i}\right]^{-1}}_{\substack{\text{Reactive Power} \\ \text{Sensitivity Index} \\ (RSI)}} \cdot \Delta V_i = \Delta V_i / \, RSI_i$$

$$\begin{aligned} &\text{if } V_i > V_{Max}, \ \Delta Q_{DG,i} = -(V_i - V_{Max}) / \, RSI_i \\ &\text{if } V_i < V_{Min}, \ \ \Delta Q_{DG,i} = (V_{Min} - V_i) / \, RSI_i \end{aligned} \tag{11}$$

where $-\left|Q_{DG,i}\right| \leq \Delta Q_{DG,i} \leq \left|Q_{DG,i}\right|$

The reactive power sensitivity index (*RSI*) can be obtained from the power flow Jacobian or using local measurements. The regulated local node voltages acts as observation points for the voltage regulator control strategy.

### 3.3 Overall Voltage Control Approach

In this paper, the suboptimal control will be used to formulate the distributed voltage control technique. The overall voltage control technique to minimize the whole system voltage violations is formulated according to a simple hierarchical control. The control procedures are as follows.

**Step 1**: Local control using the DGs reactive power.
**Step 2**: Area control using the voltage.
**Step 3**: Overall system voltage control using the suboptimal control rule.

These control steps can realize the objective of minimizing the entire system voltage violations while decreasing the stress that lies on the voltage regulators.

# 4 Simulation Results

The performance of the distributed voltage control technique is demonstrated using the IEEE 123-bus as a test system. The test system has four voltage regulators, where several photovoltaic (PVs) are introduced as DGs. In addition, the system is classified into four control areas, as shown in Fig. 2.

Different simulations are conducted based on MATLAB software, described in the following.

**Case 1**: Optimal control strategy (voltage regulators only).
**Case 2**: Suboptimal control strategy (voltage regulators only).
**Case 3**: Comparison between the optimal and suboptimal control (voltage regulators only).
**Case 4**: Long-term simulation (voltage regulators only).
**Case 5**: Overall control method (voltage regulators and DGs' reactive power).

The practical load and the PV profiles daily and monthly time intervals are presented in Fig. 3.

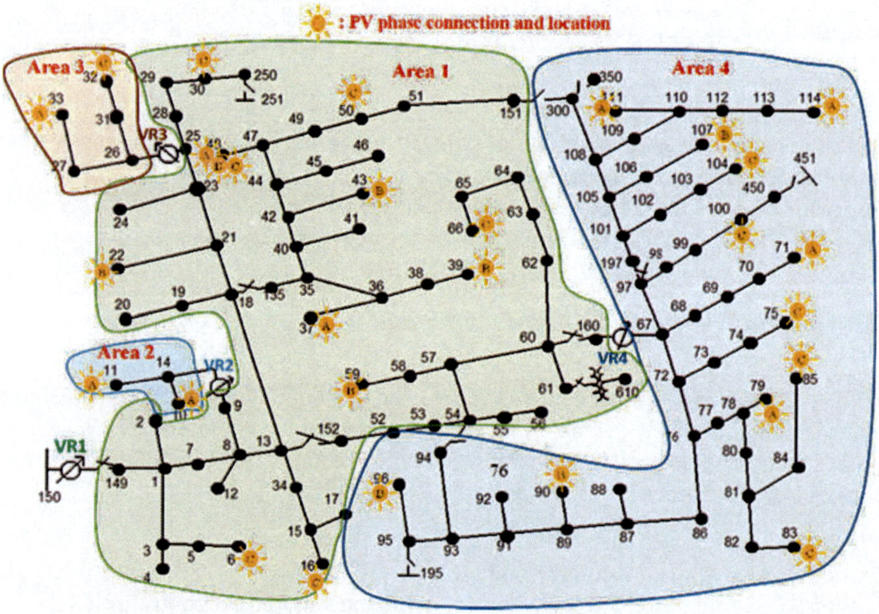


**Fig. 2** IEEE 123-bus system

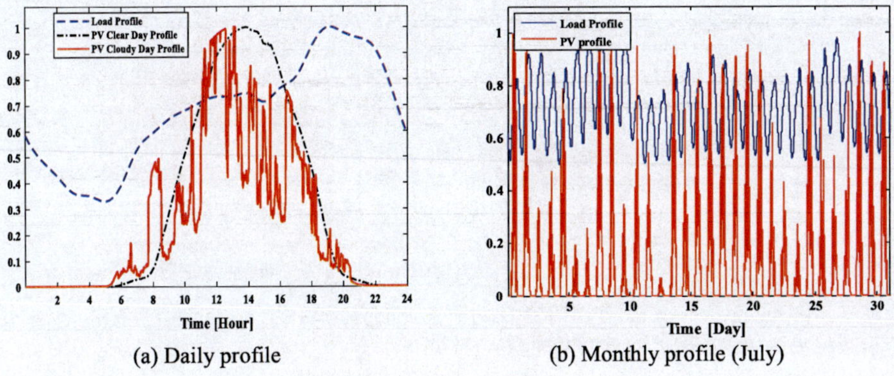


**Fig. 3** Load and PV generation profiles

## 4.1 Case 1: Voltage Control Using Optimal Control Rule

The daily voltage profiles for some nodes of the IEEE 123-bus system for sunny and cloudy PV generation without performing voltage regulation are illustrated in Fig. 4. The high PV penetration increases the voltage violations, where an overvoltage problem appears at the PV peak generation. In addition, fluctuations appear in the voltage profiles in the cloudy PV generation due to the intermittence of the PV sources, as shown in Fig. 4b. Furthermore, a voltage drop appears at some system nodes for sunny and cloudy PV generations.

After performing the optimal control technique, Fig. 5 (solid line) shows the system's voltages for sunny and cloudy PV generations. The voltage profiles are regulated within the desired limits by selecting the most effective voltage regulators at time instant *t*, as seen in Fig. 6. The simulation results demonstrate the effectiveness of the optimal technique in mitigating voltage violations while avoiding frequent tap operations.

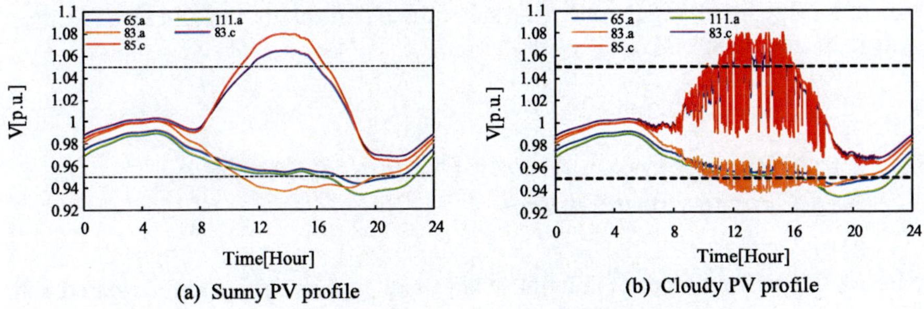


**Fig. 4** Voltage profiles of the IEEE 123-bus without voltage regulation

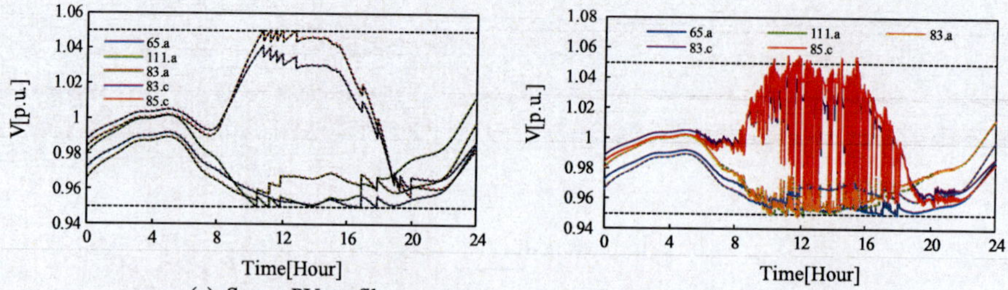


(a) Sunny PV profile (b) Cloudy PV profile

**Fig. 5** Voltage profiles of the IEEE 123-bus for the optimal and the suboptimal control. Case 2: optimal (solid lines); case 3: suboptimal (dashed lines)

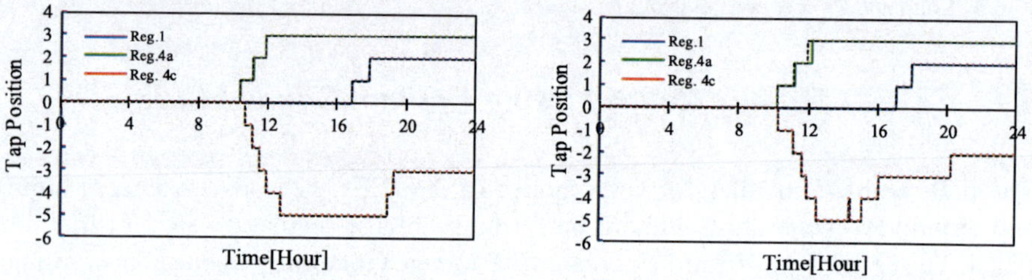


(a) Sunny PV profile (b) Cloudy PV profile

**Fig. 6** Tap positions of the IEEE 123-bus for the optimal and the suboptimal control. Case 2: optimal (solid lines); case 3: suboptimal (dashed lines)

## 4.2 Case 2: Voltage Control Using Suboptimal Control Strategy

The results of the suboptimal control role are presented in Figs. 5 and 6 (dashed lines). It is observed from the results that the regulators respond individually, which demonstrates the suboptimal rule performance. In addition, as we use a common threshold value for all regulators in the suboptimal control, its performance is almost equivalent to the optimal.

## 4.3 Case 3: Comparison Among Optimal, Suboptimal, and Conventional Control

In Fig. 7, the performances of optimal, suboptimal proposed rules are compared with a conventional voltage control technique (line drop compensator [24]) by changing the threshold value for the proposed rules and the time delay for the conventional technique. It is observed from Fig. 7 that the control performances of the optimal and the suboptimal proposed methods are almost equivalent for a wide range of

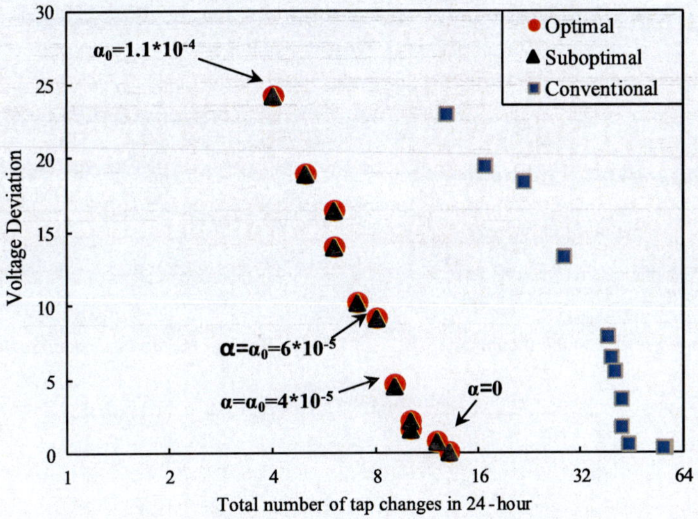


**Fig. 7** Total voltage deviations and the number of tap positions at different threshold values

threshold values. Also, the proposed approach has a better performance compared with the conventional.

## *4.4 Case 4: Long-Term Simulation*

The validation of the proposed voltage control technique for long-term operation is checked using practical load and PV profile data for one month described in Fig. 3. The proposed method successfully manages the voltage regulators and maintains the system voltages within limits, as seen in Fig. 8. Furthermore, the total amount of voltage deviations is significantly decreased.

## *4.5 Case 5: Overall Control Approach*

The overall proposed voltage control is checked by utilizing the reactive power of the PV inverter in local voltage regulation. Figure 9 illustrates that the hierarchical voltage regulation technique minimizes the voltage violations and decreases the number of control operations of voltage regulators compared with cases 1 and 2.

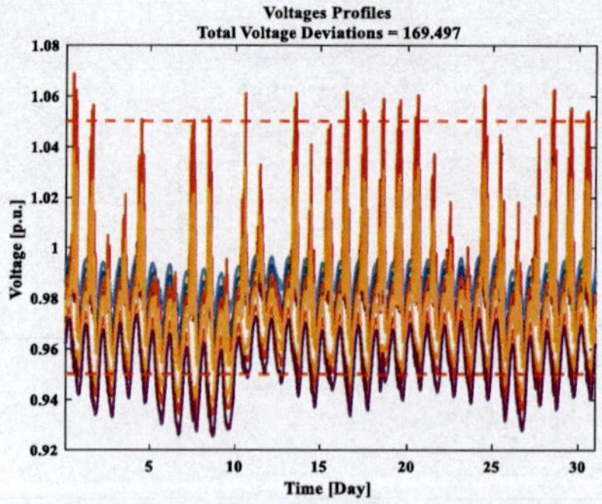


(a)Voltage Profiles without control

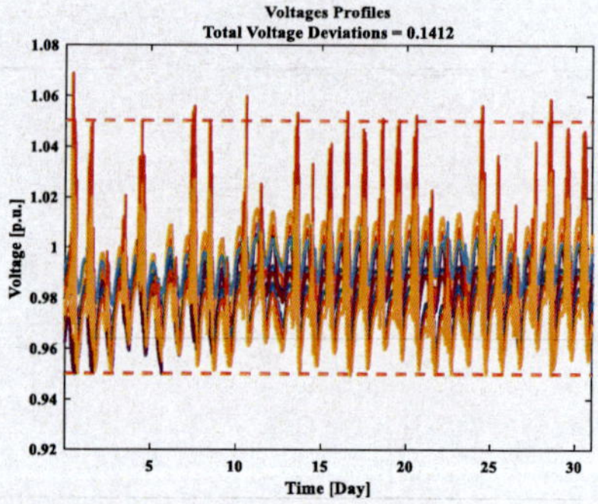


(b) Voltage Profiles with suboptimal control

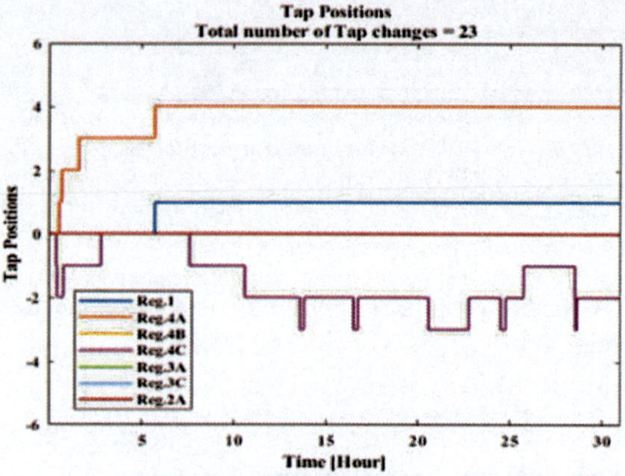


(c) Tap positions

**Fig. 8** Case 4: long-term simulation (one month: July)

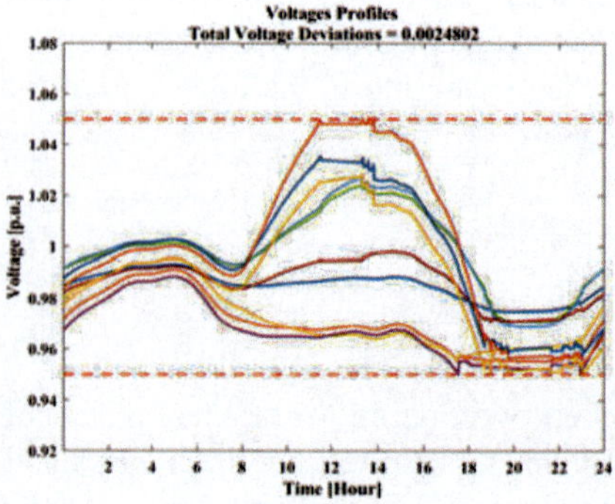


(a)Voltage Profiles

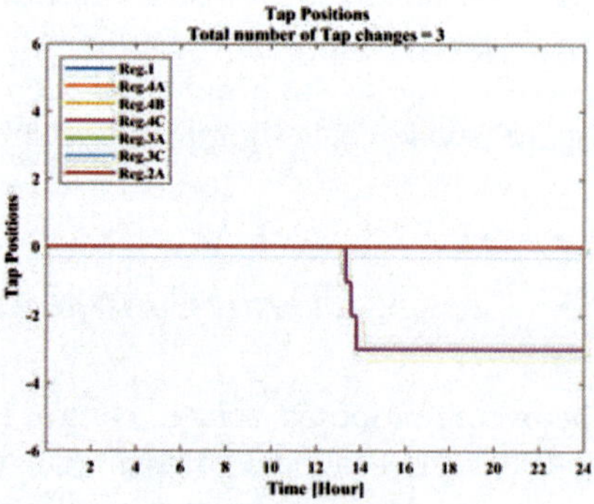


(b)Tap Positions

**Fig. 9** Result of case 5: overall control technique

## 5 Conclusions

This paper proposes a hierarchical voltage regulation approach for the LVDSs using the multi-agent system. The main objective of the approach is to minimize the voltage violations and decrease the number of control actions of the voltage regulators. In this study, the reactive power of the inverter-based DGs is utilized as local voltage controllers but also different local approaches can be incorporated in the developed approach. Several case studies are conducted using the IEEE 123-bus system to demonstrate the performance of the developed technique, including long-term simulations. Based on the simulations, the proposed technique performs in an efficient way to decrease the voltage violations for sunny and cloudy PV profiles, even in the case of long-term operation. Furthermore, the proposed voltage regulation technique can work independently with a low investment cost.

Ongoing work involves a variety of voltage regulation approaches to the developed technique. In addition, introduce the technique for microgrid systems operation.

## References


1. REN21 (2021) Renewables 2021: global status report
2. Elkhatib ME, El-Shatshat R, Salama MMA (2011) Novel coordinated voltage control for smart distribution networks with DG. IEEE Trans Smart Grid 2(4):598–605. https://doi.org/10.1109/TSG.2011.2162083
3. Ranamuka D, Agalgaonkar AP, Muttaqi KM (2017) Examining the interactions between DG units and voltage regulating devices for effective voltage control in distribution systems. IEEE Trans Ind Appl 53(2):1485–1496. https://doi.org/10.1109/TIA.2016.2619664
4. Muttaqi KM, Le ADT, Negnevitsky M, Ledwich G (2015) A coordinated voltage control approach for coordination of OLTC, voltage regulator, and DG to regulate voltage in a distribution feeder. IEEE Trans Ind Appl 51(2):1239–1248. https://doi.org/10.1109/TIA.2014.2354738
5. Long C, Ochoa LF (2016) Voltage control of PV-rich LV networks: OLTC-fitted transformer and capacitor banks. IEEE Trans Power Syst 31(5):4016–4025. https://doi.org/10.1109/TPWRS.2015.2494627
6. Abessi A, Vahidinasab V, Ghazizadeh MS (2016) Centralized support distributed voltage control by using end-users as reactive power support. IEEE Trans Smart Grid 7(1):178–188. https://doi.org/10.1109/TSG.2015.2410780
7. Sekhavatmanesh H, Cherkaoui R (2019) Analytical approach for active distribution network restoration including optimal voltage regulation. IEEE Trans Power Syst 34(3):1716–1728. https://doi.org/10.1109/TPWRS.2018.2889241
8. Efkarpidis N, De Rybel T, Driesen J (2016) Optimal placement and sizing of active in-line voltage regulators in Flemish LV distribution grids. IEEE Trans Ind Appl 52(6):4577–4584. https://doi.org/10.1109/TIA.2016.2599148
9. Simmhan Y, Kumbhare AG, Cao B, Prasanna V (2011) An analysis of security and privacy issues in smart grid software architectures on clouds. In: Proceedings—2011 IEEE 4th international conference on cloud computing, CLOUD 2011, pp 582–589. https://doi.org/10.1109/CLOUD.2011.107
10. Nikam V, Kalkhambkar V (2021) A review on control strategies for microgrids with distributed energy resources, energy storage systems, and electric vehicles. Int Trans Electr Energy Syst 31(1). https://doi.org/10.1002/2050-7038.12607

11. Tsikalakis AG, Hatziargyriou ND (2008) Centralized control for optimizing microgrids operation. IEEE Trans Energy Convers 23(1). https://doi.org/10.1109/TEC.2007.914686
12. Cintuglu MH, Youssef T, Mohammed OA (2016) Development and application of a real-time testbed for multiagent system interoperability: a case study on hierarchical microgrid control. IEEE Trans Smart Grid 9:1759–1768. https://doi.org/10.1109/TSG.2016.2599265
13. McArthur SDJ, Davidson EM, Catterson VM, Dimeas AL, Hatziargyriou ND, Ponci F, Funabashi T (2007) Multi-agent systems for power engineering applications—Part I: Concepts, approaches, and technical challenges. IEEE Trans Power Syst 22(4):1743–1752. https://doi.org/10.1109/TPWRS.2007.908471
14. McArthur SDJ, Davidson EM, Catterson VM, Dimeas AL, Hatziargyriou ND, Ponci F, Funabashi T (2007) Multi-agent systems for power engineering applications—Part II: Technologies, standards, and tools for building multi-agent systems. IEEE Trans Power Syst 22(4):1753–1759. https://doi.org/10.1109/TPWRS.2007.908472
15. Bedawy A, Yorino N, Mahmoud K, Zoka Y, Sasaki Y (2020) Optimal voltage control strategy for voltage regulators in active unbalanced distribution systems using multi-agents. IEEE Trans Power Syst 35(2):1023–1035. https://doi.org/10.1109/TPWRS.2019.2942583
16. Yorino N, Zoka Y, Watanabe M, Kurushima T (2015) An optimal autonomous decentralized control method for voltage control devices by using a multi-agent system. IEEE Trans Power Syst 30(5):2225–2233. https://doi.org/10.1109/TPWRS.2014.2364193
17. Bedawy A, Yorino N, Mahmoud K (2018) Management of voltage regulators in unbalanced distribution networks using voltage/tap sensitivity analysis. In: Proceedings of 2018 international conference on innovative trends in computer engineering, ITCE 2018, pp 363–367. https://doi.org/10.1109/ITCE.2018.8316651
18. Bedawy A, Yorino N (2018) Reactive power control of DGs for distribution network voltage regulation using multi-agent system. IFAC-PapersOnLine 51(28):528–533. https://doi.org/10.1016/j.ifacol.2018.11.757
19. Bedawy A, Yorino N, Mahmoud K, Lehtonen M (2021) An effective coordination strategy for voltage regulation in distribution system containing high intermittent photovoltaic penetrations. IEEE Access 9:117404–117414. https://doi.org/10.1109/ACCESS.2021.3106838
20. Bedawy A, Mahmoud K, Sasaki Y, Zoka Y, Yorino N, Lehtonen M (2022) A cooperative voltage control approach for distribution systems based on voltage regulators and PV inverters. In: 2021 22nd international Middle East power systems conference (MEPCON), pp 366–371. https://doi.org/10.1109/mepcon50283.2021.9686262
21. Carvalho PMS, Correia PF, Ferreira LAF (2008) Distributed reactive power generation control for voltage rise mitigation in distribution networks. IEEE Trans Power Syst 23(2):766–772. https://doi.org/10.1109/TPWRS.2008.919203
22. Zeraati M, Hamedani Golshan ME, Guerrero J (2018) Distributed control of battery energy storage systems for voltage regulation in distribution networks with high PV penetration. IEEE Trans Smart Grid 9(4):3582–3593. https://doi.org/10.1109/TSG.2016.2636217
23. Wang P, Liang DH, Yi J, Lyons PF, Davison PJ, Taylor PC (2014) Integrating electrical energy storage into coordinated voltage control schemes for distribution networks. IEEE Trans Smart Grid 5(2):1018–1032. https://doi.org/10.1109/TSG.2013.2292530
24. Kersting WH (2017) Distribution system modeling and analysis, 4th edn. https://doi.org/10.1201/9781315120782

# Application Mapping of Fully Connected 3D NoC Using Latency Prediction Model

**Ramesh Sambangi, B. Hari Krishnan, Kanchan Manna, Santanu Chattopadhyay, and Sudipta Mahapatra**

**Abstract** For Network-on-Chip (NoC) latency estimation, current analytical models do not provide reliable results. As a result, they cannot be used for optimization of design space exploration. In this paper, we propose a DNN-based learning model for predicting the latency of a 3D fully connected NoC. The features needed for the DNN model are gathered from both the analytical model and from the Booksim simulator. The resulting DNN model has been applied to the mapping optimization loop for predicting the best mapping in conjunction with the parameters of an application and the NoC. In both synthetic and application-specific traffic simulations, we have found that using our proposed DNN model, prediction error is less than 8%. Furthermore, the mapping optimization using a prediction model predicts better solutions than the mapping optimization using communication cost.

**Keywords** Network-on-Chip · Analytical models · Neural networks · DPSO

## 1 Introduction

In present and future multiprocessor systems-on-chip (MPSoCs), Network-on-Chip (NoC) has proven to be a promising communication backbone. High-performance computing applications may be based on MPSoCs with hundreds of processing elements. Small-scale designs commonly use bus-based communication, but such systems produce more delay for parallel processing applications. Dally and Towles [1] first proposed an NoC architecture to overcome the drawbacks of bus-based implementations. NoCs are composed of PEs, routers, and links, with information exchanged between PEs through these routers and links. The overall performance of an NoC-based system is greatly influenced by the application mapping. Differ-

R. Sambangi (✉) · B. Hari Krishnan · S. Chattopadhyay · S. Mahapatra
Indian Institute of Technology Kharagpur, Kharagpur, West Bengal 721302, India
e-mail: sambangi.ramesh86@iitkgp.ac.in

K. Manna
Birla Institute of Technology and Science, Pilani, Goa Campus, Pilani, Goa 403726, India

 
C. Giri et al. (eds.), *Emerging Electronic Devices, Circuits and Systems*,
Lecture Notes in Electrical Engineering 1004,
https://doi.org/10.1007/978-981-99-0055-8_30

ent tasks in an application task graph are assigned to the cores in an NoC during application mapping. During the mapping process, one of the main goals is to reduce communication latency. For this reason, simulating the overall network with the traffic pattern defined by the application is necessary in order to determine the merit of a candidate mapping. This simulation takes a considerable amount of time. Thus, in some studies, indirect metrics such as communication costs have been used as the metric to optimize mapping [2, 3]. In [4–6], analytical models are used in the optimization loop. However, these models are often inaccurate in predicting the latency value. This has led to the development of a machine learning-based prediction model [7]. It is incorporated into the mapping optimization loop to estimate the optimal mapping solution.

Based on our previous work [8], we propose an analytical model for fully connected 3D NoCs. We propose a deep neural network-based prediction model to enhance the performance of the analytical model. In the end, we used this prediction model in mapping optimization loops to identify the optimal mapping solutions.

## 2 Previous Work

Application-specific NoCs are typically designed as constrained optimization problems. Thus, performance analysis, which can be implemented in optimization loops, is of great importance. A model based on queuing theory is presented in [9] which considers the buffer sizing problem. The authors of [5] assume that the header buffer arrival process is a Poisson process, utilizing the M/G/1 queuing model for router buffers. Assumptions about the router buffers' capacity are unrealistic, as they had infinite capacity.

An G/G/1 queuing model was used in [4] to model router buffers. It may not be applicable in situations where fair arbitration is necessary since the paper assumes that input channels in routers have a fixed priority. The G/G/1/K queuing model has been used in [10] in order to increase the accuracy of an analytical model for routers. The area/power modeling accuracy and the runtime of the applications have been improved by using machine learning-based techniques in the NoC design community. To estimate area and power based on regression models for NoC routers, the multivariate adaptive regression spline technique has been used in [11]. In [7], a ML-based NoC latency model has been developed. It is important to note that accuracy depends on the similarity between test cases and training data. Researchers in [10] created a training dataset with both an analytical model and a simulator [12] to improve the prediction model's accuracy for unknown test cases.

Many researchers have published application mapping algorithms for NoCs over the years [13]. For large-scale IP applications, heuristic search-based methods are efficient at producing the most effective mapping solutions among all the existing approaches [13]. ML models, on the other hand, can automatically extract heuristics, which reduces the need for human intervention and guides the search process to discover optimal solutions [14–17].

Most of the research in this area is focused on 2D NoCs, which is driven by recent advances in analytical models and machine learning. The result of this inspired us to develop 3D NoC latency prediction models.

## 3 Analytical Model

Analytical latency models are based on queuing theory, which treats routers as service stations and buffers and channels as queues. Latency models for NoCs should consider the congestion and contention in the network. An analytical model of two-dimensional NoCs has been proposed in our previous work [8]. This work extends the approach to fully connected 3D NoCs. The router model used in this work is shown in Fig. 1. In the analytical latency model, the router model is based on the following assumptions.

- A buffered input router with $P$ channels (Fig. 1).
- Wormhole flow control with deterministic routing.
- A Poisson process is followed by incoming header flits reaching the router input channels.
- FIFO queues are used as input channel buffers, and a central switch fabric is used for communication. The service time and the waiting time for each queue are independent.

Flow latencies are influenced by packet waiting times at each router and header flit switching times at each router in its path between source and destination. In addition,

**Fig. 1** 3D router model

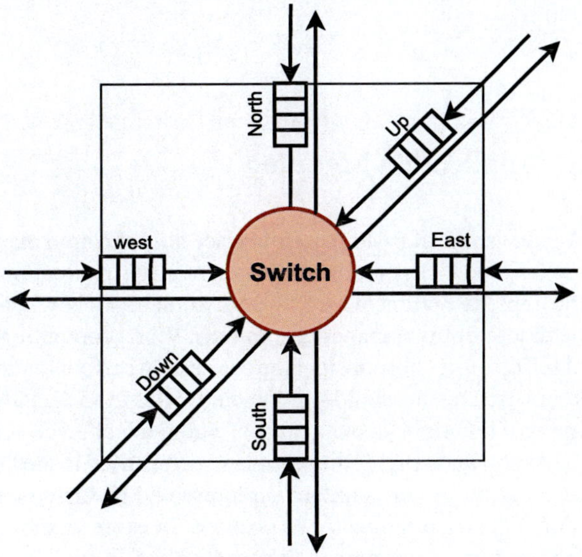

3D fully connected NoCs use the well-known XYZ routing algorithm to solve the IP mapping problem [18]. Consider $\prod_{sd}$ to be a set of all routers encountered during the route from source IP $s$ to destination IP $d$. As $W_s$, the queueing delay is how long it takes for a packet to be injected into the network from the router's output buffer. The flow $s \rightarrow d$ has an average packet latency of

$$L_{sd} = W_s + \sum_{(j,m)\in\prod_{sd}} (W_j^m + H_s) + \left\lceil \frac{S}{\Omega} \right\rceil \tag{1}$$

The calculations of various terms present in Eq. (1) are discussed in Appendix.

If $P_{s\rightarrow d}$ is the probability of occurrence of flow $s \rightarrow d$, then the overall average packet latency in the network is given by

$$L = \sum_{\forall s} \sum_{\forall d} P_{s\rightarrow d} L_{sd} \tag{2}$$

## 4 LPNet Model for Fully Connected 3D NoCs

Latency prediction neural network (LPNet) is our adopted deep neural network for latency prediction. An analytical model generates input feature vectors for this learning model. It is represented by $X_{\text{AQ}} = [X_{\text{AQ}}[1], X_{\text{AQ}}[2], \ldots, X_{\text{AQ}}[n]]$. The feature vector for a $4 \times 4 \times 2$ fully connected 3D mesh is shown in Table 1. LPNet is trained by generating feature vectors for each input parameter based on an analytical model. Any new traffic input can be predicted using Eq. (3). Here, $f_{\text{AQ}}(.)$ represents the trained prediction model. The latency value predicted by the model is given by

$$L_{\text{p}} = f_{\text{AQ}}(X_{\text{AQ}}) \tag{3}$$

## 5 Application Mapping

A vital aspect of system performance is application mapping [2, 3, 19]. Applications are mapped by assigning tasks present in the application task graph to the NoCs. With the implementation of a DPSO algorithm in [2], NoC-based applications are now able to choose optimal mapping solutions. When computing the communication cost, the algorithm does not consider the congestion and contention on the network. As a result, simulation results could significantly differ from the ideal results. We used the following cost functions in our analysis, communication cost, and LPNet prediction model.

As shown in Fig. 2, the particle structure used in the DPSO algorithm is represented by an array. A core number is represented by the contents of the array, and a router number is represented by its position. In essence, the particle fitness will depend on the position of its cores in the array shown in Fig. 2.

**Table 1** Training feature vector ($X_{AQ}$) corresponding to a fully connected 3D mesh of $4 \times 4 \times 2$

| Parameter | Element | Description | Applicable for |
|---|---|---|---|
| $\lambda$ | $X_{AQ}[1]$ | The rate at which packets are injected for the link with the highest bandwidth requirement | Whole network |
| $H_s$ | $X_{AQ}[2]$ | Number of stages in a router pipeline or the time it takes for a header flit | |
| $L_b$ | $X_{AQ}[3]$ | Serialization delay | |
| $B$ | $X_{AQ}[4]$ | Input channel buffer size | |
| $\Lambda$ | $X_{AQ}[5:11]$ | Vector with $i$th entry representing the average arrival rate at input channel $i$ for the router | Each router |
| $F$ | $X_{AQ}[12:48]$ | Forwarding probability matrix, whose each entry represents $f_{ij}$. The probability of packet arrival at $i$th channel to leave from $j$th channel | |
| $N$ | $X_{AQ}[49:55]$ | Vector represents the average number of packets present at a router's input channel buffer, whose $i$th entry represents the number of packets present there | |
| $T$ | $X_{AQ}[56:62]$ | An entry in the vector of service time represents the packet service time on input channel $i$ for the router | |
| $L$ | $X_{AQ}[63]$ | Latency value obtained from analytical model | Whole network |

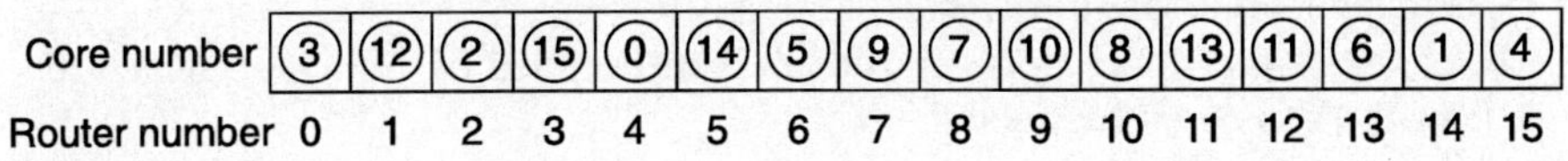


**Fig. 2** Particle structure

The initial population of the DPSO algorithm is generated randomly [2]. Afterward, the particles are updated using a global and particle local best solution. As the authors explained in [20], doping is used to update the particles in binary particle swarm optimization (BPSO). Based on the fitness value, they separate the particles in a particular generation into good particles and bad particles. When compared to bad particles, the good particles are closer to the optimal solution. Hence, bad particles are heavily doped over good particles. Using the same approach as BPSO, we created the DPSO framework. For the evolution of population generations, we use the following equation:

$$p_{k+1}^{i} = \left[\mathrm{SS}_1\left(p_k^i \rightarrow \mathrm{gbest}_k\right) \oplus \mathrm{SS}_2\left(p_k^i \rightarrow \mathrm{pbest}^i\right)\right] \times p_k^i \tag{4}$$

In Eq. (4), $a \rightarrow b$ is the swap sequence obtained by converting particle $a$ into particle $b$. The $i$th particle at $k$th iteration has been represented by $p_k^i$. We shall consider $\text{pbest}^i$ as the best local solution achieved by particle $i$ and $\text{gbest}_k$ as the best global solution for iteration $k$. A doping process is described here by swapping core positions (see Fig. 2). In [8], the procedure for evaluating swap sequences is described. Combining $a$ and $b$ with the operator $\oplus$ creates a new swap sequence [2]. A random selection of swap sequences is made by $\text{SS}_i(a \rightarrow b)$ from the total set of swap sequences $a \rightarrow b$. Depending on the concentration of doping, $\text{SS}_i(\cdot)$ selects swap sequences in a variable quantity. In every generation, particles with fitness values below average are referred to as good particles, while particles with fitness values exceeding average are considered bad particles. As a result, good particles require modest doping. Likewise, bad particles are far from an optimal solution, requiring heavy doping [8]. As a result, more swap sequences will be picked from $p_k^i \rightarrow \text{gbest}_k$ and $p_k^i \rightarrow \text{pbest}^i$.

In this paper, we examine two different scenarios for determining the particle's fitness. We considered communication costs in the first case and packet latency in the second case. The following expression has been used to calculate the communication cost:

$$\text{comm\_cost} = \sum_{\forall c_i, c_j \in C} \text{BW}(c_i, c_j) \cdot D(c_i, c_j), \tag{5}$$

where

- In the application task graph, $C$ represents all of the cores.
- A $\text{BW}(c_i, c_j)$ is the amount of communication bandwidth required between two tasks mapped to $c_i$ and $c_j$ cores.
- The Manhattan distance between routers is determined by $D(c_i, c_j)$, which represents the hop count between them.

## 6 Results and Analyses

In contrast to Booksim 2.0 [12], which supports only synthetic traffic, the simulator has been updated to support custom application-specific traffic for fully connected 3D NoCs. Evaluation of the analytical and LPNet models has been carried out using updated Booksim 2.0 [21] simulator. For evaluation and preparation of data, we assume a four-stage pipelined NoC router architecture. A $4 \times 4 \times 2$ fully connected 3D mesh NoC has been taken. The training dataset was generated using synthetic traffic patterns [22] and traffic patterns corresponding to random application task graphs [23]. We generated 220,000 training vectors, of which 80% was used as the training set and 20% as the evaluation set for validation. The injection rates are taken over a range of 0.001–0.04.

In addition to bit-complement, tornado, and transpose traffic patterns, the trained model was also tested with bit-reversal traffic patterns. All traffic patterns are con-

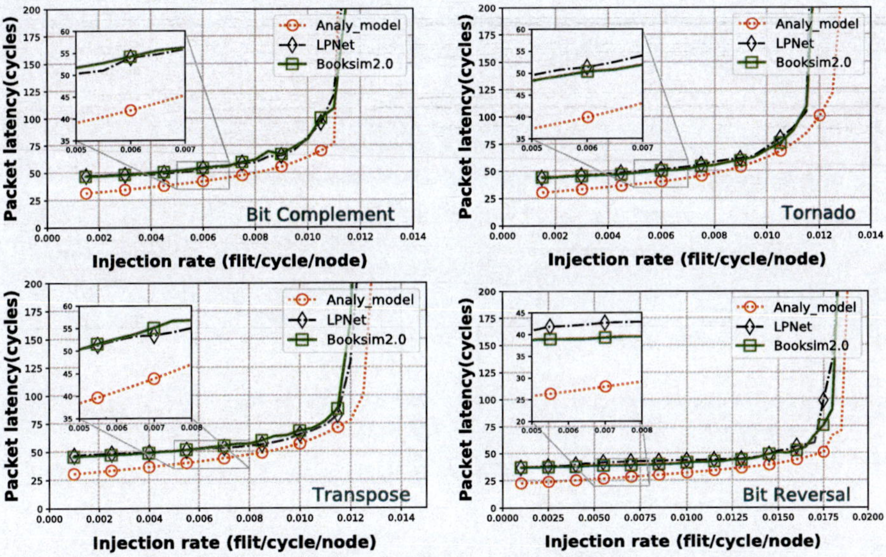


**Fig. 3** Comparison of average packet latency between LPNet, the analytical model, and the simulator for synthetic traffic patterns such as bit-complement, tornado, transpose, and bit-reversal

sidered to have a buffer width of 10 and a packet size of 16. The results of the trained model were compared with the results from the simulator. Along with the simulator and LPNet prediction model, the analytical model values are plotted as well. As depicted in Fig. 3, we would observe that LPNet model latency values almost line up with simulator values. However, we noticed a significant difference between the analytical model and the simulator. This study illustrates the importance of prediction models in minimizing analytical model error compared to simulator values.

## *6.1 LPNet Model Performance Analysis*

The trained model was tested at various injection rates with traffic patterns other than training data. We have generated random task graphs TG1, TG2, TG3, and TG4 using [23]. We have compared results from simulation and the LPNet model. In our simulations, we assumed a packet size of 16 and a buffer depth of 10. LPNet model predicted values that are very close to simulator values, as shown in Fig. 4. This study provides a strong foundation for the application of the proposed prediction model in mapping optimization loops.

Performance of the trained model against the simulator for random application task graph (TG2) has been analyzed. To do so, 500 randomly chosen mappings at varying injected rates were generated. In the next step, the relative error of the prediction

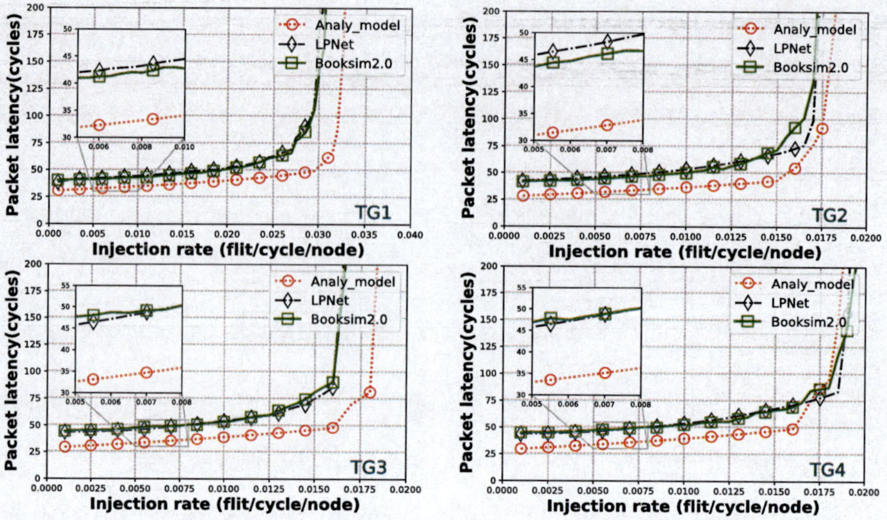


**Fig. 4** Comparison of average packet latency between LPNet, the analytical model, and the simulator for random application traffic patterns such as TG1, TG2, TG3, and TG4

**Table 2** Relative error of the analytical model and LPNet for $4 \times 4 \times 2$ fully connected 3D mesh at different injection rates

| Injection rate (flits/cycle/node) | Relative error (%) | |
|---|---|---|
| | $4 \times 4 \times 2$ mesh | |
| | Analytical model | LPNet model |
| 0.001 | 16.896 | 2.097 |
| 0.002 | 15.896 | 1.702 |
| 0.004 | 17.589 | 4.256 |
| 0.006 | 14.589 | 4.289 |
| 0.008 | 20.589 | 6.827 |
| 0.01 | 22.987 | 7.67 |

model is calculated using Eq. (6). Our 3D mesh is fully connected with a packet size of 16 and a buffer size of 10. We analyzed the relative error for TG2 at varying injection rates, and we have given it in Table 2. In comparison with the simulator, the LPNet prediction model has a minimal relative error. In contrast, analytic models produce more relative error than LPNet prediction models.

$$E_{\mathrm{rel}} = \frac{1}{500} \sum_{i=1}^{500} \frac{|L_{\mathrm{S}}(i) - L_{\mathrm{A}}(i)|}{L_{\mathrm{S}}(i)} \tag{6}$$

### 6.2 Complexity Analysis of DPSO Using Communication Cost and Prediction Model

The complexity of the DPSO model with communication costs and the LPNet model assumes the following notation. Let $G(C, E)$ represent the application task graph containing $C$ cores and $E$ edges. The 3D NoC contains $n$ routers and $L$ number of layers. It runs for $g$ number of generations and has a population size of $K$.

In DPSO, the fitness of the particle is determined by determining the source and destination routers for each edge of $G$. If the particle is presorted, the determination takes $O(\log n)$ time. Thus, overall fitness evaluation takes $O(E \log n)$ time. Therefore, the initialization phase has a complexity of $O(KE \log n)$. During the evolution part, identifying each swap sequence takes $O(n \log n)$ time; modifying takes $O(n)$ time, and evaluating fitness takes $O(E \log n)$ time. As a result, one generation of PSO takes $O(KE \log n)$ of time. Thus, the overall complexity of the algorithm becomes $O(gKE \log n)$.

The fitness of the particle has been estimated by deep neural network in DPSO with LPNet model. A deep neural network with $l$ number of layers and input feature vector size is $x \in \mathbb{R}^{(K\times(P^2+3P)n+5)}$ which has a complexity of $O(P^2Kln)$. Then the overall complexity of the DPSO algorithm using LPNet becomes $O(gnKlP^2)$.

### 6.3 DPSO Using Prediction Model

In this paper, we have mapped a randomly generated task graph onto a $4 \times 4 \times 2$ fully connected 3D mesh. We have Initialized DPSO with 500 particles and made it run up to 5000 generations. We have defined termination criteria as when the global best solution has not been improved for 200 successive generations. In all cases, LPNet takes packet sizes 16 and buffer sizes 10 with injection rates of 0.004. DPSO is run five times for each random task graph, and the best solution is selected. Two different fitness metrics are used by DPSO: communication cost and average packet latency. In Table 3, the best mapping solution for each random task graph is evaluated using DPSO. After that, the best mapping solution based on two approaches is applied to a simulator and their corresponding latency values are tabulated. Based on our results, we note that DPSO coupled with LPNet predicts a better solution compared to DPSO with communication cost as fitness function.

## 7 Conclusion

The paper aims to present a model that predicts latency for 3D NoCs with fully connected nodes using a deep neural network. The prediction model has been incorporated into the DPSO mapping optimization loop. In comparison with simulator

**Table 3** Performance comparison of DPSO using LPNet model and communication cost

| Task graph | DPSO using Comm_cost | | | DPSO using LPNet | | |
|---|---|---|---|---|---|---|
| | Best fitness achieved | Latency value from simulator | Runtime of DPSO (s) | Best fitness achieved | Latency value from simulator | Runtime of DPSO (s) |
| TG1 | 9869 | 35.87 | 55.09 | 30.21 | 32.58 | 120.87 |
| TG2 | 14,515 | 38.79 | 59.78 | 31.06 | 30.57 | 130.78 |
| TG3 | 10,761 | 33.94 | 48.05 | 32.85 | 30.39 | 87.48 |
| TG4 | 12,541 | 40.46 | 56.87 | 34.86 | 35.86 | 86.78 |
| TG5 | 13,084 | 46.92 | 77.912 | 39.68 | 41.54 | 92.78 |

results, the prediction model has a minimal relative error. In comparison with DPSO using communication cost as fitness function, better mapping solutions are obtained using DPSO with prediction model that has minimal latency.

## Appendix

The router service time for the header flits is given by $H_s$.

$$T_i = H_s + \left\lceil \frac{S}{\Omega} \right\rceil \tag{7}$$

$$H_s = t_R + t_L \tag{8}$$

$$t_R = t_{RC} + t_{VA} + t_{SA} + t_{ST} \tag{9}$$

The average waiting of an incoming packet in queue $i$ is

$$W_i = T_i \cdot N_i + \sum_{j=1, i \neq j}^{P} T_i \cdot C_{ij} \cdot N_j + R_i \tag{10}$$

Here, the term $C_{ij}$ denotes the contention probability, that is, the probability that the $i$th and $j$th input channels of a router compete for the same output channel. The average number of packets in a queue, mean arrival rate and average waiting time are related by the well-known Little's equation

$$N_i = W_i \times \lambda_i \tag{11}$$

On rearranging Eqs. (10) and (11), the average number of packets at each input buffer of a router is given in Eq. (12).

$$N_i = \lambda_i \sum_{j=1}^{P} T_j C_{ij} N_j + \lambda_i R_i \quad \text{where} \quad C_{jj} = 1 \tag{12}$$

Equation (12) describes the equilibrium condition for only the $i$th input channel of a router. Closed form expression of Eq. (12) for the entire router is given in Eq. (13).

$$N = (I - T\Lambda C)^{-1} \Lambda R \tag{13}$$

using Little's theorem, the average waiting time at the $j$th input buffer of the $m$th router is given by

$$W_j^m = \frac{N_j^m}{\lambda_j^m} \tag{14}$$

If $P_{s\to d}$ be the probability of occurrence of flow $s \to d$ then the overall average packet latency in the network can be given by

$$L = \sum_{\forall s} \sum_{\forall d} P_{s\to d} L_{sd} \tag{15}$$

## References


1. Dally WJ, Towles B (2001) Route packets, not wires: on-chip interconnection networks. In: Proceedings of the 38th design automation conference (IEEE Cat. No. 01CH37232), pp 684–689. https://doi.org/10.1109/DAC.2001.156225
2. Sahu PK, Shah T, Manna K, Chattopadhyay S (2014) Application mapping onto mesh-based network-on-chip using discrete particle swarm optimization. IEEE Trans Very Large Scale Integr (VLSI) Syst 22(2):300–312. https://doi.org/10.1109/TVLSI.2013.2240708
3. Sahu PK, Chattopadhyay S (2013) A survey on application mapping strategies for network-on-chip design. J Syst Archit 59(1):60–76. https://doi.org/10.1016/j.sysarc.2012.10.004
4. Kiasari AE, Lu Z, Jantsch A (2013) An analytical latency model for networks-on-chip. IEEE Trans Very Large Scale Integr (VLSI) Syst 21(1):113–123. https://doi.org/10.1109/TVLSI.2011.2178620
5. Ogras UY, Bogdan P, Marculescu R (2010) An analytical approach for network-on-chip performance analysis. IEEE Trans Comput-Aided Des Integr Circuits Syst 29(12):2001–2013. https://doi.org/10.1109/TCAD.2010.2061613
6. Lai M, Gao L, Xiao N, Wang Z (2009) An accurate and efficient performance analysis approach based on queuing model for network on chip. In: 2009 IEEE/ACM international conference on computer-aided design—digest of technical papers, pp 563–570
7. Qian Z, Juan D-C, Bogdan P, Tsui C-Y, Marculescu D, Marculescu R (2013) SVR-NoC: a performance analysis tool for network-on-chips using learning-based support vector regression model. In: 2013 design, automation & test in Europe conference & exhibition (DATE), pp 354–357. https://doi.org/10.7873/DATE.2013.083
8. Sambangi R, Manghnani H, Chattopadhyay S (2021) LPNet: a DNN based latency prediction technique for application mapping in network-on-chip design. Microprocess Microsyst 87:104370. https://doi.org/10.1016/j.micpro.2021.104370

9. Hu J, Ogras UY, Marculescu R (2006) System-level buffer allocation for application-specific networks-on-chip router design. IEEE Trans Comput-Aided Des Integr Circuits Syst 25(12):2919–2933. https://doi.org/10.1109/TCAD.2006.882474
10. Qian Z, Juan D, Bogdan P, Tsui C, Marculescu D, Marculescu R (2016) A support vector regression (SVR)-based latency model for network-on-chip (NoC) architectures. IEEE Trans Comput-Aided Des Integr Circuits Syst 35(3):471–484. https://doi.org/10.1109/TCAD.2015.2474393
11. Kahng AB, Lin B, Samadi K (2010) Improved on-chip router analytical power and area modeling. In: 2010 15th Asia and South Pacific design automation conference (ASP-DAC), pp 241–246. https://doi.org/10.1109/ASPDAC.2010.5419887
12. Jiang N, Becker DU, Michelogiannakis G, Balfour J, Towles B, Shaw DE, Kim J, Dally WJ (2013) A detailed and flexible cycle-accurate network-on-chip simulator. In: 2013 IEEE international symposium on performance analysis of systems and software (ISPASS), pp 86–96
13. Sahu PK, Chattopadhyay S (2013) A survey on application mapping strategies for network-on-chip design. J Syst Archit 59(1):60–76
14. Vinyals O, Fortunato M, Jaitly N (2017) Pointer networks. arXiv:1506.03134
15. Bello I, Pham H, Le QV, Norouzi M, Bengio S (2017) Neural combinatorial optimization with reinforcement learning. arXiv:1611.09940
16. Kalakanti AK, Verma S, Paul T, Yoshida T (2019) RL SolVeR Pro: reinforcement learning for solving vehicle routing problem. In: 2019 1st international conference on artificial intelligence and data sciences (AiDAS), pp 94–99. https://doi.org/10.1109/AiDAS47888.2019.8970890
17. Chen Q, Huang W, Peng Y, Huang Y (2021) A reinforcement learning-based framework for solving the IP mapping problem. IEEE Trans Very Large Scale Integr (VLSI) Syst 29(9):1638–1651. https://doi.org/10.1109/TVLSI.2021.3097712
18. Ye T, Benini L, De Micheli G (2002) Analysis of power consumption on switch fabrics in network routers. In: Proceedings 2002 design automation conference (IEEE Cat. No. 02CH37324), pp 524–529. https://doi.org/10.1109/DAC.2002.1012681
19. Murali S, De Micheli G (2004) Bandwidth-constrained mapping of cores onto NoC architectures. In: Proceedings design, automation and test in Europe conference and exhibition, vol 2, pp 896–901. https://doi.org/10.1109/DATE.2004.1269002
20. Khatua K et al (2019) A deep neural network augmented approach for fixed polarity AND-XOR network synthesis. In: TENCON 2019—2019 IEEE region 10 conference (TENCON), pp 2189–2193. https://doi.org/10.1109/TENCON.2019.8929289
21. Manghnani H, Ramesh S (2020) Booksim 2.0 for custom traffic. https://github.com/Mags17/Booksim_custom_traffic
22. Dally W, Towles B (2003) Principles and practices of interconnection networks. Morgan Kaufmann Publishers Inc., San Francisco, CA
23. Dick RP, Rhodes DL, Wolf W (1998) TGFF: task graphs for free. In: Proceedings of the sixth international workshop on hardware/software codesign (CODES/CASHE'98), pp 97–101

# Smart Device and Mobile Application for Remote Health Monitoring and Alarming

**Mrittika Ghosh, Ankan Ghosh, Aditi Bhattacharya, Abir Kapat, Soumyadeep Mandal, Sambit Prasad, and Ananya Banerjee**

**Abstract** The Internet of Things (IoT) application is an important technology to enhance medical performances. This paper focuses on the practical use of the Internet of Things (IoT) based on real-time remote health monitoring systems, real-time medical resource availability, real-time hospital bed booking as well as night emergency services with an e-commerce platform for Medicare equipment. Our initiative provides a step forward in the field of remote health monitoring and advanced Medicare facilities. The population is increasing at a high rate and a proper signal monitoring system must need the presence of a particular patient within and outside of a health center to work with. In case of any emergency or pandemic-like situation, it becomes very difficult to manage every patient and track records of them. So, this kind of technology and our modern health sector devices can provide us the promising solutions in this field. Based on that, the paper proposed a mobile application with a smart device based on the Internet of Things (IoT) platforms to remotely monitor signal, heart rate, SpO2 (define as oxygen saturation level of an individual, also it can be said as the measure of the amount of hemoglobin carrying-oxygen in the blood), blood pressure, temperature of the body in patients. The symptoms are analyzed or detected by using a microcontroller. Our main motive is to send the health records of an individual to their respective smartphone with a proper data of analysis for a further medical examination if needed. Finally, the results received from the device are being displayed in the Smart phone within our custom application, providing other facilities with it such as live doctor consultation, ambulance facilities, doctor recommended solutions or providing emergency resources through e-commerce services.

**Keywords** The Internet of Things (IoT) · Temperature sensor · ECG sensor · Pulse oximetry sensor · Emergency medical resources · Blood pressure sensor · JIASH mobile application

M. Ghosh · A. Ghosh · A. Bhattacharya · A. Kapat · S. Mandal · S. Prasad · A. Banerjee (✉)
Dr. Sudhir Chandra Sur Institute of Technology and Sports Complex, Dumdum, Kolkata, India
e-mail: ananya.banerjee@dsec.ac.in

 
C. Giri et al. (eds.), *Emerging Electronic Devices, Circuits and Systems*,
Lecture Notes in Electrical Engineering 1004,
https://doi.org/10.1007/978-981-99-0055-8_31

## 1 Introduction

Healthcare is invisible and unpredictable, nowadays, people are at high risk for their health problems as because it is not possible for them to keep track of their health in this busy life. Our health care sector mainly depends on the maintenance, support and monitoring of proper health through few steps like consulting, diagnosis, treatment and prevention. Throughout the world, such type of health care system is still widely accessible instead of many advanced technology available, which are expensive to afford or difficult to arrange for any individual for diagnosing their minimal health issues [1].

To provide advanced facilities to such issues the Internet of Things (IoT) can play an important role. It is a system of connected devices that can exchange data between electronic devices over the Internet. The useful approach of our system allows them to make wise decisions about growth and development in the Internet of Things. Under the action of such a system, better and better healthcare is now possible and easily accessible. Implementation of the Internet of Things (IoT) in the health care industry can gradually lower down the mortality rate and ensure patient care with more flexibility [2].

A smart phone usually consists of different types of sensors and more to be added in the near future. We are using such sensors by optimizing it with accuracy and providing the user a reliable data with comparison of normal health readings in their smart phones to get notified about any health issues. By using the application of such devices, we came across the smart health solution by our wearable device and application, which also provides information about the users nearest resources like hospital beds, oxygen cylinder availability, etc. And also it provides an e-commerce platform for various health products [3].

## 2 Problem Statement

A remote health monitoring system can provide useful health information at home within a short interval of time. This kind of precaution is helpful for elderly or chronically ill persons in our family who might wish to avoid staying in hospital for checkup. This kind of remote system should be able to diagnose the type of problem like heart, blood pressure, temperature and few other parameters. Nowadays, people tend to ignore and pass-on if they see a person is suffering or in danger. Also, in case of urgent and sudden health issues, people find no one beside them to help them get a proper medical consultation without intervening in serious condition. Our growing stressful life has a heavy impact on public health. As the crowd grows in hospitals and the number of patient's increases, doctor's fees have also skyrocketed, especially some patients who cannot afford proper fees are unable to pay. It is difficult to remember a schedule checkup once we are suffering from chronic diseases like diabetes, which includes old-aged persons mainly. Facing the same problem with buying medicines too. It is important to complete a dose of each medicine to get cured properly but it is as difficult for people having to spare time to go out and buy them or mainly old-aged people not having that much strength to walk to a medicine store and have medicine for them. Local ambulance services are often found to be inappropriate and ignorant. In some cases, delayed services could bring endangerment to a patient's health, causing a life-scarce situation. Also, poor condition and maintenance of ambulances also adds to the disadvantage of the situation.

## 3 Methodology

Our proposed system is mainly to track various health information and present it to our users in an easy-to-use interface through a simple mobile application. After that, the tracked data is then saved to cloud and can be accessed through our provided app and the result will be shown by comparing the readings within the normal range or any step needed to be taken, respectively. The application itself will send a monthly, weekly or daily remainder (set by the user itself) to them and their shortlisted family contacts regarding checkup or medication requirements such as a scheduled checkup dosage, meal time, etc. The device is very handy to use as it is a simple watch but more effective to our life [4].

On wearing the device and turning it on, it will automatically start our health diagnosing process and collect all the data like BP, SpO2 level, heart rate, body temperature, etc. Necessary readings are required to analyze any individual's health condition. An Internet of Things (IoT) gateway will play a key role between the sensors and the cloud. It will then communicate with the sensors in the wearable device over a Bluetooth connectivity type and the filtered data being processed. Finally, the received data is transferred into an important standard protocol named MQTT which is to be sent to the cloud through internet. The collected readings will be shown to our connected mobile device with a proper comparison between normal readings and the user's one. Usually, all data of sensors, records and databases are stored in the form of Big Data (Big Data is the next step after the Internet of Things (IoT). All of this data is retrieved according to the requirement. An analysis is performed on the structure of the data and is then provided to any user). To handle the security of the user data, Big Data has the ability to manage itself which will keep out an unauthorized user by using the protection layer of firewalls, end-user training and intrusion protection systems (IPS), strong user authentication, intrusion detection systems (IDS) [5].

We are using Bluetooth module to establish connection between the device and the mobile application, so that Internet will not play any role in the process of transferring data. If there is some abnormalities noted in the readings and one wish to take some consultancy, they can use the "Panic Button" attached with the device, and at that very moment, an alert will be created that will state the following: An option to consult a doctor (which will be suggested by our application if necessary for the patient) or get some information from previously suffered patients suggestions about the same

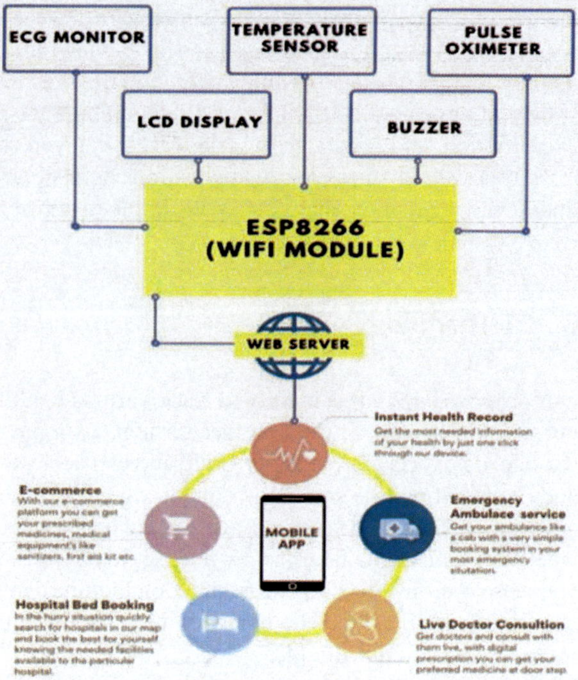


**Fig. 1** Block diagram of working of remote health monitoring system

and get cured domestically. The app also has the features of live doctor consultation and ambulance services which will help people get their emergency needs hassle free and they do not have to wait in queues for getting their requirements [6] (Fig. 1).

# 4 Hardware and Software Requirements

## *4.1 Hardware Requirements*

The hardware and software needed for the above proposed system consist of heart rate, ECG sensor, blood pressure sensor, temperature sensor and JIASH (Jivaan Ashash has been named to justify the action of an assurance to bring silver lining at the time of extremity). The wearable device will be active always as health issue does not come with notification.

### 4.1.1 Blood Pressure Sensor

Blood pressure sensor (BPS) is mainly used to detect human blood pressure. It has the ability to measure pulse pressure, systolic pressure and also mean arterial pressure.

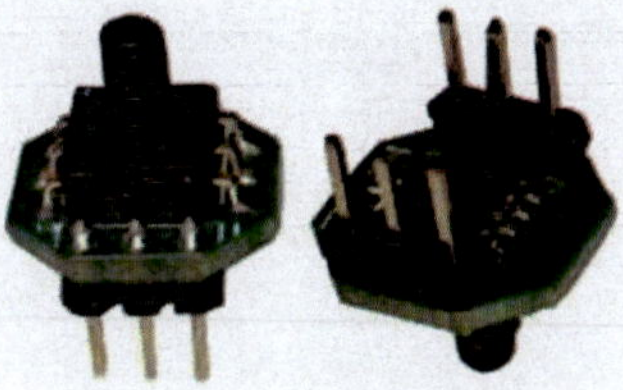

**Fig. 2** Blood pressure sensor

**Fig. 3** Temperature sensor pulse oximeter

When the blood pressure rises, it builds up in the blood and thus causes weakness, and that is why, it is important to keep an eye on the blood pressure (Fig. 2).

### 4.1.2 Temperature Sensor and Pulse Oximeter

Pulse oximeter and the temperature sensor, they are mainly used to measure oxygen saturation and temperature, respectively. To measure saturation level of oxygen in our blood, the pulse oximeter works like it uses a probe and it can be protected by a finger. In general, the normal oxygen saturation level is considered to be in range of 95–100%, and levels below 90% can cause cells to become depressed and damaged (Fig. 3).

### 4.1.3 ECG Monitoring Sensor

Electrocardiogram (ECG) sensor mainly used to detect a cardiac output at rest. This device provides us the details of patient important problems such as heart rate. Also it provides us the information on increased heart rate cause due to hypertension (high blood pressure) or any symptoms like supply of less oxygen supply to the heart (Fig. 4).

**Fig. 4** ECG monitoring sensor

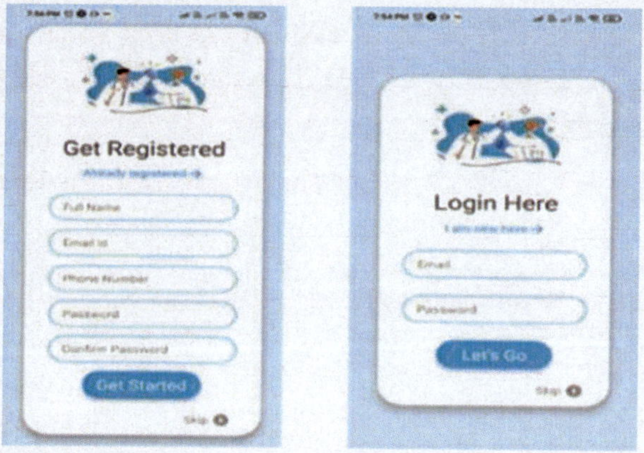


**Fig. 5** Register/login screen

## *4.2 Software Requirements*

**Jivaan Ashash (JIASH) Application**

JIASH (an assurance to bring silver lining at the time of extremity) is an application which is iOS and Android platform independent, developed using React Native technology for front-end and node-Js for back-end. This app will receive all the data from the device and notify the user of his/her health and compare it to send an alert in case of any abnormalities. This application also has other features like, scheduling monthly/weekly health checkup, collecting blood samples from the user's house in case any lab test has been prescribed by the doctor, supplying needful health commodities like masks, sanitizer, medicine, live doctor consultation and ambulance booking facilities. The lab reports of the user can be received through our app and can be consulted with the doctor by scheduling a checkup date. The glimpse of the application has been attached below:

### 4.2.1 Register/Login Screen

An user can get registered/login inside the app to keep a track of all the actions or services taken by them (Fig. 5).

### 4.2.2 Home Screen

This is the main screen of the app where all the services will be listed and can be availed by a click. It will also help the user to navigate to other useful screens (Fig. 6).

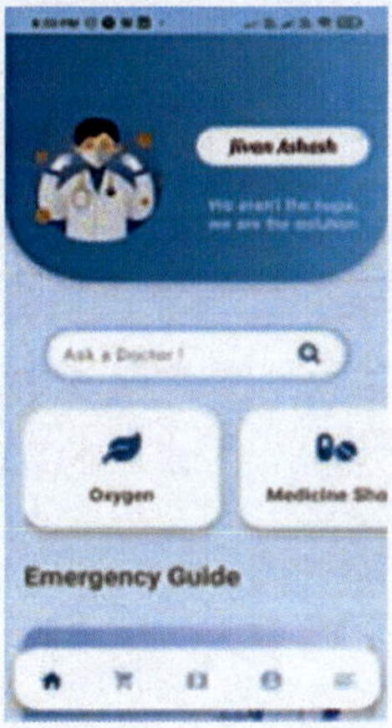


**Fig. 6** Home screen

### 4.2.3 Map Screen

This screen will enable users to search emergency resources. On entering one's current location, the user can get all the information of important (Fig. 7).

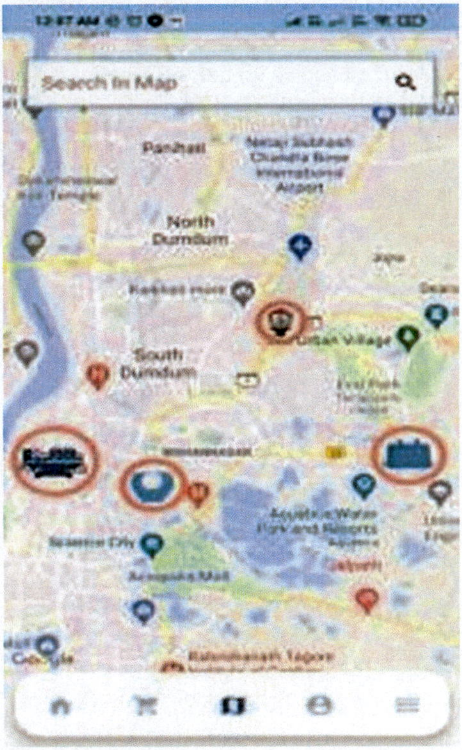


**Fig. 7** Map screen

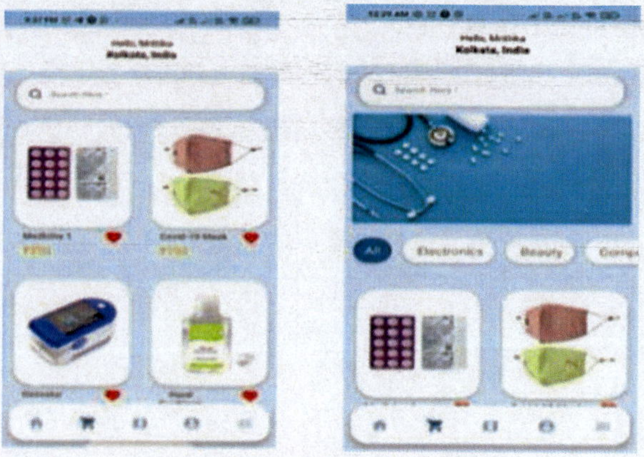


**Fig. 8** Shop screen

### 4.2.4 E-Shop Screen

This screen focuses on e-commerce and enables the user to order their useful health products at very ease with the comfort of home using the application (Fig. 8).

### 4.2.5 User Profile Screen

This screen shows all the information about the user account, and the user can track all the orders and can change app settings according to their convenience (Fig. 9).

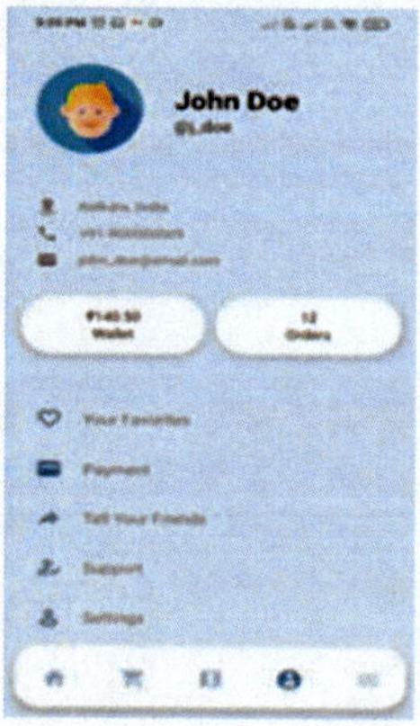


**Fig. 9** Profile screen

## 5 Future Scope

Our future scope includes upgrading the device with a more compact design, faster connectivity with devices using the Internet, more accuracy with data or readings and auto-analyze vital diseases and predict upcoming issues in the health using artificial intelligence. This will help us to detect or analyze rare cases of heart diseases in early stage or other health issues and prevention of its adverse effects. Our ultimate goal is to provide user-friendly, effective, reasonable health care services to the people which would enlighten their life with new hopes.

## 6 Conclusion

Remote health monitoring system is mainly based on the Internet of Things (IoT), which is cost-efficient, energy efficiency, portable, user-friendly and has given a satisfactory result. This system makes the use of sensors along with microcontrollers and applications. All the health parameters of an individual are being sensed by the sensors in our device and sent to the cloud with a proper firewall secure data protection layer. The health details are shown on the JIASH application preinstalled on the smartphone phone. This proposed system is very valuable as well as useful to our society, especially for the senior citizens as they suffers a lot in comparison with the rest young generation and can avail a regular checkup or a schedule checkup at their homes.

## References

1. Rahaman A, Noouddin S (2019) Sensor based health monitoring system
2. Bathilde JB, Chameera R, Zaidel DNA (2018) Continuous heart rate monitoring system using the Internet of Things (IoT)
3. Pawar S, Deshmukh HR (2018) A survey on e-health care heart monitoring for heart care using the Internet of Things (IoT)
4. Vijay Kumar G, Bharadwaj A, Nikhil Sai N (2017) Temperature and heartbeat monitoring system using the Internet of Things (IoT)
5. Said MS, Islam MM, Islam MR (2019) Microcontroller based health monitoring system
6. Rahman MA, Barai A et al (2016) Development of a device for remote monitoring of heart rate and body temperature

# Designing a Silicon-on-Insulator (SOI) Waveguide with an Aim of Studying Nonlinear Pulse Reshaping

**Hemant, Somen Adhikary, and Mousumi Basu**

**Abstract** In recent years, silicon photonics has attracted a lot of attention. On the other hand, pulse reshaping has been a fascinating research subject in recent years. In this paper, we describe a silicon-on-insulator (SOI) waveguide that has been precisely designed for pulse reshaping to generate a triangular pulse (TP). By matching the *v*-parameters in the *x*- and *y*-direction, we were able to create a rectangular buried SOI waveguide. The effective index method was used to calculate group velocity dispersion, and the modal distribution and effective area were used to calculate nonlinearity. Even though two-photon absorption and free-carrier generation contribute significantly to lose parameters, the highly nonlinear waveguide is shown to be capable of reshaping the Super-Gaussian pulse input into a triangular pulse shape. In addition, the input chirp, pulse width, and peak power values are also important for the formation of a high-quality triangular pulse. When compared to our result for the optical fibers, the required length is less for the triangular pulse reshaping. Our waveguide has potential in pulse generation, signal processing, and many other fields.

**Keywords** Buried waveguide · Effective index method · Pulse reshaping · Triangular pulse

## 1 Introduction

Nowadays, there are many technology platforms to work on photonic integrated circuits using different types of systems such as high-index glass, semiconductors, polymers, and silicon [1]. Silicon attracts the attraction of researchers due to its high refractive index and high confinement of the field distribution in the waveguide with minimum propagation loss. The motivation research is the availability of high-quality Si wafers, for creating buried waveguides [2]. The optical waveguide plays a vital role in optical communication engineering and integrated optical electronics. Thus, the so-designed waveguide is capable of pulse reshaping due to high nonlinearity

Hemant (✉) · S. Adhikary · M. Basu
Department of Physics, IIEST Shibpur, Howrah 711103, India
e-mail: hemant11061998@gmail.com



C. Giri et al. (eds.), *Emerging Electronic Devices, Circuits and Systems*,
Lecture Notes in Electrical Engineering 1004,
https://doi.org/10.1007/978-981-99-0055-8_32

and generating a triangular pulse with maximum stability [3]. We use different types of pulses for input as Gaussian pulse, Super-Gaussian, and hyperbolic secant pulse to generate a triangular pulse with the maximum stability as compared to the optical fibers. In our work, we design a rectangular buried waveguide using the inputs to get the triangular pulse.

Now, pulse reshaping has become a more fascinating research interest over the years. Thus, the pulse propagates in an optical waveguide in the presence of nonlinearity without any loss of propagation [4] and does not suffer from an optical wave-breaking effect [5]. One of the key concerns in optical pulse regeneration is pulse reshaping. Waveguide intrinsic features such as dispersion and nonlinearities are responsible for pulse distortion in the temporal and spectral domains during the transmission of an optical pulse through an optical waveguide communication system. As a result, having all-optical pulse reshaping capabilities would be ideal [6]. The modified pulse has a variety of applications in signal processing and quantum optics, as well as ultra-high-speed optical systems. Many practical applications, however, necessitate the use of a wider range of pulse waveforms, including flat-top (rectangular-like), parabolic, and triangular pulses. Changing the pulse waveform from the well-known Gaussian, Super-Gaussian, and hyperbolic secant shapes to more exotic triangular pulses could be useful for a variety of optical signal processing and modification applications [7]. Triangular pulses' basic intensity curve is ideal for a variety of photonic applications. The influence of the beginning pulse width and initial pulse chirp on the propagation of triangular pulse pairs is also investigated.

## 2 Designing a Buried Waveguide Using the Effective Index Method

We use silicon (Si) as the core and silica ($SiO_2$) as the cladding of the waveguide for designing and optimizing a buried waveguide in single mode using the effective index method. The refractive indices of the core ($n_1$) and cladding ($n_2$) materials are 3.477 [7] and 1.450 [8], respectively, from Sellmeier's formula at the propagation wavelength of 1.55 m. Numerically calculating the following equation [6, 9], the propagation constant () and the field associated with it can be derived. The buried waveguide is separated into two waveguides I in the *x*-direction and 2D waveguide II in the *y*-direction using this method [3]. Modal analysis is an important design procedure because it offers information on the modes that propagate on their propagation constant. The basic goal is to figure out what wavelength range the waveguide propagates in as a single mode. The number of confined modes depends on waveguide size, core–cladding refractive index difference, and operating frequency, which can be estimated by applying effective index method (Fig. 1).

First, determine the normalized frequency of the slab waveguide I so that mode is guided in the slab.

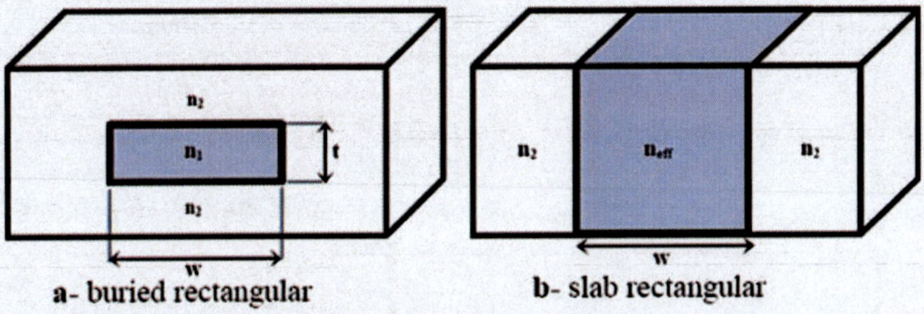


**Fig. 1** Simplifying the waveguide structure by using the effective index method

$$V_1 = K_0 T \sqrt{n_{\text{core}}^2 - n_{\text{clad}}^2} \tag{1}$$

Using the cutoff condition, the normalized frequency should be in a range of:

$$\tan^{-1}\sqrt{a} < V_1 \leq \pi + \tan^{-1}\sqrt{a} \tag{2}$$

where $a$ is the asymmetry measure of the waveguide:

$$a = \frac{\left(n_{\text{sub}}^2 - n_{\text{clad}}^2\right)}{\left(n_{\text{core}}^2 - n_{\text{clad}}^2\right)} \tag{3}$$

For a symmetric waveguide, $n_{\text{sub}} = n_{\text{clad}}$

$$0 < V_1 \leq \pi \quad \text{and} \quad 0 < V_2 \leq \pi$$

where

$$V_2 = k_0 W \sqrt{N_{\text{eff1}}^2 - n_{\text{clad}}^2} \tag{4}$$

represents the normalized frequency of the slab waveguide II (Fig. 2).

$N_{\text{eff1}}$ can be obtained by solving the eigenvalue equation.

$$k_x - 2\tan^{-1}\frac{\mu}{k_x} - m\pi = 0 \tag{5}$$

where $m$ is the integer (0, 1, 2, 3 …) which identifies the number of modes.

$$k_x = \sqrt{(k_0 n_{\text{core}})^2 - (k_0 n_{\text{clad}})^2} \tag{6}$$

$$\mu = \sqrt{(k_0 N_{\text{eff1}})^2 - (k_0 n_{\text{clad}})^2} \tag{7}$$

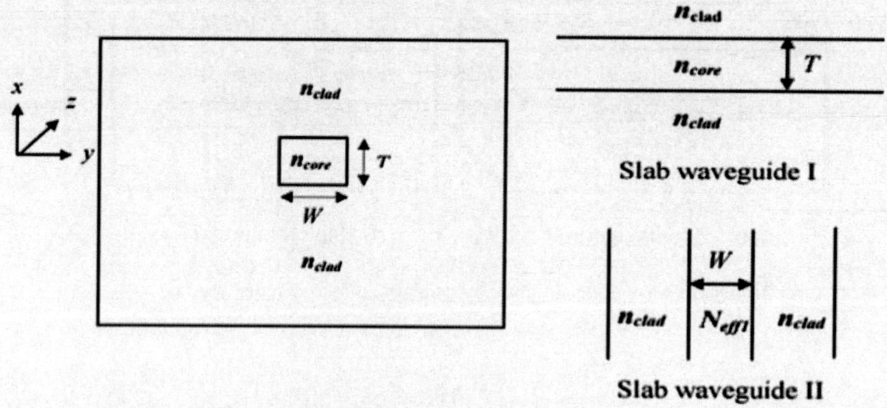


**Fig. 2** Structure division for calculation of the effective index calculation

Thus, putting the values from (6) and (7) in Eq. (5), we can get a transcendental equation which can be solved analytically to estimate $N_{\text{eff1}}$. Similarly, we can obtain a similar set of equation to calculate $N_{\text{eff2}}$ which will give us the effective index of our waveguide. From this effective index, one can easily estimate the propagation constant ($\beta$) and group velocity dispersion ($\beta_2$) by using the relations:

$$N_{\text{eff2}} = \beta / k_0 \tag{8}$$

$$\beta_2 = \frac{d^2\beta}{d\omega^2} \tag{9}$$

To ensure single-mode propagation inside our so-designed waveguide, we must ensure the fulfillment of condition $0 < V < \pi$ for both $V_1$ and $V_2$ [10]. To achieve that, we have optimized the height ($T$) and width ($W$) of the waveguide in such a way that $V_1 \approx V_2$ at the operating wavelength of 1.55 μm. We have checked for a large set of values and found that for $T$ ~ 0.24 μm and $W$ ~ 0.3 μm, $V_1 \approx V_2$ condition is satisfied as shown in Figs. 3 and 4.

Waveguide parameters such as height ($T$) and width ($W$) are optimized for the generation of triangular pulse at an operating wavelength ($\lambda$) of 1.55 μm. When the height of the waveguide is ~ 0.24 μm and the width is ~ 0.3 μm, the group velocity dispersion is estimated to be ($\beta_2$) ~ 2.432 ($ps^2/m$) and nonlinear coefficient ($\gamma$) is ~ 441.9 $(Wm)^{-1}$.

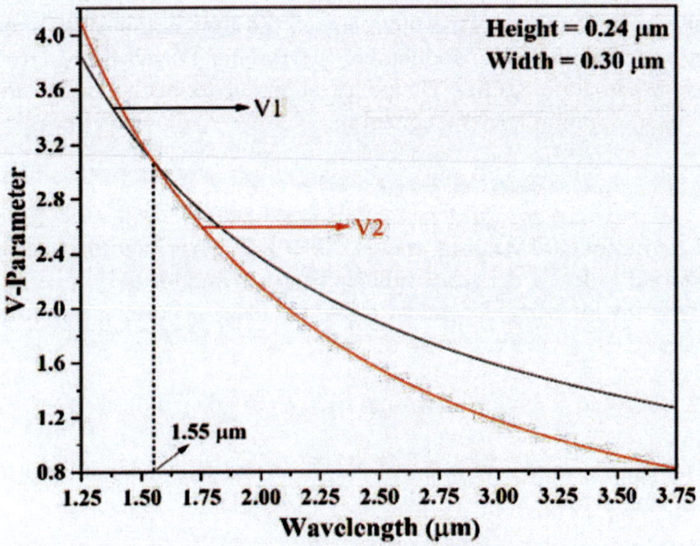


**Fig. 3** Variation of *V*-parameter with operating wavelength

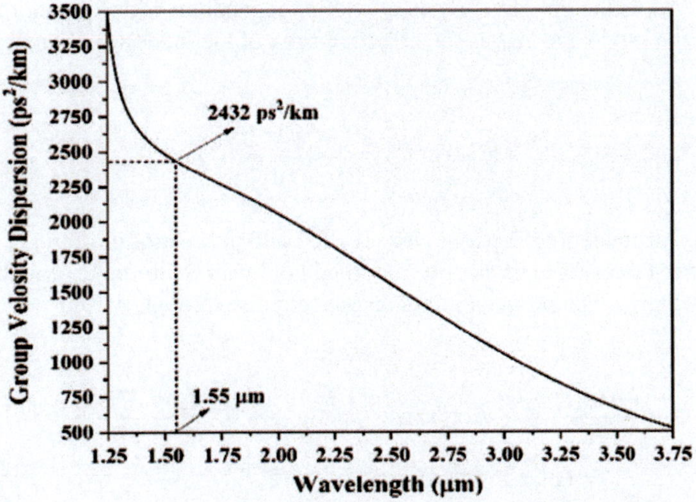


**Fig. 4** Variation of group velocity dispersion of buried waveguide with operating wavelength

## 3 Nonlinear Pulse Reshaping in the Proposed SOI Buried Waveguide

The nonlinear effects occur due to intensity dependence of the refractive index of the medium or due to the inelastic-scattering phenomenon, which is responsible for the Kerr effect. Depending on the input signal, the Kerr nonlinearity shows three different effects as Self-Phase Modulation (SPM), Four-Wave Mixing (FWM), and Cross-Phase Modulation (CPM). The polarization due to this bound electron is not linear and satisfies a more general equation:

$$P = \varepsilon_0 \chi^{(1)} E + \varepsilon_0 \chi^{(2)} E^2 + \varepsilon_0 \chi^{(3)} E^3 + \cdots \tag{10}$$

$\varepsilon_0$ is the permittivity of vacuum, and $\chi^{(k)}$ ($k = 1, 2, \ldots$) is $k$th order susceptibility. In silica, second order of the susceptibility vanishes due to its symmetry, and the nonlinearity is the intensity-dependent refractive index according to the following equation:

$$n = n_0 + n_2' I \tag{11}$$

where $n_0$ is the linear refractive index of the silica, and $n_2$ is the nonlinear refractive index coefficient, defined as:

$$n_2 = \frac{3}{4} \frac{\chi^{(3)}}{c \varepsilon_0 n_0^2} \tag{12}$$

$I = P/A_{\text{eff}}$ is the effective intensity within the medium with $P$ being the power carried by the mode and $A_{\text{eff}}$ is the **effective area** of the mode which is defined [9] as follows:

$$A_{\text{eff}} = \frac{\left[\int_0^\infty E_y^2(x) x \mathrm{d}x \mathrm{d}y\right]^2}{\int_0^\infty E_y^4(x) x \mathrm{d}x \mathrm{d}y} \tag{13}$$

Intensity-dependent refractive index is called **self-phase modulation** [11] (SPM). For our waveguides, we can choose modified nonlinear Schrodinger equation with higher order term. The propagation of an optical pulse through an SOI waveguide is given by:

$$\begin{aligned} &\frac{\partial A(z,t)}{\partial z} + \frac{\alpha}{2} A(z,t) + \frac{i\beta_2}{2} \frac{\partial^2 A(z,t)}{\partial T^2} - \frac{\beta_3}{6} \frac{\partial^3 A(z,t)}{\partial T^3} \\ &= i\kappa^{|} A(z,t)^{2|} A(z,t) - \frac{\alpha_{\text{f}}}{2} A(z,t) \end{aligned} \tag{14}$$

where $\alpha$ is the linear loss, and $\kappa = \gamma + i\frac{\Gamma}{2}$. The imaginary part of $\kappa$ is related to two-photon absorption [12] coefficient ($\beta_{\text{TPA}}$), where $\Gamma = \frac{\beta_{\text{TPA}}}{A_{\text{eff}}}$ and FCA is included by $\alpha_{\text{f}}$ (in unit of $\text{m}^2$) by the relation $\alpha_{\text{f}} = \sigma N$, and this $\sigma$ has the dimension of cross-section and $N_{\text{c}} \sim \frac{2\beta_{\text{TPA}} P_0^2 T_0}{3h\nu_0 A_{\text{eff}}^2}$.

Here, $P_0$ and $T_0$ are the peak power and pulse width, respectively. In order to calculate the quality of reshaped triangular pulses (TPs), a misfit parameter ($\mu$) is estimated as given by [12–15]:

$$\mu^2 = \frac{\int\left[|A|^2 - |A_{\mathrm{TP}}|^2\right]^2 \mathrm{d}T}{\int |A|^4 \mathrm{d}T} \tag{15}$$

where |*A*| is the intensity of a perfect triangular pulse and $A_{\mathrm{TP}}$ is the intensity of the triangular pulse at the output end. The optimum waveguide length ($L_{\mathrm{opt}}$), for which the misfit parameter is less than 2% ($\mu \leq 0.02$) and when misfit parameter remains stable within the above range for a sufficiently long length of the waveguide, is known as waveguide length as the stability length ($L_{\mathrm{st}}$) [16] as shown in Fig. 5. As shown in Fig. 6, the Super-Gaussian pulse with $m = 3$ order is suitable for input pulse to get triangular pulse in reshaping.

After determining the suitable input pulse for reshaping to triangular pulse, we have optimized the input parameters such as input power pulse ($P_0$), width ($T_0$), and pre-chirp. For optimization, input peak power of a Super-Gaussian pulse with $m =$ 3 order is varied between 0.9 and 1.5 W keeping the pulse width at a constant value of 1.1 ps. As shown in Fig. 7a, then the stability becomes maximum when the input peak power is 1.1 W, so it is considered as optimum power. Similarly in Fig. 7b, by keeping power of the pulse constant at the value of 1.1 W, the pulse width of the

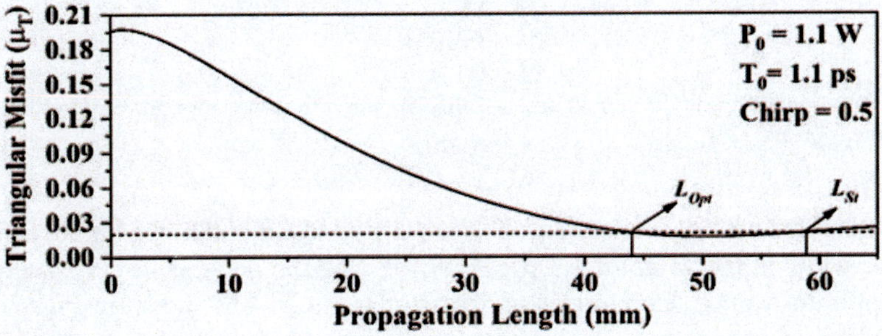


**Fig. 5** Variation of triangular misfit parameter for reshaped pulse with different lengths of propagation

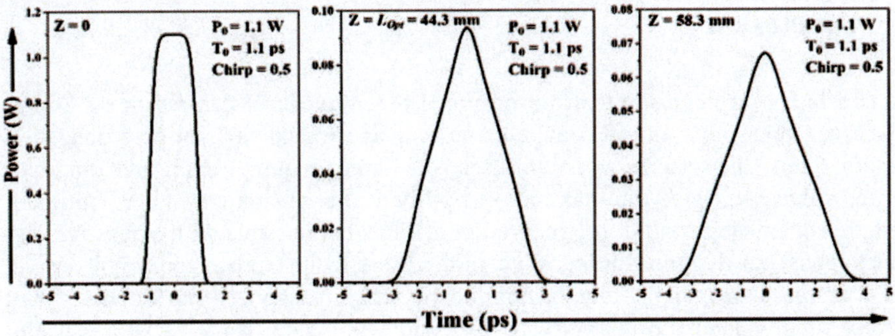


**Fig. 6** Input pulse, reshaped output pulse at optimum length and at the end of stability length (left to right) for triangular pulse generation

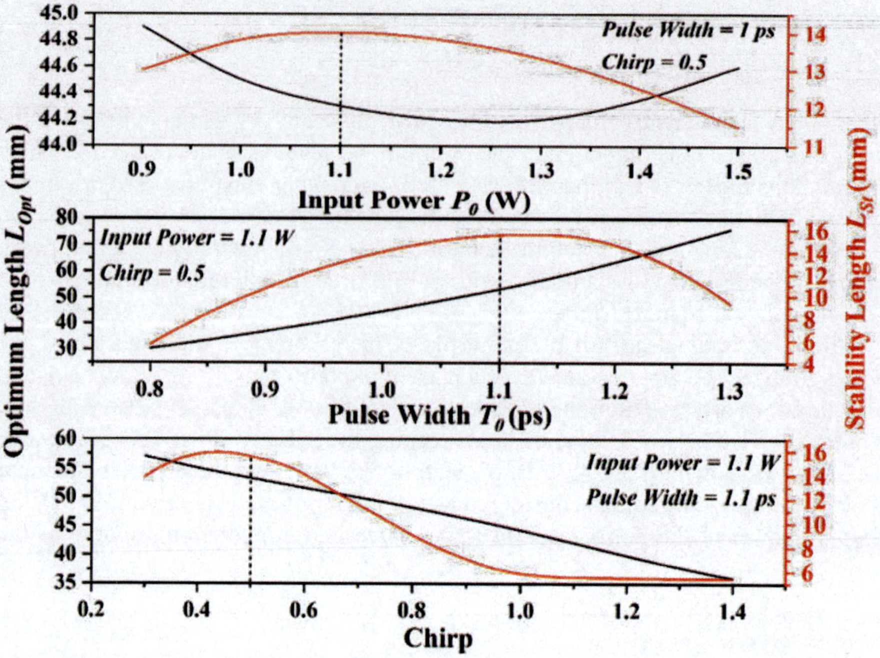


**Fig. 7** Variations of optimum length and stability length with (top) input power, (middle) pulse width, and (bottom) chirp

Super-Gaussian pulse is varied from 0.8 to 1.3 ps and we obtain the higher stability of pulse width at the value of 1.1 ps. Finally, fixing the power and pulse widths at their optimum values, the chirp is varied from 0.3 to 1.4 as shown in Fig. 7c, but the stability of generated TP is highest for a chirp value of 0.5. Generation of the triangular pulse is not possible with negative value of the chirp.

## 4 Conclusion

In conclusion, we can say that our designed SOI waveguide can reshape a Super-Gaussian input pulse into a triangular pulse shape within few cm of propagation length. Thus, this can be possible after calculating required data, and our optimized values will give the maximum stability at a comparatively lower optimum length. Under such conditions, generation of TP is not possible with a negative pre-chirp. Thus, our designed buried waveguide of core width 0.3 $\mu$m and height around 0.24 $\mu$m has a very high value of nonlinearity coefficient $\gamma \sim 441.19\ (\text{Wm})^{-1}$ and a very small value of group velocity dispersion $\beta_2 \sim 2.43\ (\text{ps}^2/\text{m})$ which is useful for the development of triangular pulses at lower optimal length of the waveguide, but the effect of nonlinearity is high as compared to dispersion; then, the stability range becomes moderate. Thus, this type of waveguides can be very useful for the applications in pulse production, signal processing, and a variety of other fields.

## References


1. Bogaerts W, Chrostowski L (2018) Silicon photonic circuit design: methods, tools and challenges. Laser Photonics Rev 12:1700237
2. Jalali B (2006) Silicon photonics. IEEE 24:4600–4615
3. Yulianti I, Supa'at ASM, Idrus SM (2008) Optimization of buried type waveguide for single mode operation. In: 5th international conference, AsiaCSN, Malaysia
4. Pavesi L, Lockwood DJ (eds) (2004) Silicon photonics. Springer, New York
5. Reed GT, Knights AP (2004) Silicon photonics: an introduction. Wiley, Hoboken, NJ
6. Chowdhury D, Bose N et al (2017) Performance of different normal dispersion fibers to generate triangular optical pulses. Opt Quant Electron 49:294
7. Lin Q, Painter OJ et al (2007) Nonlinear optical phenomena in silicon waveguides: modeling and applications. Opt Exp 15:16604–16644
8. Boscolo S, Latkin AI (2008) Passive nonlinear pulse shaping in normally dispersive fiber systems. IEEE J Sel Top Quant Electron 44:1196–1203
9. Bose N, Ghosh D et al (2013) Nonlinear pulse reshaping in a designed erbium-doped fiber amplifier with a multi cladded index profile. Opt Eng 52:086104
10. Agrawal GP (2008) Applications of nonlinear fiber optics, 2nd edn. Academic Press
11. Yin L, Lin Q, Agrawal GP (2006) Dispersion tailoring and soliton propagation in silicon waveguides. Opt Lett 31:9
12. Yin L, Agrawal GP (2007) Impact of two-photon absorption on self-phase modulation in silicon. Opt Lett 32:2031–2033
13. Ghatak A, Thyagarajan K (1999) Introduction to fiber optics. Cambridge University Press, Cambridge
14. Adhikary S, Ghosh BK (2020) Triangular pulse generation by using chalcogenide fibers and creation of tunable high frequency oscillations from the interaction of reshaped pulse pair. Optik 204:164208
15. Adhikary S, Basu M (2021) Nonlinear pulse reshaping in a typically designed silicon-on-insulator waveguide and its application to generate a high repetition rate pulse train. J Opt 23:125506
16. Liang TK, Nunes LR et al (2006) High speed logic gate using two-photon absorption in silicon waveguides. Opt Commun 265:171–174

# FaceDig: A Deep Neural Network-Based Fake Image Detection Scheme

**Simantini Ghosh, Suman Kayal, Manab Malakar, Anirbit Sengupta, Supriyo Srimani, and Abhijit Das**

**Abstract** Social media plays a significant role in contemporary societies in this modern era. The majority of users use social media to exchange text, photographs, and videos on a regular basis. Images are one of the most widely shared forms on social platforms and also the most vulnerable medium for tempering. Small and medium sized businesses may now easily create and distribute these images in a short frame of time, compromising news authenticity and public faith in social networking sites. This work proposes an approach for extracting the content of image, classifying it and verifying the authenticity (i.e., real or fake) of digital images. The algorithm uses convolutional neural network (CNN) to detect real and forged images. In addition, efficiency and refinement of training data are required to support its use in everyday life. The concept of error rate bid data and the deep learning method are used for further solutions. The result shows that the proposed method using convolutional neural network (CNN) technology and error level analysis (ELA), fake image detection can reach more than 96%. The findings of this work will aid in the detection of anomalous material and forged photographs on social media.

**Keywords** Image forgery · Classification · Convolution neural network (CNN) · Rectified linear unit (ReLU) · Sigmoid function

## 1 Introduction

Social networking websites have become a popular media trend in recent years, attracting a considerable percentage of users. The number of users [1–3] has now surpassed over five billion worldwide. The development of smart devices such as smartphones plays an important role in uploading and downloading images to those

S. Ghosh · S. Kayal · M. Malakar · A. Das
RCC Institute of Information Technology, Kolkata, India

A. Sengupta · S. Srimani (✉)
Dr. Sudhir Chandra Sur Institute of Technology and Sports Complex, Kolkata, India
e-mail: supriyo.srimani@dsec.ac.in

 
C. Giri et al. (eds.), *Emerging Electronic Devices, Circuits and Systems*,
Lecture Notes in Electrical Engineering 1004,
https://doi.org/10.1007/978-981-99-0055-8_33

networks. With emerging social networking platforms such as WhatsApp, Facebook, and Instagram, there has been an increased volume of photograph content created over the past few decades. The use of image processing software such as Pixelmator, Inkscape, Fireworks, GNU Gimp, and Adobe Photoshop to create embedded images is a major concern for social media companies such as Instagram and Facebook. In a sense, it has created a culture of impeccable perfection and non-existent splendor. With modified models, slim abdomen, fair skin, and beautiful hair, this has been described as a direct blow to the self-esteem of many and which leads people to behave responsibly. These images are a major source of false stories and are often used in a vicious way, such as to arouse public interest. To the bare eyes, the real and fake image is indistinguishable. In this technological age, around the world, most crimes and harassment or blackmail involve the use of photograph forgery. Many researchers in last decade tried to detect photograph forgery using different algorithms [4–15].

According to work by Zheng et al. [16], recognizing non-photographic stories is challenging, and news identification is essentially an open problem that can be solved using a few existing models. It has been recommended that the problem of 'fake news detection' can be investigated. Many valuable structures are determined by the text and visuals used in fake news after a thorough investigation. There are hidden elements in non-fiction texts and pictures that may be determined by a combination of hidden elements chosen from this model in different layers. The pattern called TI-CNN suggested. TI-CNN is developed in text and picture details concurrently by displaying a clear vision and embedded elements in a compact space. Raturi [17] reported an approach to identify a fake account in social platform. In this work, a machine learning technique has been employed to anticipate fraud accounts based on their activities on their social platform. In this method, complement Naive Bayes (CNB) and support vector machine (SVM) were used to evaluate material based on text classification and analysis of date. They have shown that, SVM has a 96% rate in detecting fraudulent accounts on Facebook, whereas CNB has a 95% accuracy rate in detecting fraudulent accounts with a wallet. The findings of this study indicate that the fundamental issue with social network security is that data is not properly investigated before being published.

The aim of this work is to develop a method that receives an input image and identifies it using CNN model. CNNs are highly good feature extractors for a whole new task/problem. CNN can perform automated feature extraction, and it removes the requirement for individual feature extraction. Weight sharing is another important aspect of CNN, which means that the same weight is applied for two layers in the model. CNN is selected in this research instead of other deep learning algorithms due to the features and functionality mentioned previously. The outcomes of this suggested study will be useful in identifying and tracking social media material as well as detecting fraud on social networking sites.

In this work, a methodology for fake image classification in social media is depicted. The remaining of this work is structure as follows: Sect. 2 discusses methodology and the problem formulation. Section 3 covers the results and observations. Section 4 concludes the work.

## 2 Methodology

The purpose behind this research work is to investigate a supervised machine learning classification issue with two classes: the actual image and the fake image. Conventional neural network (CNN) is used in this work to classify these images. Figure 1 shows the fake image detection algorithm in detail.

### *2.1 Dataset Preparation*

An image database is created containing high resolution (2048 × 1536 pixels), unencrypted, and abstract images from Kaggle, and for predictive purposes, a synthetic database is created using Adobe Photoshop CC. After importing all the images from the merged images were first scaled to 224 × 224 images. Images in the database are digitized using acquisition conditions. Each image has a label and one of two classes: (1) fake image, (2) real picture. Figure 2 shows the representation of real and fake images. The dataset is made of an extended training set of 1866 images, and a separate test set of 800 images. In the training datasets, the two classes are balanced, i.e., number of fake images and real images are 998 and 868, respectively.

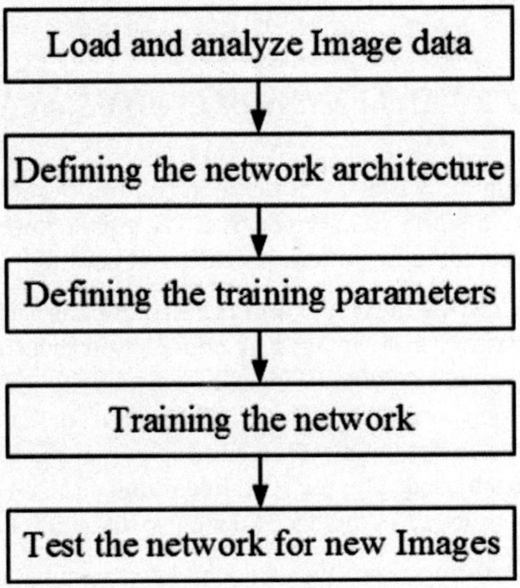


**Fig. 1** Flowchart of fake image detection

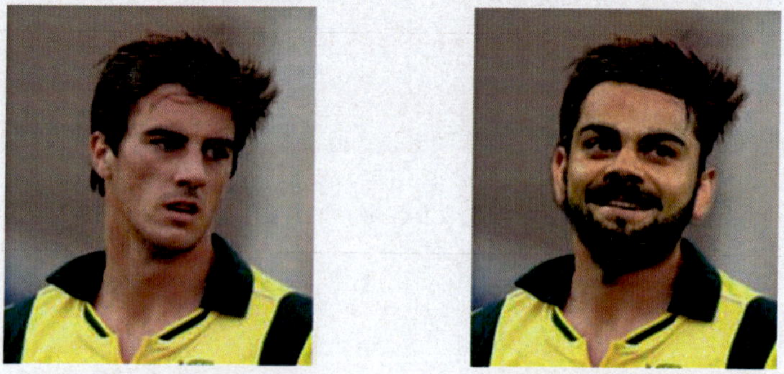

(a) Real Image (b) Fake Image

**Fig. 2** Example of training images

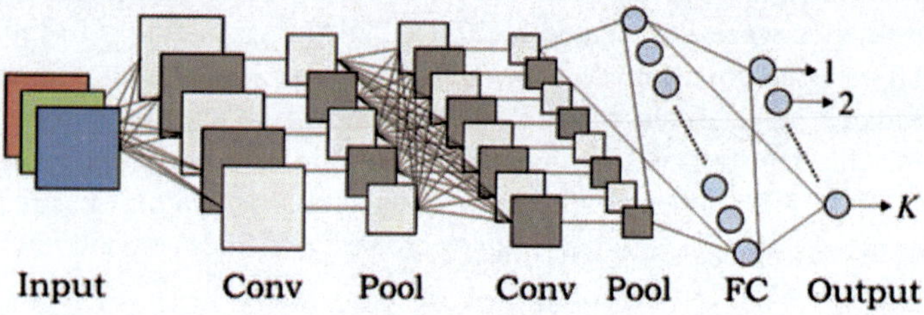


**Fig. 3** Convolution neural network (CNN)

## 2.2 *Architecture of Convolution Neural Network (CNN)*

Figure 3 illustrates the architecture of convolution neural network (CNN) used in this work. Each layer of CNN architecture consists of four major parts known as convolution, padding, stride, and pooling.

**Convolution Operation** Convolution is a mathematical approach that extracts information from images. A convolution layer cross-correlates the input and kernel to produce an output. Initially, the kernel goes horizontally, then moves downward and goes horizontally again. The output matrix is the summation of the dot product of the image pixel value and the kernel pixel value. In this work, a $3 * 3$ kernel has been used. The rectified linear unit (ReLU) is a nonlinear activation function that is employed in this work owing to the nonlinear behavior of image data.

**Padding** From Fig. 4, it can been seen hat the image size shrinks due to the loss in the pixels on the perimeter of the image. A small kernel size in the convolution layer will lead to reduction in loss in the pixel but it requires many successive convolution layers to extract the information from the image. Also, the kernel reaches the edge of the picture less frequently in the convolution layer, but it touches the center of the image more frequently, and it overlaps in the middle. As a result, the corner features of any image loss the prominence. So, extra pixels of filler around the boundary of

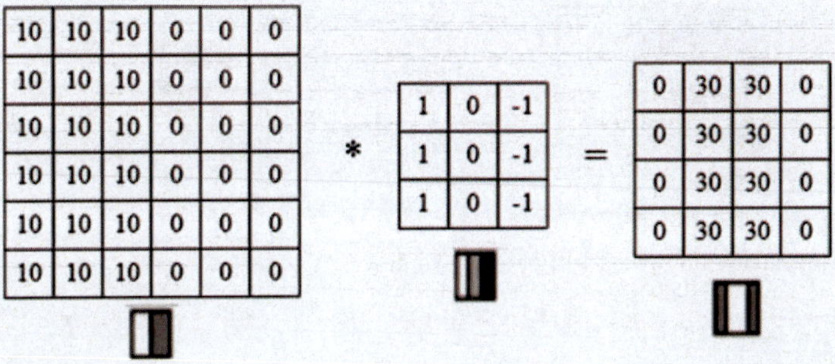


**Fig. 4** Convolution operation

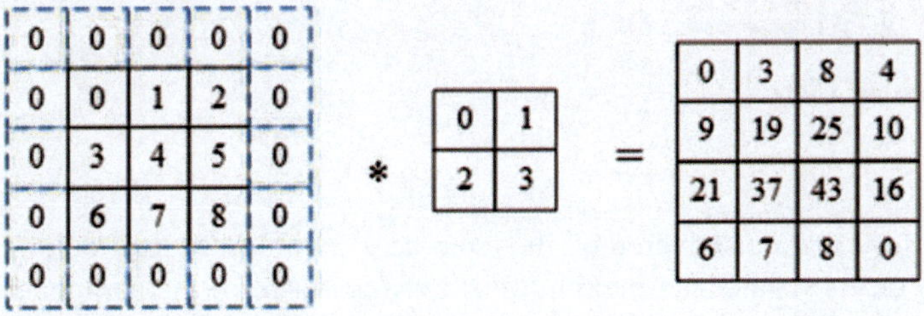


**Fig. 5** Padding with 2 $*$ 2 kernel

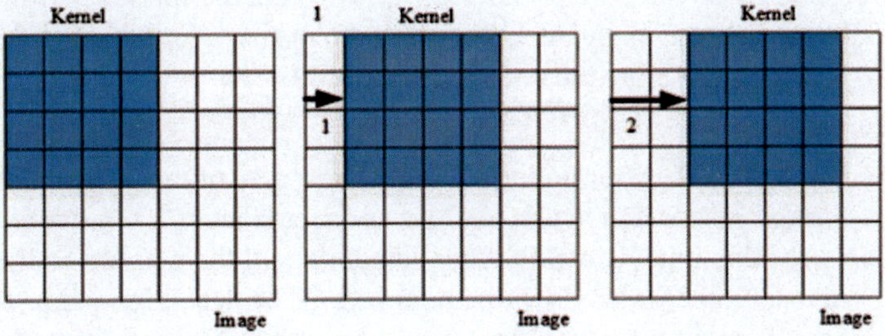


**Fig. 6** Stride = 0, stride = 1, stride = 2

the input image are added, i.e., increasing the effective size of the image. Figure 5 shows padding operation for 2 $*$ 2 kernel.

**Stride** Stride is denoted as the number of shift in the pixels over the input image. Higher stride is used either for computational efficiency or because of the need of downsample. The convolution window is moved more than one element at a time, skipping the intermediate locations for higher stride. Figure 6 shows stride for position 0, 1 and 2.

**Pooling** The pooling layer is also utilized to dramatically reduce the spatial size of the representation, cutting network complexity, and processing costs. In pooling operators, a fixed-shape window is slid across all areas in the input, generating a sin-

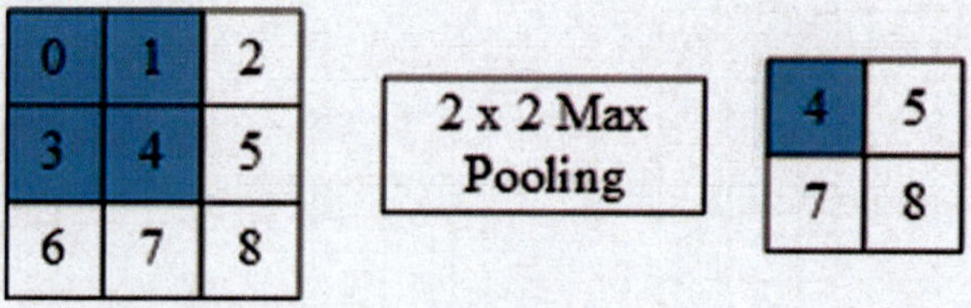


**Fig. 7** Maximum pooling with 2 × 2 pooling window

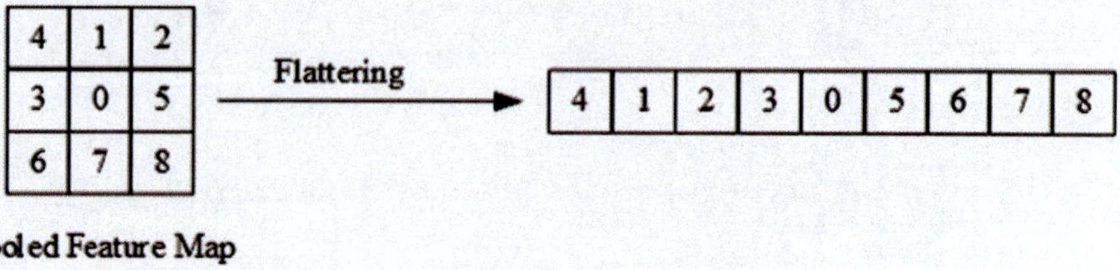


**Fig. 8** Flattening process

gle output for each area covered by the fixed-shape window or the pooling window, depending on its stride. The maximum or average value of the components in the pooling window is often calculated via pooling operators, which are deterministic in nature. In this work, maximum pooling is used as it selects most important features from the image. Figure 7 shows a pooling window, where the shaded portions are the first output element as well as the input tensor elements used for the output computation: $\max(0, 1, 3, 4) = 4$. Before the construction of fully connected network, the 'flattening' process converts the pooled feature data to the single column, shown in Fig. 8.

**Full Connected Network** After flattering, the feature map of the compact element is transmitted through a neural network. A fully connected layer is similar to a hidden layer in ANNs but this time, it is fully connected, i.e., all the neurons in the cell are connected with each other to those in the next layer. The activation parameter helps to implement the element activation function in a single layer. In this work, rectified linear unit (ReLU) is used as a activation function (Fig. 9).

**Dropout Layer** Dropout is a form of regularization, in which a fraction of the nodes feeding into a fully linked layer is dropped at random (shown in Fig. 10). It is a process used to prevent a model from overfilling. Dropout indicates that the activation function's contribution is set to zero. The gradients for lost nodes become 0 since there is no activation contribution. This effects the sensitiveness of the cost function to neighboring neurons that change the process of updating the weights, during the back propagation process.

**Output Layer** The output layer of CNN as mentioned earlier is a fully integrated layer, where input from other layers is downloaded and sent to convert the output to the number of classes as required by the network. The 'Sigmoid' function is used as an activation function in the output layer of neural network model that predicts and formulates the distribution of possibilities of the possible outcome.

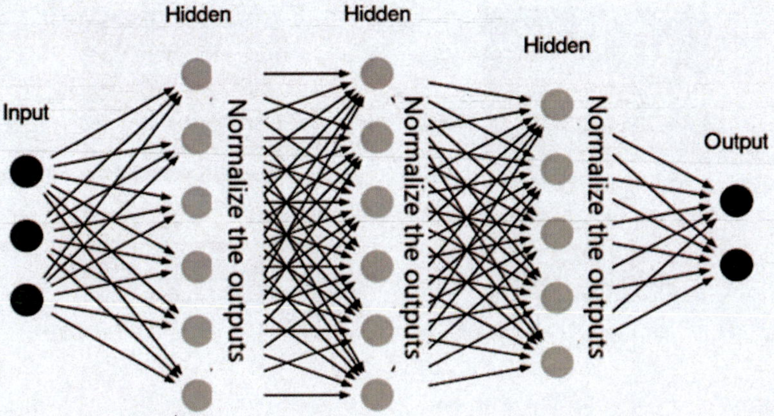


**Fig. 9** Architecture of a fully connected network

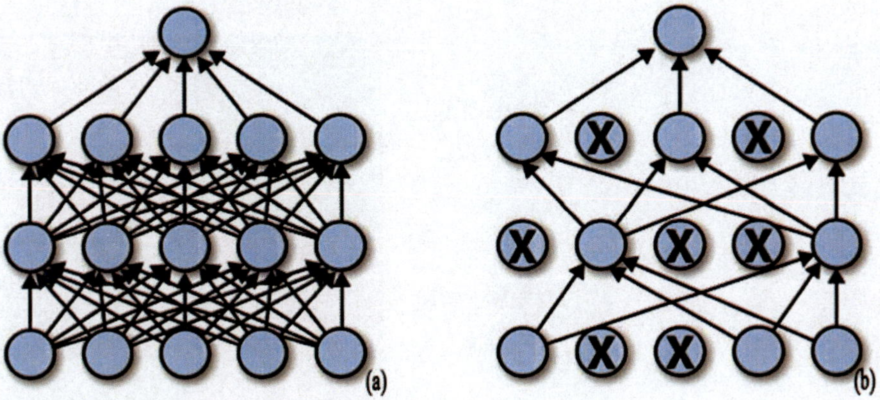


**Fig. 10** **a** Fully connected network, **b** connected network with dropout layers

## 3 Results and Discussion

The main purpose of this work is to accurately distinguish between fake and real photographs. For this objective, a convolution neural network is implemented in this work. Python 3.7 and TensorFlow 2.2.0 are used in this work for training and testing procedures, which are being conducted on the Google Colab platform. The parameter 'accuracy' is not a well-defined metric for predictive algorithms used for classification. Four other metrics like specificity, sensitivity, precision, and false positive rate have been evaluated in this work to assess the efficacy of the proposed image classification algorithm.

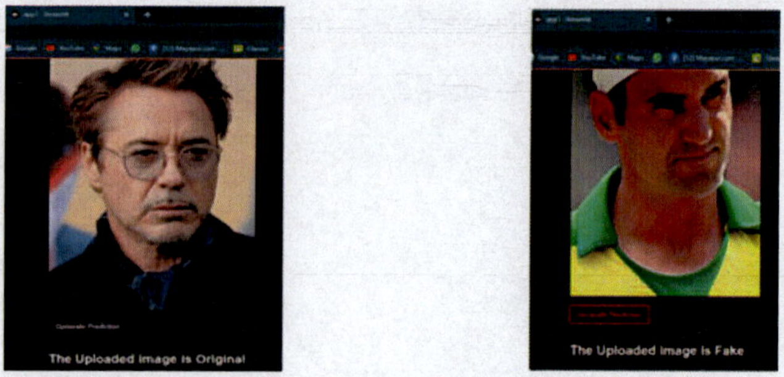


(a) Classified Real Image (b) Classified Fake Image

**Fig. 11** Output images

**Fig. 12** Confusion matrix for image classification

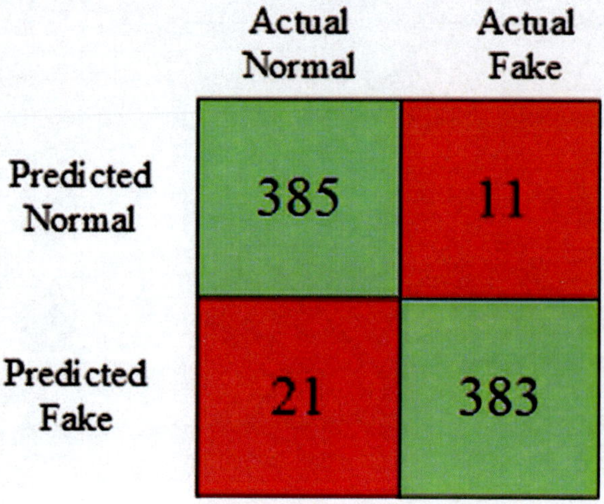


## *3.1 Dataset Preparation*

The implementation of the algorithm in Google Colab platform determines the originality of the image as shown in Fig. 11a, b. The confusion matrix shown in Fig. 12 shows the effectiveness of the algorithm for fake image classification. From the above confusion matrix, the values are calculated where accuracy is 96, precision is 97.22, specificity is 97.21, sensitivity is 94.83, and FPR is 0.0279. The proposed image classification method is compared (shown in Table 1) with some of the image classification techniques described in the previous works [18] to validate the performance. Depending on the results, it is concluded that the suggested CNN-based methodology detects fake pictures more effectively than earlier methods. This demonstrates the excellence of the implemented techniques.

**Table 1** Comparison of image classification with the previous works

| Performance parameters | Previous work [18] | This work |
|---|---|---|
| Accuracy | 93.25 | 96 |
| Specificity | 95.5 | 97.21 |
| Sensitivity | 91 | 94.83 |
| Precision | 95.29 | 97.22 |
| FPR | 0.0450 | 0.0279 |

## 4 Conclusion

Social media, nowadays, plays a very important role in our daily lives. Globally, there are an estimated 3.96 million people using social media, according to platform reports of the current number of active users. Social media is an useful platform for sharing, communicating, and spreading information. In the recent years, it has reduced the cost, time, effort, and capabilities of modern photography. Through social media, the spread of fake news and the spread of fake images are changing day by day. Especially in the political arena, counterfeit images can make or break the credibility of politicians. Therefore, there is no protection from threats to privacy and privacy. In this work, a method for extracting image content, classifying it, verifying digital image authenticity, and detecting fraud is suggested. This work has contributed to the discovery of counterfeit information, photographs, and social media, thus solving the problem of reporting false stories and false accounts through fake images. From the results, it is shown that the convolutional neural network model developed using in-depth learning is capable of achieving results with an accuracy of 96%. There are some limitations of the proposed algorithm like training dataset size and processing unit to train the model. The application of this algorithm for other types of images, for example, gray images would be the focus of future research.

## References

1. Mohamed SG. 100 social media statistics you must know. https://statusbrew.com/insights/social-media-statistics/
2. Blogger G. Saudi Arabia social media statistics. https://www.globalmediainsight.com/blog/saudi-arabia-social-media-statistics/
3. Ansari MD, Ghrera SP, Tyagi V (2014) Pixel-based image forgery detection: a review. IETE J Educ 55(1):40–46
4. Krizhevsky A, Sutskever I, Hinton GE (2012) ImageNet classification with deep convolutional neural networks. In: Advances in neural information processing systems, vol 25
5. Bunk J, Bappy JH, Mohammed TM, Nataraj L, Flenner A, Manjunath B, Chandrasekaran S, Roy-Chowdhury AK, Peterson L (2017) Detection and localization of image forgeries using resampling features and deep learning. In: 2017 IEEE conference on computer vision and pattern recognition workshops (CVPRW). IEEE, pp 1881–1889

6. Aphiwongsophon S, Chongstitvatana P (2018) Detecting fake news with machine learning method. In: 2018 15th international conference on electrical engineering/electronics, computer, telecommunications and information technology (ECTI-CON). IEEE, pp 528–531
7. Kim DH, Lee HY (2017) Image manipulation detection using convolutional neural network. Int J Appl Eng Res 12(21):11640–11646
8. Hsu CC, Lee CY, Zhuang YX (2018) Learning to detect fake face images in the wild. In: 2018 international symposium on computer, consumer and control (IS3C). IEEE, pp 388–391
9. Hsu CC, Hung TY, Lin CW, Hsu CT (2008) Video forgery detection using correlation of noise residue. In: 2008 IEEE 10th workshop on multimedia signal processing. IEEE, pp 170–174
10. Farid H (2009) Image forgery detection. IEEE Signal Process Mag 26(2):16–25
11. Ahmed SRA, Sonuç E (2021) DeepFake detection using rationale-augmented convolutional neural network. Appl Nanosci 1–9
12. Dang LM, Hassan SI, Im S, Lee J, Lee S, Moon H (2018) Deep learning based computer generated face identification using convolutional neural network. Appl Sci 8(12):2610
13. Marra F, Gragnaniello D, Cozzolino D, Verdoliva L (2018) Detection of GAN-generated fake images over social networks. In: 2018 IEEE conference on multimedia information processing and retrieval (MIPR). IEEE, pp 384–389
14. Dang LM, Hassan SI, Im S, Moon H (2019) Face image manipulation detection based on a convolutional neural network. Expert Syst Appl 129:156–168
15. He K, Zhang X, Ren S, Sun J (2016) Deep residual learning for image recognition. In: Proceedings of the IEEE conference on computer vision and pattern recognition, pp 770–778
16. Yang Y, Zheng L, Zhang J, Cui Q, Li Z, Yu PS (2018) TI-CNN: convolutional neural networks for fake news detection. arXiv preprint arXiv:1806.00749
17. Raturi R (2018) Machine learning implementation for identifying fake accounts in social network. Int J Pure Appl Math 118(20):4785–4797
18. AlShariah NM, Khader A, Saudagar J (2019) Detecting fake images on social media using machine learning. Int J Adv Comput Sci Appl 10(12):170–176

# Self-heating Effects on Power Loss of SiC-Based General-Purpose Inverter-Stack Circuit

**Subhajit Das, Takahiro Iizuka, and Mitiko Miura-Mattausch**

**Abstract** This investigation presents a simulation-based analysis of the self-heating effect on power loss of a general-purpose inverter-stack circuit for high-power applications. This study addresses the self-heat consequence due to the instantaneous temperature increase in the circuit due to the individual device temperature increase. It has been observed that the insignificant change in switching loss with the original material-specific physical quantities is in contradiction to the common observation. It is found that the most significant reason for the high-temperature increase of circuits is due to the hindrance of the heat propagation to the outside of the devices. The investigation is done to simulate the power loss change under such conditions. In order to simulate the real circuit in a package, an equivalent device thermal network-based simulation setup has been developed. It is found that the realistic power loss for long-term circuit operation increases linearly with the thermal resistance of the heat conduction medium, such as the junction to ambient via the case and the junction to ambient through the case and the heatsink. One-dimensional thermal distribution network of GPIS considering different heatsink thermal resistances has also been demonstrated in the present study.

**Keywords** Self-heating · General-purpose inverter-stack circuit · SiC-MOSFET · Thermal network · Power device · Compact model · HiSIM_HSiC

S. Das (✉) · T. Iizuka · M. Miura-Mattausch
HiSIM Research Center, Hiroshima University, Hiroshima 739-8530, Japan
e-mail: sjdsubha@hiroshima-u.ac.jp

T. Iizuka
e-mail: iizuka@hiroshima-u.ac.jp

M. Miura-Mattausch
e-mail: mmm@hiroshima-u.ac.jp

 
C. Giri et al. (eds.), *Emerging Electronic Devices, Circuits and Systems*,
Lecture Notes in Electrical Engineering 1004,
https://doi.org/10.1007/978-981-99-0055-8_34

## 1 Introduction

In the ever-increasing requirements in the energy budget, the adoption of silicon carbide (SiC) power transistor for high-voltage circuits is following a growing trend across different high-power applications, which includes modern electrical vehicles (EVs) and their charging infrastructure, power generation (wind, photovoltaics, etc.), uninterruptible power supplies, high-power inductive applications, aircraft, deep-space exploration, etc. [1]. Subsequently, this emerging acceptance of the SiC technology in the field of high-temperature, as well as high-power applications, the thermal problem of heat generation, and dissipation become extremely important with comprehensive understanding and analysis of the SiC system performance [2]. The multiphysics modeling tools capturing the fundamental physics of the SiC material are widely used both in academia and industry to assess the benefits of SiC power devices to optimize the power electronic system performance. However, the undesirable effects such as the formation of the hotspot, thermal runaway, electrothermal coupling of the switching devices, which adversely impact the reliability or the performance of power electronic (PE) integrated circuits, require a careful electrothermal design of SiC-based PE systems [3, 4]. As the switching elements of PE systems can experience high-temperature transients during the switching events, a large amount of switching power produces considerable self-heating in the switching transistors [5]. The study on the reliability of PE devices has revealed that it is closely related to the thermal loading profile of the system [6]. The power switching device can be damaged in a single switching event if the junction temperature reaches more than the allowable temperature while in operation. The SiC has high thermal conductivity and subsequently possesses more easy thermal propagation capability than that of the silicon (Si) and gallium nitride (GaN) counterparts. It is demonstrated in the literature that the SiC die is capable of operating with junction temperatures being above 200 °C [7, 8]. Figure 1 depicts the comparative heat propagation performance of Si, GaN, and SiC-based devices. Therefore, it could be presumed that the self-heating effect is expected to be smaller in SiC-based power electronic circuits. Subsequently, to ensure the sustained and reliable operation of the SiC-based device, a computationally efficient thermal model capturing the real-time PE system performance and predicting the high-temperature thermal profile is highly demanded.

The thermal modeling based on the thermal profile of the PE switching devices using resistor–capacitor (RC)-based electrothermal model includes very high computational efficiency for extensive circuit simulation [10]. Consequently, the Foster and Cauer thermal model has extensive use in the thermal profiling of the system [11]. In the Foster configuration, the RC extraction is carried out from the transient thermal experimental measurement, while the device geometry and material property build the RC model in the Cauer counterpart. The Cauer model in electrothermal modeling has been presented in several studies in the literature [12, 13]. However, in realistic

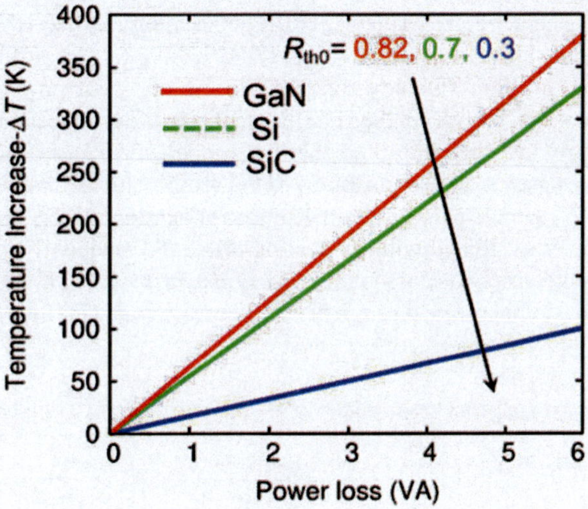


**Fig. 1** Self-heat as a function of DC power loss for different materials. $R_{th0}$ presents the thermal resistivity ($W^{-1}$ K) of SiC, Si, and GaN [9]

circuit operation, a considerable switching loss increase is found due to the self-heating in the PE circuits, which in turn alters the thermal profile of the power electronic circuit. The interfacial thermal network of the transistor plays a significant role in heat conduction to the ambient. The heat generation in the circuit and, as a consequence, the generated power loss necessitate careful study on the realistic circuit to shape a practical thermal design. Considering the aforementioned issue, in this work, the self-heating effects on the general-purpose inverter stack (GPIS), an essential module of extensive PE application, have been studied. Subsequently, the study on the thermal network and its effect on the real-time circuit operation has been investigated.

## 2 General-Purpose Inverter-Stack Circuit

A 12 kVA GPIS circuit has been built with two half-bridge inverters, as depicted in Fig. 2a [14]. Individual bridge of the GPIS design spites the DC buses so that they can be connected externally; subsequently, it is useful for different electromagnetic (EM) applications, high-speed motor drives, high-temperature induction welding applications, etc. The SiC power MOSFET C2M0160120D from Cree [15] has been adopted as the device for the GPIS circuit. This SiC device allows the maximum junction temperature, $T_{jmax} = 150$ °C. Subsequently, the maximum allowed temperature rise ($T_{jrise}$) with the safety factor of 15% is 85 °C. Consequently, the switching loss, resulting self-heating effect, and optimized thermal design play a significant role in building a stable and reliable system. In this simulation-based investigation, the model card for the compact model of world standard HiSIM-HSiC [16] has been developed utilizing the datasheet provided by Cree. Table 1 summarizes the GPIS system features. HiSIM_HSiC is based on HiSIM_HV, the industry-standard model for Si high-voltage devices [17–19].

In order to study the self-heating and heat propagation in the GPIS circuit, a compact model in the HiSIM_HSiC [16] for self-heating has been applied. Figure 2b presents the simplified thermal model of HiSIM_HV [17–19], where the thermal resistance ($R_{th}$) contributes to the temperature increase and the thermal capacitance ($C_{th}$) took part in the thermal propagation. First, the device characteristics have been evaluated without heating effect. Subsequently, the temperature due to instantaneous power loss is updated, considering the self-heating impact by incorporating the temperature through the temperature node ($T_j$) into the device characteristics [17].

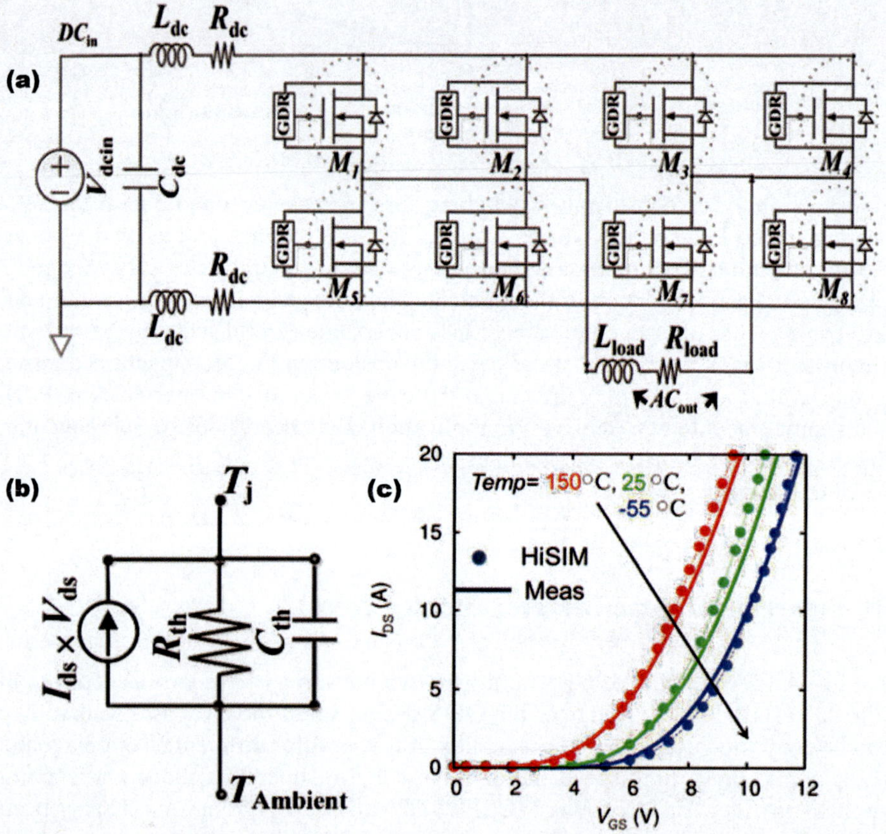


**Fig. 2** **a** Circuit diagram of the 12 kVA general-purpose inverter-stack (GPIS) circuit [14], based on the SiC power MOSFET C2M0160D from Cree [15]; GDR is the ideal gate driver for pulsating switch control. **b** Equivalent thermal network for incorporating the self-heating effect into a HiSIM-HV, using a thermal resistance $R_{th}$ and a thermal capacitance $C_{th}$. **c** Comparison of the $I_{ds}$–$V_{gs}$ characteristics of the Cree C2M0160120D SiC power MOSFET with simulation results with HiSIM_HSiC. The dotted line represents Cree device characteristics, and the solid line represents the HiSIM result

**Table 1** Rating of GPIS and sub-systems

| Item | Quantity/range |
|---|---|
| Power | 15 kVA |
| DC-bus voltage $V_{DC}$ | 800 V |
| Output voltage $V_{LL}$ | 460 V |
| Output RMS current | 22 A |
| Nominal power factor | 1 |
| Switching frequency $f_{sw}$ | Up to 500 kHz |
| SiC-MOSFET | C2M0160120D |

Then, the re-evaluation of the device characteristics is realized with the calculated temperature $T$. The $R_{th}$ and $C_{th}$ are dimension-oriented entities and function of heat resistivity ($R_{th0}$) and thermal capacity ($C_{th0}$) of the transistor's junction to ambient, respectively. The model accuracy is verified by comparing transfer characteristics for three different temperatures for a fixed $V_{ds}$ of 20 V as presented in Fig. 2c. In the GPIS circuit, the transistors M1, M2, M7, and M8 work to attribute the rising edge of the load current, where the M3, M4, M5, and M6 contribute to the falling edge of the converted AC signal of the GPIS. The gate pulses of the switch transistors are driven through the gate driver network with − 5 to 20 V range with a current sink rating of 5 A, in such a fashion when M1, M2, M7, and M8 are turned 'ON,' the rest of the groups M3 M4, M5, and M6 are turned 'OFF.' This complementary nature of the switching causes delay to dissipate power, which has been studied later. The power line LPF protects the powerline harmonics generated during the high-frequency switching of the power transistors. The body diode is embedded with the SiC device for reverse current surge protection. The power loss during the circuit in operation can be evaluated as the sum of all devices included, where some devices calculate the switching loss and some conduction loss of the switching devices. The switching power loss ($P_{sw\_loss}$) is given as:

$$P_{sw\_loss} = 0.5 \times V_{ds} \times I_{ds} \times f_{sw} \times t_{sw} \quad (1)$$

where switching frequency ($f_{sw}$) is the switch driving frequency by the gate driver with an actual time of switching event ($t_{sw}$). Subsequently, to study the thermal propagation from case to ambient through the external thermal network, the Cauer model [20] for a thermal network of the circuit has been adopted for this work, as shown in Fig. 3. It should be noted that the individual transistor has been connected to the ambient via the case to ambient through the $R_{th}$ and $C_{th}$, as shown in Fig. 2a. In the external thermal network, the RC elements of base plates and other heat-spreading components (e.g., heatsink) are connected to individual transistors in the system as presented in Fig. 3. In this figure, thermal RC connected for each transistor has been presented as $R_{nm}$–$C_{nm}$, where $n$ represents the position of the transistor and $m$ is the subsequent thermal interfaces of it. In order to demonstrate the model, the experimental board-level setup has been referenced [21, 22] in Fig. 4. From this

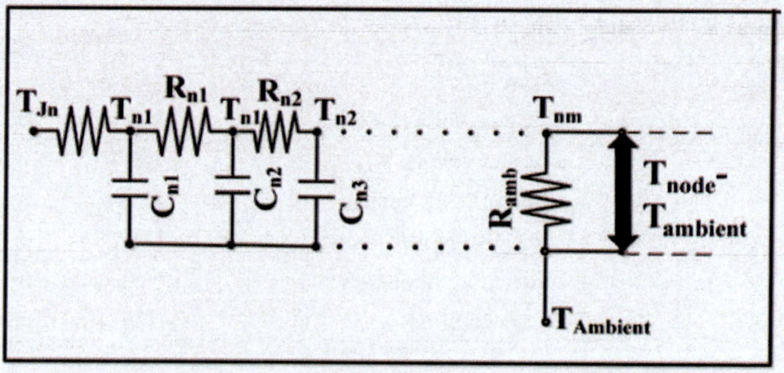


**Fig. 3** Cauer model of electrical transmission line equivalent circuit for modeling external thermal network of GPIS. $T_{jn}$ presents the $T_j$, as shown in Fig. 2b, of the transistors of the GPIS circuit

**Fig. 4** Experimental board-level setup of SiC MOS-based GPIS system with the thermal arrangement [21, 22]

figure, it is evident to show that the back surface of the package is connected to the ambient via heatsink, where the front part of the package is suspended in the air to carry the heat to the ambient.

## 3 Self-heating Effects in GPIS System

The 12 kVA GPIS circuit depicted in Fig. 2a has been fabricated with C2M0160120D, packaged with the TO-247-3, presented in Fig. 4 [21, 22]. Each (16 mm × 21 mm) package contains the vertical N-channel power MOSFETs with a planar gate [23]. The heatsink (300 mm × 80 mm × 33 mm) arrangement is installed with the eight transistors in a 2-H-bridge configuration. The GPIS arrangement has been simulated to investigate the self-heating effect on the circuit. It is evident that the junction temperature propagates to the ambient through the case and the case interface temperature reaches the steady-state value after the thermal time constant period ($\tau = R_{th} \times C_{th}$).

In order to simulate the thermal behavior, the 4-T configuration [17] of the transistor has been considered, where the effective internal thermal resistance considering the contacts and dielectric of the transistor on SiC substrate ($R_{th} = 0.055\ W^{-1}\ K$, and $C_{th} = 30\ \mu J\ K^{-1}$) [12] has been considered. In this setup, the self-heat generated in the device can be measured; however, the same can be connected to an external thermal network with the 5-T configuration [17] demonstrated later. In the study, the simulations have been performed with various thermal capacitive settings keeping the $R_{th}$ constant as well as the magnitude of the self-heating considering the different $R_{th0}$ in the circuit. It should be noted that with an optimized design of half-bridge GPIS as presented in [14], all the transistors generate almost identical steady-state self-heat during the circuit operation. Subsequently, the self-heating characteristics of any transistor represent the performance of each transistor individually in the H-bridge of the GPIS. In this study, it has been verified from Fig. 5a–d that the magnitude of temperature rise in steady state in the individual transistor (study on transistor M1 has been portrayed in subsequent studies) is dominantly controlled by the thermal resistance, where the time to reach steady-state ($T_{sat}$) is the thermal time constant, which is analogous to the effect of electrical resistance and capacitance. Figure 5a, b present a significant observation of the thermal capacitance on heat propagation. The heat capacitance obstructs the heat from propagating and puts forth a hindrance to reaching the steady-state temperature. Subsequently, the design of a circuit is essential to achieve steady-state thermal conditions by proper selection of the heat capacity of the transistor interface.

The self-heating phenomenon puts forth a significant impact on the switching loss of the circuit. This excess switching loss again rises the operational temperature. Subsequently, it is evident that the heat-conductive medium holds a vital role to propagate heat to the ambient and attaining the switching loss in safe operating margin. It can be studied from Fig. 5d that switching power loss rises linearly with the thermal resistance between the junction and ambient. The excess switching loss generation with the rise of the operating temperature is due to the fact that the device reaction time during the transient switch in medium-to-high frequency circuits increases with the delay in discharging current to drain out capacitive charges from the large drift region during the turnoff transition, which contributes to the drain current as:

$$I(t) = I(V) + \mathrm{d}Q(t)/\mathrm{d}t \tag{2}$$

The discharging current shows a relative increase due to the rise in minority carrier lifetime with temperature, resulting in the delay of the expected $I_{gs}$ and $I_{ds}$, thus causing, in particular, the $V_{ds}$-response delay. This slowed-down response of the individual devices causes the total power loss increase [14, 24]. Consequently, the discharging delay rises, causing an additional switching loss in the switching transistors in the GPIS circuit which can be inferred from Fig. 6a. However, it is

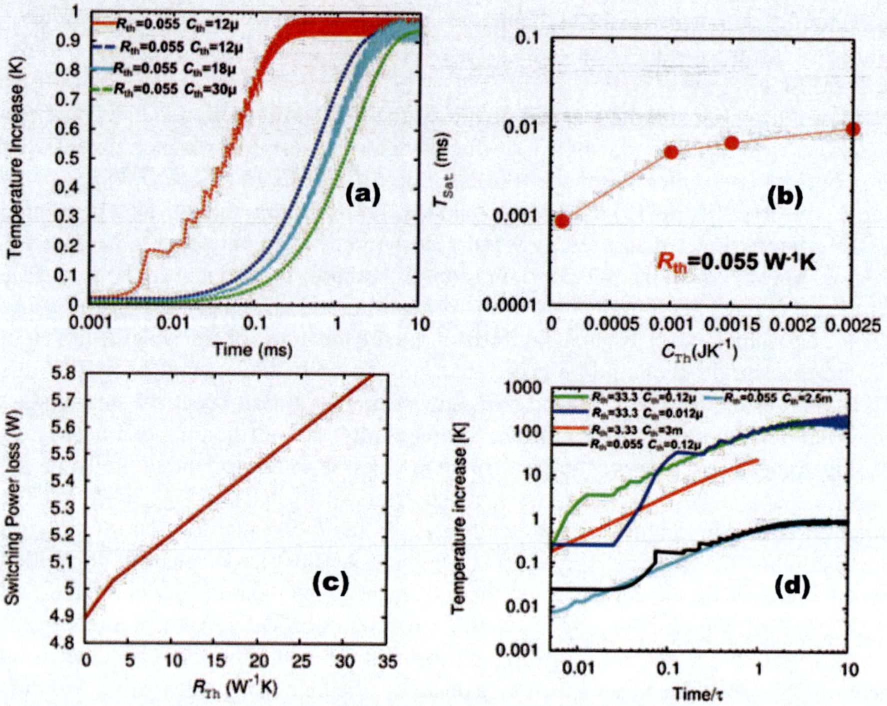


**Fig. 5** **a** $\Delta T$ converges to a fixed steady state for fixed $R_{\text{th}}$ at a different time for different $C_{\text{th}}$ values, **b** time to reach steady-state different $C_{\text{th}}$ (in J K$^{-1}$) values of individual transistors of the GPIS, **c** simulated power loss as a function of thermal resistivity of the heat conduction medium $R_{\text{th}}$ (W$^{-1}$ K), **d** simulated switching loss as a function of time/$\tau$, where $\tau$ is the thermal time constant of the heat propagation medium of the transistors

worth mentioning that this additional switching loss varies significantly with the design of the body diode of the power device. The complementary MOSFETs in the circuit experience the reverse inductive potential during the falling gate pulse of the active MOSFETs resulting in the turning on of the body diode. The turnon time of the body diode of the complementary MOSFETs further boosts the discharging duration of the MOSFETs, which are in turnoff transition.

In order to investigate the thermal behavior of the power module together with the interfacial components, the Cauer network depicted in Fig. 3 represents the actual physical layer of the system from the heat source to the chip ambient, as shown in Fig. 4. This network represents the RC elements of the semiconductor chip, substrates, base plates, and other heat-spreading components in the system. This Cauer network for interfacial thermal design of the circuit has a significant role in defining the system's final thermal state. Subsequently, to simulate the thermal condition in the GPIS system developed with C2M0160120D, the simulation setup as 'Approach-I' with 5T configuration [17] of the device has been developed as presented in Fig. 7a. The 'Approach-I' depicts the simulation setup with the devices

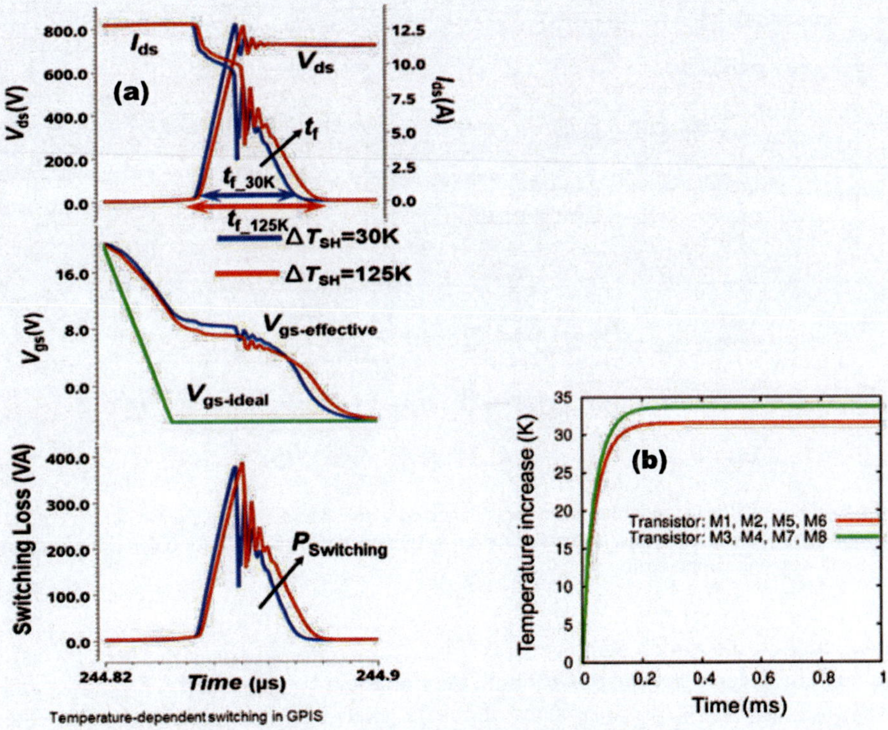


**Fig. 6** **a** Switching response of GPIS circuit considering two different self-heating conditions, the red line indicated a higher self-heating condition, and the blue is the lower counterpart. $V_{\text{gs-ideal}}$ is the ideal gate-to-source voltage applied from ideal gate driver, **b** increasing self-heat in switching transistors of GPIO circuit considering $R_{\text{th}} = 40\ \text{W}^{-1}\ \text{K}$ [15]

connected to the external thermal network as demonstrated in the experimental board-level setup in Fig. 4, keeping the junction to case thermal resistance ($R_{\text{j-c}} = 1\ \text{W}^{-1}\ \text{K}$) as per the device datasheet [15]. As demonstrated in [21, 22] the back surface of the case of every transistor is connected to the heatsink; subsequently, in order to design the thermal resistance of the case to ambient ($R_{\text{c-a}}$) of the front part of the case which is directly connected to the ambient, the $R_{\text{c-a\_effective}}$ is calculated as $2 * R_{\text{c-a}}$, where $R_{\text{j-c}}$ is obtained as $39\ \text{W}^{-1}$ K from the datasheet [15].

It is quite an obvious fact that due to the substantial thermal conductivity of the heatsink, the steady-state temperature of the surface of the heatsink connected to all the transistors in the network preserves almost constant temperature. The heatsink temperature for the different thermal resistances of the heatsink to ambient ($R_{\text{heatsink-amb}}$) has been tabulated in Table 2. As a consequence of this fact, the external heatsink network has been replaced with a single thermal voltage, and the simulation setup has been obtained for the simulation with device thermal network only through 'Approach-II,' as demonstrated with 'Approach-II' in Fig. 7b. Now, with both 'Approach-I' and 'Approach-II,' the self-heat rises ($\Delta T_{\text{approach-I}}$, $\Delta T_{\text{approach-II}}$), and

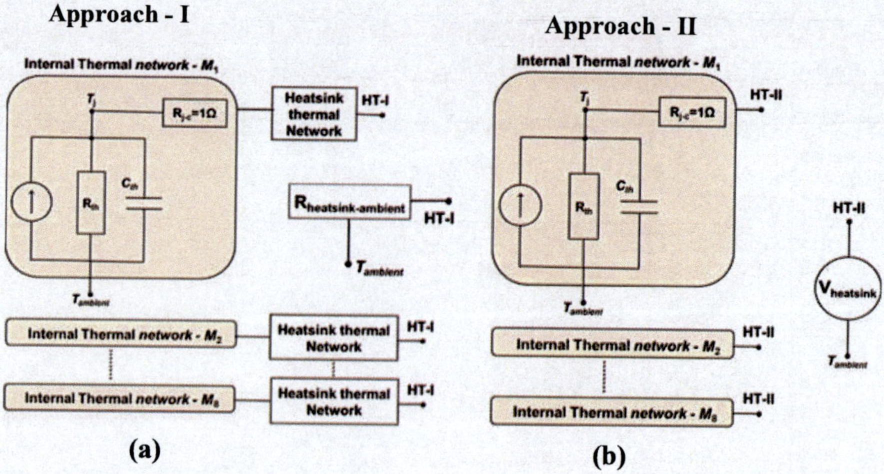


**Fig. 7** **a** Simulation setup with the devices connected to the external thermal network in a circuit, **b** equivalent simulation setup for the simulation with device thermal network only with a studied constant heatsink temperature

subsequently, the switching power loss ($P_{\text{Sw-approach-I}}$, $P_{\text{Sw-approach-II}}$) has been investigated. It has been observed that for both the studies, $\Delta T_{\text{approach-I}}$ and $\Delta T_{\text{approach-II}}$ and $P_{\text{Sw-approach-I}}$ and $P_{\text{Sw-approach-II}}$ show almost an exact match as portrayed in Fig. 8a, b. It should be noted that the second approach can now offer an opportunity to study and analyze any power device in a PE circuit with the measured heatsink temperature with infrascope.

It has been analyzed that the switching power loss as well as the self-heat increase the linear function of the heatsink temperature, as shown in Fig. 8a, b. A one-dimensional temperature distribution post-steady state in the GPIS circuit has been studied, which is a very significant study on the placement and routing of the integrated PE transistor of any PE circuit considering the self-heating effect in the package. It has been found that the transistors work synchronously in GPIS, leading to an integral thermal impact so that the system can reach up to a specific temperature higher than that of the ambient counterpart. In this study, the case temperature of the

**Table 2** $R_{\text{heatsink-amb}}$ to heatsink temperature

| $R_{\text{heatsink-amb}}$ ($W^{-1}$ K) | Heatsink temperature (°C) |
|---|---|
| 25 m | 21.72 |
| 35 m | 23.19 |
| 50 m | 25.26 |
| 70 m | 28.04 |
| 100 m | 32.25 |
| 130 m | 36.49 |
| 160 m | 48.42 |

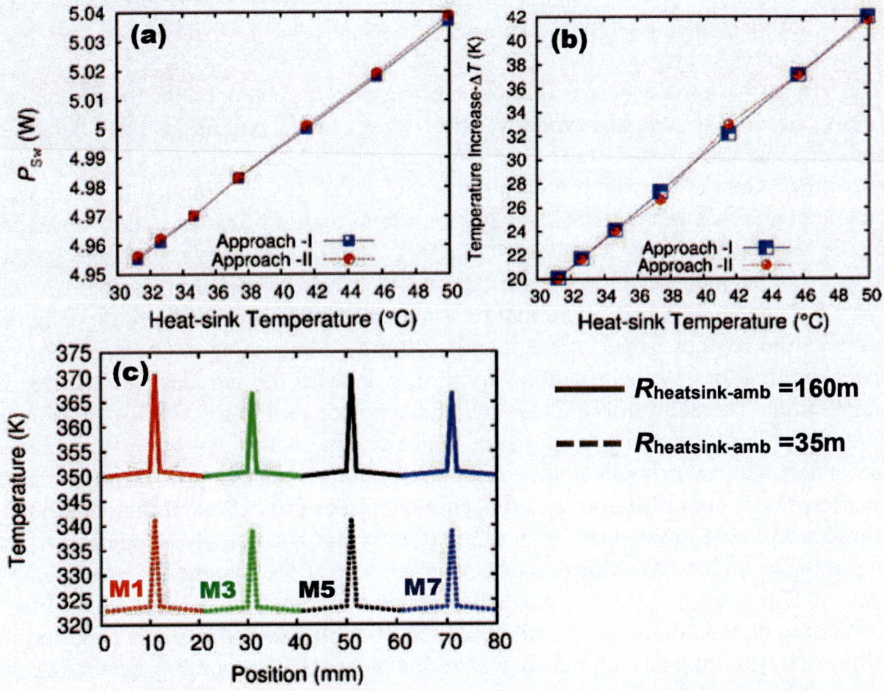


**Fig. 8** **a** Calibration of switching power loss as a function of the heatsink temperature with 'Approach-I' by 'Approach-II,' **b** calibration of operating temperature increases as a function of the heatsink temperature with 'Approach-I' by 'Approach-II.' The square lines represent the results of the circuit simulation result device thermal network through Approach-II and circle lines represent the circuit simulation result with the thermal network through Approach-I, **c** one-dimensional thermal distribution in the GPIS system with $R_{\text{heatsink-amb}} = 160\text{E}{-}3$ and $35\text{E}{-}3\ \text{W}^{-1}\ \text{K}$. The heat distribution has been portrayed as per the position of the transistor in the GPIS system

individual transistor has been obtained. Subsequently, the temperature from the case to several points of the heatsink has been obtained through the thermal resistance as a function of distance in the 1D thermal as shown in Fig. 8c. It can be inferred from the figure that the temperature would not reach the ambient temperature due to the integral thermal effect of the transistors of the circuit. It should be further noted from the figure that the heatsink surface temperature is almost constant over the distance, in which assumption was imperative to the design of 'Approach-II' presented in Fig. 7b. However, this gives the first stage of the thermal model, leading to incorporating all the heat conduction effects in a real-time switching environment.

## 4 Conclusion

The authors have mainly investigated the effect of self-heating on the switching loss in SiC circuits. A general-purpose inverter-stack (GPIS) circuit has been designed for this simulation-based study. The compact model HiSIM_HSiC was applied to solve the potential distribution within the SiC-MOSFET for simulating the device characteristics accurately. The investigation shows that self-heating in the transistors of the SiC-based circuit is generated due to switching, and the temperature propagates from the junction to the ambient through the case as well as via heatsink in the circuit. Depending upon the thermal resistance of the circuit, the operating temperature of the devices varies; subsequently, the rise in operating temperature boosts the switching loss by incorporating additional delay in the switching transistors of the circuits. The study reveals that switching power loss in the GPIS circuit rises linearly with the operating temperature. Since device characteristics are a nonlinear linear dependence on applied bias conditions, circuit simulation provides a powerful tool to predict circuit performances accurately under critical operating conditions. Simulation-based thermal study on such circuitry has been presented to offer a fast and reliable method to achieve the thermal effects of the integrated system proficiently. However, the 1D thermal network has been applied to perform the electrothermal circuit simulation in this study. The ongoing investigation is expected to improve by designing a comprehensive thermal network incorporating heat radiation and heat physics-of-failure emission models.

**Acknowledgements** Spectre®-based circuit simulation section has been partly supported through the activities of VDEC, the University of Tokyo, in collaboration with Cadence Design Systems. The authors wish to express special thanks to Mr. Abhishek Kar for his valuable comments and permission to portray the experimental board-level setup of C2M0160120D-based GPIS for this work.

## References

1. Lin H, Villamor A (2018) Power SiC 2018: materials, devices, and applications. Yole Development, France
2. Székely V, Rencz M, Courtois B (1998) Tracing the thermal behavior of ICs. IEEE Des Test Comput 15(2):14–21
3. Gao GB, Wang MZ, Gui X, Morkoc H (1989) Thermal design studies of high-power heterojunction bipolar transistors. IEEE Trans Electron Devices 36(5):854–863
4. Rinaldi N (2000) Thermal analysis of solid-state devices and circuits: an analytical approach. Solid-State Electron 44(10):1789–1798
5. Kadambi V, Abuaf N (1985) An analysis of the thermal response of power chip packages. IEEE Trans Electron Devices 32(6):1024–1033
6. Tega N, Sato S, Shima A (2019) Comparison of extremely high-temperature characteristics of planar and three-dimensional SiC MOSFETs. IEEE Electron Device Lett 40(9):1382–1384
7. Wang H, Liserre M, Blaabjerg F, de Place Rimmen P, Jacobsen JB, Kvisgaard T, Landkildehus J (2013) Transitioning to physics-of-failure as a reliability driver in power electronics. IEEE J Emerg Sel Top Power Electron 2(1):97–114

8. Wang Z, Shi X, Tolbert LM, Wang F, Liang Z, Costinett D, Blalock BJ (2014) A high-temperature silicon carbide MOSFET power module with integrated silicon-on-insulator-based gate drive. IEEE Trans Power Electron 30(3):1432–1445
9. Kimoto T, Cooper JA (2014) Fundamentals of silicon carbide technology: growth, characterization, devices, and applications. Wiley
10. Koh R, Iizuka T (2012) Self-heating parameter extraction of power MOSFETs based on transient drain current measurements and on the 2-cell self-heating model. In: 2012 IEEE international conference on microelectronic test structures. IEEE, pp 191–195
11. Bahman AS, Ma K, Ghimire P, Iannuzzo F, Blaabjerg F (2016) A 3-D-lumped thermal network model for long-term load profiles analysis in high-power IGBT modules. IEEE J Emerg Sel Top Power Electron 4(3):1050–1063
12. Ceccarelli L, Kotecha RM, Bahman AS, Iannuzzo F, Mantooth HA (2019) Mission-profile-based lifetime prediction for a SiC MOSFET power module using a multi-step condition-mapping simulation strategy. IEEE Trans Power Electron 34(10):9698–9708
13. Acharya S, She X, Todorovic MH, Datta R, Mandrusiak G (2018) Thermal performance evaluation of a 1.7-kV, 450-A SiC-MOSFET based modular three-phase power block with wide fundamental frequency operations. IEEE Trans Ind Appl 55(2):1795–1806
14. Kar A, Miura-Mattausch M, Sengupta M, Navaroo D, Kikuchihara H, Iizuka T, Rahaman H, Mattausch HJ (2021) Simulation-based power-loss optimization of general-purpose high-voltage SiC MOSFET circuit under high-frequency operation. IEEE Access 9:23786–23794
15. Datasheet of C2M0160120D silicon carbide power MOSFET. [Online]. Accessed 23 Sept 2020. Available: https://www.wolfspeed.com/power/products/sic-mosfets
16. Webpage of HiSIM Research Center. [Online]. Accessed 2022. Available: https://www.hisim.hiroshima-u.ac.jp/
17. HiSIM_HV 2.4.0 (2018) User's manual. Hiroshima University, Higashihiroshima, Japan
18. Tanaka A, Oritsuki Y, Kikuchihara H, Miyake M, Mattausch HJ, Miura-Mattausch M, Liu Y, Green K (2011) Quasi-2-dimensional compact resistor model for the drift region in high-voltage LDMOS devices. IEEE Trans Electron Devices 58(7):2072–2080
19. Mattausch HJ, Miyake M, Iizuka T, Kikuchihara H, Miura-Mattausch M (2012) The second-generation of HiSIM_HV compact models for high-voltage MOSFETs. IEEE Trans Electron Devices 60(2):653–661
20. Bagnoli PE, Casarosa C, Ciampi M, Dallago E (1998) Thermal resistance analysis by induced transient (TRAIT) method for power electronic devices thermal characterization. I. Fundamentals and theory. IEEE Trans Power Electron 13(6):1208–1219
21. Kar A, Manna S, Banerjee G, Sengupta M (2021) Design, analysis, fabrication & testing of SiC device-based high frequency synchronous DC-DC converters. In: 2021 national power electronics conference (NPEC). IEEE, pp 01–06
22. Kar A, Sengupta M (2021) Design, analysis and experimental validation of a variable frequency silicon carbide-based resonant-converter for welding applications. Sādhanā 46(2):1–7
23. Li H, Wang J, Ren N, Xu H, Sheng K (2019) Investigation of 1200 V SiC MOSFETs' surge reliability. Micromachines 10(7):485
24. Miura-Mattausch M (2008) The physics and modeling of MOSFETS: surface-potential model HiSIM. World Scientific

# A Novel Approach to Model and Analyze Wafer–Wafer Hybrid Bonding

**Debika Chaudhuri, Hafizur Rahaman, and Tamal Ghosh**

**Abstract** In wafer–wafer hybrid bonding, maintaining vigorous reliability of bonding and minimal deformation is essential. This paper probes the methodology for different design and process parameters for a two-layer wafer–wafer bonding using the finite element analysis (FEA) methodology. The shortcoming of poor bonding reliability allied with IC design and process parameters is discussed to endorse a better understanding on parameters that could be used to establish strategies for the wafer–wafer hybrid bonding techniques.

**Keywords** Wafer–wafer hybrid bonding · Stress · CTE mismatch · FEA

## 1 Introduction

The advantages of three-dimensional integrated circuits (3D ICs) are the low form factor, good performance, less power consumption, and more integration density. So, it can be regarded as the semiconductor technology for the next generation. In 3D integration, wafer bonding is the most important fabrication techniques, and for the vertical interconnection of 3D ICs, the stacked bonding is one of the importantly used technologies [1]. In 3D integration, the used primary level wafer bonding techniques include (i) eutectic bonding, (ii) polymer or adhesive bonding, (iii) metal diffusion bonding, and (iv) fusion or direct silicon bonding. Though metal diffusion bonding and eutectic bonding offer direct interconnection, the unbounded area with air gap make their reliability questionable. Fusion (direct silicon) bonding offers effective via density and alignment, but the dirt-free surface and bonding environment requirements are very important. Polymer (adhesive) bonding, on the other hand, is a low-temperature method suitable for patternable 3D IC micro-electromechanical

D. Chaudhuri (✉) · T. Ghosh
School of VLSI Technology, IIEST, Shibpur, Howrah, India
e-mail: 2020vlp001.debika@students.iiests.ac.in

H. Rahaman
Department of Information Technology, IIEST, Shibpur, Howrah, India

 
C. Giri et al. (eds.), *Emerging Electronic Devices, Circuits and Systems*,
Lecture Notes in Electrical Engineering 1004,
https://doi.org/10.1007/978-981-99-0055-8_35

systems (MEMSs) very large-scale integration (VLSI) and microsensor packaging. Additional to these approaches, another promising technique with high reliability and yield is the hybrid bonding technique [1–5]. It unites metal–metal and wafer bonding either with organic adhesives or with inorganic dielectrics. Among recently emerging packaging techniques, wafer–wafer hybrid bonding is a unique technique as it allows multi-stacking of the device wafers. As compared with the planar-integrated conventional wafer level packaging, the wafer–wafer hybrid bonding technique enables ultrafine (as low as 1 μm) vertical dimensional pitch interconnections through cost-effective methods.

Hybrid bonding combines a dielectric bond with embedded metal to form interconnects. A conventional 2.5-dimentional packaging process needs interposer, but this bonding reduces complex TSV fabrication. Conversely, during both the front and back end scenario, wafer–wafer hybrid bonding technique is compatible. However, the prime benefit is that the direct placement of Cu connection pad above the CMOS wafer through back-end-of-line (BEOL) process radically enhances the design flexibility of the circuit. This helps to achieve reduced die size and excellent pitch. Better adhesion, minimum residual stress, better chemical resistance, and excellent and thermomechanical stability are the basic requirements for the adhesives to be used for this purpose. Apart from satisfactory collocation with metal, adhesives should also tolerate the rigorous metal bonding conditions and consequent follow-up stages. This helps to recognize the mechanical solidity of the stacked ICs by imposing intrinsic metal interconnects with adhesive support. Filling and performing as a bonding material, adhesives or dielectrics effectively elevate the strength of the bonding and the reliability of the device. Also, as the electrical interconnections and micro-gap filling can be fabricated simultaneously, it makes simpler the process flow and reduces the challenge of micro-gap filling to optimize the yield and reliability issues.

During the last few decades, a significant number of fundamental studies [4–8] were conducted on wafer–wafer hybrid bonding. In this paper, a 3D FEA modeling approach is developed to study the bonding quality. The stress for various design parameters like the diameter, depth, and pitch of Cu pad and the process temperature are achieved to set up the design and process strategies.

## 2 Three-Dimensional Integration Using Wafer–Wafer Hybrid Bonding

Many 3D integration platforms have been developed and utilized worldwide by various industries and research organizations. To carry out these platforms, the main necessary technologies are the TSV, slim wafer handling, wafer bonding, and interconnection. Although numerous thin wafer handling potential technologies can be applied, they are still under assessment and evaluation. By the application of electric field, pressure or temperature multi-layer substrates can be permanently bonded together. By increasing the number of stacked layers, to meet the ever-increasing demand in 3D packaging for high-performance applications, wafer–wafer hybrid bonding technology provides a significant cost-effective solution.

However, due to the exceptionally rigorous requirements like finer ($\leq 6\ \mu m$) pitch, such method is facing major challenges in terms of the yield and precision issues and is still under progress. One such challenging issue is the mismatch due to thermal expansion which leads to the deformation in the wafer. Also, if the numbers of stacked layers are more, a high stress is developed there. This arises due to complexity in the process parameters and the different thermal expansion coefficients of the used materials.

Besides this, significant stress can be generated due to unnecessary volume of Cu or insufficient Cu dishing [9]. This may also cause delamination of bonded layers during annealing with the increase in temperature. As a result of this, unwanted delamination or poor bonding may be generated, and hence, the yield may be radically reduced.

## 3 Model Description

In order to mitigate the poor bonding performance risk, one practicable localized model is developed. In wafer–wafer hybrid bonding techniques, for the process engineers, bonding performance on the Cu pad interfaces and dielectric surfaces is of huge concern.

The model includes two identical structures containing four Cu pads and surrounding materials on each face, which are bonded collectively side by side. Figure 1 represents the schematic of two identical wafers bonded together by wafer–wafer hybrid bonding technique. In Fig. 1, $L$ represents the Cu pad pitch, $x$ is the depth of the $SiO_2$ layer, $h$ is the Cu pad depth, and $d$ is the diameter of the Cu pad. Tables 1 and 2, respectively, describe the chosen material properties and the design parameters along with their units.

## 4 Results and Discussions

In this study, Cu–Cu bonding area developments are simulated. Cu stress is generated during annealing process. Based on various design parameters like the pitch, the diameter and the depth of Cu pad results on the stresses on both the Cu and dielectric interface are analyzed. Annealing temperatures chosen for this study are 300, 350, and 400 °C. Figure 2 shows the stress profile as obtained from the simulation platform for the structure in Cu–$SiO_2$ interface. Stress on the interface increases with increasing annealing temperature and decreases with decrease in Cu pad design density.

Figure 3 shows the effect of Cu diameter in different annealing temperatures.

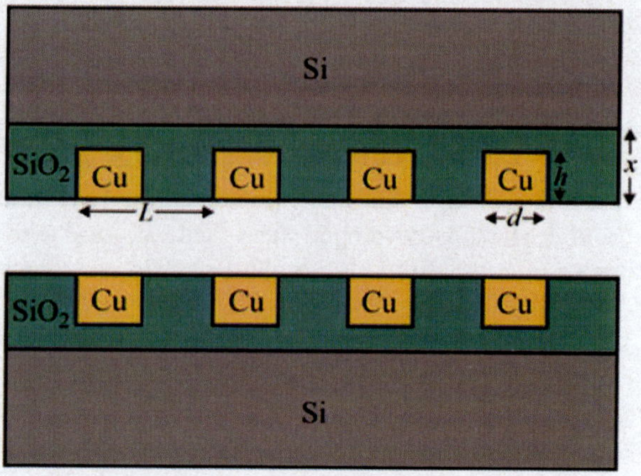


**Fig. 1** Schematic illustrating two identical wafers bonded together by wafer–wafer hybrid bonding technique

**Table 1** Material properties

| Materials | Young's modulus (GPa) | Poisson's ratio | Thermal expansion coefficient (ppm/°C) |
|---|---|---|---|
| Si | 131 | 0.28 | 2.6 |
| Cu | 91.8 | 0.34 | 17.6 |
| $SiO_2$ | 73 | 0.17 | 0.5 |

**Table 2** Design parameters

| | |
|---|---|
| Diameter (*d*) of Cu pad | 3, 5, 7 μm |
| Pitch (*L*) of Cu pad | 6, 9, 12 μm |
| Depth (*h*) of Cu pad | 5, 7, 9 μm |
| Depth (*x*) of $SiO_2$ layer | 0.2 μm |

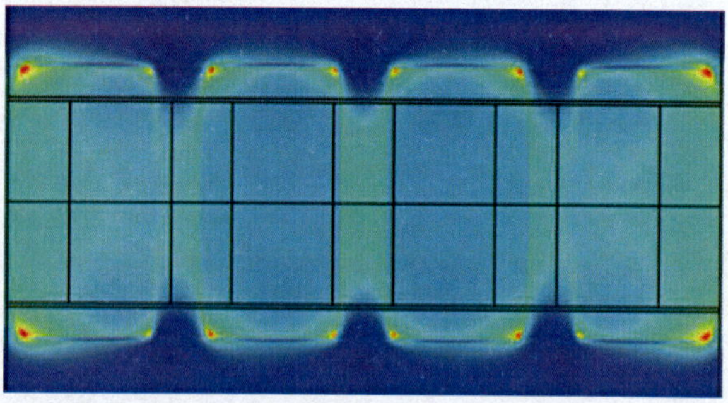

**Fig. 2** Stress profile for the structure in Cu–$SiO_2$ interface as derived from the simulation platform

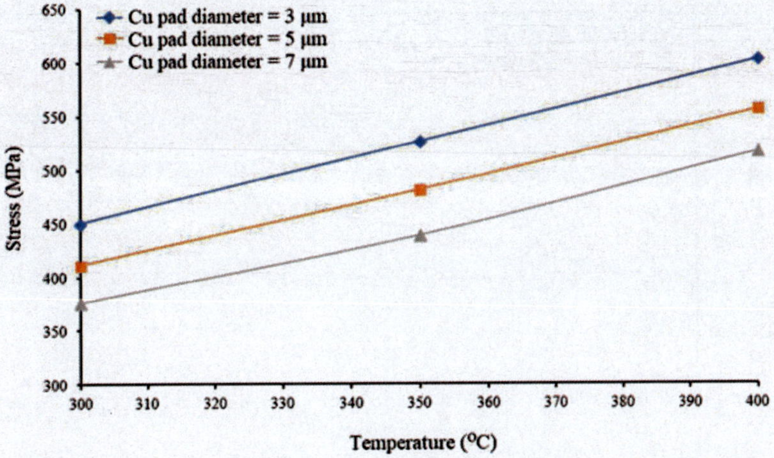


**Fig. 3** Effects of Cu diameter in different annealing temperatures

This reveals that the stress decreases with the increase in Cu diameter. Figure 4 shows the effect of Cu depth in different annealing temperatures. This reveals that the stress increases with the increase in Cu depth.

Figure 5 on the other hand shows the effects of pitch length in different annealing temperatures. This reveals that the stress also increases with the increase in pitch length. But, the rate of increase is not linear.

Based on the performed parametric studies, the strategies to aid the wafer–wafer hybrid bonding techniques have thus been effectively set up. In the future growth of wafer–wafer hybrid bonding to assist sustain the wafer deformation and stress at suitable levels, this strategy could thus be implemented.

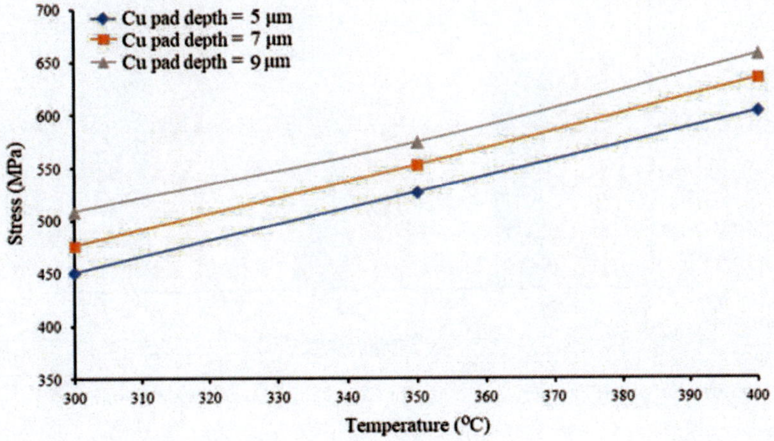


**Fig. 4** Effects of Cu pad depth in different annealing temperatures

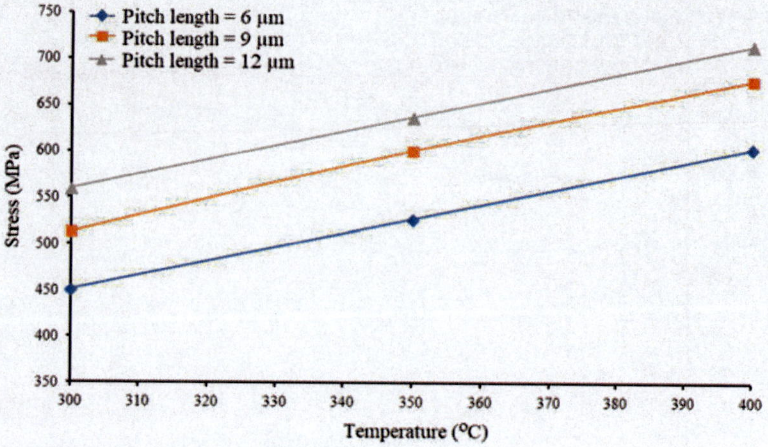


**Fig. 5** Effects of pitch length in different annealing temperatures

Figure 6 shows the comparative stress–temperature study for Cu pad and Cu pad with Ti liner (0.1 μm), from which it is clear that although Ti has low thermal expansion coefficient as compared to Cu, there is no improvement in the stress values for different annealing temperatures with the addition of Ti as lining material.

The sectional view of the Ti lining is illustrated in Fig. 7.

In future studies, different lining materials and liner width can be incorporated to find better stress values.

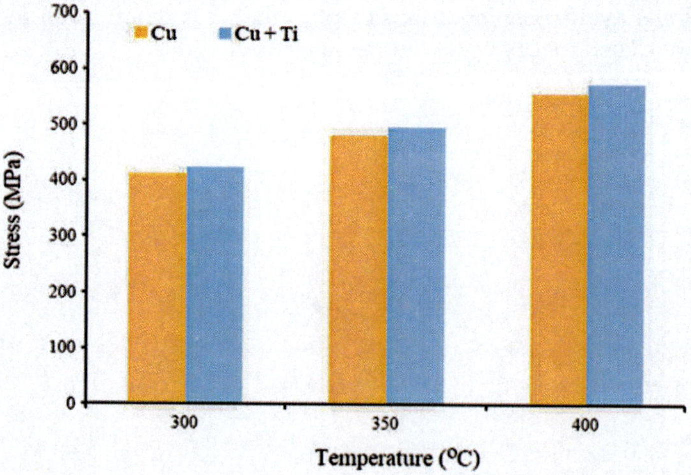


**Fig. 6** Comparative stress–temperature study for Cu pad and Cu pad with Ti lining

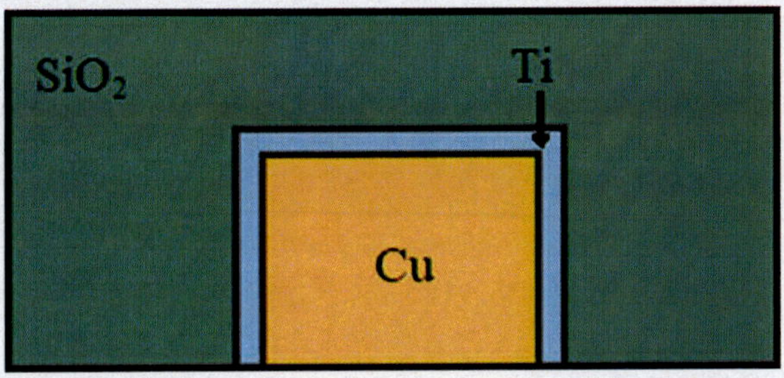


**Fig. 7** Sectional view of the Ti lining (0.1 μm) with Cu pad

## 5 Conclusion

In this work, hybrid bonding between two wafers at different annealing temperatures is analyzed for different design parameters. Simulation results show that lower stress in the structure is achievable using higher Cu diameter and lower depth. Pitch length also has a significant role in bonding accuracy for hybrid bonding mechanism, and its lower value is to be used for a better result. These findings can be useful to implement good-quality bonding with low thermal stress between IC layers.

## References

1. Ji L, Che FX, Ji HM, Li HY, Kawano M (2020) Wafer-to-wafer hybrid bonding development by advanced finite element modeling for 3-D IC packages. IEEE Trans Compon Packag Manuf Technol 10(12):2106–2117
2. Ko CT, Hsiao ZC, Fu HC, Chen KN, Lo WC, Chen YH (2010) Wafer-to-wafer hybrid bonding technology for 3D IC. In: 3rd electronics system integration technology conference ESTC, pp 1–5
3. Che FX, Putra WN, Heryanto A, Trigg A, Zhang X, Gan CL (2013) Study on Cu protrusion of through-silicon via. IEEE Trans Compon Packag Manuf Technol 3(5):732–739
4. Bayrashev A, Ziaie B (2002) Silicon wafer bonding with an insulator interlayer using RF dielectric heating. In: Fifteenth IEEE international conference on micro electro mechanical systems, pp 419–422
5. Chidambaram V, Lianto P, Wang X, See G, Wiswell N, Kawano M (2021) Dielectric materials characterization for hybrid bonding. In: 2021 IEEE 71st electronic components and technology conference (ECTC), pp 426–431
6. Lim SPS, Chong SC, Chidambaram V (2021) Comprehensive study on chip to wafer hybrid bonding process for fine pitch high density heterogeneous applications. In: 2021 IEEE 71st electronic components and technology conference (ECTC), pp 438–444
7. Workman T, Mirkarimi L, Theil J, Fountain G, Bang KM, Lee B, Uzoh C, Suwito D, Gao G, Mrozek P (2021) Die to wafer hybrid bonding and fine pitch considerations. In: 2021 IEEE 71st electronic components and technology conference (ECTC), pp 2071–2077

8. Lhostis S, Farcy A, Deloffre E, Lorut F, Mermoz S, Henrion Y, Berthier L, Bailly F, Scevola D, Guyader F, Gigon F, Besset C, Pellissier S, Gay L, Hotellier N, Berrigo ALL, Moreau S, Balan V, Fournel F, Jouve A, Cheramy S, Arnoux M, Rebhan B, Maier GA, Chitu L (2016) Reliable 300 mm wafer level hybrid bonding for 3D stacked CMOS image sensors. In: 66th proceedings of electronic packaging technology conference, pp 870–876
9. Beilliarda Y, Estevez R, Parrya G, McGarrye P, Di Cioccio L, Coudrain P (2017) Thermomechanical finite element modeling of Cu–$SiO_2$ direct hybrid bonding with a dishing effect on Cu surfaces. Int J Solids Struct 117:208–220

# Successive Approximation Register Analog-to-Digital Converter—A Tutorial

**Shruti Konwar, Utkarsh Jaiswal, and Bibhu Datta Sahoo**

**Abstract** This tutorial paper covers two crucial aspects of successive approximation register (SAR) analog-to-digital converter (ADC). First, a mathematical analysis for two commonly used SAR ADC topologies is presented, and then the power dissipation aspect of SAR ADC is discussed for different switching schemes. The paper also provides some design criteria for practical implementation of SAR ADCs.

**Keywords** SAR ADC · Capacitive-DAC · Dynamic switching · Split SAR · Power dissipation · Charge sharing

## 1 Introduction

Successive approximation register (SAR) analog-to-digital converters (ADCs) find widespread applications requiring medium-to-high resolution and/or low-to-medium sampling speeds [1]. Pipelining [2] and time-interleaving [3] have been used to enhance the sample rates of SAR ADCs. Different variants of SAR ADCs have been reported in literature, viz. noise shaping SAR [4], time-interleaved [3] SAR, asynchronous SAR [5], pipelined SAR [2], and multibit SAR [6], that give either high sample rate or high resolution. Literature also shows the amalgamation of two or more categories for further performance improvement.

Figure 1 shows the basic block diagram of a conventional SAR ADC architecture. The major blocks include a capacitive sampling network, a high precision comparator, and a successive approximation register (SAR) logic network.

The SAR ADC conversion cycle can be broken into the sampling or acquisition phase and the bit-cycling or conversion phase. The capacitive-DAC (CDAC) used in the architecture serves the dual purpose of both sample/hold circuit in sampling phase and DAC in bit-cycling phase to convert the digital output to analog equivalent

S. Konwar (✉) · U. Jaiswal · B. D. Sahoo
Department of Electronics and Electrical Communication Engineering, Indian Institute of Technology Kharagpur, Kharagpur, West Bengal, India
e-mail: shruti24@iitkgp.ac.in



C. Giri et al. (eds.), *Emerging Electronic Devices, Circuits and Systems*, Lecture Notes in Electrical Engineering 1004,
https://doi.org/10.1007/978-981-99-0055-8_36

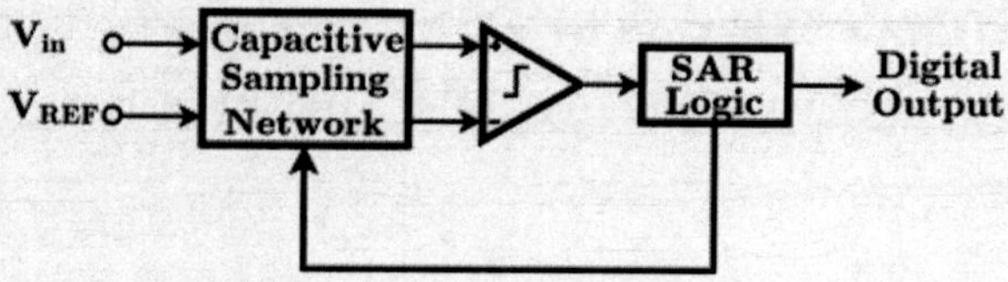


**Fig. 1** Block diagram of a basic SAR ADC

and fed back to the comparator for the next bit cycle using a binary search algorithm. The input value is sampled to the capacitive network during the acquisition phase. During the bit-cycling phase, $N$-bit DAC generated mid-scale voltage is compared with the sampled analog voltage. The output of the comparator is used to select the subsequent DAC output which is used as the threshold of the comparator in the next cycle. For example, if the input is larger than the DAC output, the output of the comparator is set to "high", the MSB bit is set to logic "1", and (MSB − 1) bit is also set to logic "1", resulting in the DAC output to be set to the midpoint of the upper half of the input range. On the other hand, if the input is smaller than the DAC output, the output of the comparator is set to "low", the MSB bit is set to logic "0", and (MSB − 1) bit is set to logic "1", resulting in the DAC output to be set to the midpoint of the lower half of the input range. Repeating the above process $N$ times results in the input signal getting digitized to an $N$-bit digital word.

This paper is organized as follows: Sect. 2 shows the commonly used SAR ADC architectures, and Sect. 3 gives a design approach of the ADC along with comparison of various energy switching schemes.

## 2 SAR ADC Architectures

The type of the SAR ADC is defined by the way in which the capacitor-based charge redistribution network (CDAC) is realized. The most commonly used architectures are binary-weighted capacitor array and split capacitor array. These two architectures are discussed below.

### *2.1 Binary-Weighted SAR*

The charge sampling network in this topology consists of binary-weighted capacitors and MOS switches, as shown in Fig. 2a along with timing diagram in Fig. 2b for a single-ended SAR ADC. An $N$-bit conventional binary-weighted SAR ADC consists of capacitors of values $C_i = 2^{(i-1)} C_u$, for $i = 1$ to $N$, where $C_u$ is unit capacitance of the CDAC, and a dummy capacitance $C_0 = C_u$, giving a total CDAC capacitance of $2^N C_u$.

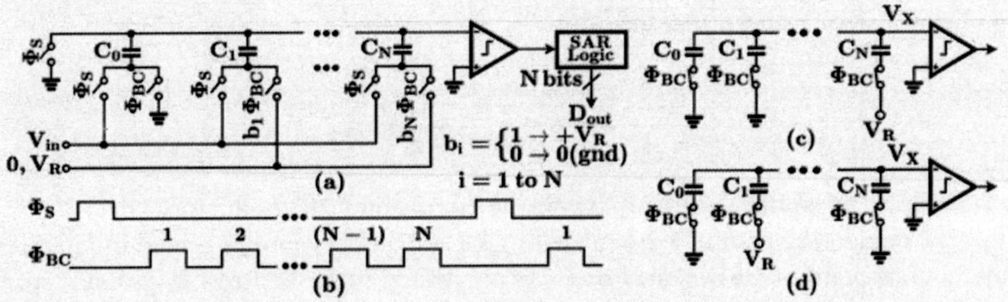


**Fig. 2** Binary-weighted SAR ADC **a** block diagram, **b** timing diagram, **c** bit-cycling for MSB decision, and **d** bit-cycling for LSB decision

From the principle of charge conservation for the two phases of operation, sampling (also called acquisition) and bit cycling ($N$ bit cycles for an $N$-bit SAR), the DAC output voltage can be calculated individually and superposition can be used to obtain the total DAC voltage ($V_X$) [7].

During sampling phase, the bottom plate of all the capacitors is connected to input voltage $V_{in}$ giving a total charge given by (1).

$$Q_S = \sum_{i=0}^{N-1} 2^i C_u = 2^N C_u V_{in} \tag{1}$$

During bit-cycling phase, the bottom plates of the capacitors are connected to either reference voltage $V_R$ or ground, depending on the bit pattern given by the SAR logic. For example, Fig. 2c shows the MSB capacitor switching for an $N$-bit binary DAC where the MSB capacitor is connected to reference voltage $V_R$ and rest to ground. Charge during bit-cycling phase is given by (2)

$$Q_{BC} = 2^{N-1} C_u V_R - \sum_{i=0}^{N} 2^i C_u V_X \tag{2}$$

From (1) and (2) using principle of charge conservation, i.e., $Q_S = Q_{BC}$, the voltage $V_X$ is given by (3) for Fig. 2c.

$$V_X = \frac{-2^N}{\sum_{i=0}^{N} 2^i} V_{in} + \frac{2^{N-1}}{\sum_{i=0}^{N} 2^i} V_R. \tag{3}$$

The MSB capacitor ($C_N = 2^{N-1} C_u$) connected to $V_R$ causes $V_X$ to settle to (4)

$$V_X = -V_{in} + \frac{V_R}{2} \tag{4}$$

giving the comparator output as (5).

$$b_{\mathrm{N}}\ (\mathrm{MSB}) = \begin{cases} 1 & V_{\mathrm{in}} < \frac{V_{\mathrm{R}}}{2} \\ 0 & V_{\mathrm{in}} > \frac{V_{\mathrm{R}}}{2} \end{cases} \tag{5}$$

Thus, at the end of the first bit cycle the comparator makes the MSB decision at $V_{\mathrm{R}}/2$. Considering another extreme case for LSB bit decision, as shown in Fig. 2d, the bottom plate of LSB capacitor $C_1$ is connected to $V_{\mathrm{R}}$ and rest to ground, thus giving,

$$Q_{\mathrm{BC}} = C_{\mathrm{u}} V_{\mathrm{R}} - \sum_{i=0}^{N} 2^i C_{\mathrm{u}} V_{\mathrm{X}} \tag{6}$$

Similarly, at the end of LSB bit cycle, the bit decision is taken at comparator threshold as (7).

$$b_1\ (\mathrm{LSB}) = \begin{cases} 1 & V_{\mathrm{in}} < \frac{V_{\mathrm{R}}}{2^N} \\ 0 & V_{\mathrm{in}} > \frac{V_{\mathrm{R}}}{2^N} \end{cases} \tag{7}$$

Thus in the SAR ADC conversion cycle, the latching comparator output controls the next switching transition. If MSB bit $b_{\mathrm{N}}$ is "low", the second-largest capacitor ($C_{\mathrm{N}-1}$) is connected to $V_{\mathrm{R}}$, raising the voltage at $V_{\mathrm{X}}$, and this is called an "up" transition. If, on the other hand, $b_{\mathrm{N}}$ is high, $C_{\mathrm{N}}$ is returned to ground and $C_{\mathrm{N}-1}$ is connected to $V_{\mathrm{R}}$, and this is called a "down" transition. Repeating the above process for the remaining capacitors in the array, the value of $V_{\mathrm{X}}$ at each stage after switching transients have settled is given as (8)

$$V_{\mathrm{X}} = -V_{\mathrm{in}} + \frac{C_{\mathrm{T}}}{C_{\mathrm{T}} + C_{\mathrm{B}}} V_{\mathrm{R}} \tag{8}$$

where $C_{\mathrm{T}}$ denotes the total capacitance connected to the reference voltage $V_{\mathrm{R}}$ and $C_{\mathrm{B}}$ denotes the total capacitance connected to ground.

Thus, for any combination of comparator output or the ADC's digital bits $b_i$, $i = 1$ to $N$, we can use superposition principle so as to obtain the total DAC voltage $V_{\mathrm{X}}$,

$$V_{\mathrm{X}} = -V_{\mathrm{in}} + \sum_{i=1}^{N} b_i w_i V_{\mathrm{R}} \tag{9}$$

where $w_i$ can be defined as the weights of the DAC capacitors given as (10).

$$w_i = \frac{C_i}{\sum_{i=1}^{N} 2^i C_{\mathrm{u}}} \tag{10}$$

where $i = 1$ to $N$ for the $N$-bit binary SAR ADC.

## 2.2 Split SAR

In binary-weighted SAR, the size of the capacitors increases exponentially as the desired resolution increases. MSB capacitor can attain a value unfeasible for practical silicon implementation. This problem is alleviated by the split SAR topology in which the DAC sampling network is split into two capacitor banks connected through a bridge capacitor $C_B$ for a two-section split SAR architecture. In general, $M$ capacitor array banks are connected by $(M - 1)$ bridge capacitors. The section closest to the comparator is called the MSB-section, while the farthest is called the LSB-section. Value of $C_B$ is chosen such that series combination of LSB capacitance bank and $C_B$ gives total capacitance equal to unit capacitance $C_u$.

For two sections split DAC, the capacitance of LSB bank totals to be $2^{N/2}C_u$. In general, total capacitance $C_{LSB}$ of LSB capacitance bank will be $2^{N/s}C_u$ where "$s$" is the number of sections while total capacitance of all other sections will be $(2^{N/s}C_u - 1)$ [8]. Figure 3 shows a generic split SAR architecture for Q-sections.

To analyze the charge transfer phenomenon in a split SAR, an example for a two-section 6-bit split SAR architecture is shown in Fig. 4. The capacitor values are given as $C_0 = C_u$, $C_1 = C_u$, $C_2 = 2C_u$, $C_3 = 4C_u$, $C_4 = C_u$, $C_5 = 2C_u$, $C_6 = 4C_u$ where $C_u$ is the unit capacitance used. The bridge capacitance $C_B$ is chosen so as to have an equivalent capacitance to the left of the MSB-section equal to the unit capacitance $C_u$ as by (11). In this topology, the CDAC is affected by top-plate and bottom-plate parasitic capacitances of $C_B$ shown as $C_{TP}$ and $C_{BP}$ in Fig. 4.

$$\frac{\sum_{i=0}^{3} C_i C_B}{\sum_{i=0}^{3} C_i + C_B} = C_u \implies C_B = \frac{8}{7}C_u \tag{11}$$

The timing diagram for the different clock cycles used in the conversion procedure for the split SAR is same as that in binary-weighted SAR. The charge conservation in this case is explained for the 6-bit case as follows.

During the sampling phase, $\Phi_s$, the bottom plates of capacitors are connected to $V_{in}$ and top plates to ground giving the equivalent circuit as shown in Fig. 5a.

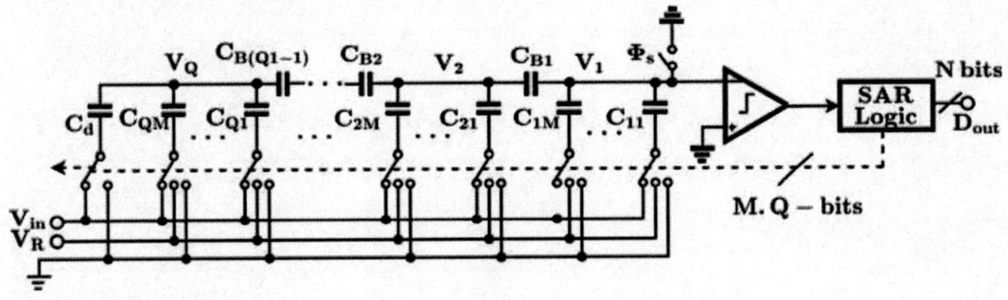


**Fig. 3** Generic split SAR architecture for single-ended implementation

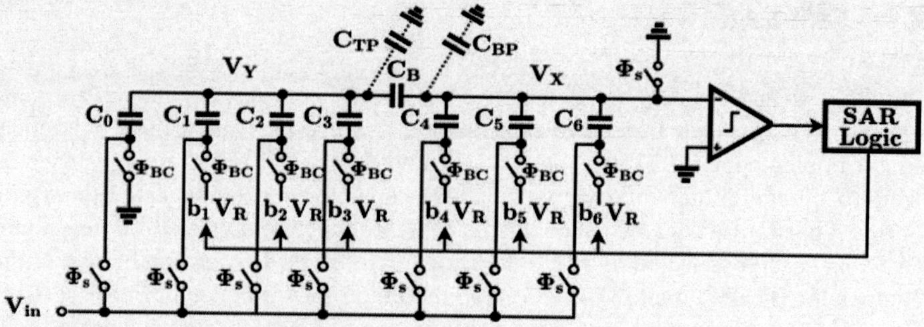


**Fig. 4** Circuit diagram of split SAR ADC for 6-bit

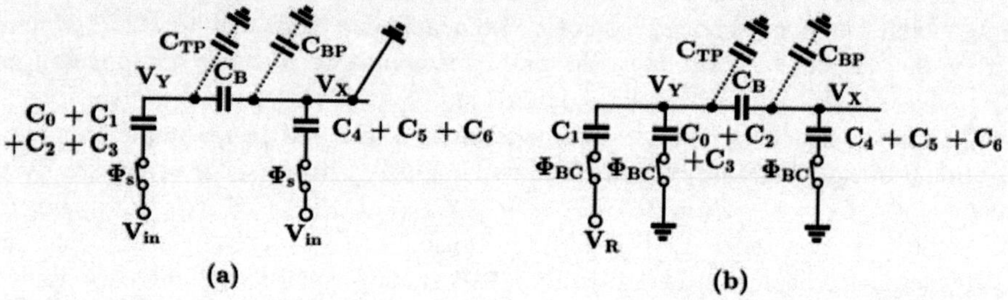


**Fig. 5** Charge transfer phases for 6-bit split SAR. **a** Sampling phase $\Phi_s$. **b** Bit-cycling phase $\Phi_{BC}$ (for weight $w_1$)

At left plate of $C_B$:

$$(V_{in} - V_Y) \sum_{i=0}^{3} C_i - V_Y C_{TP} - V_Y C_B = 0$$

$$\implies V_Y = \frac{\sum_{i=0}^{3} C_i V_{in}}{\sum_{i=0}^{3} C_i + C_B + C_{TP}} \tag{12}$$

Similarly, from right plate of $C_B$ we obtain

$$Q_S = V_{in} \sum_{i=4}^{6} C_i + V_Y C_B = \zeta V_{in} \tag{13}$$

where $\zeta$ is given by (14).

$$\zeta = \left[ \sum_{i=4}^{6} C_i + \frac{\sum_{i=0}^{3} C_i C_B}{\sum_{i=0}^{3} C_i + C_B + C_{TP}} \right] \tag{14}$$

During bit-cycling phase, $\Phi_{BC}$, the capacitance for which the weight is to be determined is connected to reference voltage $V_R$ while the rest to ground. For example, to calculate the weight corresponding to $C_1$, the bottom plate of $C_1$ is connected to $V_R$, whereas the bottom plates of the rest are connected to ground as shown by the equivalent circuit in Fig. 5b.

From left plate of $C_B$:

$$(V_R - V_Y)C_1 - V_Y(C_0 + C_2 + C_3) - V_Y C_{TP} + (V_X - V_Y)C_B = 0$$

$$\implies V_Y = \frac{C_1 V_R + C_B V_X}{\sum_{i=0}^{3} C_i + C_B + C_{TP}} \tag{15}$$

And from right plate of $C_B$:

$$Q_{BC} = -V_X \sum_{i=4}^{6} C_i - V_X C_{BP} + (V_Y - V_X)C_B$$

Using (15) and simplifying,

$$Q_{BC} = \frac{C_1 V_R}{\sum_{i=0}^{3} C_i + C_B + C_{TP}} - \gamma V_X \tag{16}$$

where $\gamma$ is given by (17).

$$\gamma = \sum_{i=4}^{6} C_i + C_B + C_{BP} - \frac{C_B^2}{\sum_{i=0}^{3} C_i + C_B + C_{TP}} \tag{17}$$

From principle of charge conservation, using (13) and (16)

$$\zeta V_{in} = \frac{C_1 V_R}{\sum_{i=0}^{3} C_i + C_B + C_{TP}} - \gamma V_X$$

$$V_X = \frac{1}{\gamma}\left[-\zeta V_{in} + \frac{C_1 V_R}{\sum_{i=0}^{3} C_i + C_B + C_{TP}}\right]$$

Thus, the weight corresponding to capacitor $C_1$ is given as

$$w_1 = \frac{C_1 V_R}{\sum_{i=0}^{3} C_i + C_B + C_{TP}} \tag{18}$$

Similarly, for calculating the weight of any other capacitor, the bottom plate of that capacitor is connected to $V_R$ during the bit-cycling phase and that of the rest are connected to ground. In general, ignoring the effect of parasitic capacitances, the corresponding weight for capacitor $C_i$ in the 2-section 6-bit Split SAR of Fig. 4 can be written as (19).

$$w_i = \frac{\sum_{i=0}^{3} b_i C_i + C_B \sum_{i=4}^{6} b_i C_i}{\sum_{i=4}^{6} C_i + C_B - \frac{C_B^2}{\left(\sum_{i=0}^{3} C_i + C_B\right)}}. \tag{19}$$

Using charge conservation principle during the sampling and bit-cycling phases, the above equation can be modified for Split SAR with any number of sections.

## 3 SAR ADC Design Approach

This section provides an initial design approach of a SAR ADC considering various design constraints of the various blocks of the ADC. The sizing of the DAC capacitors and switches are most crucial in meeting the overall linearity of the ADC. This is done by the settling time and noise considerations. The CDAC switching contributes significantly to the overall power consumption and therefore techniques to minimize this have been discussed.

### 3.1 Settling Time Consideration

To understand the time constant $\tau$ consideration, let us consider an $N$-bit SAR ADC with a sampling rate of $f_s$, i.e., sampling period of $t_s = 1/f_s$. Thus for the $N$-bit conversion, $(N+1)$ cycles have to be within $t_s$ resulting in the bit-cycling clock period given as $t_{BC} = t_s/(N+1)$ or corresponding bit-cycling rate $f_{BC} = 1/t_{BC}$. Considering switch resistance as $R_{sw}$ and total DAC capacitance as $C_{total}$, the DAC output voltage should follow input voltage and must be settled with at most an error of LSB/2 (a rule of thumb) within 50% duty cycle of $t_{BC}$ for the $R_{sw}C_{total}$ equivalent and the comparator should start comparing. Thus, the actions that should happen within this $t_{BC}/2$ time include the SAR logic taking the previous comparator value to generate the digital output, this output is converted to analog equivalent by the DAC, and the comparator is comparing this DAC output to produce the next bit for the next conversion cycle. Mathematically, with reference to Fig. 6, for full scale input voltage, the time constant $\tau = R_{sw}C_{total}$ can be written as

$$\begin{aligned} V_{out} &= V_{in}\left[1 - \exp^{-t_{DACsettling}/\tau}\right] \\ \implies \Delta V &= V_{in} \exp^{-t_{DACsettling}/\tau} = \frac{\text{LSB}}{2} \\ \implies (N+1)\tau \ln(2) &= t_{DACsettling} = \frac{t_{BC}}{2} \end{aligned}$$

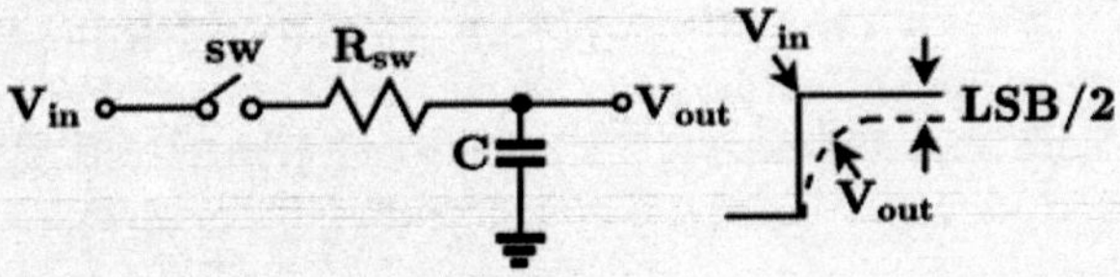


**Fig. 6** DAC settling time

$$\tau = \frac{t_{BC}}{2(N+1)\ln(2)} \tag{20}$$

Once $\tau$ is obtained, the total DAC capacitance $C_{total}$ can be calculated next considering noise as discussed in the succeeding section.

## 3.2 *Noise Consideration*

Considering the two major noise sources in the system as quantization noise ($P_N$) due to the quantization error of the ADC and thermal noise ($P_{Thermal}$) basically introduced by the MOS switches, the overall SNR can be written as

$$\mathrm{SNR} = 10 \times \log_{10}\left(\frac{P_S}{P_N + P_{Thermal}}\right) \tag{21}$$

where $P_S$ is the signal power.

Thus, considering the thermal noise and linearity specification in order to avoid degradation of SNR, the sampling capacitors are accordingly sized. For example, in the 12-bit system, considering 1 dB degradation in SNR from ideal value due to thermal noise, for a signal amplitude of $1\,V_{p-p}$, we have

$$P_S = 0.5^2/2,\ P_N = \Delta^2/12 = 1/\left(12 \times 4096^2\right),\ P_{Thermal} = 2kT/C_{total}$$

where $C_{total}$ is the total capacitance of the sampling network and factor 2 is considering differential implementation.

Thus, we can write from (21),

$$73\ \mathrm{dB} = 10 \times \log_{10}\left(\frac{0.5^2/2}{1/(12 \times 4096^2) + 2kT/C_{total}}\right) \tag{22}$$

Solving (22), we can find the size for the unit capacitance $C_{unit}$ of the sampling network as

$$\sum_{i=0}^{11} 2^i C_{unit} = C_{total} \tag{23}$$

Thus, from (20) and (23), $R_{sw}$ and therefore the sizes of the MOS switches can be obtained and the binary-weighted DAC switches are sized such that the $RC$ time constant of each switch-capacitor pair remains the same.

## 3.3 Power Consumption in CDAC

The dominant source of power dissipation is SAR ADC's comparator and the switching of the capacitor array [9, 10], which can be minimized further. With reference to Sect. 2.1, it can be observed that either an "up" or a "down" transition occurs in every bit cycle of the ADC. Five switching schemes are discussed below, where four out of five methods behave identically for "up" transition and differ only when there is a "down" transition, while the last switching scheme is different as will be explained in the succeeding sections. For ease of understanding for the readers, all calculations in this section are explained for a 3-bit capacitor array with $C_3 = 4C_0$, $C_2 = 2C_0$, $C_1 = C_0$, $C_0$, along with closed form equations for a general $N$-bit capacitor array.

### 3.3.1 One-Step Switching Method

Considering initial time, i.e., at time 0, the bottom plate of the $C_3 = 4C_0$ is switched to $V_R$. The capacitor array is charged to reach the final value in (4). The total energy drawn from $V_R$ is given by (24).

$$E_{0\to 1} = 2C_0 V_R^2 \tag{24}$$

For every $N$-bit code, $E_{0\to 1}$ can be generalized as (25).

$$E_{0\to 1} = \frac{1}{2} V_R^2 C_N \tag{25}$$

where $C_N$ is the MSB capacitor. After the end of the first period of bit cycle, if MSB bit $b_3$ is 0, $C_2 = 2C_0$ in Fig. 7 is connected to $V_R$ giving

$$V_X[2] = \frac{3}{4} V_R - V_{in} \tag{26}$$

Thus, the generalized expression for $V_X[i]$ for $i$th bit cycle for each $N$-bit code can be written as (27).

$$V_X[i] = \frac{C_{T_i}}{C_{T_i} + C_{B_i}} V_R - V_{in} \tag{27}$$

where $C_{T_i}$ and $C_{B_i}$ is the sum of all capacitors connected to the reference voltage and ground, respectively, after making alterations in connections for $i$th bit cycle for each $N$-bit code, giving the energy drawn as (28).

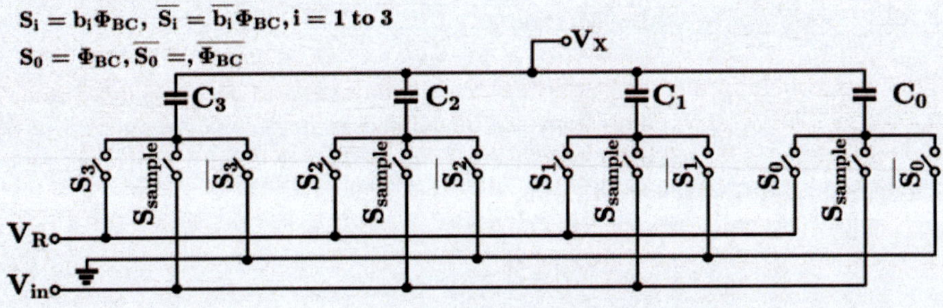


**Fig. 7** 3-bit capacitor array for one-step and two-step switching methods

$$E_{1\to 2} = \frac{C_0 V_R^2}{2} \tag{28}$$

Generalized expression for $E_{i\to(i+1)}$ for $i$th bit cycle for each $N$-bit code after "up" conversion is given as (29).

$$E_{i\to(i+1)} = \frac{C_{B_i} V_R^2}{2^{(i+1)}} \tag{29}$$

For down conversion in one-step switching scheme, it involves switching down the MSB capacitor ($C_3$) and switching up the MSB/2 capacitor ($C_2$) simultaneously. Using (27), the energy drawn from $V_R$ while the capacitor array settles is given as (30).

$$V_X[2] = \frac{1}{4}V_R - V_{in}$$

$$E_{1\to 2,\text{1-step}} = \frac{5C_0 V_R^2}{2} \tag{30}$$

Generalized expression for $E_{i\to(i+1)}$ for $i$th bit cycle for each $N$-bit code after "down" conversion can be written as (31).

$$E_{i\to(i+1),\text{1-step}} = C_{\text{Latest}} V_R^2 + \frac{C_{T_i} V_R^2}{2^{(i+1)}} \tag{31}$$

where $C_{\text{Latest}}$ is the capacitor which just got charged to $V_R$ in $i$th bit cycle during "down" conversion.

While this switching scheme is the simplest to implement in terms of number of switches and clock edges, the ratio of (30) to (28) points toward significant inefficiency, since it requires five times more energy to lower $V_X$ than to raise it by the same amount.

### 3.3.2 Two-Step Switching Method

Two-step switching method requires two switching steps to accomplish a "down" transition. The two switching steps can be defined as steps $1 \rightarrow 1.5$ and $1.5 \rightarrow 2$. $C_3 = 4C_0$ and $C_2 = 2C_0$ are connected to $V_R$ in the first switching step, resulting in a transition similar to (26)–(28).

$$V_X[1.5] = \frac{3}{4}V_R - V_{in} \tag{32}$$

$$E_{1\rightarrow 1.5} = \frac{C_0 V_R^2}{2} \tag{33}$$

Generalized expression for $E_{i\rightarrow(i+0.5)}$ for $i$th bit cycle for each $N$-bit code after "first" step of "down" conversion is given as (34).

$$E_{i\rightarrow(i+0.5)} = \frac{C_{B_i} V_R^2}{2^{(i+1)}} \tag{34}$$

Then, at second step, i.e., at time 1.5, the largest capacitor is disconnected from $V_R$ and connected to ground, drawing energy given by (35). Using (27),

$$V_X[2] = \frac{1}{4}V_R - V_{in}$$

$$E_{1.5\rightarrow 2} = V_R\,[2C_0(V_X[1.5] - V_X[2])] = C_0 V_R^2 \tag{35}$$

Thus, generalized expression for $E_{(i+0.5)\rightarrow(i+1)}$ for $i$th bit cycle for each $N$-bit code after "second" step of "down" conversion can be written as (36).

$$E_{(i+0.5)\rightarrow(i+1)} = C_{Latest} V_R^2 \tag{36}$$

where $C_{Latest}$ is the capacitor which just got discharged to ground in the second step of $i$th bit cycle during "down" conversion. The total switching energy is

$$E_{1\rightarrow 2,\text{2-step}} = E_{1\rightarrow 1.5} + E_{1.5\rightarrow 2} = \frac{3C_0 V_R^2}{2} \tag{37}$$

Generalized expression for $E_{1\rightarrow 2,\text{2-step}}$ for $i$th bit cycle for each $N$-bit code is calculated by adding (34) and (36).

### 3.3.3 Charge Sharing Switching Method

The charge from the largest capacitor can be used to charge the next largest capacitor, thereby leading to this new switching scheme [11]. Figure 8 shows the modified capacitor array, with additional switches $S_{CS}$. Once again, the switching is done in

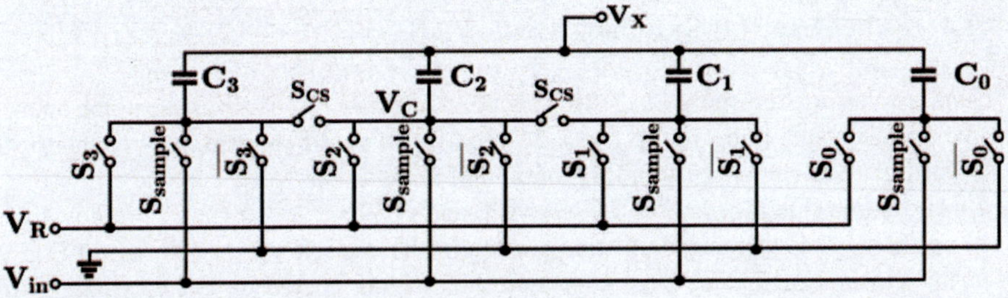


**Fig. 8** 3-bit capacitor array for charge sharing methods

two phases [12]. A new switch, $S_{CS}$, is used to connect the first two capacitors, by first disconnecting them from $V_R$ and ground, thereby drawing no energy from $V_R$. Using the principle of charge conservation, the voltage $V_C$ at the end of the first phase is given by (38).

$$V_C[1.5] = \frac{2}{3} V_R \tag{38}$$

Generalized expression for $V_C[i + 0.5]$ for $i$th bit cycle for each $N$-bit code after "down" conversion is given by (39).

$$V_C[i + 0.5] = \frac{C_{\text{prior}_{i-1}}}{C_{\text{prior}_{i-1}} + C_{\text{latest}_i}} V_R \tag{39}$$

where $C_{\text{prior}_{i-1}}$ is last capacitor that was connected to $V_R$ before $i$th bit cycle and $C_{\text{latest}_i}$ is the current capacitor value which is going to be charged to $V_R$ in the $i$th bit cycle. It is clear that the MSB/2 capacitor is effectively charged to $\frac{2}{3}$rd of the reference voltage with no energy expenditure. During the second phase, the MSB capacitor is connected to ground and the MSB/2 capacitor is connected to $V_R$. Hence, the total energy dissipated is given by (40).

$$E_{1\to 2,\text{CS}} = \frac{7 C_0 V_R^2}{6} \tag{40}$$

Generalized expression for $E_{i\to(i+1)}$ for $i$th bit cycle for each $N$-bit code after "down" conversion is given as (41).

$$E_{i\to(i+1),\text{CS}} = \frac{C_{\text{latest}_i}}{(C_{\text{prior}_{i-1}} + C_{\text{latest}_i})} C_{\text{latest}_i} V_R^2 + \frac{C_{\text{T}_i} V_R^2}{2^{(i+1)}} \tag{41}$$

### 3.3.4 Split Capacitor Switching Method

In charge sharing, the energy is getting saved in a "down" transition, and some energy must still be spent charging up $C_2 = 2C_0$ to $V_R$. To avoid charging any capacitor to $V_R$ during a "down" transition, this method [13] splits the MSB capacitor into three capacitors of value $C_{3,2} = 2C_0$, $C_{3,1} = C_0$, and $C_{3,0} = C_0$ and then switches down one of them. This capacitor splitting results in the capacitor array shown in Fig. 9. During the first bit cycle, $C_{3,2}$, $C_{3,1}$ and $C_{3,0}$ are connected to $V_R$, dissipating the energy (24). After time 1, instead of connecting $C_2$ to $V_R$, $C_{3,2}$ is simply connected directly to ground, as in Fig. 9. This requires the energy

$$E_{1\to 2,\text{split}} = \frac{C_0 V_R^2}{2} \tag{42}$$

Generalized expression for $E_{i\to(i+1)}$ for $i$th bit cycle for each $N$-bit code after "down" conversion is given as (43).

$$E_{i\to(i+1),\text{split}} = \frac{C_{T_i} V_R^2}{2^{(i+1)}} \tag{43}$$

Thus, the capacitor splitting approach requires no additional energy to charge up a capacitor from the ground to $V_R$ during a "down" transition and hence achieved the same energy for an "up" and a "down" transition.

For a $N$-bit converter, the MSB capacitor is a copy of the rest of the capacitor array, as shown in Fig. 9. During the first bit cycle, all sub-capacitors of the MSB capacitor are connected to $V_R$. For subsequent "up" transitions, the capacitor in the main array is connected to $V_R$ [14]. In contrast, for any "down" transition, the capacitor from the MSB capacitor array is connected to the ground. However, this approach requires twice as many switches as generally small switches, so the area penalty is minimal.

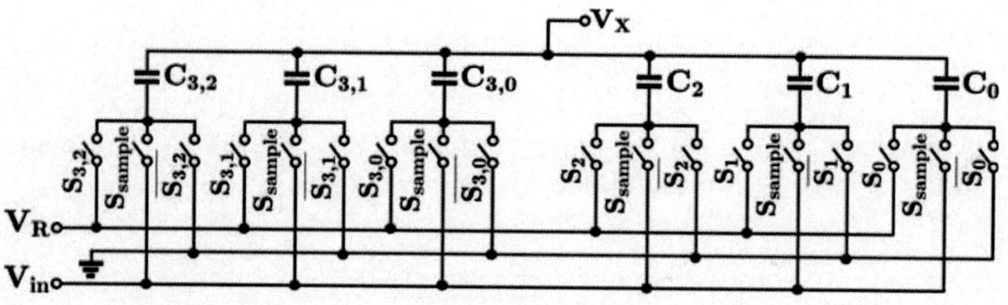


**Fig. 9** 3-bit capacitor array for capacitor splitting method

### 3.3.5 Switching Method of [15]

In this method, input signal which lies in between [0, $V_R$] is digitized by reference voltage $V_R/2$ instead of $V_R$. For this, $V_R/2$ being reference voltage, it is the largest digitizing voltage along with other digitizing voltages $[2^{-2}, 2^{-3}, 2^{-4}, \ldots, 2^{-N}]\,V_R$ which are generated by DAC.

Figure 10 shows the block diagram of SAR ADC with this [15] switching method. The problem of voltage error at DAC2's top plate due to parasitic capacitors is overcome by a dummy capacitor array DAC1, which is identical to DAC2. Switching of capacitors in DAC1 generates MSB, while that in DAC2 generates other bits. Here, the working of a 3-bit SAR ADC is shown in Fig. 11 as an example with this switching method. The main steps are first sample the input signal then invert the DAC2. After that, DAC1 generates the MSB after which all other bits are generated by capacitor switching in DAC2.

MSB is generated by connecting bottom plates of both DACs to $V_R/2$. So top-plate voltage of DAC1, $V_R/2$, is compared to that of DAC2 ($V_R - V_{in}$). If $V_{in} \geq V_R/2$, the bottom plates of DAC1 get connected to GND, otherwise, it remains connected to $V_R/2$. In both cases, the capacitor array consumes no power during generation of MSB except parasitic capacitors consume small power. However, capacitors switch between $V_R/2$ and GND in this method rather than between $V_R$ and GND.

The above five switching methods have been applied to a 10b capacitor array. Simulated energy versus codes plot is shown in Fig. 12, neglecting the bottom-plate parasitic of the capacitors. From the simulation results, it is evident that energy saving is achieved in the different switching methods and is highest for switching method of [15] where $V_R/2$ is used as a reference voltage. It is to be noted that all the schemes

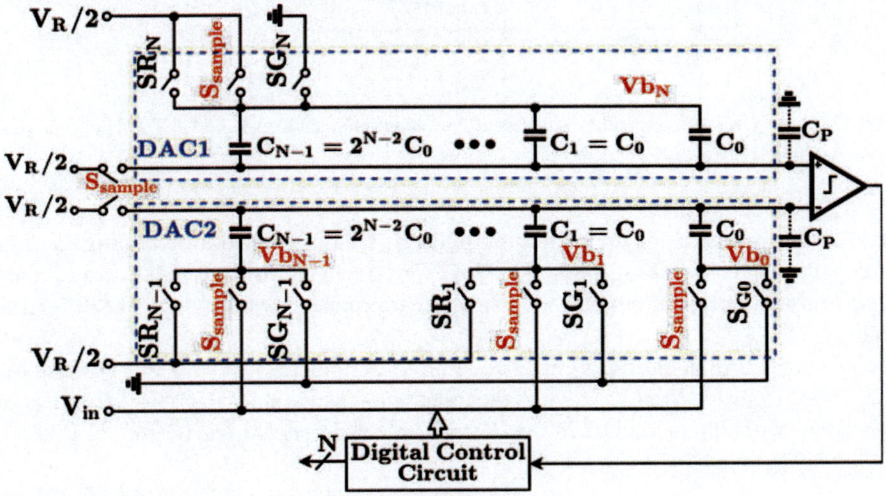


Fig. 10 3-bit capacitor array with switching scheme of [15]

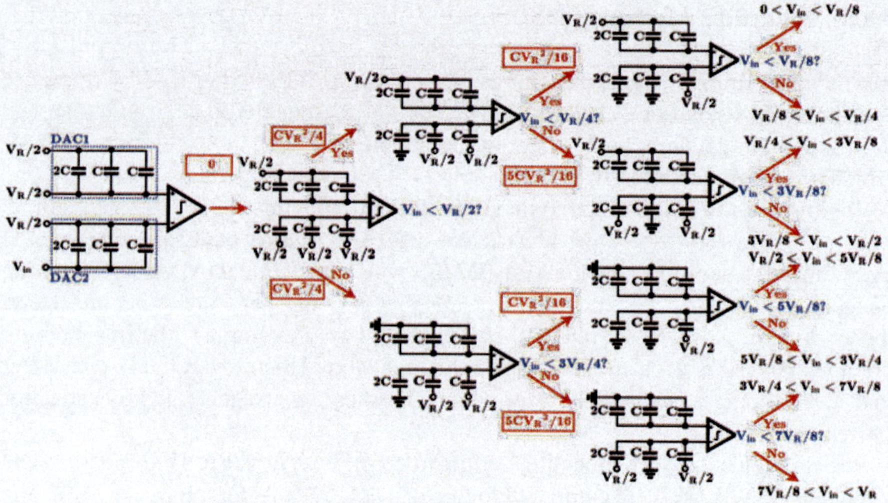

Fig. 11 Working flow of two 3-bit single-ended SAR ADC with switching method of [15]

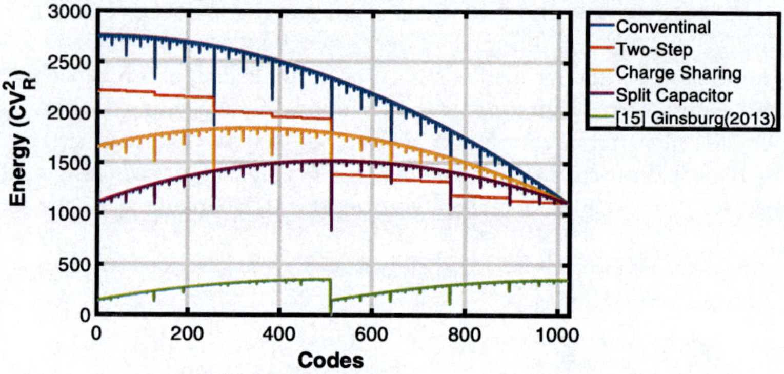


Fig. 12 Energy versus codes plot for different switching scheme when $V_R = 1.8$ V, $C_0 = 1$ unit and resolution is 10-bit

have "up" transitions for the highest output code resulting in same switching energy except for that in switching method of [15]. The "down" transitions in the array occur frequently for the lower output codes, so the energy can be saved in different-different manner.

The switching method [15] has two similar sections in the curve, for codes 0 to $(2^{N-1} - 1)$ and $2^{N-1}$ to $(2^N - 1)$, as the input voltages $V_{in}$ and $(V_R/2 + V_{in})$ undergo the same switching procedure in DAC2 with only difference in the biasing of DAC1.

## 4 Conclusion

In this paper, two SAR ADC topologies have been discussed, along with a detailed mathematical analysis. Compared to the conventional binary-weighted SAR, split SAR topology is area efficient for higher resolutions. Design constraints are discussed to assist the practical implementation of the ADC. Five different switching methodologies for the capacitive-DAC of the SAR ADC are discussed, out of which the last switching method where $V_R/2$ is used as a reference voltage proves to be the most power efficient.

## References

1. Park J, Park HJ, Kim JW, Seo S, Chung P (2000) A 1 mW 10-bit 500 KSPS SAR A/D converter. In: 2000 IEEE international symposium on circuits and systems, vol 5, pp 581–5845
2. Lim Y, Flynn MP (2015) A 1 mW 71.5 dB SNDR 50 MS/s 13 bit fully differential ring amplifier based SAR-assisted pipeline ADC. IEEE J Solid-State Circuits 50(12):2901–2911
3. Ginsburg BP, Chandrakasan AP (2007) Dual time-interleaved successive approximation register ADCs for an ultra-wideband receiver. IEEE J Solid-State Circuits 42(2):247–257
4. Guo W, Sun N (2016) A 12b-ENOB 61 μW noise-shaping SAR ADC with a passive integrator. In: 42nd European solid-state circuits conference, pp 405–408
5. Chen S, Michael W, Robert BW (2006) A 6-bit 600-MS/s 5.3-mW asynchronous ADC in 0.13-μm CMOS. IEEE J Solid-State Circuits 41(12):2669–2680
6. Hong HK, Kim W, Kang HW, Park SJ, Choi M, Park HJ, Ryu ST (2014) A decision-error-tolerant 45 nm CMOS 7 bit 1 GS/s non-binary 2b/cycle SAR ADC. IEEE J Solid-State Circuits 50(2):543–555
7. Johns DA, Martin K (2008) Analog integrated circuit design. Wiley
8. Keshattiwar A, Sahoo BD (2019) A systematic approach to sizing capacitors in split-SAR ADC to achieve optimum redundancy. In: 62nd IEEE international Midwest symposium on circuits and systems, pp 117–120
9. Hu W, Lie DY, Liu YT (2012) An 8-bit single-ended ultra-low-power SAR ADC with a novel DAC switching method. In: 2012 IEEE international symposium on circuits and systems, pp 2349–2352
10. Hariprasath V, Guerber J, Lee SH, Moon UK (2010) Merged capacitor switching based SAR ADC with highest switching energy-efficiency. Electron Lett 46(9):620–621
11. Khoo KY, Wilson AN (1995) Charge recovery on a databus. In: Proceedings of the 1995 international symposium on low power design, pp 185–189
12. Van Elzakker M, Van Tuijl E, Geraedts P, Schinkel D, Klumperink EAM, Nauta B (2010) A 10-bit charge-redistribution ADC consuming 1.9 W at 1 MS/s. IEEE J Solid-State Circuits 45(5):1007–1015
13. Ginsburg BP, Chandrakasan AP (2005) An energy-efficient charge recycling approach for a SAR converter with capacitive DAC. In: IEEE international symposium on circuits and systems, pp 184–187
14. Ginsburg BP, Chandrakasan AP (2007) 500-MS/s 5-bit ADC in 65-nm CMOS with split capacitor array DAC. IEEE J Solid-State Circuits 42(4):739–747
15. Hu W, Liu YT, Nguyen T, Lie DC, Ginsburg BP (2013) An 8-bit single-ended ultra-low-power SAR ADC with a novel DAC switching method and a counter-based digital control circuitry. IEEE Trans Circuits Syst I 60(7):1726–1739

# Origin of Hump in $I_{ds}$ for Body-Tied SOI-MOSFET and Its Influence on Circuit Performance

T. Iizuka, M. Miura-Mattausch, H. Kikuchihara, and S. Ghosh

**Abstract** In course of device scaling toward the usage of thinner semiconductor layer, body-tied SOI-MOSFETs have exhibited the onset of floating-body-like effects commonly known to floating-body SOI-MOSFETs. One of such manifestations is a hump or sudden jump in drain-to-source and body currents notably in the subthreshold bias region. This discontinuous behavior is attributed to the onset of snapback through 3D device simulation enabling a realistic placement of the body contact. Thinner SOI layer easily depletes and the consequent vanishment of neutral region hinders impact ionization-generated holes from exiting the SOI layer at the body contact, thus causing a buildup of body potential and an enhancement of carrier injection at the source/body junction. These physical insights were modeled in HiSIM_SOI, an industry-standard compact model for SOI-MOSFETs. Thanks to the slowness of its physical mechanism in comparison to device operating speed in actual circuitry, the time-dependent floating-body effect, also known as the history effect, would be mitigated.

**Keywords** SOI-MOSFET · Floating-body effects · Compact model

## 1 Introduction

SOI-MOSFET is one of the candidates both for low-power and high-speed applications. Aggressive downscaling not only for the channel length but also for the SOI substrate thickness improves the device characteristics suitability for high-performance applications. There are two types of advanced SOI-MOSFETs. One is the floating-body (FB) type, where the body contact is given to the body bottom, and the other is the body-tied (BT) type, where the bottom of the substrate is contacted to the earth, to erase non-intentional carriers induced during device operation. The carrier accumulation happens within the device due to the impact ionization which is well known for the FB SOI-MOSFET. The carrier accumulation causes a drain

T. Iizuka (✉) · M. Miura-Mattausch · H. Kikuchihara · S. Ghosh
Hiroshima University, Higashi-Hiroshima 739-8530, Hiroshima, Japan
e-mail: iizuka@hiroshima-u.ac.jp

 
C. Giri et al. (eds.), *Emerging Electronic Devices, Circuits and Systems*,
Lecture Notes in Electrical Engineering 1004,
https://doi.org/10.1007/978-981-99-0055-8_37

current hump in the subthreshold region. It has been demonstrated by Lu et al. that the $I_{ds}$ hump can be enhanced significantly by applying positive back-gate voltage $V_{bg}$ in addition to the relatively large $V_{ds}$ as shown in Fig. 1 [1]. It has been argued that the hump could improve the switching performance due to the very high transconductance.

It has been believed that BT SOI-MOSFET is free from such extraordinary phenomenon due to the body contact, which provides the generated hole-flow path. Ida et al. utilized the enhanced body contact to inject holes into the body by applying positive bias on the contact as shown in Fig. 2 [2]. However, recent measurements show the $I_{ds}$ hump even for BT SOI without $V_{bg}$ biasing as shown in Fig. 3.

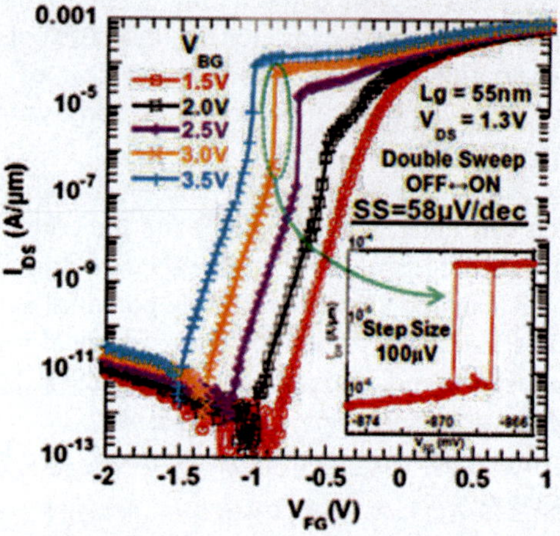


**Fig. 1** Reported example of sharp subthreshold drain current on FD SOI [1]

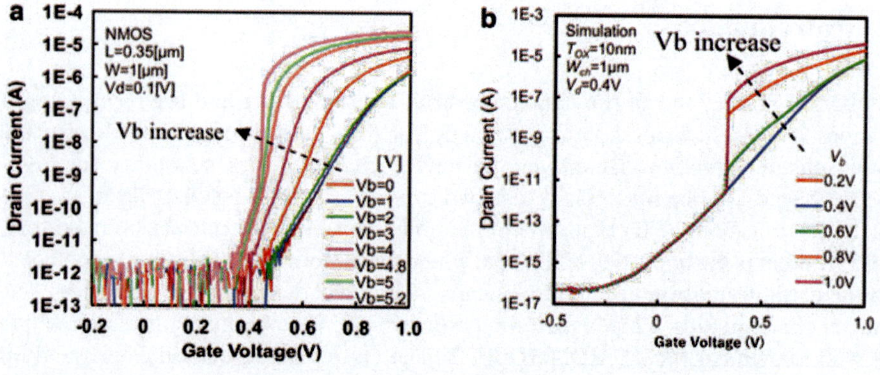


**Fig. 2** Reported example of sharp subthreshold drain current on PN body-tied SOI [2]. **a** Measurements and **b** device simulation

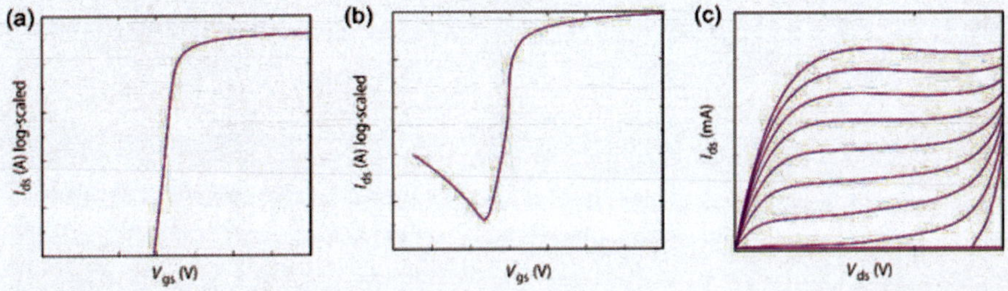


**Fig. 3** Measurement examples of BT SOI. **a** $I_{ds}$–$V_{gs}$ at a low $V_{ds}$. **b** $I_{ds}$–$V_{gs}$ at a high $V_{ds}$, where an abrupt increase of $I_{ds}$ is noticeable in the subthreshold bias region. **c** $I_{ds}$–$V_{ds}$ for multiple $V_{gs}$

Purpose of present investigation is to identify the origin for the $I_{ds}$ hump in BT SOI-MOSFET. We propose a compact model for the effect by considering the hole accumulation explicitly in the Poisson equation in the same way as FB SOI-MOSFET but in a simplified way. Influence of $I_{ds}$ hump on circuit performance will be discussed.

## 2 Investigation Based on Simulation

In our investigation, 3D device numerical simulation was performed to understand the microscopic features exhibiting during the device operation [3]. Figure 4 shows the studied device structure with its dimensions (Table 1). The body contact was placed away from the active device so that a realistic layout of the device structure was considered.

The simulated $I$–$V$ characteristics are shown in Fig. 5 schematically, which shows the same feature as observed in measurements. It is clearly seen that the $I_{ds}$ hump is exactly synchronized with that of the body current. This reveals that the impact ionization causes the hump, namely excess holes induced by the impact ionization.

**Fig. 4** Studied device structure

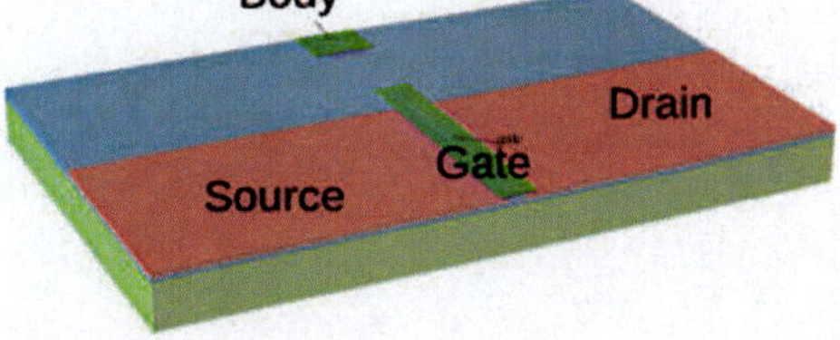


**Table 1** Device structure

| | | |
|---|---|---|
| TSOI | 100 nm | SOI body thickness |
| NSOI | $3 \times 10^{17}$ cm$^{-3}$ | SOI body doping concentration |
| Tfox | 2 nm | FOX thickness |
| Tbox | 450 nm | BOX thickness |

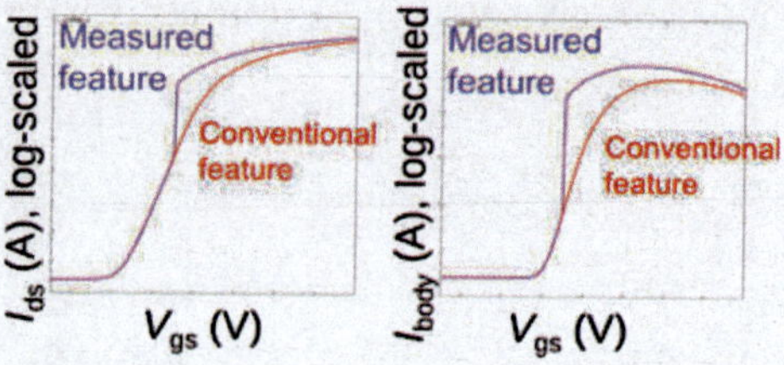


**Fig. 5** Featured characteristics of $I_{ds}$ and $I_{body}$

Simulation results with different $V_{ds}$ values are shown in Fig. 6 as a function of $V_{gs}$. $V_{gs}$ is varied from low to high (up-hill sweep) and the opposite direction (down-hill sweep). Clear hysteresis is observed with increased $V_{ds}$. For the case, $I_{ds}$ jump is obvious. Additionally, discontinuity occurs for low $V_{gs}$ sweep at high $V_{ds}$ values. Figure 7 depicts the hole density as a function of $V_{gs}$. For small $V_{ds}$, the hole density does not exceeds the impurity concentration of the SOI layer due to the depletion condition within the body. The simulated hole density, origin of the body current detected at the body contact, shows the bell shape as a function of $V_{gs}$ as can be expected from the theory. By increasing $V_{ds}$, the hole density increases but keeping the bell shape. With increased $V_{ds}$, the hole density calculation with the two different $V_{gs}$ sweeps does not show the same results but discontinuity occurs in the same way as that of $I_{ds}$. This refers to the hysteresis observed in $I_{ds}$ (see Fig. 6). This is exactly due to the snapback phenomenon as depicted in Fig. 8 as a function of $V_{ds}$.

The drain current $I_{ds}$ is simulated as a function of upward $V_{ds}$ (high sweep). When $V_{ds}$ reaches the breakdown voltage, the avalanche multiplication of carrier generation occurs and $I_{ds}$ enters a non-controllable condition. To maintain the relationship within the device described by the Poisson equation, the device system derives a new equilibrium condition through reducing the effective $V_{ds}$. Since the voltage sweep ($V$-driven) is usually performed one directionally either in monotonous increments or decrements of the external bias, the regression of the voltage sweep is out of scope during simulation. Once the device finds itself at a critical point across which in the bias plane no continuous state would be found at the vicinity in the direction of the one-directional bias sweep, a new state is only available through a turnaround at the

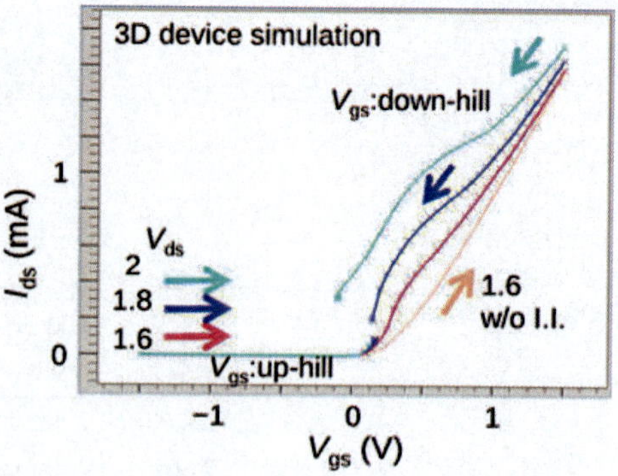


**Fig. 6** $I_{ds}$–$V_{gs}$ simulation results (3D device simulation)

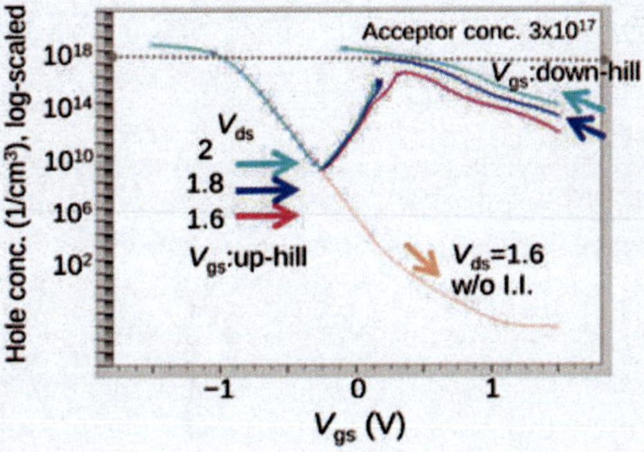


**Fig. 7** Hole concentration probed at the mid-channel surface (3D device simulation)

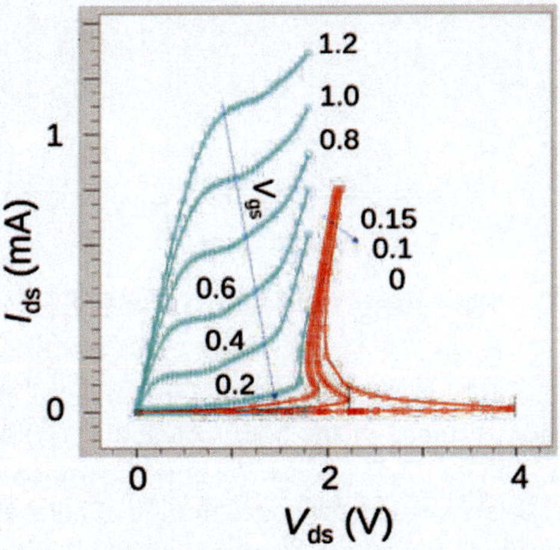


**Fig. 8** Simulated $I_{ds}$–$V_{ds}$ characteristics (3D device simulation). Lines are guide for the eye only. In the $V$-driven mode (light blue), $V_{ds}$ was swept from 0 to 1.8 V. For selected lower $V_{gs}$ cases (0, 0.1, and 0.15 V), $I_{ds}$ was swept from 0 to 0.8 mA (red) instead of $V_{ds}$, in the $I$-driven mode

critical point in the opposite direction, but not through tracing back the original path. Should a new state be found at a distance from the original point in the bias plane, a discontinuous transit to the new state results in a current jump. Simulation often gets stuck near the critical point after its futile attempts of finding a new solution nearby. If the simulation is, however, performed by sweeping the current ($I$-driven) instead, rather than $V$-driven, a smoothly responding voltage change is obtained. Figure 9 shows the microscopic insight of the condition, which is the explanation of the current jump called the snapback phenomenon.

Snapback happens when the hole concentration exceeds a certain threshold within the device as schematically shown in Fig. 7. Thus, it can be concluded that the $I_{ds}$ hump is originated by the high hole concentration accumulated within the device (1. Impact ionization in Fig. 9), which reduces the built-in potential at the source/channel junction (2. Potential increase), resulting in the current flow to the source electrode (3. Bipolar effect). This phenomenon occurs even in the BT SOI-MOSFET. The main reason is the resistance of the body contact, which prevents from the holes flowing out.

**Fig. 9** Passage to the snapback, probed by charge separation in 2D device simulation

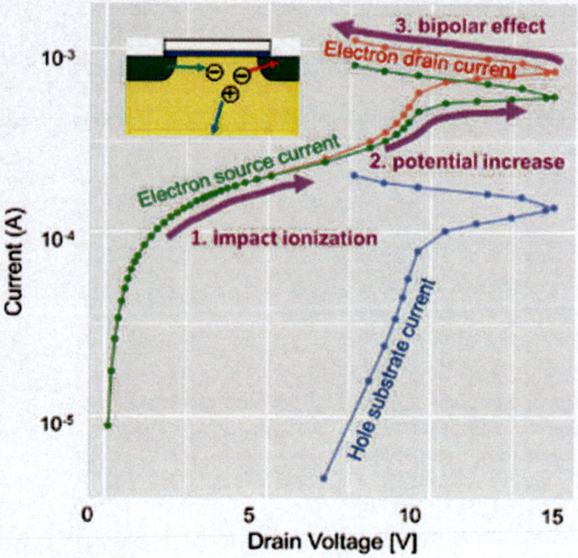


## 3 Modeling $I_{ds}$ Hump for BT SOI

The hole accumulation has been modeled in HiSIM_SOI for FB SOI. For BT SOI, the resistance of the body contact ($R_{body}$) has been considered instead of the hole accumulation [4]. Since the hole generation is faster than the hole flow out due to the resistance effect, the concentration of holes remaining within the SOI body increases, resulting in the history effect as usually observed as the same as FB SOI-MOSFET. Namely, the hole accumulation occurs continuously during device operation even for the BT SOI-MOSFET. For FB SOI-MOSFET, HiSIM_SOI solves the whole potential distribution along the device to the vertical direction with the quasi-Fermi potential distribution along the channel direction iteratively. Here, the potential value at the bulk $\phi_{bulk}$ plays an important role together with that at the front-surface potential $\phi_s$ (see Fig. 10). On the contrary, the potential value at the back-gate $\phi_b$ is the key potential, which is varied according to the accumulated hole concentration $Q_h$ (see Fig. 11).

As shown in Fig. 12, $Q_h$ is a function of both $V_{gs}$ and $V_{ds}$, which shows the same feature exactly as that of $I_{body}$. Modeling considers the time-dependent $Q_h$ value, which describes the history effect. For BT SOI-MOSFET, however, only a part of totally generated holes are stored within the device, which is dependent on the body-contact resistance $R_{body}$. However, the potential drop induced by $R_{body}$ is known from $I_{body}$ calculated. Thus, the back-gate potential $\phi_b$ is calculated iteratively to obtain consistent potential distribution under the impact ionization condition. The solving approach is shown in Fig. 13.

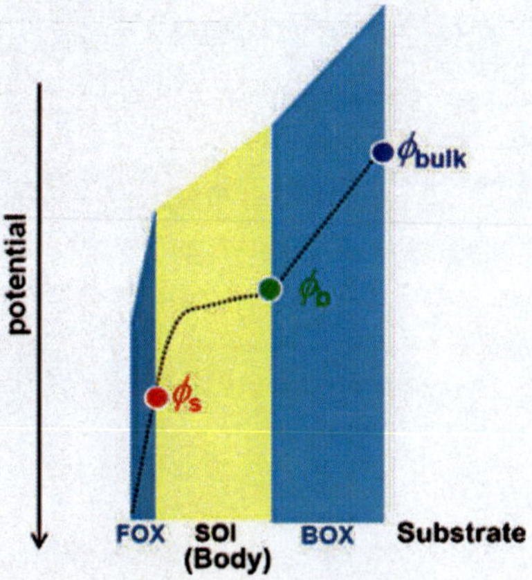


**Fig. 10** Milestone potentials, across the SOI-MOSFET from FOX to BOX, internally used for HiSIM_SOI to simulate device characteristics

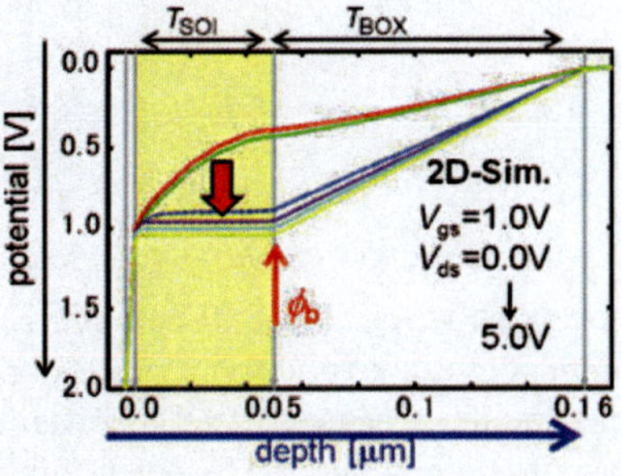


**Fig. 11** Buildup of the back-gate potential $\phi_b$ due to the accumulation of impact ionization-generated holes. Two-dimensional device simulation results

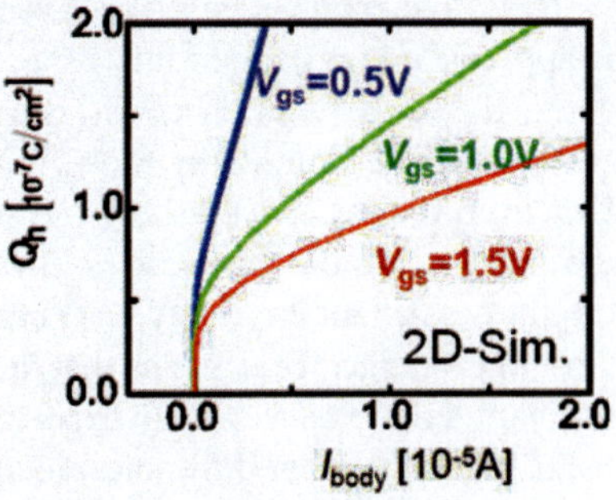


**Fig. 12** Accumulated hole concentration per unit area, as function of $I_{body}$ and $V_{gs}$, simulated by 2D device simulation

# 4 Discussions

It has been believed that the history effect could never be observed for the BT SOI-MOSFET, because the generated holes can flow out from the body contact. The reality was, however, not the case but a clear hump in $I_{ds}$ has been observed, which is the proof for the hole accumulation. Modeling can be done by considering the resistance effect of the body contact together with the accumulated charge density in

**Fig. 13** Internal procedures implemented into HiSIM_SOI for calculating the accumulated hole concentration

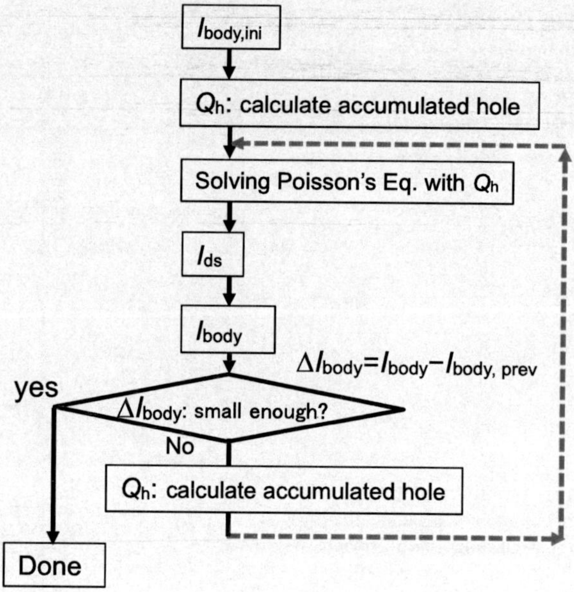


the same way as FB SOI-MOSFET. However, BT SOI-MOSFET includes no back-gate potential explicitly in modeling. Since the body current $I_{body}$ is known, the node potential can be calculated consistently by solving the Poisson equation including the accumulated charge. However, this can be done only iteratively to achieve accurate solution.

Here, a question is raised as several researchers argued, whether the history effect is appreciated on the circuitry aspect. Figure 14 shows measured transient $I_{ds}$ characteristics as a function of $V_{ds}$ sweep. Three different voltage ramping rates are compared. The slow sweep shows kink effect (red curve) due to the impact ionization. However, the fast sweep shows no such effect (blue curve). The reason reveals that the hole accumulation requires time to become observable. HiSIM_SOI considers the time constant explicitly, and our measurements show that the time constant is around μs range. This means that the $I_{ds}$ hump observed in the DC $I$–$V$ characteristics would not be necessarily expected under the fast circuit operations. Capacitance measurements under different frequencies would be required, to confirm the reality.

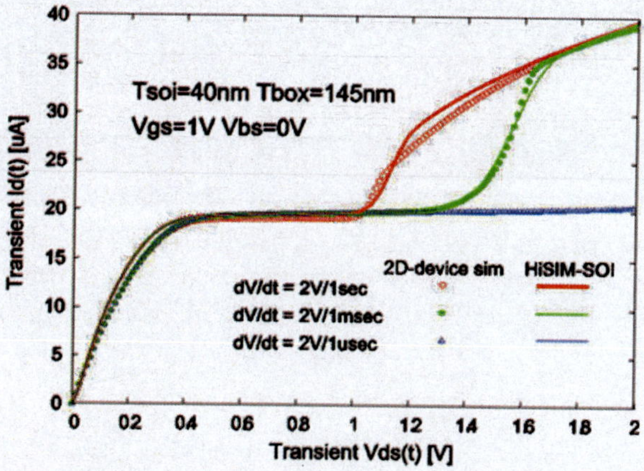


**Fig. 14** Transient drain current in response to voltage ramping of $V_{ds}$. Two-dimensional device simulation-generated characteristics are compared with the predictions from circuit simulation using HiSIM_SOI for the compact model to SOI-MOSFET. A characteristic kink is clearly visible in the slower transient (red). Its onset delays at the faster transient (green) and becomes further delayed and obscured within the time scale (up to 1 μs) during the further faster transient (blue). A characteristic time constant for the impact ionization is $t_d \propto 1/I_{body}$

## References


1. Lu Z et al (2010) Realizing super-steep subthreshold slope with conventional FDSOI CMOS at low-bias voltages. In: 2010 international electron devices meeting. IEEE, Piscataway, NJ, pp 16.6.1–16.6.3
2. Ida J et al (2015) Super steep subthreshold slope PN-body tied SOI FET with ultra low drain voltage down to 0.1 V. In: 2015 IEEE international electron devices meeting (IEDM). IEEE, Piscataway, NJ, pp 22.7.1–22.7.4
3. ATLAS, Silvaco Inc.
4. HiSIM_SOI webpage. https://hisim.hiroshima-u.ac.jp/cgi/HiSIM_SOI/public_release.cgi. Accessed 2022/2/7

# IR-LED Using Electroluminescence in PbS Quantum Dot

**Abhigyan Ganguly, Siddhartha S. Nath, and Viranjay M. Srivastava**

**Abstract** In the present report, PbS quantum dots embedded in polyvinyl alcohol (PVA) matrix have been synthesized by simple chemical method. UV/VIS spectroscopy, X-ray diffraction study (XRD), and high-resolution transmission electron microscopy (HRTEM) are used for characterization of the laboratory-synthesized lead sulfide quantum dots. To test the electroluminescence of the PbS quantum dots, a simple TCO/ZnO/QD/Al structure is been fabricated and tested for a range of applied voltage. It is observed that the variation of the EL intensity in the infrared range is almost linear for a given range of applied voltage.

**Keywords** PbS · Quantum dots · LED · Electroluminescence

## 1 Introduction

In recent times, semiconductor quantum dots (QDs) have emerged as one of the leading research topic among researchers, due to their versatile and wide range of applications in the field of electronics, optoelectronics, sensors, and photovoltaics [1–3]. In case of nano-dots or more specifically quantum dots, the properties of the material are found to be quite different from that of their bulk material properties. The main reason for this being their dimensional confinement in all the three dimensions, due to which quantum mechanical properties are dominant in quantum dots. Size-dependent bandgap variation, multiple exciton generation, higher surface-to-volume ratio, etc. are few of such properties that are exclusive to nano-dots. One of such convenient phenomena observed in case of quantum dots is electroluminescence

A. Ganguly (✉)
Dr. Sudhir Chandra Sur Institute of Technology and Sports Complex, Kolkata, India
e-mail: abhigyanganguly666@gmail.com

S. S. Nath
Cachar College, Assam University, Silchar, India

V. M. Srivastava
Howard College, University of KwaZulu Natal, Durban, South Africa

 
C. Giri et al. (eds.), *Emerging Electronic Devices, Circuits and Systems*,
Lecture Notes in Electrical Engineering 1004,
https://doi.org/10.1007/978-981-99-0055-8_38

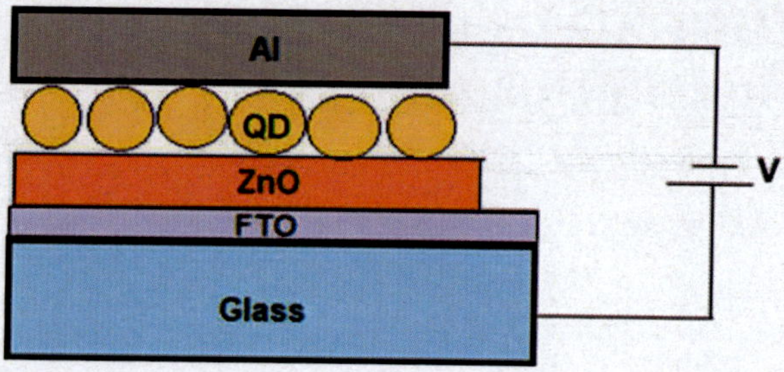


**Fig. 1** Schematic diagram of the ZnO/ZnS-QD-based light-emitting device

(EL), which makes quantum dots one of the chosen materials for fabrication of light-emitting devices (LEDs) [4].

The electroluminescence in a material can be defined as the phenomenon of emission of light when voltage is applied to it. In the present work, PbS quantum dots are prepared on polyvinyl alcohol (PVA) matrix by a simple chemical method of one-pot synthesis. The quantum dots are characterized by using standard techniques such as UV/VIS spectroscopy, X-ray diffraction study (XRD), and high-resolution transmission electron microscopy (HRTEM), to confirm the structural and optical properties of the laboratory-made PbS. For the next part of the experiment, the PBS quantum dots are deposited on a TCO/ZnO structure to test for electroluminescence. While the TCO acts as one of the electrodes, aluminum plate is used the other electrode. Voltage is applied across the two electrodes, and the emission intensity is measured by a electroluminescence spectroscope. The schematic representation of the fabricated TCO/ZnO/QD/Al structure is shown in Fig. 1.

## 2 Experimental

To prepare the polymer capping matrix, 5 g of polyvinyl alcohol (PVA) is added to 100 ml of distilled water and stirred with mild heating for four hours. When the solution PVP turns transparent and viscous, the solution is kept in the dark chamber and given one-day standing time. The role of the capping matrix is vital in the formation of quantum dots. As because, during the chemical synthesis process, the quantum dots are formed in the interstitial gaps of the polymer matrix, which prevents the growth of QDs beyond a certain point. The reason for choosing PVA is because the PVA polymer remains itself inert during the reaction, but only controls the size of the quantum dots. Also, the PVA is easily soluble in water, and hence after completion of the synthesis, they can mostly be washed out from the precipitate quite easily [5].

To prepare the Pb precursor, 11.38 g of lead acetate (PbC2H3O2) is mixed with 100 ml of distilled water to form a 0.3 M solution. The lead acetate solution was then added to the PVA polymeric solution, and the mixture was stirred at 600 °C for 3 h with a magnetic stirrer. By adding 2 M NaOH solution, the pH of the solution was kept at 6. A viscous translucent solution was formed after a one-day standing period was given to the solution. After one day, 0.3 M sodium sulfide was prepared by dissolving 2.34 g Sodium Sulfide ($Na_2S$) in 100 ml distilled water, then added to Pb /PVA matrix solution [6].

Zinc Acetate and Sodium Hydroxide (NaOH) were combined in ethanol to make ZnO, which was then uniformly deposited on conductive FTO-coated glass (resistivity 10 Ω/sq.) using the tape template method and the doctor's blade technique. The thin film is then solidified and the surface of FTO is planarized by heating it to 80 °C and air annealing it to 1000 °C. The ZnO-coated glass plate is then dipped coated for roughly 60 s in each of the previously prepared CdS quantum dot solutions to form a PbS QD layer on the oxide via chemical bath deposition (CBD). Then, a thin aluminum plate is put between them and kept together with the FTO [4, 6].

The samples have been characterized by optical absorption spectroscopy (using PerkinElmer Lambda 35 1.24), X-ray diffraction (XRD) (Bruker AXS, X-ray source: Cu Ka), high-resolution transmission electron microscopy (using JEM 1000 C XII), scanning, and electroluminescence spectroscopy (using HITACHI-F-2005).

## 3 Results and Discussion

Figure 2 shows the optical absorption spectrum for the laboratory-prepared PbS quantum dots. From the figure, strong absorbance edge can be observed at about 335 nm of wavelength. As the absorption edge of bulk PbS is in and around 3000 nm, thus one can confer that there has been a large blue shift in the absorption spectra of the prepared PbS nanoparticles. This blue shift is the indication of formation of nano-sized particles. For initial approximation, the average particle size of the samples can be estimated by using the hyperbolic approximation model, which is given as [7]:

$$R = \sqrt{\frac{2\pi^2 h^2 Egb}{m * (Egn^2 - Egb^2)}} \tag{1}$$

where $R$ is the quantum dot radius, $E_{gb}$ is the bulk band gap, $E_{gn}$ is the quantum dot band gap, $h$ is Planck's constant, and $m^*$ is the effective mass of electron of the specimen. Here, the bulk band gap ($E_{gb}$) for PbS is 0.41 eV and electron effective mass at room temperature is 0.175 $m_0$ [8], where $m_0$ is the electron rest mass. The quantum dot band gap ($E_{gn}$) of the prepared PbS QD sample as determined from the absorption edge wavelength (335 nm) is 3.70 eV. Thus, the radius ($R$) of the quantum dot obtained from the HBM is 3.2 nm, which implicates the QD (2$R$) size to be 6.4 nm, as presented in Table 1.

Figure 3 provides the obtained X-ray diffraction (XRD) pattern for the synthesized PbS nano-dots. One can observe that the XRD PbS at 28 (111), 30 (200), and 41 (220), which matches perfectly with the data for PbS (JCPDS no. 78–1901). Thus, we can also conclude that PbS crystal is pure cubic. From X-ray diffraction study,

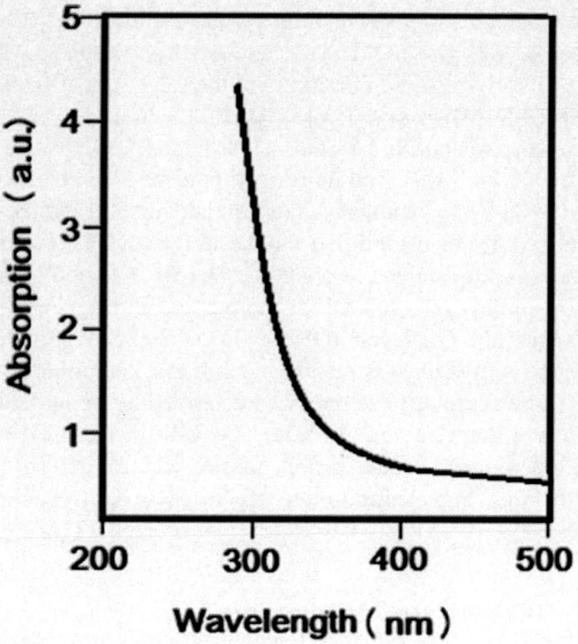


**Fig. 2** UV-visible absorption spectroscopy of PbS quantum dots

**Table 1** Data from UV-V is spectra and XRD

| QD sample | Absorption edge (nm) | Bandgap (eV) | Size (nm) from HBM | $2\theta$ (degree) | Average size (nm) from Scherrer formula |
|---|---|---|---|---|---|
| PbS | 335 | 3.70 | 6.40 | 28(111)<br>30 (200)<br>41 (220) | 6.60 |

average particle size (crystallite size) can again be estimated for initial analysis by using Scherrer formula [9]:

$$D = 0.9\lambda/W\cos\theta \quad (2)$$

where '$\lambda$' is the wavelength of X-ray (0.1541 nm), '$W$' is full width at half maxima (FWHM), and '$\theta$' (theta) and '$D$' are the glancing angle and the particle diameter (crystallite size), respectively. Considering all the three peaks ($2\theta$ in degree) in the X-ray diffractogram, the average crystallite (quantum dot) size has been assessed and found to be about 6.6 nm for PbS QDs. All data from absorption spectra and XRD are presented in tabulated form in Table 1.

The high-resolution transmission electron microscopy (HRTEM) images of the PbS quantum dots are shown in Fig. 4a, which confirms the formation of quantum

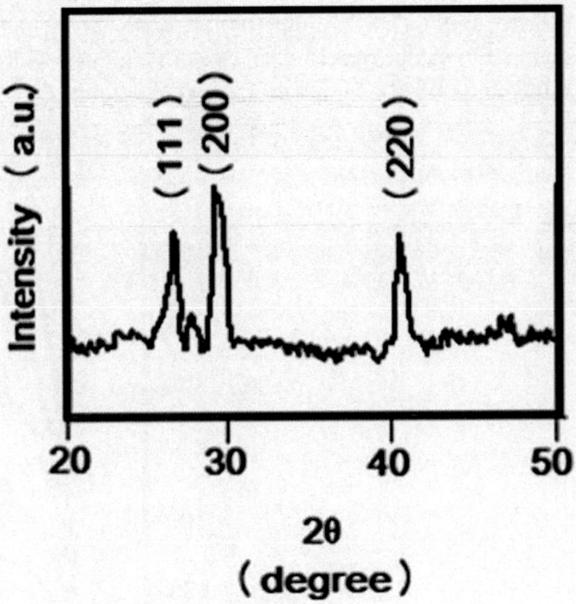


**Fig. 3** XRD of PbS quantum dots

dots within the size 10 nm. The scanning electron microscopy (SEM) images of the PbS QDs are shown in Fig. 4b. The image shows the surface morphology of the fabricated quantum dots, and it can be observed that the fabricated quantum dots are arranged in ordered arrays.

Figure 5 shows the EL plot of the PbS at different bias voltages. The EL intensity is observed at around 1400 nm wavelength in case of PbS QDs, which is in the IR region with increasing the applied voltage the emission intensity increases [10]. But, the main disadvantage of nano-LED is that if the device is operated for a longer period at higher voltage, the device may get damaged [11, 12]. Also, beyond in Fig. 6, we have plotted the EL intensity versus voltage obtained for the device, and we can see that for the specific voltage range, the plot is almost linear [13]. The EL response

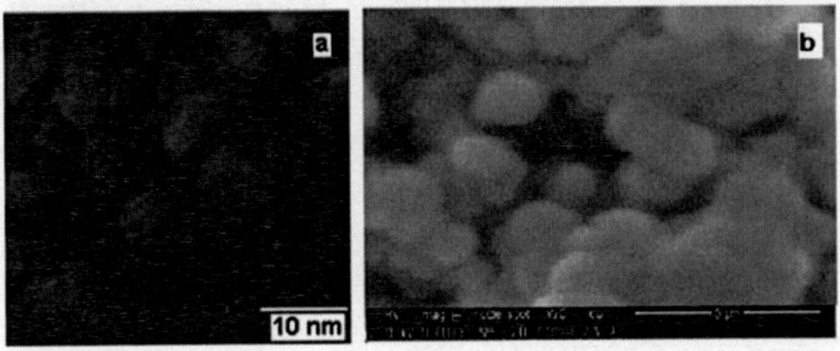


**Fig. 4** **a** HRTEM and **b** SEM images of fabricated PbS quantum dots

time for PbS QDs is around $10^{-9}$ secs. Hence, it can be inferred that PbS QD-based nano-LED can be operated successfully at lower voltage range. The EL date for the fabricated PbS QD nano-LED is given in Table 2.

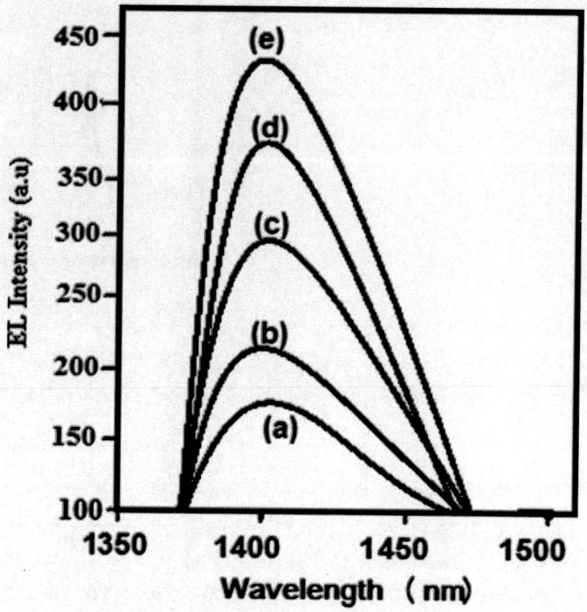


**Fig. 5** Electroluminescence (EL) of the synthesized PbS nano-dots at applied voltage of, **a** 3 V, **b** 6 V, **c** 9 V, **d** 12 V, and **e** 15 V, at room temperature 300 K

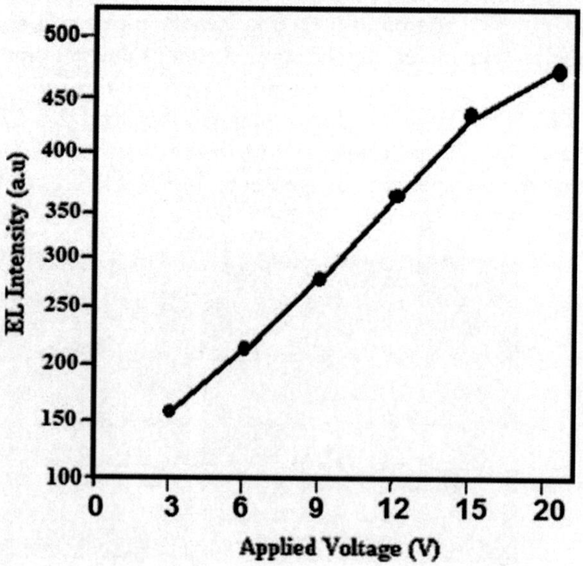


**Fig. 6** EL intensity V/S applied voltage for PbS QDs at 300 K

**Table 2** EL intensity data for the fabricated PbS QD LED

| Applied voltage (V) | EL intensity (a.u.)at 300 K | Response speed |
|---|---|---|
| 3 | 154 | Of the order of $10^{-9}$ s |
| 6 | 210 | |
| 9 | 271 | |
| 12 | 363 | |
| 15 | 446 | |
| 20 | 467 | |

## 4 Conclusions

Lead sulfide (PbS) quantum dots have been prepared by using low-cost chemical method of one-pot synthesis. The nanoparticle samples are characterized by standard techniques and they are deposited on an TCO/ZnO electrode, with aluminum as the other electrode. The fabricated device is tested for electroluminescence for a low range of applied voltage at room temperature. The emission is observed in the IR range and is almost linear for a small range of applied voltage. The IR-LED works good for the given range of voltage, but beyond 15 V the linearity is lost.

## References


1. Nath SS, Choudhury M, Chakdar D, Gope G, Nath RK (2010) Acetone sensing property of ZnO quantum dots embedded on PVP. Sens Actuators: B 148:353–357
2. Semonin OE, Luther JM, Beard MC (2012) Quantum dots for next-generation photovoltaics. Mater Today 15:508–515
3. Zaban A, Micic OI, Gregg BA, Nozik AJ (1998) Photosensitization of nanoporous TiO2 electrodes with InP quantum dots. Langmuir 14:3153–3156
4. Ganguly A, Nath SS, Choudhury M (2018) Effect of Mn doping on multilayer PbS quantum dots as sensitized solar cell. IEEE J Photovoltaics 8(6):1656–1661
5. Nath SS, Ganguly A, Gope G, Kanjilal MR (2017) SnO2 quantum dots for nano light emitting devices. Nanosyst: Phys Chem Math 8(5):661–664
6. Ganguly A, Nath SS, Choudhury M (2018) Cu doped PbS quantum dots as sensitizer in solar cells. J Nanoelectron Optoelectron 13(6):906–9011
7. Dandia A, Parewa V, Rathore KS (2012) Synthesis and characterization of CdS and Mn doped CdS nanoparticles and their catalytic application for chemoselective synthesis of benzimidazoles and benzothiazoles in aqueous medium. Catal Commun 28:90–94
8. Walton AK, Moss TS, Ellis B (1962) Determination of effective mass in the lead salts by infra-red Farady Effect. Proc Phys Soc 79(5)
9. Chukwuocha EO, Onyeaju MC, Harry TST (2012) Theoretical studies on the effect of confinement on quantum dots using the Brus equation. World J Condens Matter Phys 2:96–100
10. Ganguly A, Nath SS, Srivastava VM (2021) Swift heavy ion irradiated SnO2 quantum dot based light emitting device. Optoelectron Adv Mater Rapid Commun 15(3–4):120–123
11. Anikeeva PO, Halpert JE, Bawendi MG (2009) Quantum dot light-emitting devices with electroluminescence tunable over the entire visible spectrum. Nano Lett 9(7):2532–2536

12. Nath SS, Ganguly A, Gope G, Kanjilal MR (2018) ZnS quantum dots based voltage sensing light emitting device. IEEE Sens Lett 2(3). https://doi.org/10.1109/LSENS.2018.2862915
13. Anikeeva PO, Halpert JE, Bawendi MG, Bulovic V (2007) Electroluminescence from a mixed red-green-blue colloidal quantum dot monolayer. Nano Lett 7(8):2196–2200

# 100X Increase in Industrial and Personal Productivity Augmenting the State-of-the-Art Technologies AI/ML, Edge Computing, and 5G Network

**Biswajit Patra**

**Abstract** Artificial Intelligence (AI) has become a major innovative technology driver and crucial pillar in next-generation industrial revolution and major force for new digital world. This trend has been accepted by all major industries and acknowledged by the European Commission, who has very clearly pointed out that how a high-performance, ultra-low latency, intelligent, and trustworthy safety networks are fundamental for the evolution of the multiservice Next-Generation Internet (NGI). Excellent progress has been done in the accuracy and performance of AI-enabled platforms. This is now super critical that AI/ML integration in autonomous decision-making and mission critical systems with end-to-end quality assurance with ultra-low latency has advanced cellular network. 5G is the next-generation cellular network that aspires to achieve substantial improvement on quality of service, such as higher throughput and lower latency. Edge computing is an emerging technology that enables the evolution to 5G by bringing cloud capabilities near to the end users (or user equipment, UE) to overcome the intrinsic problems of the traditional cloud, such as high latency and the lack of security. In this paper, 100X increase in efficiency is demonstrated deploying the state-of-the-art technologies AI/ML, edge computing, and 5G network in highly secure environment. Other important aspects including the key requirements for its successful deployment in 5G and the applications of edge computing in 5G are also described. Then, we explore, highlight, and categorize recent advancements in edge computing for 5G. By doing so, we reveal the salient features of different edge computing paradigms for 5G. Finally, open research challenges for the research community are mentioned.

**Keywords** Artificial intelligence · Next-generation internet · AI-enabled platforms · 5G · Edge computing

B. Patra (✉)
Bangalore, India
e-mail: biswajitpatra@ieee.org



C. Giri et al. (eds.), *Emerging Electronic Devices, Circuits and Systems*,
Lecture Notes in Electrical Engineering 1004,
https://doi.org/10.1007/978-981-99-0055-8_39

# 1 Introduction

Artificial Intelligence (AI) platforms, to take human like decisions, are not only irreversibly sets on the evolutionary path of every future business and also of every object and technology service we human will interact in coming days. This trend is influenced and driven by the need to support demanding real-world use cases such as autonomous driving, online-health, collaborative gaming to play games in a group, online retail business. In every such scenarios, it is acknowledged that the 5G operators will have the opportunity to fill a central role in providing innovative solutions for application and service developers that want to combine the advanced capabilities of 5G with cloud-based application development processes emerged in the last decade, such as, for example, the Platform as the Service (PaaS) and the microservice/serverless models to the edge in secured manner by near-edge trust zone computing. A significant example of this ecosystem is AI-enabled applications running on a high-performance compute either co-located at edge or at cloud, which have become a major innovative force in almost any business vertical and are being foreseen as one of the pillars boosting the next-generation industrial revolution. This is mainly due to the fact that cloud and mobile networks are already converging at several technological levels. From one side, cloud technologies, such as cloud-native and virtualization, are making their way into the telecom operators' domain. At the same time, networking use cases and requirements, such as access-aware operations and service function chaining, are influencing the evolution of cloud technologies. Despite such advances, today the cloud operators' approach remains centralized with few large data centers deployed at key locations, while telecom operators manage a distributed infrastructure based on a hybrid multi-cloud approach and very localized service need.

In this context, 5G is a preferred technology due to its high performance in terms of latency, data rate, and reliability, a great future cellular technology for a technological and business convergence between the cloud computing and the telecom worlds. 5G features like slicing, multi-access edge computing (MEC), and more flexible radio connectivity can be used to support qualitatively different applications and to deliver a richer user experience, faster interactions, large-scale data processing, and machine-to-machine communications in industrial environment spanning across multiple locations across different parts of the word to all in a same premises inside a factory. This is a great challenge and an opportunity to develop solutions from entry level to large corporate business. Nevertheless, the challenges to be overcome to realize this connectivity/computing convergence are still notable. In particular, the increasing number of control and optimization dimensions of the end-to-end 5G infrastructure may result in an overly complex network that operators and vendors may find it difficult to operate, manage, and evolve.

AI and machine learning (ML) technologies will be crucial in the cloud-network convergence process and will help operators achieve a higher level of automation, increase network performance, and decrease the time-to-market of new features. Early attempts at applying AI/ML in the cellular domain can be found in several academic works [1–4]. Nevertheless, it cannot be expected that each subsystem of future access, edge, core, and cloud segments will employ distinct and separated AI tools and datasets. Such an approach would lead to AI-silos, slowing down advances vital to achieving sustainable networking and ultra-scale complex services relying on distributed compute-connect fabric.

The approach of AI@EDGE to answer the above-mentioned challenges has two lines of action. First, we will design, prototype, and validate a network and service automation platform able to support flexible and programmable pipelines for the creation, utilization, and adaptation of secure and privacy-aware AI/ML models. Second, we will use this platform to orchestrate AI-enabled end-to-end applications. Here, we introduce the novel concept of Artificial Intelligence Functions (AIFs) to refer to the AI-enabled end-to-end applications sub-components that can be deployed across the *AI@EDGE* platform. Finally, the *AI@EDGE* platform will be validated using four well-chosen use cases with specific requirements that cannot be satisfied by current 5G networks according to 3GPP R15 and 3GPP R16, in terms of support for latency-sensitive and highly dynamic AI-enabled applications.

The rest of the paper is structured as follows. In Sect. 1.1, we discuss the challenges addressed by the AI@EDGE project. Section 1.3 covers the AI@EDGE concept for beyond 5G networks. The four reference use cases are described in Sect. 1.4. Finally, Sect. 2 concludes this paper.

.

## *1.1 Challenges*

The main objective of AI at the edge is to build a secure connect–compute platform capable of enabling the automated roll-out and management of large-scale heterogeneous edge and cloud computing infrastructures of hardware and software. To this end, the platform encompasses the required APIs to enable the deployment of large-scale virtual compute overlays (e.g., containers and serverless instances, etc.) across a multi-connected heterogeneous infrastructure able to support a range of future critical applications. This is illustrated in Fig. 1, which represents a typical functional overview of the platform across AI/ML-driven multi-connected applications at massive scale and AI/ML-centered security. However, this ambitious objective involves several mission critical challenges. Such challenges are described in detail in the following subsections:

(1) Network Automation Platform Leveraging Flexible and Reusable Ai Pipelines

5G is a full paradigm shift where high performance in terms of latency, data speed, and reliability calls for a technological and business convergence between cloud computing and networking. The increasing number of control and optimization dimensions of the 5G infrastructure may lead to a very complex network that operators and vendors may find it very challenging to operate, manage, and continuously

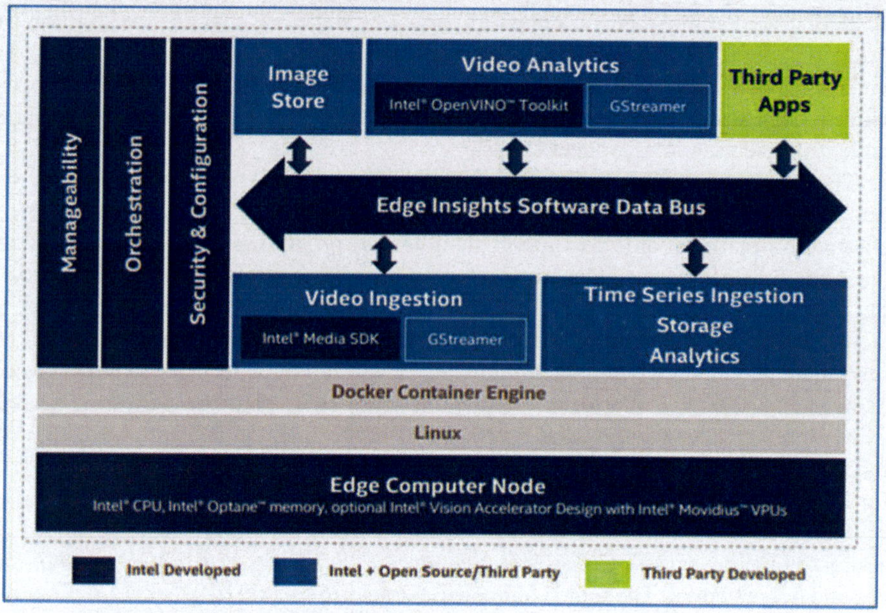


**Fig. 1** Functional overview of the AI at the edge platform

improve upon. AI technologies will be key in this roadmap, and although early attempts to address this issue can be already found [1–4], they focus on specific training and inference types that cannot be extrapolated to other network segments easily. However, in a fully automated system, having distinct and independent AI tools and datasets would make management of networking and highly scalable services on a distributed connect–compute platform a very challenging task. Therefore, the challenge is to implement a general-purpose network automation framework capable of supporting flexible and reusable end-to-end AI pipelines. Scalable AI/ML models, fast data pipelines, and effective data dissemination models are crucial to realize automation at scale. Some of the key pillars of AI at the edge computing are the potential of multi-access edge computing and the powerful mechanisms of scalable distributed and federated learning in the 5G context. Based on this, AI@EDGE addresses this challenge by developing a platform for closed-loop automation that allows the deployment of AI/ML compute infrastructures over the edge, which also accounts for secure isolation of co-located AI/ML algorithms by multiple stakeholders running on shared MEC resources. The progress on this challenge will enable two main results:

- Scaling of AI/ML distributed algorithms to ensure application performance and model reliability under varying resource availability.
- Zero-touch end-to-end network and service management including the creation, utilization, and adaptation of reusable AI/ML pipelines in a connect-compute platform.

## 1.2 Secure AI in Multi-Stakeholder Environments

From a security perspective, there are several relevant aspects for the success of 5G and beyond systems. In production networks, the potential risks for tangible assets, such as servers and human beings victims of attacks, are significantly increasing. Consequently, intrusion detection systems are a native part of the current 5G security architecture. However, they are usually under proprietary licenses, which highlights the need to enable an open exchange of models and parameters for intrusion detection, especially in multi-stakeholder environments. It must be noted that this issue worsens when the platforms are driven by AI/ML models, since a new and dangerous attack surface is added. The ability to achieve resilience and service continuity requires simple and information-effective data-driven models, suitably designed for running on Internet of Things and MEC devices with limited resources. Therefore, the challenge is to provide lightweight, secure, and resilient ML systems that are robust to evasion and poisoning attacks.

With the advent of ML, privacy techniques have been recently revisited to better accommodate the trade-off between privacy risk and data usefulness in the construction of ML pipelines. Federated learning (FL) [5–7] and adversarial networks [8] have recently been used to cover the security aspects. FL allows assembling a common model combining local models built from edge devices without disclosing any data. However, this approach poses numerous problems such as local biases, temporal offsets. Security is one of the cornerstones of the architecture of *AI@EDGE*. Consequently, *AI@EDGE* aims to define and implement an ML methodology following a distributed paradigm, allowing the development, implementation, and evaluation of effective intrusion detection algorithmic framework, accounting for security and isolation of co-located AI/ML algorithms. Tackling this challenging research, question will mainly lead to three key results:

- Increase in attack detection speed and intrusion resiliency, including early detection and automated configuration and speedup of countermeasures.
- Model propagation and computational efficiency to optimize the set of exchanged parameters, e.g., the weights of a neural network, and their aggregation methods.
- Privacy security of the model parameters exchanged between edge devices.

## 1.3 The AI@EDGE Connect–Compute Fabric for Beyond 5G Networks

The design of the *AI@EDGE* platform envisions the automated roll-out of adaptive and secure compute overlays and a new generation of AI-enabled end-to-end applications. Such applications are made possible by *AI@EDGE* through the introduction of the novel concept of AIFs, which refer to the AI-enabled applications sub-components that can be deployed and chained across the various levels of the architecture. This vision is presented in Fig. 2, in which *AI@EDGE* combines a

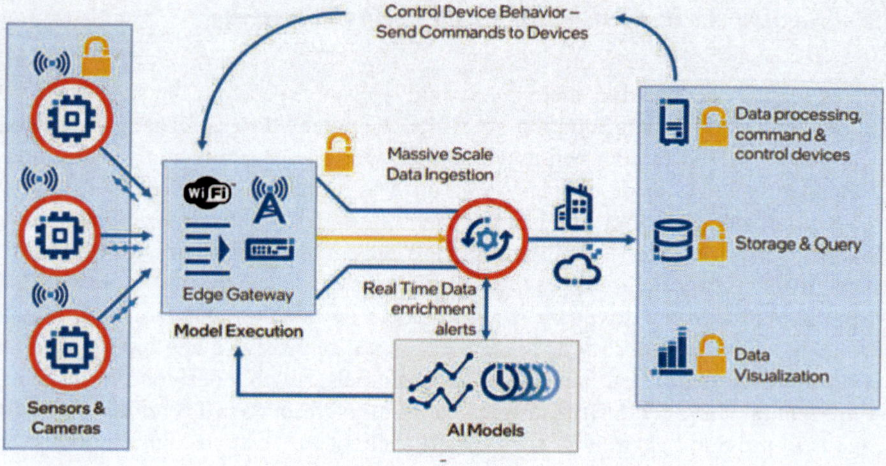


**Fig. 2** AI@EDGE AI-enabled connect-compute platform

set of cutting-edge cloud computing and 5G concepts with a reusable, secure, and privacy preserving AI/ML layer to enable an innovative network automation platform supporting all aspects of network and service management including the deployment and scaling of AIFs of different natures (i.e., latency-critical, low-latency, and latency-tolerant AIFs) over a distributed facility and the various tasks needed to deploy such applications, e.g., the creation of a new network slice.

The remainder of this section describes the technological enablers on which *AI@EDGE* builds to convert the connect–compute platform into a reality and that composes the functional blocks of the AI-driven platform, as sketched in Fig. 3.

## *1.4 A. Distributed and Decentralized Connect–Compute Platform*

**Enabler.** *AI@EDGE* combines the function as a service (FaaS) paradigm with serverless computing, hardware acceleration (GPU, FPGA, and CPU), and a cross-layer, multi-connectivity-enabled disaggregated radio access network (RAN) into a single connect-compute platform to allow over-the-top providers to fully use the 5G capabilities though well-established cloud-native paradigms to develop and run applications. The serverless and FaaS approaches are gaining attention as cloud computing models, in which the infrastructure provider manages on-demand infrastructure and resources, while the stakeholders (e.g., service providers) can focus only on their core activities. Building on this, *AI@EDGE* intends to define a set of open APIs, by which network operators, vertical industries, service providers, and users can interact with the network on a neutral host model. Moreover, the platform encompasses the

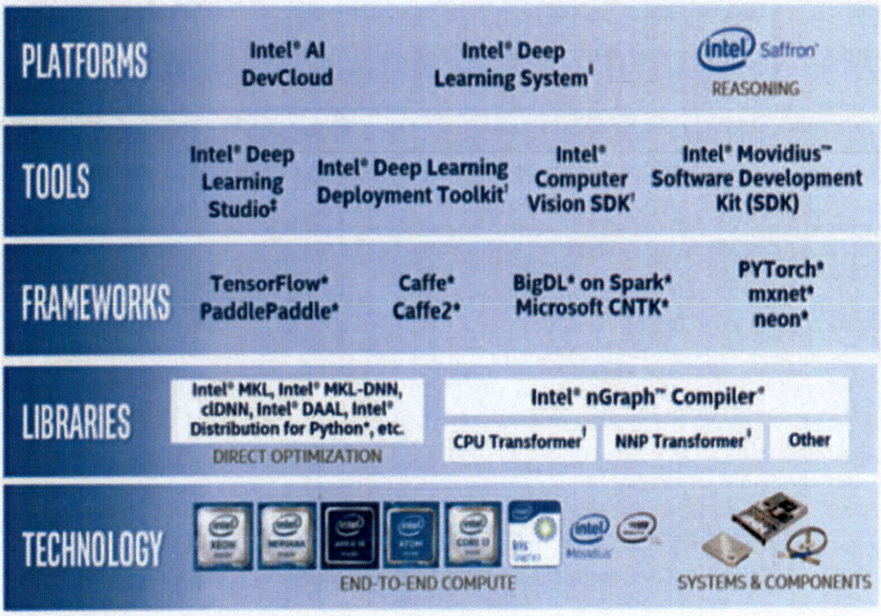


**Fig. 3** Conceptual architecture and functional blocks of AI@EDGE

path toward a hybrid multi-cloud-native deployment supporting Virtual Machines (VMs) and containers and their integration with the serverless paradigm.

**Innovation.** *AI@EDGE* aims to account for this mixture by extending the current ETSI MEC/NFV architectures with application and application-intent models able to capture the huge heterogeneity in the application building domain. For this purpose, *AI@EDGE* takes as a basis the Cloud Native Application Bundling (CNAB) initiative and will propose its extension to support for serverless technologies (besides VMs and containers). In addition, context and metadata from application and application-intent modeling studies are meant to be used for realizing intelligent control and management of applications and services deployed over the serverless decentralized and distributed *AI@EDGE* platform. This can be observed for a wide range of verticals in Fig. 2.

### 1.4.1 Orchestration of Artificial Intelligence Functions

**Enabler**. The provisioning of AI-enabled applications over a distributed computing platform requires reference models and standards, especially in heterogeneous and complex scenarios as edge computing platforms spanning across multiple domains. Defining these AI-enabled applications involves the representation of their AIFs (i.e.,

AI-enabled applications sub-components). To this end, *AI@EDGE* leverages standard knowledge representation languages and well-known state-of-the-art ontology engineering methodologies, to represent AIFs, as well as their relationships and status at the different levels of the technology stack. Conversely, *AI@EDGE* considers "defacto" standards for cloud and edge services orchestration and from their emergent variants (e.g., FaaS) to the end-to-end orchestration and chaining of AIFs.

**Innovation**. *AI@EDGE* aims to build on the above ontologies to propose a reference model that provides the tools to describe AIFs, their requirements (e.g., storage, hardware acceleration, etc.), and the necessary metadata for their orchestration, which will also compose a catalog of available AIFs. Furthermore, *AI@EDGE* envisions innovative solutions for end-to-end orchestration to partition AIFs across different segments attending to their requirements (as shown in Fig. 2) considering the heterogeneity and complexity of the underlying edge computing platforms and the collection of valuable quality of service indicators to create complex AI-enabled applications and detect abnormal situations.

### 1.4.2 Hardware-Accelerated Serverless Platform for AI/ML

**Enabler**. The combination of serverless computing and virtualization augmenting CPU, GPU, AI accelerators introduces an efficient and cost-effective event-driven accelerated computing approach that can be applied for a wide variety of scientific applications. The most recent hardware acceleration solutions (FPGA, GPU, and CPU) with privacy preserving ML techniques allow and speed up the execution of sensitive and computing-intensive workloads over the same platform. The deployment of heterogeneous acceleration platforms at the edge enables advanced processing scenarios to be exploited in far more complex processing functions. In addition to GPUs, which are currently the prime solution for AI/ML processes' acceleration, FPGA are gaining momentum for deployments at the edge due to their ability to ensure optimal performance to execute specialized functions (e.g., real-time network intensive processing), in an energy and cost-efficient manner.

**Innovation**. *AI@EDGE* makes resource-aware hardware acceleration techniques a key point of its design with the goal of increasing resource efficiency across the computing need. As depicted in Fig. 2, aims to go a step further by exploring approaches to tame accelerators' heterogeneity and enable their integration (both GPUs and FPGAs) with the serverless computing concept, in order to offer a unified platform able to allocate resources and migrate functionality between accelerators on different edge devices or between edge and cloud infrastructures.

### 1.4.3 Cross-Layer, Multi-Connected, Disaggregated Radio Access

**Enabler**. Supporting beyond 5G use cases requires relying on different communication technologies to increase reliability, as exposed in Rel15 and Rel16 through dual-connectivity techniques using data duplication at the **PDCP** layer. However, besides reliability, various use cases demand greater flexibility and openness in the RAN to implement more advance multi-connectivity layers and to enable a higher degree of automation. This demand is being promoted in O-RAN specifications [9], which propose an open architecture where RAN control and management functions are divided into near-Real-Time (nRT) and non-Real-Time (nonRT) RAN Intelligent Controller (RIC). *AI@EDGE* aims to rely on both multi-connectivity options and O-RAN specifications to deliver a flexible, open, and unified platform including 3GPP and non-3GPP radio access technologies.

## 2 Conclusions

In this paper, the challenges and conceptual architecture of the *AI@EDGE* is presented. The paper aims converging and evolving AI/ML and 5G at the network edges with the goal of providing a flexible platform on top of which the next-generation AI-enabled application and services can be deployed. The paper also describes the reference use cases that may be used to validate the concept, namely: cooperative perception for vehicular networks, secure, multi-stakeholder AI for Industrial Internet of Things, aerial infrastructure inspections, and in-flight entertainment. Leveraging the 5G, edge computing, and data analytics (AI/ML), the productivity of industrial automation can be more than 100X of today's average efficiency with more than 99.99% accuracy by 2025. A great amount of research is needed by both academic community and industry to make best of these technology in human life, industrial, and home automation.

**Acknowledgements** The views expressed do not necessarily represent with any project work ongoing in Intel. Intel is not liable for any use that may be made of any of the information contained therein. The authors would also like to acknowledge A) CERCA Program/Generalitat de Catalunya for great insights shared in public domains, Jin Yuntong, Hu Xiaopao, Yih Leong Sun, and Du Yongfeng from Intel organized the design of this reference architecture, Intel's Open Source Technology Center team manager Ding Jianfeng, Michael Kadera, Senalka McDonald, Wang Qing, Intel China Strategic Cooperation and Innovation Business Unit (Ecosystem Development Office) Zhang Zhibin, Intel DCG, and Li Hua, CTO of AWCloud Software Co.Ltd, Chen Ke, Li Xiaoyan, Feng Shaohe, Shang Dehao, Yang Yanguo, Wen Wei, Xia Lei, Xu Maorong, Xu Yihua, Wang Ruxin, Zhang Xin, and Hu Chao from Intel, and AWCloud Software Co., Ltd.., AWCloud Software Co., Ltd. contributed to the reference deployment case of the China AI Open Platform. Liang Bing from Intel wrote a case study of the China AI Open Platform

## References


1. Letaief KB, Chen W, Shi Y, Zhang J, Zhang YA (2019) The roadmap to 6G: AI empowered wireless networks. IEEE Commun Mag 57(8):84–90
2. David K, Berndt H (2018) 6G vision and requirements: is there any need for beyond 5G? IEEE Veh Technol Mag 13(3):72–80
3. Calvanese Strinati E, Barbarossa S, Gonzalez-Jimenez JL, Ktenas D, Cassiau N, Maret L, Dehos C (2019) 6G: The next frontier: from holographic messaging to artificial intelligence using subterahertz and visible light communication. IEEE Veh Technol Mag 14(3):42–50
4. Tariq F, Khandaker MRA, Wong KK, Imran MA, Bennis M, Debbah M (2020) A speculative study on 6G. IEEE Wirel Commun 27(4):118–125
5. Abeshu A, Chilamkurti N (2018) Deep learning: the frontier for distributed attack detection in fog-to-things computing. IEEE Commun Mag 56(2):169–175
6. Lim WYB, Luong NC, Hoang DT, Jiao Y, Liang YC, Yang Q, Niyato D, Miao C (2020) Federated learning in mobile edge networks: a comprehensive survey. IEEE Commun Surv Tutorials 22(3):2031–2063
7. Nguyen TD, Marchal S, Miettinen M, Fereidooni H, Asokan N, Sadeghi A (2019) DÏoT: a federated self-learning anomaly detection system for IoT. In: 2019 IEEE 39th international conference on distributed computing systems (ICDCS), pp 756–767
8. Tripathy A, Wang Y, Ishwar P (2019) Privacy-preserving adversarial networks. In: Proceedings of IEEE Allerton, Monticello, IL, USA
9. O-RAN Allianc (2020) O-RAN architecture description v1.0
10. Calabrese FD, Frank P, Ghadimi E, Challita U, Soldati P (2020) Enhancing RAN performance with AI. ERICSSON, Technology Review
11. Hassan N, Yau KLA, Wu C (2019) Edge computing in 5G: a review
12. Riggio R, Coronado E, Linder N, Jovanka A, Mastinuk G, Goratti L, Rosa M, Schotten H, Pistore M (2021) AI@EDGE: a secure and reusable artificial intelligence platform for edge computing